Hybrid Feedback Control

Hybrid Feedback Control

Ricardo G. Sanfelice

PRINCETON UNIVERSITY PRESS

PRINCETON AND OXFORD

Requests for permission to reproduce material from this work
should be sent to permissions@press.princeton.edu

Published by Princeton University Press
41 William Street, Princeton, New Jersey 08540
6 Oxford Street, Woodstock, Oxfordshire OX20 1TR

press.princeton.edu

All Rights Reserved

Library of Congress Control Number: 2020943855

ISBN 978-0-691-180229
ISBN (e-book) 978-0-691-18953-6

British Library Cataloging-in-Publication Data is available

Editorial: Susannah Shoemaker and Kristen Hop
Production Editorial: Nathan Carr
Production: Jacquie Poirier
Publicity: Matthew Taylor and Katie Lewis
Jacket/Cover Credit: Kayakers controlling their flow and jumps while avoiding crashing into boulders within the Soča River, Slovenia / Janossy Gergely / Shutterstock

The publisher would like to acknowledge the author of this volume for providing the camera-ready copy from which this book was printed.

This book has been composed in LaTeX

Printed on acid-free paper. ∞

Printed in the United States of America

10 9 8 7 6 5 4 3 2 1

This book is dedicated to Andy Teel, for his advising, mentoring, and friendship throughout the years. Without his deep insight and inspiring ideas, this book would have not been possible. Furthermore, without his courage, which started my academic career, not only would this book not have been possible, but I also would not have met my wife. And without the help from my wife closing the feedback loop, I would not have been able to finish this book either.

As a corollary of the statement above, it follows that this book is also dedicated to my soul mate and love of my life, Christine, my wife, for her unconditional love, friendship, and encouragement. She has always been my biggest supporter and encourager, and allowed me to write this book for quite some time. All of it while raising our N children, with N converging (in finite time) to the number of equilibrium configurations of the pendubot treated in this book, plus one.

The above statements imply that this book is also dedicated to my children, Ariana, Melia, Gabriel, Nicolas, and Giovanni, for their joy and excitement, and their inquiries about my work and the content of this book. They designed a hybrid control algorithm that robustly and recurrently triggers an arbitrarily large number of events causing very welcomed distractions to my writing. They claim to have contributed by allowing this book to get better with hybrid time, as $t + j$ attempted to tend to infinity.

Contents

Preface

Why hybrid feedback control?

Classical control theory features powerful tools for analysis of systems and design of feedback controllers. These tools are applicable in scenarios where systems and controllers are given in continuous time or in discrete time. *Hybrid feedback control* can lead to controllers that surpass the capabilities of controllers designed using classical control theory due to featuring

- Variables that change continuously – or, equivalently, *flow*; and

- Variables that are instantaneously updated to new values – or, equivalently, *jump*.

In this book, this behavior is referred to as *hybrid*, and the system generating it is said to be a *hybrid system* and to have *hybrid dynamics*.

Hybrid dynamics confers unique capabilities to hybrid controllers. Hybrid controllers can implement feedback strategies that combine behavior that is typical of continuous-time controllers and of discrete-time controllers. A hybrid controller can combine multiple state-feedback laws to solve a complex problem by dividing and conquering it. It can reset its variables when certain events occur, so as to reconfigure itself or to accommodate sporadic availability of information. Logic variables, timers, and memory states as part of the state vector of a hybrid controller enable such unique features. Very importantly, the evolution of these variables can be modeled mathematically and systematically designed using tools that resemble the classical ones.

Relative to classical control theory tools, hybrid control theory notably broadens the class of systems that can be analyzed and designed. The system to be controlled, which is typically referred to as *the plant*, can exhibit hybrid dynamics. Arguably, the combination of continuous and discrete behavior in systems of today is prevalent. Technology has transformed the classical feedback paradigm involving two systems – the plant and the controller – into a complex interconnection between physics, digital devices and computing systems, and interfaces between them. The digital and computing systems define the "cyber" component of the entire system, and are mainly in charge of executing algorithms. The interfaces between the physical and the cyber provide the means to exchange information between the continuous-time and the discrete-time processes. With the physics being naturally described by continuous-time models, and the digital and computing component by discrete-time models, the resulting system exhibits hybrid dynamics.

Goal of this book and its structure

The goal of this book is to present a self-contained introduction to hybrid feedback control and to design tools that make hybrid control a powerful design method. To accomplish its goal, the book starts with an introduction to hybrid control systems, in Chapter 1. Several examples featuring plants and algorithms with hybrid behavior are provided. After formulating the general hybrid control problem, this introductory chapter illustrates the power of hybrid feedback control, by showing that it solves problems that continuous-time and discrete-time feedback control cannot solve. Chapter 2 provides a brief introduction to the modeling framework used to capture the dynamics of hybrid plants and of hybrid controllers. Models of hybrid systems in terms of *hybrid equations* or, more generally, in terms of *hybrid inclusions* are introduced and illustrated in numerous examples. The concept of solution used in this modeling framework is also introduced in this chapter. An overview to a tool for numerical simulation of hybrid equations is also included. This tool is used throughout the book for validation of the designs. Chapter 2 also includes a discussion on when a mathematical model exhibits true hybrid behavior, in the sense that it has variables that may evolve both continuously and discretely. Determining if a model is truly hybrid is critical as it defines the toolset for analysis and design that must be employed – in fact, treating a classical continuous-time or discrete-time system as a hybrid system would be an overkill. The introductory portion of the book concludes in Chapter 3 where formal notions and tools used throughout the book are presented. This chapter is concise, as readers can find more details and proofs in the 2012 book *Hybrid Dynamical Systems: Modeling, Stability, and Robustness* by R. Goebel, R. G. Sanfelice, and A. R. Teel [1].

The next eight chapters present hybrid feedback control strategies. For easy readability, the presentation in some of the chapters is for continuous-time plants, but insight on how the strategies can be extended to the hybrid case are provided. Chapter 4 presents a hybrid control strategy that uses two feedback control strategies and a logic-based algorithm to select the feedback law that should be applied. This strategy unites two feedback controllers, where one of the strategies is considered to work well locally and the other one globally. It carries the name *uniting control*. Chapter 5 introduces a general hybrid control strategy for *event-triggered control*. The strategy allows for different types of events, such as those triggered by the state of the plant, the state of the controller, time, inputs, outputs, or external signals. Chapter 6 presents a hybrid control strategy that extends the one in Chapter 4 by allowing for the use of more than two controllers, some of which might be of feedback or of open loop type. This strategy is particularly suitable for settings where it is possible to design state-feedback control laws that locally asymptotically stabilize isolated points in the state space and open-loop control laws capable of steering the solutions to the plant between those points, following an order that steers the solutions to the desired equilibrium point or set. This strategy is referred to as *throw-catch control*. The hybrid control strategy in Chapter 7 employs the Lyapunov function associated to each controller available in a family of feedback controllers, and appropriately selects the feedback law that should be applied to the plant to assure asymptotic stabilization. The basic idea stems from problems where several gradient-like feedback laws are available and are such that at least one of them can drive the state away from points where others may get stuck, namely, points at which the gradient vanishes, and towards a desired set-point. Such feature leads to the term *synergistic feedback*. Chapter 8 presents a general supervisory algorithm that coordinates a family of general hybrid controllers. This is the most general logic-based strategy in this book. It subsumes the hybrid control strategies in Chapter 4, Chapter 6, and Chapter 7. Due to such generality, it is the most powerful one, though somewhat abstract.

The next three chapters of this book strive for generality and present feedback control strategies for general hybrid plants. Chapter 9 introduces passivity notions for hybrid systems useful for the design of asymptotically stabilizing static state-feedback and output-feedback controllers. Passivity notions that are tailored to hybrid plants are presented. Passivity-based control is quite powerful as it leads to a very elegant feedback control strategy. In Chapter 10 the concept of *control Lyapunov function* for hybrid plants is introduced. Control Lyapunov functions have the nice feature that, under appropriate assumptions, they lead to a constructive feedback design method. This chapter presents different methods to systematically design an asymptotically stabilizing feedback control law. In Chapter 11 tools to certify forward invariance of a set using barrier functions and to design feedback controllers inducing forward invariance are presented. Forward invariance is a key property that has to be satisfied when safety is desired. In the final chapter of this book, Chapter 12, operators and semantics to formulate as well as tools to certify linear temporal logic specifications are introduced. Temporal logic is an expressive language that permits the mathematical formulation of high-level specifications, such as "eventually reach a target and always avoid an obstacle." Guaranteeing that such specifications are satisfied by solutions to hybrid systems is critical in many applications.

Teaching and learning plan

This book was written for graduate level education, primarily for courses in M.S. and Ph.D. programs. Background in real analysis, classical and modern control theory, as well as nonlinear differential equations/nonlinear systems is recommended. The intention is to influence researchers interested in topics related to hybrid systems, cyber-physical systems, control, and automation. To reach out to different levels of expertise, each chapter starts with a detailed overview of the material to be delivered. Each such overview section is written in a high-level, not (overly) technical manner, so that readers without a strong background on control theory can understand. A short "primer" on modeling and mathematics summarizing basic background is in Appendix A. As the reader gets deeper into each chapter, the use of mathematics increases. Each chapter includes exercises that give the reader an opportunity to put test the acquired knowledge in concrete problems.

The following teaching plans are proposed to teach from this book:

1. For a short ≈20-hours-long course for beginners, one could cover Chapters 1-5 and Chapters 9-12, except the design sections. The design sections, along with the overview sections of Chapters 6-8, can be assigned as suggested reading.
2. For a 10-week long course – which is a typical quarter-long course with ≈3 hours of instruction per week – for beginners, covering the entirety of Chapters 1-5 and Chapters 9-12 should be possible. Chapters 6-8 can be assigned as suggested reading in advanced courses.
3. For a 15-week long course, which is a typical semester-long course with ≈3 hours of instruction per week, one should be able to cover all chapters. The structure and content of each chapter were designed to be able at cover one chapter per week, roughly.

In addition to the material in this book, the readers are provided with the following resources to supplement their learning experience:

- Book website located at

  https://hybrid.soe.ucsc.edu/HybridControlBook

  The label @BookSite is used throughout the book in place of the url above.

- The 2012 Princeton University Press book *Hybrid Dynamical Systems: Modeling, Stability, and Robustness* [1].

- The Coursera MOOC *Cyber-Physical Systems: Modeling and Simulation.*

- The YouTube Channel for the *Hybrid Systems Laboratory* at the University of California, Santa Cruz.

- The webinar and MATLAB/Simulink simulation tool *HyEQ: A Toolbox for Simulation of Hybrid Dynamical Systems.*[1]

Acknowledgments

First and foremost, as stated in the dedication of this book, I am deeply indebted to Andy Teel for his advising, mentoring, and friendship throughout the years. Back in 2010, he recommended that I teach a Ph.D. course at the European Embedded Control Institute in Paris. That journey led to the very first draft of the chapters included in this book. And this is how this book project started. The material included in this book was molded over the years through graduate courses I have taught while at the University of Arizona and at the University of California, Santa Cruz, and through short Ph.D. courses I taught in Paris, Buenos Aires, Beijing, Bologna, L'Aquila, and Shanghai. The feedback I have received while teaching these courses helped immensely in shaping this book.

I would like to gratefully acknowledge the continuous financial support by the *Air Force Office of Scientific Research, CITRIS and the Banatao Institute* at the University of California, and the *National Science Foundation*, which partially supported the generation of the research results that are reported in this book.

I would also like to acknowledge my gratitude to the following colleagues and students who, in many different ways, contributed to the material presented in this book (listed in alphabetical order by last name): Berk Altin, Adam Ames, Pauline Bernard, Jun Chai, Francesco Ferrante, Masoumeh Ghanbarpour, Rafal Goebel, Hyejin Han, Maurice Heemels, Dawn Hustig-Schultz, Ryan Johnson, David Kooi, Santiago Jimenez Leudo, Yuchun Li, Mohamed Maghenem, Lorenzo Marconi, Roberto Naldi, Pablo Nanez, Iman Nodozi, Sean Phillips, Romain Postoyan, Laurent Praly, Adnane Saoud, Mario Spirito, Nan Wang, Nathan van de Wouw, and Nathan Wu.

Ricardo G. Sanfelice
Santa Cruz, California
September 2020

---

[1]MATLAB© and Simulink© are registered trademarks of The MathWorks Inc. and are used with permission. The MathWorks does not warrant the accuracy of the text or exercises in this book. This book's use or discussion of MATLAB©, Simulink©, or related products does not constitute an endorsement or sponsorship by The MathWorks of a particular pedagogical approach or particular use of the MATLAB© and Simulink© software.

List of Symbols

$\dot{x}$	The derivative, with respect to time, of the state of a hybrid closed-loop system.
x^+	The state of a hybrid closed-loop system after a jump.
$\mathbb{R}$	The set of real numbers.
$\mathbb{R}^n$	The n-dimensional Euclidean space.
$\mathbb{R}_{\geq 0}$	The set of nonnegative real numbers, that is, $\mathbb{R}_{\geq 0} = [0, \infty)$.
$[0, 1]^n$	All vectors in $\mathbb{R}^n$ such that each component is the interval $[0, 1]$.
$\mathbb{Z}$	The set of all integers.
$\mathbb{N}$	The set of nonnegative integers, that is, $\mathbb{N} = \{0, 1, \ldots\}$.
$\mathbb{N}_{\geq k}$	$\{k, k+1, \ldots\}$ for a given $k \in \mathbb{N}$.
$\varnothing$	The empty set.
$\mathbb{B}$	The closed unit ball, of appropriate dimension, in the Euclidean norm.
$\mathbb{B}^\circ$	The open unit ball, of appropriate dimension, in the Euclidean norm.
$\mathbb{S}^n$	The set $\{x \in \mathbb{R}^{n+1} : |x| = 1\}$.
$\subset$	The (strict or nonstrict) subset symbol. See Remark A.2.
$:=$	The *defined as* symbol.
$\in$	The *belongs to* symbol.
$\forall$	The *for all* symbol.
$\backslash$	The *set subtraction* or *complement* symbol.
I	The identity matrix.
1_N	The vector of dimension N with all of its entries equal to one.
Id	The identity function.
$\mathbf{1}$	The vector with first entry equal to one and all of its other entries equal to zero.
$\overline{S}$	The closure of the set S.
int S	The interior of the set S.
∂S	The boundary of the set S.
con$S$	The convex hull of the set S.
$\overline{\text{con}}S$	The closure of the convex hull of the set S.

$S_1 \setminus S_2$	The set of points in S_1 that are not in S_2.
$S_1 \times S_2$	The set of ordered pairs (x_1, x_2) with $x_1 \in S_1$, $x_2 \in S_2$.
$\sup S$	The supremum of the set $S \subset \mathbb{R}$.
$\sup_t S$	The supremum of the set $S \subset \mathbb{R}_{\geq 0} \times \mathbb{N}$ on the first component; i.e., $\sup_t S := \sup \{t : (t, j) \in S\}$.
$\sup_j S$	The supremum of the set $S \subset \mathbb{R}_{\geq 0} \times \mathbb{N}$ on the second component; i.e., $\sup_j S := \sup \{j : (t, j) \in S\}$.
$x^\top$	The transpose of the vector x.
(x, y)	Equivalent notation for the vector $[x^\top \ y^\top]^\top$.
$\langle x, y \rangle$	The inner product between vectors x and y. For vectors x and y in $\mathbb{R}^n$, $\langle x, y \rangle = x^\top y$.
$\|x\|$	The Euclidean norm of the vector x.
$\|x\|_S$	The distance from x to the set S, which is given by $\inf_{y \in S} \|x - y\|$ for a nonempty set $S \subset \mathbb{R}^n$ and a point $x \in \mathbb{R}^n$.
$x + \mathbb{B}$	The closed unit ball, of appropriate dimension in the Euclidean norm centered at x, where $x + \mathbb{B} = \{x + \chi : \|\chi\| \leq 1\}$.
$\lambda_{\max}(A)$	The largest eigenvalue of the positive definite matrix A.
$\lambda_{\min}(A)$	The minimum eigenvalue of the positive definite matrix A.
$A \otimes B$	The Kronecker product between the matrices A and B.
$f : \mathbb{R}^m \to \mathbb{R}^n$	A single-valued map from $\mathbb{R}^m$ to $\mathbb{R}^n$. See § A.2.
$x \mapsto f(x)$	A single-valued map mapping x to $f(x)$. See § A.2.
$F : \mathbb{R}^m \rightrightarrows \mathbb{R}^n$	A set-valued map from $\mathbb{R}^m$ to $\mathbb{R}^n$. See § A.2.
$f(x)$	The value of the map f at x.
$\mathrm{dom}\, f$	Domain of the map f. See Definition A.1.
$\nabla f(x)$	The value of the gradient of the map f with respect to x at x.
$\frac{\partial f}{\partial x}(x)$	The value of the Jacobian of the map f with respect to x at x.
$\partial f(x)$	The value of the Clarke generalized gradient of f at x.
$\mathrm{rge}\, f$	The range of the map f. See Definition A.3.
$\mathrm{gph}\, f$	The graph of the map f. See Definition A.4.
$R(\cdot)$	The rotation matrix given, for each $\phi \in \mathbb{R}$, as $$R(\phi) = \begin{bmatrix} \cos \phi & -\sin \phi \\ \sin \phi & \cos \phi \end{bmatrix}.$$
$F(S)$	$\cup_{x \in S} F(x)$ for the set-valued map $F : \mathbb{R}^m \rightrightarrows \mathbb{R}^n$ and the set $S \subset \mathbb{R}^m$.
$T_S(\chi)$	The *tangent cone* to the set $S \subset \mathbb{R}^n$ at $\chi \in \mathbb{R}^n$. See Definition A.8.
$\mathcal{PD}$	The class of positive definite functions. See Definition A.25.
$\mathcal{K}$	The class of functions from $\mathbb{R}_{\geq 0}$ to $\mathbb{R}_{\geq 0}$ that are continuous, zero at zero, and strictly increasing. See Definition A.16.

$\mathcal{K}_\infty$	The class of functions from $\mathbb{R}_{\geq 0}$ to $\mathbb{R}_{\geq 0}$ that are continuous, zero at zero, strictly increasing, and unbounded. See Definition A.17.
$V^{-1}(r^*)$	The r^*-level set of the function $V : \operatorname{dom} V \to \mathbb{R}$ restricted to X, which is the set of points $\{x \in X : V(x) = r^*\}$.
$\dot{V}^{-1}(0)$	The 0-level set of the function $\dot{V}$ restricted to C, which is the set of points $\{x \in C : \dot{V}(x) = 0\}$.
$\Delta V^{-1}(0)$	The 0-level set of the function ΔV restricted to D, which is the set of points $\{x \in D : \Delta V(x) = 0\}$.
$L_V(r^*)$	The r^*-sublevel set of the function $V : \operatorname{dom} V \to \mathbb{R}$, which is the set of points $\{x \in \operatorname{dom} V : V(x) \leq r^*\}$.
$\mathcal{I}(r^*)$	The complement closed of the r^*-sublevel set of the function $V : \operatorname{dom} V \to \mathbb{R}$, which is the set of points $\{x \in \operatorname{dom} V : V(x) \geq r^*\}$.
s.t.	The abbreviation of *such that*.
:	The mathematical symbol for *such that*.
sign	The single-valued map $\operatorname{sign}(r) = -1$ if $r < 0$, $\operatorname{sign}(r) = 1$ if $r > 0$, and $\operatorname{sign}(0) \in \{-1, 1\}$.
$\overline{\operatorname{sign}}$	The set-valued function $\overline{\operatorname{sign}}(r) = -1$ if $r < 0$, $\overline{\operatorname{sign}}(r) = 1$ if $r > 0$, and $\overline{\operatorname{sign}}(0) = \{-1, 1\}$.
$\Pi(S)$	$\{z \in \mathbb{R}^n : \exists u \text{ s.t. } (z, u) \in S\}$ for the set $S \subset \mathbb{R}^n \times \mathbb{R}^m$.
$\Pi_0(S)$	$\{z \in \mathbb{R}^n : (z, 0) \in S\}$ for the set $S \subset \mathbb{R}^n \times \mathbb{R}^m$.
$\Pi_c(C_P)$	$\{z \in \mathbb{R}^{n_P} : (z, u) \in C_P\}$ for the set C_P that is part of the data of $\mathcal{H}_P$.
$\Pi_{c,0}(C_P)$	$\{z \in \mathbb{R}^{n_P} : (z, 0) \in C_P\}$ for the set C_P that is part of the data of $\mathcal{H}_P$.
$\Pi_d(D_P)$	$\{z \in \mathbb{R}^{n_P} : (z, u) \in D_P\}$ for the set D_P that is part of the data of $\mathcal{H}_P$.
$\Pi_{d,0}(D_P)$	$\{z \in \mathbb{R}^{n_P} : (z, 0) \in D_P\}$ for the set D_P that is part of the data of $\mathcal{H}_P$.
$\Pi_c^w(C)$	$\{x \in \mathbb{R}^{n_P} : (x, w_c) \in C\}$ for the set C that is part of the data of $\mathcal{H}_w$.
$\Pi_d^w(D)$	$\{x \in \mathbb{R}^{n_P} : (x, w_d) \in D\}$ for the set D that is part of the data of $\mathcal{H}_w$.
$\Psi_c^u(z)$	$\{u_c \in \mathbb{R}^{m_{P_c}} : (z, u_c) \in C_P\}$ or $\{u_c \in \mathbb{R}^{m_{P_c}} : (z, u_c, w_c) \in C_P\}$.
$\Psi_d^u(z)$	$\{u_d \in \mathbb{R}^{m_{P_d}} : (z, u_d) \in D_P\}$ or $\{u_d \in \mathbb{R}^{m_{P_d}} : (z, u_d, w_d) \in D_P\}$.
$\Psi_c^w(z)$	$\{w_c \in \mathbb{R}^{s_{P_c}} : (z, w_c) \in C\}$ or $\{w_c \in \mathbb{R}^{s_{P_c}} : (z, u_c, w_c) \in C_P\}$.
$\Psi_c^w(z)$	$\{w_d \in \mathbb{R}^{s_{P_d}} : (z, w_d) \in D\}$ or $\{w_d \in \mathbb{R}^{s_{P_d}} : (z, u_d, w_d) \in D_P\}$.

Hybrid Feedback Control

Chapter One

Introduction

Hybrid dynamical systems are ubiquitous in science and engineering as they permit capturing the complex, intertwined continuous and discrete behavior of a myriad of systems. Over the years, the popularity of systems combining physical and software components has propelled the development of tools that can systematically handle such a complex combination. As a result, several frameworks for modeling, analysis, and design of such systems have emerged in the literature. A key challenge imposed by hybrid dynamical systems is the fact that some of its variables – for example, those describing the temporal evolution of the system – change continuously, while others – for example, those associated to the algorithms – change impulsively, upon the triggering of events. In particular, this combination of continuous and discrete behavior emerges when physical systems are controlled by algorithms that are implemented digitally and that, through the use of appropriate interfaces, exchange analog and digital information. In such a control setting, which is depicted in Figure 1.1, the physical process to control, *the plant*, and the digital elements need to be jointly modeled, analyzed, and designed.

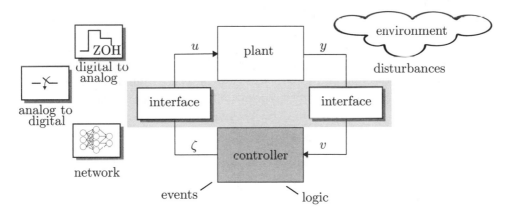

Figure 1.1: A hybrid control system: a feedback system with a plant, a controller, and interfaces – along with environmental disturbances – as subsystems featuring variables that flow and, at times, jump.

In this book, a *hybrid control system* is a hybrid dynamical system whose variables may evolve continuously – namely, *flow* – and at times, they change instantaneously – namely, *jump*. Such a hybrid behavior can be present in one or more of the subsystems of a hybrid control system:

- Hybrid behavior in the plant is prevalent in many systems. These include impulsive oscillators, walking robots, as well as robotic manipulators and vehicles interacting with the environment.

- Hybrid behavior in the algorithm used for control – namely, *the controller* – is typically unavoidable due to information being available only at isolated time instances. Hybrid behavior in the controller is needed to steer the variables of the plant, rapidly and robustly, to a set-point or region.

- Hybrid behavior in the subsystems interconnecting the plant and the controller – namely, *the interfaces* – is present due to the interaction between the two. Typically, the plant portrays continuous behavior, while the controller portrays discrete behavior, leading to hybrid behavior in the interfaces.

Most hybrid control systems involve *feedback* since their inputs are assigned by an algorithm that uses measurements of some of the system variables. Figure 1.1 depicts a hybrid control system in the so-called *closed-loop configuration* with feedback involving its subsystems: the plant, the controller, and interfaces.

1.1 OVERVIEW

In this book, hybrid control systems are modeled as[1] *hybrid equations* or, more generally, as *hybrid inclusions*. Within this framework, the continuous dynamics of the system are modeled using a differential equation – or, respectively, a differential inclusion – while the discrete dynamics are captured by a difference equation – or, respectively, a difference inclusion. To this end, the variables that describe the evolution of the system are collected in a *state* vector. The external signals that affect the values of the state are collected in an *input* vector. A state trajectory to such a system can *flow* over intervals of time with nonzero length and *jump* at certain time instants. Figure 1.2 depicts the evolution of a state variable in a state trajectory to such a system. From an initial condition denoted ×, the state variable flows, then instantaneously jumps to a different value, from where it flows, and so on. This state variable also experiences two consecutive jumps without flow in between. In Figure 1.2, a solid trace represents flow and a dashed trace represents jumps.

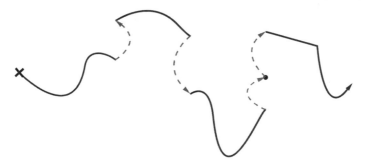

Figure 1.2: A state variable of a hybrid system evolving from an initial condition denoted by ×: it flows (solid trace) and, at times, it jumps (dashed trace).

---

[1]The literature is rich in frameworks for the study of hybrid systems. See § 2.6 for an in-depth discussion of many of those frameworks and their relationships with the one used here.

To informally introduce the hybrid equations/inclusions framework, a model of the plant depicted in Figure 1.1 is first presented.[2] Denoting its state as the vector z taking values from an Euclidean space $\mathbb{R}^{n_P}$ (for some positive integer n_P), its evolution over ordinary time $t \in \mathbb{R}_{\geq 0}$ is governed by a constrained differential equation – or, more generally, inclusion. One particular such case is when z is governed by a plant with linear and time-invariant dynamics, in which case z flows according to

$$\dot{z} = Az + Bu$$

where $u \in \mathbb{R}^{m_P}$ is the control input, the matrices A and B have appropriate dimension, and $\dot{z}$ denotes derivative with respect to t. When the dynamics of the plant are nonlinear, the change of z can be captured by a general nonlinear continuous-time system of the form

$$\dot{z} = F_P(z, u)$$

where F_P is a function that defines the evolution of z over ordinary time. In particular, for the linear time-invariant case, the function F_P is given as $F_P(z, u) := Az + Bu$. See § A.2 for a discussion about functions.

It is important to note that in a real-world setting, the state and the input to the plant are typically subjected to constraints emerging from the physics or from the design specifications. One way to incorporate state and input constraints in the model is by restricting z and u to a subset of $\mathbb{R}^{n_P} \times \mathbb{R}^{m_P}$ capturing the constraints. Denoting such a set as C_P, state and input constraints for the plant are given as

$$(z, u) \in C_P$$

The resulting plant model consists of a constrained nonlinear continuous-time system. Given an input signal, a state trajectory (or solution) is given the time function

$$t \mapsto z(t)$$

that, as ordinary time evolves, flows according to $\dot{z} = F_P(z, u)$ as long as the constraint defined by C_P is satisfied. Such evolution is depicted in Figure 1.2 in solid line, by the continuous, uninterrupted paths. Very importantly, the constraint has the effect of restricting the (continuous) evolution of z and the input u to the set C_P.

The following example illustrates how to define C_P and F_P to model the flows of a continuous-time system with constraints. This system has only input constraints. Later, in Example 1.2, a system with flows subjected to state-dependent constraints is provided.

Example 1.1 (DC/AC inverter). *Consider the single-phase DC/AC inverter with a low-pass RLC filter at its output shown in Figure 1.3. The control objective is to convert the input DC voltage, which is denoted as V_{DC}, into a pseudo-sinusoidal output signal that approximates a given reference sinusoidal voltage with constant angular frequency. The output signal is denoted v_{C_a} and is the voltage on the capacitor C_a. The current through the inductor L is denoted i_L. The position of the*

---

[2]Background on modeling is given in § A.1.

switches $S_1 - S_4$ are the available "control knobs" to accomplish the control objective. Each one of these switches can be set to either the ON or to the OFF position by the control algorithm.

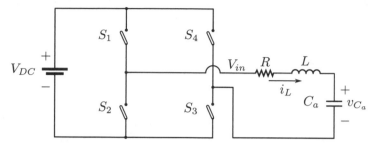

Figure 1.3: Single-phase DC/AC inverter circuit diagram.

A suitable plant model has state vector $z = (i_L, v_{C_a}) \in \mathbb{R}^2$ and control input u taking values that correspond to the positions of the controllable switches. Without loss of generality, the possible values of u are defined as follows:

- *$u = 1$ corresponds to $S_1 = S_3 = ON$ and $S_2 = S_4 = OFF$;*

- *$u = 0$ corresponds to $S_2 = S_3 = ON$ and $S_1 = S_4 = OFF$;*

- *$u = -1$ corresponds to $S_1 = S_3 = OFF$ and $S_2 = S_4 = ON$.*

Kirchhoff's laws are used to determine the continuous change of z for each possible value of u. Circuit laws indicate that the voltage on the capacitor satisfies the differential equation

$$C_a \dot{v}_{C_a} = i_L$$

When $u = 1$, Kirchhoff's voltage law[3] implies that the input voltage V_{DC} is equal to the sum of the voltage on the resistor, on the inductor, and on the capacitor, leading to

$$V_{DC} = i_L R + L \dot{i}_L + v_{C_a}$$

This equality leads to the differential equation

$$\dot{i}_L = \frac{V_{DC}}{L} - \frac{R}{L} i_L - \frac{1}{L} v_{C_a}$$

When $u = 0$, Kirchhoff's voltage law gives

$$0 = i_L R + L \dot{i}_L + v_{C_a}$$

The differential equation governing i_L is

$$\dot{i}_L = -\frac{R}{L} i_L - \frac{1}{L} v_{C_a}$$

Finally, when $u = -1$, the voltages satisfy

$$-V_{DC} = i_L R + L \dot{i}_L + v_{C_a}$$

---

[3]This law states that the (signed) net sum of the voltages in a closed loop is equal to zero.

The associated differential equation is

$$\dot{i}_L = -\frac{V_{DC}}{L} - \frac{R}{L}i_L - \frac{1}{L}v_{C_a}$$

Combining these derivations, the evolution of $z = (i_L, v_{C_a})$ is governed by

$$\dot{z} = \begin{bmatrix} \dot{i}_L \\ \dot{v}_{C_a} \end{bmatrix} = F_P(z, u) := \begin{bmatrix} \frac{V_{DC}}{L}u - \frac{R}{L}i_L - \frac{1}{L}v_{C_a} \\ \frac{1}{C_a}i_L \end{bmatrix} \quad (z, u) \in C_P := \mathbb{R}^2 \times \{-1, 0, 1\}$$

This is a continuous-time system with a discrete-valued input – it is not a hybrid system![4] In contrast, an algorithm that controls this plant to accomplish the control objective would necessarily assign the input u to either a function of the state z or, more generally, to a variable of the algorithm itself. In the latter case, since u is restricted to take values from the discrete set $\{-1, 0, 1\}$, the resulting closed-loop system would be hybrid due to having variables that change continuously – in particular, the state z – and variables that jump – in particular, the variables of the control algorithm. This continuous-time system is revisited in Example 1.4 and Example 2.37, where a particular control algorithm leading to a hybrid closed-loop system is provided.

When in addition to flows, the state of the plant exhibits instantaneous jumps in the state (as depicted by the dashed lines in Figure 1.2), the plant model needs to be augmented with a law resetting the state and a mechanism triggering the jumps. The general difference equation model

$$z^+ = G_P(z, u)$$

is suitable to model resetting laws: when a jump occurs, the state z is instantaneously reset[5] (or reinitialized) to the value $G_P(z, u)$, where the arguments of G_P are the current value of the state z and of the input u. The new value of z after the jump is denoted as z^+. A mechanism that triggers the jumps can be formulated as (z, u) belonging to a subset of $\mathbb{R}^{n_P} \times \mathbb{R}^{m_P}$, similar to the set C_P for the flows. Denoting this set as D_P, when the state and the input satisfy

$$(z, u) \in D_P$$

the state z may exhibit a jump. The following example illustrates this construction.

Example 1.2 (Walking robot). *Consider the planar walking robot shown in Figure 1.4. The figure shows a leg being in contact with the ground. This leg is referred to as the stance leg. The other leg on the air is referred to as the swing leg. It also shows the angles θ_p and θ_s associated to the stance and swing legs, respectively, relative to an axis that is perpendicular to the flat ground. Each leg has mass equal to m_ℓ. The walking robot includes a torso with mass equal to m_t. The angle of the torso is denoted θ_t. The hip has mass equal to m_h. Torque actuators are assumed to be attached to the legs and to the torso: inputs u_p and u_s are available at the stance and swing legs, respectively, and an input u_t is available at the torso.*

---

[4]Indeed, it is not a hybrid system in the sense of the term *hybrid* as employed in this book; see § 1.2 for more details.

[5]In computer algorithms, such an assignment is typically written as $z \leftarrow G_P(z, u)$.

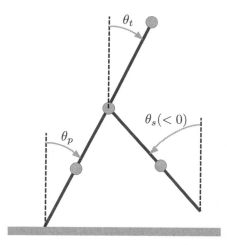

Figure 1.4: A compass model of a walking robot in two dimensions. The leg in contact with the ground is the stance leg and the leg up in the air is the swing leg.

The Euler-Lagrange method can be employed to derive the equations of motion of the walking robot in between impacts with the ground.[6] For the case of walking motion of the robot, the variation of the angles over time is given by the angular velocity associated with each angle. Denoting by ω_p, ω_s, and ω_t the angular velocity associated with θ_p, θ_s, and θ_t, respectively, and defining the vectors

$$\theta := \begin{bmatrix} \theta_p \\ \theta_s \\ \theta_t \end{bmatrix}, \qquad \omega := \begin{bmatrix} \omega_p \\ \omega_s \\ \omega_t \end{bmatrix}$$

the change of θ in between impacts is governed by the linear differential equation

$$\dot{\theta} = \omega$$

and the change of ω is governed by the nonlinear differential equation

$$\mathfrak{M}(\theta)\dot{\omega} + \mathfrak{C}(\theta, \omega)\omega + \mathfrak{N}(\theta, \omega) = \mathfrak{B}u$$

where $\mathfrak{M}$ is the (invertible) inertia matrix of the walking robot, $\mathfrak{C}$ is the Coriolis matrix, $\mathfrak{N}$ includes gravity terms and other forces acting at the joints, $\mathfrak{B}$ is the torque input matrix, and u is the vector of torque inputs with components u_p, u_s, and u_t. The two differential equations above lead to the nonlinear differential equation with state $z = (\theta, \omega) \in \mathbb{R}^3 \times \mathbb{R}^3$ and input $u = (u_p, u_s, u_t)$ given by

$$\dot{z} = \begin{bmatrix} \dot{\theta} \\ \dot{\omega} \end{bmatrix} = \begin{bmatrix} \omega \\ \mathfrak{M}^{-1}(\theta)\left(\mathfrak{B}u - \mathfrak{C}(\theta, \omega)\omega - \mathfrak{N}(\theta, \omega)\right) \end{bmatrix} =: F_P(z, u)$$

In walking motion, impacts correspond to both legs being in contact with the ground. In a widely accepted model in the literature known as the compass model,

---

[6] See Exercise 4.

this impact condition is defined as follows. Given an angle parameter ϕ, the impact occurs when the stance angle θ_p is equal to the parameter ϕ, the swing angle θ_s is equal to $-\phi$, and the stance leg is not moving backwards. This triggering mechanism is captured by the set of points $(z, u) \in \mathbb{R}^6 \times \mathbb{R}^3$ defined as

$$D_P := \left\{ (z, u) \in \mathbb{R}^6 \times \mathbb{R}^3 : \theta_p = -\theta_s = \phi, \omega_p \geq 0 \right\}$$

When an event is triggered according to condition captured by D_P, the angle θ and the angular velocity ω are updated according to the impact model given as follows:

- *After the angle of the swing leg reaches the desired angle $-\phi$, the swing leg remains planted while the other leg (the stance leg, which was planted) swings forward. To capture this role exchange of the legs, at the impacts, the angle of the swing leg θ_s is reset to the value of the angle of the stance leg θ_p right before the impact. Similarly, θ_p is reset to θ_s before the impact. Then, at jumps, the state θ is updated according to*

$$\left. \begin{array}{c} \theta_p^+ = \theta_s \\ \theta_s^+ = \theta_p \\ \theta_t^+ = \theta_t \end{array} \right\} \quad \Rightarrow \quad \theta^+ = \begin{bmatrix} 0 & 1 & 0 \\ 1 & 0 & 0 \\ 0 & 0 & 1 \end{bmatrix} \theta$$

- *The angular velocities after the impact are governed by a contact model. For the particular case of the compass model, it is typically assumed that there is no slipping, no rebounding at impacts, that the impacts occur instantaneously, that the external forces during the impact are represented by impulses resulting in instantaneous changes of the angular velocities only (that is, the angles are not affected), and that the signals applied to the torque inputs are impulse free. The resulting reset law for ω is of the form*

$$\omega^+ = \Gamma(z)$$

where the function Γ determines ω after the impact.

Then, when there is an impact – or, equivalently, a jump – due to (z, u) belonging to the set D_P defined above, z is updated according to

$$z^+ = \begin{bmatrix} \begin{bmatrix} 0 & 1 & 0 \\ 1 & 0 & 0 \\ 0 & 0 & 1 \end{bmatrix} \theta \\ \Gamma(z) \end{bmatrix} =: G_P(z, u)$$

Necessarily, the function Γ vanishes when ω is zero. Hence, the time at which jumps occur may accumulate if both θ and ω converge to zero. In fact, from $(\theta, \omega) = (0, 0)$ the system can exhibit a solution that has consecutive jumps without flow in between.

As already pointed out in Example 1.1, the condition $(z, u) \in D_P$ can be interpreted as a state-input constraint that the difference equation is restricted to. It is natural to expect that the flow set C_P for this system involves conditions on the state. The derivation of that set, of the function Γ governing the jumps, and of the differential equations governing (θ, ω) are left as an exercise to the reader; see Exercise 4.

In this book, the system to control is referred to as the *hybrid plant* as it might have hybrid dynamics, as illustrated in Example 1.2. Denoted $\mathcal{H}_P$, a hybrid plant has a state vector z, an input u, and an output y.

The variables of $\mathcal{H}_P$ collected by the state z can evolve as follows:

- **Flow**: z is allowed to evolve continuously – namely, to flow – when, for the current input u, the condition

$$(z, u) \in C_P$$

is satisfied. The set C_P is a subset of the state and input space. This set can include constraints that the state and the input have to satisfy during flows; see Example 1.1. During flows, z changes over "time" according to the differential equation

$$\dot{z} = F_P(z, u)$$

or, more generally, according to the differential inclusion

$$\dot{z} \in F_P(z, u)$$

In simple words, given an input u, the state z evolves continuously over "time," with a velocity defined by F_P when (z, u) is in the set C_P. Differential inclusion models are introduced in § A.1.

- **Jump**: The state z might also experience instantaneous changes – namely, jumps. Jumps are allowed when the condition

$$(z, u) \in D_P$$

is satisfied. The set D_P is a subset of the state and input space; see Example 1.2. Similar to C_P, the set D_P can include constraints that the state and the input have to satisfy for jumps to occur. When a jump occurs at a specific "time," the new value of the state, which is denoted z^+, is assigned via the difference equation

$$z^+ = G_P(z, u)$$

or, more generally, via the difference inclusion

$$z^+ \in G_P(z, u)$$

Indeed, given an input u, the state z changes instantaneously, to a value given by G_P when (z, u) belongs to D_P.

The reason that "time" is written is to emphasize that the notion of time has to be carefully defined. In fact, the following notion of "hybrid time" is employed to parameterize the state, the input, and the output during both flows and jumps:

Hybrid time is defined by pairs

$$(t, j)$$

where

- $t \in \mathbb{R}_{\geq 0}$ captures the duration of flows, and

- $j \in \mathbb{N}$ indicates the number of jumps.

In simple words, the hybrid time instant (t, j) indicates the amount of flow and the number of jumps in the state, input, and output of the system. In particular, the value of z after it has flowed for t seconds and jumped j times is given by $z(t, j)$.

This hybrid parameterization of time gives meaning to the change of z during flows, which in $\mathcal{H}_P$ is denoted as $\dot{z}$ and corresponds to the derivative of z with respect to t. As stated earlier, the change of z at jumps is denoted as z^+. Indeed, the notation z^+ corresponds to the new value of z at jumps.

A *hybrid equation* model of a plant with hybrid dynamics is given by

$$
\mathcal{H}_P : \begin{cases} (z, u) \in C_P & \dot{z} = F_P(z, u) \\ (z, u) \in D_P & z^+ = G_P(z, u) \\ & y = h(z) \end{cases}
$$

where z is the state of the plant and takes values from the Euclidean space $\mathbb{R}^{n_P}$, u is the input to the plant and takes values from the Euclidean space $\mathbb{R}^{m_P}$, and y is its output and takes values from the output space $\mathbb{R}^{r_P}$. At times, for convenience, X_P and U are introduced to denote possible values for z and u, respectively. In general, these allowed ranges are already enforced by the definitions of C_P and D_P, as the examples in this section illustrate.

At times, it is convenient to allow the maps F_P and G_P to be set valued. In such a case, the evaluation of F_P and of G_P at (z, u) may result in a collection of more than one point, rather than just a singleton, as the following model captures.

A *hybrid inclusion* model of a plant with hybrid dynamics is given by

$$
\mathcal{H}_P : \begin{cases} (z, u) \in C_P & \dot{z} \in F_P(z, u) \\ (z, u) \in D_P & z^+ \in G_P(z, u) \\ & y = h(z) \end{cases}
$$

The symbol $\in$ in the *differential inclusion* $\dot{z} \in F_P(z, u)$ indicates that the velocity of z is an element taken from the set $F_P(z, u)$. Similarly, the symbol $\in$ in the *difference inclusion* $z^+ \in G_P(z, u)$ indicates that the new value of the state z is an element taken from the set $G_P(z, u)$.

By allowing F_P and G_P to be set-valued maps, a hybrid system model can conveniently capture multiple possibilities at a given time, in particular, to model uncertainty on the parameters, disturbances on the system, or lack of knowledge of the actual behavior of the system. See § A.2 for details.

The *data* of the hybrid plant is defined by

$$(C_P, F_P, D_P, G_P, h)$$

Due to the roles they play, the following terminology is employed:

- The set C_P is referred to as the *flow set*;

- The map F_P is referred to as the *flow map*;

- The set D_P is referred to as the *jump set*;

- The map G_P is referred to as the *jump map*; and

- The map h is referred to as the *output map* of the hybrid plant.

The general model for $\mathcal{H}_P$ captures the dynamics of (constrained or unconstrained) continuous-time nonlinear systems as given earlier when $D_P = \emptyset$ and G_P is arbitrary. In such a case, the data of $\mathcal{H}_P$ is given by

$$(C_P, F_P, \emptyset, \star, h)$$

where $\star$ indicates that the jump map is arbitrary. Similarly, the general model for $\mathcal{H}_P$ captures the dynamics of (constrained or unconstrained) discrete-time systems when $C_P = \emptyset$ and F_P is arbitrary. Note that while the output equation does not explicitly include a constraint on (z, u), the output map is only evaluated along solutions, so it inherits the conditions involving z and u imposed by C_P and D_P during flows and jumps, respectively. These conditions can be considered to be constraints that the state and the input have to satisfy during flows and at jumps, respectively.

To simplify the notation, at times, the input u may not be written when defining objects that do not depend on it explicitly. For instance, if F_P depends only on z, then $F_P(z)$ may be written, while if C_P only depends on z, it might be defined as a subset of $\mathbb{R}^{n_P}$.

To control a hybrid plant $\mathcal{H}_P$, control algorithms that can cope with the nonlinearities induced by the flow and jump equations/inclusions are required. In general, feedback controllers designed using classical techniques from the continuous-time and discrete-time domains may fall short. Due to this limitation, hybrid feedback controllers are considered. Following the hybrid plant model above, hybrid controllers are denoted $\mathcal{H}_K$, have state η, input v, and output ζ. Similar to $\mathcal{H}_P$, the state η is allowed to evolve continuously when (v, η) is in the flow set, and jump when in the jump set of the controller. More precisely, a hybrid inclusion model of a hybrid controller is given by

$$\mathcal{H}_K : \begin{cases} (v, \eta) \in C_K & \dot{\eta} \in F_K(v, \eta) \\ (v, \eta) \in D_K & \eta^+ \in G_K(v, \eta) \\ & \zeta = \kappa(v, \eta) \end{cases}$$

Its state takes values from the subset X_K of the Euclidean space $\mathbb{R}^{n_K}$, its input takes values from $\mathbb{R}^{r_P}$, and its output takes values from the output space $\mathbb{R}^{m_P}$. Note that instead of (η, v), (v, η) is written in $\mathcal{H}_K$ so as simplify the forthcoming definition of the closed-loop system. The objects

$$(C_K, F_K, D_K, G_K, \kappa)$$

define the *data* of the hybrid equation – or hybrid inclusion when F_K or G_K are set valued – defining the hybrid controller. As with the hybrid plant, C_K is the flow set, F_K is the flow map, D_K is the jump set, G_K is the jump map, and κ is the output map of the controller. Hybrid controllers are designed throughout this book using different techniques.

A concrete example of a hybrid controller modeled as $\mathcal{H}_K$ is sample-and-hold control. The control of dynamical systems based on discrete samples of the output of the plant and discrete updates of its input is a widely used technique in practice. Unlike purely discrete-time models, a hybrid system model for sample-and-hold control allows to capture the evolution of the variables in between sampling times.[7] It also allows to model aperiodic sampling. The following example provides hybrid models of a sample-and-hold controller with these features.

Example 1.3 (Sample-and-hold control). *Consider the control of a continuous-time plant by a sample-and-hold controller that implements the static feedback law*

$$u = \kappa_c(y)$$

where κ_c is a function designed to guarantee a particular property of interest for the closed-loop system. One such controller includes the following state variables to implement the sample-and-hold mechanism:

- *A timer state τ that triggers the sampling and hold events; and*

- *A memory state ℓ_v that assigns the input to the plant using a zero-order hold mechanism:*

    - *At each event, the memory state ℓ_v is reset to a value given by the evaluation of the feedback κ_c at the current value of the output y;*

    - *In between events, the value of the memory state ℓ_v is kept unchanged.*

In the standard periodic sample-and-hold implementation, the events are triggered when the timer reaches a positive constant T^* defining the sampling period. A jump set D_K of the controller triggering such events includes the condition

$$\tau \geq T^*$$

so as to trigger one such event every T^* seconds. At such events, the timer is reset to zero via

$$\tau^+ = 0$$

and the memory state ℓ_v is updated to the current value of the state-feedback law via

$$\ell_v^+ = \kappa_c(y)$$

---

[7]See § 1.2.5 for more details.

In between events, the memory state is kept constant so as to apply a constant input to the plant in between events.

The hybrid controller $\mathcal{H}_K$ has state η defined as $(\ell_v, \tau) \in \mathbb{R}^{r_P} \times \mathbb{R}_{\geq 0}$. Its input is denoted $v \in \mathbb{R}^{r_P}$ and is assigned to the output $y = h(z)$ of the plant. The data $(C_K, F_K, D_K, G_K, \kappa)$ of $\mathcal{H}_K$ is given as follows:

$$C_K = \{\eta \in \mathbb{R}^{r_P} \times \mathbb{R}_{\geq 0} : \tau \in [0, T^*]\}$$

$$F_K(\eta) = \begin{bmatrix} 0 \\ 1 \end{bmatrix} \qquad \forall \eta \in C_K$$

$$D_K = \{(y, \eta) \in \mathbb{R}^{r_P} \times \mathbb{R}^{r_P} \times \mathbb{R}_{\geq 0} : \tau \geq T^*\}$$

$$G_K(y, \eta) = \begin{bmatrix} \kappa_c(y) \\ 0 \end{bmatrix} \qquad \forall (y, \eta) \in D_K$$

$$\kappa(\eta) = \ell_v$$

Note that the definition of C_K is such that flows are allowed when $\tau \in [0, T^*]$, and that the plant output y does not play a role; hence, y is omitted in the definition of C_K and F_K. Since, from the definition of F_K above, $\dot{\ell}_v = 0$ and $\dot{\tau} = 1$, flows from $\tau \geq T^*$ are not possible, as flowing from such points would violate the allowed range for τ imposed by C_P. The definition of G_K resets, at each event, the timer to zero and the memory state ℓ_v to $\kappa_c(y)$. In this way, ℓ_v stores the current value of the feedback in between events. The output function of the controller is defined as $\kappa(\eta) = \ell_v$ so that the memorized output feedback law is applied to the plant following the sample-and-hold control paradigm.

The construction of the hybrid controller implementing sample-and-hold control proposed above can be easily extended to incorporate other mechanisms. For instance, aperiodic sampling can be modeled by enlarging the flow set C_K and the jump set D_K. For instance, the choices

$$C_K = \{\eta \in \mathbb{R}^{r_P} \times \mathbb{R}_{\geq 0} : \tau \in [0, T^*_{\max}]\}$$

$$D_K = \{(y, \eta) \in \mathbb{R}^{r_P} \times \mathbb{R}^{r_P} \times \mathbb{R}_{\geq 0} : \tau \geq T^*\}$$

with $T^*_{\max} \geq T^*$ allows for sampling and hold events to occur no sooner than T^* seconds and no later than $T^*_{\max}$ seconds. In fact, due to the overlap between C_P and D_P, the hybrid controller may exhibit a jump at any value of τ in $[T^*, \infty)$. More general constructions inspired by state-triggered control techniques are possible. Such hybrid control algorithms are the focus of Chapter 5.

Though the model of the plant in Example 1.1 is not hybrid as the term *hybrid* is employed in this book (see §2.2), as argued in the following example, the use of a hybrid control algorithm arises naturally to solve the associated control objective.

Example 1.4 (DC/AC inverter, revisited). *Consider the power conversion control problem formulated in Example 1.1, which consists of inverting DC input voltage to a desired AC output voltage. Due to the fact that no particular position of the switches $S_1 - S_4$ in the diagram shown in Figure 1.3 leads to an AC output voltage, it is mandatory to recurrently toggle the switches to be able to generate such an output signal. Fortunately, it is possible to design a hybrid control algorithm that generates pseudo-sinusoidal state trajectories as shown in Figure 1.5. The hybrid control strategy leading to the desired state evolution shown there is as follows:*

- *Define an ideal periodic trajectory, referred to as the reference, which is to be approximated as close as possible. One such reference is the perfectly circular trajectory in Figure 1.5 (solid trace, left).*

- *Construct switching sets that trap the reference. These sets are denoted in dash-dot line style in the same figure and define a tracking band around the set of points described by the reference.*

- *When the state $z = (i_L, v_{C_a})$ of the circuit reaches one of the switching sets, choose an appropriate value of u in $\{-1, 0, 1\}$ that not only keeps the state within the band but also keeps it moving in the same general (clockwise) direction.*

The reference is defined as the steady-state trajectory of the RLC filter, which is a current-voltage pair that makes

$$V(z) := \left(\frac{i_L}{a}\right)^2 + \left(\frac{v_{C_a}}{b}\right)^2$$

constant, where $a = \frac{1}{\sqrt{R^2 + (L\omega - \frac{1}{C_a\omega})^2}}$ and $b - \frac{a}{C_a\omega}$. Denoting by c the constant value of V along the chosen reference trajectory, the switching sets are defined as subsets of level sets of V: for constants c_i and c_o such that $0 < c_i < c < c_o$, define

$$S_i := \{z \in \mathbb{R}^2 : V(z) = c_i\}, \qquad S_o := \{z \in \mathbb{R}^2 : V(z) = c_o\}$$

Then, when z reaches S_i or S_o, the hybrid controller selects a value for u that keeps z within the tracking band. The set of states z in the tracking band satisfy

$$\{z \in \mathbb{R}^2 : c_i \leq V(z) \leq c_o\}$$

A key feature of this set is that the design parameters c_i and c_o adjust the precision of the approximation provided by the sinusoidal output; as $c_o - c_i$ converges to zero while satisfying $0 < c_i < 1 < c_o$, the z components of the resulting closed-loop trajectories are "closer" to the reference trajectory. Unlike pulse-width modulation techniques, this algorithm uses state-dependent conditions to toggle the value of the (discrete-valued) input u of the model in Example 1.1. The details of this algorithm are given later on in this book, in Example 2.37 and Example 11.12, in the context of invariant sets. Associated simulation files are at @BookSite/Simulation/DCAC.

The control of the plant $\mathcal{H}_P$ via the controller $\mathcal{H}_K$ defines an interconnection through the input/output assignment

$$u = \zeta, \qquad v = y$$

Such an interconnection is depicted in Figure 1.1 for the case when the interfaces are omitted or included in the models for $\mathcal{H}_P$ and $\mathcal{H}_K$. The resulting closed-loop system is a hybrid dynamical system given in terms of a hybrid equation/inclusion. This system is denoted $\mathcal{H}$. Its state is given by x, which is the stack of the plant state z and the controller state η. As shown in Chapter 2, its data can be constructed from

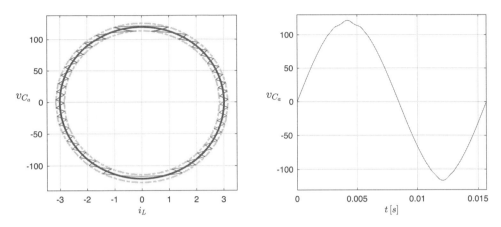

Figure 1.5: A closed-loop solution and output voltage using the hybrid controller for power conversion outlined in Example 1.4. *Source: Chai and Sanfelice, 2014 [2]. Reproduced by permission of IEEE.*

the data (C_P, F_P, D_P, G_P, h) and $(C_K, F_K, D_K, G_K, \kappa)$ of each of the subsystems. In principle, when the inputs of $\mathcal{H}_P$ and of $\mathcal{H}_K$ are all assigned, the closed-loop system can be described by a hybrid system without inputs of the form

$$\mathcal{H} \; : \; \begin{cases} x \in C & \dot{x} \;\; \in F(x) \\ x \in D & x^+ \in G(x) \end{cases}$$

Inputs and outputs can be easily added to the model. In either case, the general hybrid control problem is as follows:

Hybrid Control Problem*: Design a controller $\mathcal{H}_K$ for the given plant $\mathcal{H}_P$ such that desired properties are conferred on the closed-loop system $\mathcal{H}$.*

Desired properties of common interest include the following:

- Global asymptotic stability of an equilibrium point, a set, or a time-varying trajectory;

- Robustness of the behavior of the closed-loop system, in particular, robustness of asymptotic stability;

- Safety of the closed-loop system or of the individual systems, which is the property that particular state components of the systems do not take on unsafe values.

Fortunately, hybrid feedback control algorithms have the capability of not only assure these properties but also of solving problems that classical feedback control algorithms cannot solve. The next section describes control problems for which classical feedback control falls short of conferring a desired property on the system. On the other hand, the problems can be solved via hybrid feedback control.

1.2   WHY HYBRID CONTROL?

Classical control theory provides powerful tools for the analysis of the closed-loop system and the design of feedback controllers. These tools are suitable when the dynamics of the system under study are of continuous-time or of discrete-time nature, but do not apply to systems that combine both continuous and discrete behavior. Additionally, the closed-loop system properties obtained with feedback controllers designed within those frameworks are limited by the capabilities of continuous-time and discrete-time control. For example, continuous-time controllers are not always capable of robustly and globally stabilizing a nonlinear system to an equilibrium point (or to a set), while discrete-time feedback controllers are unable to react to the behavior of the plant in between samples. Hybrid feedback control theory has the power to remove the limitations of the classical frameworks and solve a broad range of problems that are not solvable within the classical frameworks. Key reasons for why using hybrid control algorithms are explained next.

1.2.1   Hybrid Models Capture Rich Behavior

Model-based control techniques require a model of the plant that is accurate enough, yet mathematically tractable. Hybrid system models have the remarkable capability of capturing the evolution of variables that flow – as in continuous-time systems – and, at times, jump – as in discrete-time systems. A mathematical model with such a capability can describe the evolution of a variable that changes continuously for some time and, under certain conditions, is reset to a new value as Figure 1.2 illustrates. Flows intertwined with jumps are required in control algorithms that perform advanced tasks, such as the supervision of multiple controllers and the self-reconfiguration of the controller upon events. Such tasks can be accomplished using a hybrid controller $\mathcal{H}_K$ that includes logic variables, timers, and memory states in its state η.

The capability of incorporating such variables in an algorithm is a unique feature of hybrid control, making it a powerful and versatile tool for the solution of a myriad of problems that classical control theory is unable to solve.

The state variables employed by the algorithms presented in this book include the following:

1. *Logic variables*: these state variables take values from a discrete set, remain constant during flows, and, at jumps, are reset to a value in a discrete set. A logic variable that is part of the state η of a hybrid controller $\mathcal{H}_K$ is denoted q. This variable takes values from the discrete set Q. It flows according to the trivial differential equation

$$\dot{q} = 0$$

   and at jumps, is reset via

$$q^+ \in G_{K,q}(v, \eta)$$

   where $G_{K,q}$ is the q-th component of the jump map of $\mathcal{H}_K$ – the map $G_{K,q}$ collects all possible values of q allowed after jumps.

A particular scenario using a logic variable is when multiple feedback laws are available for stabilization. In such a case, one of the components of η, say the first one, would be a logic variable, in which case $\eta_1 = q$. Then, the first component of G_K should be defined to appropriately select the control law after jumps. In the case of static state-feedback laws, the output of the controller would be $\zeta = \kappa(z, \eta) := \kappa_q(z)$, where κ_q denotes the feedback law associated to each possible value of the logic variable q. In Example 1.1, which features a DC/AC inverter, the input is only allowed to take on discrete values, which, as argued in Example 1.4, would lead to a controller with a logic variable that keeps track of the position of the switches of the inverter. Hybrid control algorithms with logic variables that define multiple modes of operation on which different control laws are used are presented in Chapter 4, Chapter 6, Chapter 7, and Chapter 8.

2. *Timers*: these state variables count the ordinary time elapsed and get reset to zero when they reach a threshold. A timer τ typically takes values from a set of the form $[0, T^*]$, where $T^* \in \mathbb{R}_{>0}$ is a constant defining the threshold at which the timer is reset. A timer flows according to the differential equation

$$\dot{\tau} = 1$$

whose right-hand side defines the τ component of the flow map F_K of the hybrid controller. When

$$\tau = T^* \qquad (\text{or } \tau \geq T^*)$$

the timer τ is reset to zero via

$$\tau^+ = 0$$

This choice leads to a τ component of the jump map of $\mathcal{H}_K$ that is equal to zero.

The events in the sample-and-hold control algorithm in Example 1.3 can be triggered when a timer expires, using a triggering condition that involves timers. Other control algorithms using timers are introduced in Chapter 5 and Chapter 6.

3. *Memory states*: these variables act as data buffers that store values of the state, inputs, or functions of the two. Similar to logic variables, they remain constant until a jump in the controller updates their value. A memory variable ℓ flows according to the differential equation

$$\dot{\ell} = 0$$

and is updated via

$$\ell^+ \in G_{K,\ell}(\eta, v)$$

at jumps, where $G_{K,\ell}$ is the ℓ component of the jump map G_K of $\mathcal{H}_K$.

The sample-and-hold controller in Example 1.3 uses a memory state denoted ℓ_v to record the output feedback law at each sampling event. Other control algorithms using memory states are introduced in Chapter 5 and Chapter 6.

The following example introduces a system of practical relevance for which a hybrid controller with logic variables and a timer can achieve the control objective.

Example 1.5 (Multi-link pendulum). *Consider the two-link pendulum shown in Figure 1.6, known as the pendubot. It consists of two rigid links with torque actuation only at one of the joints – the lower joint of link 1 – and angular encoders at both joints. The dynamics of the angular positions ϕ_1 and ϕ_2, and their time derivatives defining the angular velocities ω_1 and ω_2, respectively, can be obtained via the Euler-Lagrange method. To this end, define the state vector collecting these variables as $z = (\phi_1, \omega_1, \phi_2, \omega_2) \in \mathbb{R}^4$ and the torque input as $u \in \mathbb{R}$. Using the*

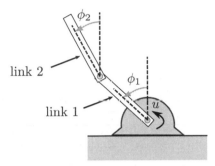

Figure 1.6: The pendubot system: two links with torque actuation only at the lower joint of link 1. The pendulum angles are denoted by ϕ_1 and ϕ_2 for link 1 and link 2, respectively. The torque input is denoted by u.

Euler-Lagrange method,[8] it can be shown that the dynamics of the pendubot are given by

$$\dot{z} = F_P(z, u) := \begin{bmatrix} \omega_1 \\ F_{P,2}(z, u) \\ \omega_2 \\ F_{P,4}(z, u) \end{bmatrix}$$

where $F_{P,2}$ and $F_{P,4}$ are smooth functions defining the change of angular speed as a function of the state and input. The interest is in a control algorithm that swings up the pendubot from every initial condition, with robustness to small perturbations. In other words, an algorithm that globally and robustly asymptotically stabilizes the configuration defined by the two links "up" with zero angular velocities is desired.

One of the key challenges in swinging up the pendubot is the fact that it accepts multiple equilibrium configurations. As such, when an energy-based feedback controller is used, the controller may get stuck at equilibria other than the desired one. These equilibrium configurations are depicted in Figure 1.7. To cope with multiple equilibria and overcome the issue of the controller getting stuck, a hybrid controller that features multiple modes of operation to perform the following tasks is proposed:

1. *Catch nearby z_u^*: From points nearby z_u^*, the controller applies a static state-feedback controller that locally asymptotically stabilizes the point z_u^*;*

2. *Throw from z_{ur}^* and z_{ru}^*: From points nearby z_{ur}^* or z_{ru}^*, the controller applies a constant input that drives the state away from nearby those points;*

---

[8]Exercise 4 poses a modeling problem for which this method can be used.

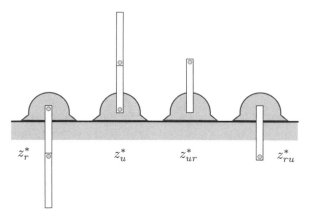

Figure 1.7: Equilibrium configurations for the pendubot: z_r^* is the resting equilibrium with $\phi_1 = \pi$, $\omega_1 = 0$, $\phi_2 = \pi$, $\omega_2 = 0$; z_u^* is the upright equilibrium with $\phi_1 = 0$, $\omega_1 = 0$, $\phi_2 = 0$, $\omega_2 = 0$; z_{ur}^* is the upright/resting equilibrium with $\phi_1 = 0$, $\omega_1 = 0$, $\phi_2 = \pi$, $\omega_2 = 0$; and z_{ru}^* is the resting/upright equilibrium with $\phi_1 = \pi$, $\omega_1 = 0$, $\phi_2 = 0$, $\omega_2 = 0$. (Hence, the value of N in the dedication page is revealed.)

3. *Throw from z_r^*: From points nearby z_r^*, the controller applies an input signal that steers the pendubot to a neighborhood of z_u^*;*

4. *Catch to some equilibria: From any other point, the controller applies a static state-feedback controller that removes energy from the system so as to steer the state to nearby one of the equilibrium configurations;*

5. *Recover from throw: If when performing the throw from z_r^* in item 3 above, the pendulum does not reach a neighborhood of z_u^* in the expected amount of time, then the algorithm executes the catch operation in item 4. This logic enables the algorithm to retry the throw in item 3 until the state of the pendubot reaches a neighborhood of z_u^*.*

These tasks can be coordinated by the hybrid controller through the use of logic variables that keep track of the task being executed, and a timer that keeps track of the time while an input (open-loop) signal is applied to the plant. Figure 1.8 depicts a hybrid automaton[9] implementing the control logic outlined in items 1-5 above. Complete details abouts the structure and design of this hybrid controller are given in Example 6.1.

Although they are needed in the implementation of most feedback control algorithms, logic variables are not typically included in the model used for feedback control design. Rather, they are heuristically added after the design – typically, during implementation. On the other hand, logic variables can be systematically incorporated in the model of a hybrid controller. In fact, the data of $\mathcal{H}_K$ can be defined to implement the dynamics emerging from the rules governing the logic variables, in that way, capturing the entire behavior of the controller in a compact mathematical model.

---

[9]See § 2.6 for more details about modeling hybrid automata.

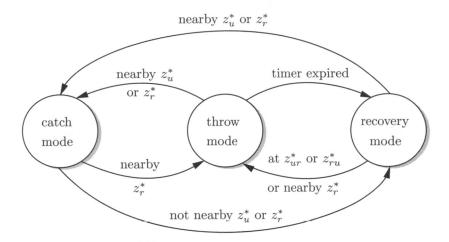

Figure 1.8: Outline of the hybrid automaton associated to the hybrid control strategy for the pendubot system in Example 1.5.

1.2.2 Continuous-Time Systems not Stabilizable via Continuous State-Feedback Can Be Stabilized via Hybrid Control

Many continuous-time plants cannot be stabilized to a point (or to a set) via continuous static state-feedback laws. The challenge in stabilizing a continuous-time system to a point via continuous feedback is that local controllability is not enough, and additional properties of the right-hand side are required. To illustrate this point, consider the problem of stabilizing the position and orientation of a mobile robot given by the unicycle model with state $z = (\xi, \theta)$ and dynamics

$$\dot{\xi} = \begin{bmatrix} \cos(\theta) \\ \sin(\theta) \end{bmatrix} \vartheta, \qquad \dot{\theta} = \omega$$

where $\xi \in \mathbb{R}^2$ denotes the planar position and $\theta \in \mathbb{R}$ its orientation. The inputs $\vartheta \in \mathbb{R}$ and $\omega \in \mathbb{R}$ denote the forward velocity and angular velocity inputs, respectively. The mobile robot is oriented towards the right of the planar coordinate system when $\theta = 0$. Without loss of generality, it is assumed that θ is positive when its orientation changes counterclockwise from $\theta = 0$ and that the control design objective is to asymptotically stabilize the point $(\xi, \theta) = (0, 0) \in \mathbb{R}^2 \times \mathbb{R}$. Unfortunately, this control system fails Brockett's well-known condition for (local) asymptotic stabilization via *a continuous feedback*, that is, via a static state-feedback law that is continuous.

Nevertheless, asymptotic stability of the desired configuration can be achieved using a hybrid controller. The details of one such hybrid control strategy are given in Example 8.11 in the context of supervisory control (see Chapter 8). In simple words, the idea consists of supervising two controllers designed as follows. The first controller is a local hybrid controller that is able to locally asymptotically stabilize θ to zero while simultaneously driving the position to zero. Such a property is possible if the controller induces a persistency-of-excitation type behavior on the orientation of the vehicle. The second controller consists of a hybrid controller capable of steering the vehicle to a neighborhood of the origin – both in position and orientation – that is contained in the region of operation of the first controller. A supervisory algorithm with state $q \in Q := \{0, 1\}$ is then employed to select the controller to be applied to the vehicle.

It can be shown that with the hybrid controller outlined above, the set $\{0\} \times Q$ is *asymptotically stable* for the resulting hybrid closed-loop system; see Example 8.11. Moreover, the stability property guaranteed by this controller is both global and robust to small perturbations. The analysis of controllers accomplishing such useful properties through the use of logic variables are presented in Chapter 4, Chapter 6, Chapter 7, and Chapter 8.

1.2.3   Almost Global Asymptotic Stability Turns Global

Even if the stabilization of a point is possible via continuous feedback, the controller might not guarantee the property to hold globally. Global asymptotic stability is particularly difficult to attain when the state of the plant is restricted to a manifold. In fact, global asymptotic stabilization of an isolated point for a dynamical system with state that lives on a manifold may not be possible via a continuous static state-feedback control law.

To illustrate this challenge to global stabilization, consider the problem of globally stabilizing the following system with state $z \in \mathbb{S}^1$ and input $u \in \mathbb{R}$ to a set-point z^*:

$$\dot{z} = u \begin{bmatrix} 0 & -1 \\ 1 & 0 \end{bmatrix} z$$

This system models a particle traveling on the unit circle centered at the origin of the plane. The state z denotes the position of the particle, or its angle, which can be extracted from the orientation of z on the plane. The unit circle is a compact manifold. It is denoted by $\mathbb{S}^1$. For this system, the sign of the control input u determines the direction of motion of the particle. Now, for simplicity, let the point to stabilize be $z^* = (1, 0) \in \mathbb{S}^1$. A globally asymptotically stabilizing controller has to guarantee the following properties of solutions starting from $\mathbb{S}^1$:

- Solutions stay in $\mathbb{S}^1$;

- Solutions converge to z^* – or, equivalently, z^* is globally attractive;

- If a solution starts nearby z^* then it stays nearby z^* (and within $\mathbb{S}^1$).

It is easy to show that regardless of the choice of the input u, this manifold is forward invariant. Hence, the first property in this list holds for free. To satisfy the other two properties, a natural continuous static state-feedback controller to consider is

$$\kappa(z) = -z_2 \qquad \forall z \in \mathbb{S}^1$$

Then, let the function V be defined as

$$V(z) = 1 - z_1 \qquad \forall z \in \mathbb{S}^1$$

The state-feedback law κ defined above leads to the following change of V over time:

$$\left\langle \nabla V(z), -z_2 \begin{bmatrix} 0 & -1 \\ 1 & 0 \end{bmatrix} z \right\rangle = -(1 - z_1^2) \qquad \forall z \in \mathbb{S}^1$$

This expression is equivalent to the time derivative of V along the solutions to the system, which is typically written as $\dot{V}$. Being defined on $\mathbb{S}^1$, note that V only vanishes at $z = z^*$ and is positive everywhere else. A function with such a property is said to be positive definite with respect to z^*.[10] The quantity $\dot{V}$ is negative at points z in $\mathbb{S}^1$ satisfying $z_1^2 \neq 1$. Using Lyapunov stability arguments (see §3.3.1), solutions to the closed-loop system that start from such points converge to z^*. However, solutions that start from $-z^*$ do not converge to z^* since they stay at $-z^*$ for all time. As a consequence, the continuous static state-feedback law $z \mapsto \kappa(z) := -z_2$ guarantees stability of z^* but not global attractivity of that point. It only assures convergence to z^* for solutions starting from $\mathbb{S}^1 \setminus \{-z^*\}$.

> The property induced by κ outlined above is an *almost global asymptotic stability property*, in the sense that there exist solutions from isolated points (precisely, $-z^*$) that do not converge to z^*. Attractivity of z^* holds from almost every point in $\mathbb{S}^1$. Unfortunately, this almost global asymptotic stability property on $\mathbb{S}^1 \setminus \{-z^*\}$ cannot be extended to the entire manifold $\mathbb{S}^1$ by preserving continuity of the state-feedback law. This limitation motivates the design of a hybrid controller that, unlike continuous static state-feedback controllers, induces global attractivity.

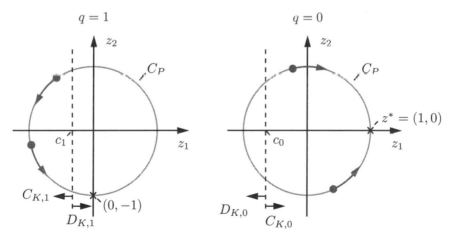

Figure 1.9: The modes of operation and associated data of the hybrid control strategy outlined in §1.2.3 for global (and robust) asymptotic stabilization of z^*.

A hybrid controller can be designed to globally (and robustly) asymptotically stabilize the point z^*. The analysis above suggests that a modification around the point $-z^*$ of the proposed continuous state-feedback control law could lead to a globally stabilizing controller. The intuition behind such a strategy is as follows:

- In a connected region of $\mathbb{S}^1$ not including $-z^*$ but including z^* (in its interior), define the feedback law to be equal to $-z_2$;

---

[10]See Definition A.25.

- At the other points in $\mathbb{S}^1$, define the new feedback law so as to drive the solutions to the region where the control law is equal to $-z_1$.

A particular construction of the sets of a hybrid controller implementing such strategy is depicted in Figure 1.9. With c_0 and c_1 being constants satisfying $-1 < c_0 < c_1 < 0$, the set of points in $\mathbb{S}^1$ that belong to C_P with $z_1 \leq c_1$ define $C_{K,1}$, which is where z is allowed to flow with the new feedback law – the original feedback law is only used on $C_{K,0}$. Such a mechanism can be implemented by a hybrid controller $\mathcal{H}_K$ with state given by a logic variable $\eta = q \in Q := \{0, 1\}$ that is toggled based on the value of the state z. More details about this implementation are given in Exercise 15.

1.2.4 Nonrobust Stability Becomes Robust

The lack of continuous state-feedback laws stabilizing systems like the mobile robot in §1.2.2 or the system on the unit circle in §1.2.3 motivate the use of discontinuous feedback laws for global stabilization. It turns out that discontinuous state-feedback laws can be designed to solve those stabilization problems, globally. However, the resulting global stability property may not necessarily be robust to measurement noise, no matter how small the noise is.

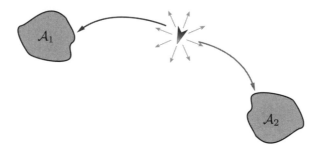

Figure 1.10: A stabilization problem in robotics involving two sets: a mobile robot needs to pick a destination between two different locations.

To illustrate this point in a concrete example, let $z^* > 0$ and consider the stabilization of the two-point set

$$\mathcal{A} = \{-z^*\} \cup \{z^*\}$$

for the linear time-invariant system

$$\dot{z} = u$$

where $z \in \mathbb{R}$ is the state and $u \in \mathbb{R}$ is the input. Two-point set (or two-set) stabilization problems emerge in robotics, in particular, when robotic manipulators are used to transport objects between locations or when a mobile vehicle needs to pick a destination between two different options. Figure 1.10 depicts one such scenario.

It is easy to verify that there is no continuous state-feedback law that globally asymptotically stabilizes $\mathcal{A}$, but that the discontinuous state-feedback law

$$\kappa(z) := -z + \text{sign}(z)z^* \qquad \forall z \in \mathbb{R}$$

does indeed render $\mathcal{A}$ globally asymptotically stable. In fact, when $z > 0$, the closed-loop system resulting from using κ is

$$\dot{z} = -z + z^*$$

whose solutions remain positive and converge to z^* exponentially. Similarly, when $z < 0$, the closed-loop solutions remain negative and converge to $-z^*$. However, this attractivity property is not robust to noise in the measurements of z. Indeed, when the feedback law measures $z + w_y$ instead of z, with w_y being a disturbance, there exist solutions that do not converge to a small neighborhood of z^*, no matter how small the magnitude of w_y is.

To visualize the undesirable effect of w_y on the nominal stability property of the closed-loop system, let $\varepsilon \in (0, z^*)$ and define the disturbance $t \mapsto w_y(t)$ as the periodic signal with period equal to $\delta_1 + \delta_2$ and values $w_y(t) = -\varepsilon$ for all $t \in [0, \delta_1]$, $w_y(t) = 0$ for all $t \in (\delta_1, \delta_2]$, and so on, where $\delta_1 = -\ln\left(\frac{z^* - \frac{3}{4}\varepsilon}{z^* - \frac{\varepsilon}{2}}\right)$ and $\delta_2 = -\ln\left(\frac{z^* - \frac{\varepsilon}{2}}{z^* - \frac{\varepsilon}{4}}\right)$. Note that the magnitude of the noise w_y is $\sup_{t \in [0, \infty)} |w_y(t)| = \varepsilon$. In particular, for the initial condition $z(0) = \frac{\varepsilon}{2}$, the unique solution to the closed-loop system with the discontinuous control law κ above satisfies

$$|z(t)| \in \left[\frac{\varepsilon}{4}, \frac{\varepsilon}{2}\right] \qquad \forall t \in \mathbb{R}_{\geq 0}$$

Convergence to a neighborhood of $\mathcal{A}$ is not possible, even when the magnitude ε of the noise w_y is arbitrarily small. Fortunately, a hybrid feedback controller can be designed to globally and robustly asymptotically stabilize the two-point set $\mathcal{A}$, where the robustness property shields the closed-loop system from bounded noise signals w_y like the one above.

One such hybrid controller combines two static state-feedback laws using a simple logic implementing binary hysteresis.[11] The controller has a logic variable as state, $\eta = q \in Q = \{0, 1\}$, which defines two modes of operation. To outline the control logic, let c be a constant satisfying $c \in (0, z^*)$, where z^* is the constant in the definition of the two-point set $\mathcal{A}$. The hybrid controller enforces the following behavior:

- If the initial value of z is larger than $-c$, and q is initialized at mode 0, then a control law that steers z to z^* is applied.

- If the initial value of z is smaller than c, and q is initially equal to mode 1, then q is reset to 0 and the control law that steers z to z^* is applied.

Similarly, if the initial value of z is smaller than c and q is set to 1 initially, then

---

[11]The lines with arrows in the cover of this book define a binary hysteresis function.

the controller steers z to $-z^*$. With an appropriately defined notion of solution, which is given in § 2.3.3, the other initial conditions also lead to z converging to $\mathcal{A}$. Furthermore, solutions starting nearby $\mathcal{A}$ stay nearby $\mathcal{A}$. Hence, the set $\mathcal{A}$ is globally asymptotically stable. More importantly, as long as the measurement noise $t \mapsto w_y(t)$ is such that $\sup_{t \in [0,\infty)} |w_y(t)| < c$, solutions to the closed loop would not get stuck at points away from $\mathcal{A}$ as they would when using the discontinuous control law given above.

To provide a glimpse to the forthcoming mathematical models of hybrid controllers $\mathcal{H}_K$ and resulting hybrid closed-loop systems $\mathcal{H}$, the mechanism outlined above is implemented by the following hybrid controller $\mathcal{H}_K$ modeled as a hybrid equation with data $(C_K, F_K, D_K, G_K, \kappa)$. The algorithm measures the plant state z and its data is defined as follows:

$$C_K = \bigcup_{q \in Q} (C_{K,q} \times \{q\}), \quad \begin{cases} C_{K,0} = \{z \in \mathbb{R} : z \geq -c\} \\ C_{K,1} = \{z \in \mathbb{R} : z \leq c\} \end{cases}$$

$$F_K(z,q) = 0 \qquad \forall (z,q) \in C_K$$

$$D_K = \bigcup_{q \in Q} (D_{K,q} \times \{q\}), \quad \begin{cases} D_{K,0} = \{z \in \mathbb{R} : z \leq -c\} \\ D_{K,1} = \{z \in \mathbb{R} : z \geq c\} \end{cases}$$

$$G_K(z,q) = 1 - q \qquad \forall (z,q) \in D_K$$

$$\kappa(z,q) = \kappa_q(z), \quad \begin{cases} \kappa_1(z) = -z + z^* \; \forall z \in C_{K,0} \\ \kappa_2(z) = -z - z^* \; \forall z \in C_{K,1} \end{cases}$$

where $\zeta = \kappa(z,q)$ is its output, which, when controlling the plant $\dot{z} = u$, is assigned to the plant input u. Furthermore, the hybrid closed-loop system $\mathcal{H}$ resulting from using this hybrid controller is given by

$$\mathcal{H} : \begin{cases} (1-2q)z \geq -c & \dot{z} = -z + (1-2q)z^*, & \dot{q} = 0 \\ (1-2q)z \leq -c & z^+ = z, & q^+ = 1 - q \end{cases}$$

Note that when $q = 0$, z converges to z^* exponentially during flows, while when $q = 1$, z converges to $-z^*$ exponentially. This property follows from the fact that, for each $q \in Q$, the solution to $\dot{z} = \kappa_q(z)$ is given by the function $t \mapsto z(t) = \exp(-t)z(0) + (1 - 2q)(1 - \exp(-t))z^*$. Tools for the design of similar hybrid controllers for global and robust stabilization are presented in Chapter 8.

1.2.5   Controlled Intersample Behavior and Aperiodic Sampling

As pointed out above Example 1.3 and therein, a method used for the design of discrete-time controllers for a continuous-time plant consists of discretizing the model of the plant and, within a discrete-time systems framework, of designing a controller for it. The properties induced by a controller designed with such a *direct design method* are only guaranteed for the solutions to the discrete-time closed-loop system. However, the controller is to be applied to a continuous-time plant, via the use of sample-and-hold devices, and one would be interested in characterizing the behavior of the state in between samples. For instance, a static state-feedback controller $\zeta = \kappa(z)$ designed for a discrete-time model of the continuous-time plant

$$\dot{z} = F_P(z, u) \qquad z \in \mathbb{R}^{n_P}, u \in \mathbb{R}^{m_P}$$

guarantees that there exist positive constants α and β such that every solution $k \mapsto \widetilde{z}(k)$ to the resulting discrete-time closed-loop system satisfies

$$|\widetilde{z}(k)| \leq \alpha \exp(-\beta k)|\widetilde{z}(0)| \qquad \forall k \in \{0, 1, 2, \ldots\}$$

The quantity $\widetilde{z}(k)$ is the value of the solution to the discrete-time closed-loop system after k sampling events, typically associated with ordinary time kT^* with $T^* > 0$ being the sampling period. Note that without further analysis, the bound above does not necessarily hold for the solution to the continuous-time plant in between sampling events, i.e., the property induced by the controller does not characterize the intersample behavior of the plant.

Another approach for the design of a control algorithm is *emulation*. It consists of determining a discrete-time controller that approximates, in an appropriate sense, a continuous-time controller designed for the continuous-time plant. When the information used for control is not available periodically, determining an algorithm that accomplishes the stabilization goal might be challenging and, if not properly done, lead to unstability. To illustrate this potential issue, consider the problem of asymptotically synchronize the family of continuous-time systems with dynamics

$$\dot{z}_i = Az_i + Bu_i, \qquad y_i = Mz_i, \qquad i \in \mathcal{V} := \{1, 2, \ldots, N\}, \qquad N > 1$$

where z_i is the state, u_i is the input, and y_i is the output of the i-th system. The matrices A, B, and M have appropriate dimension. At each (not necessarily periodic) communication event, the i-th system in the network measures its own output, y_i, as well as the information received from its neighbors, that is, y_r with $r \in \mathcal{N}_i$, where $\mathcal{N}_i \subset \mathcal{V}$ denotes the neighbors of the i-th system. For this system, the goal is to design a feedback controller assigning the inputs u_i to drive the solutions to the family of linear time invariant systems to synchronization, asymptotically. More precisely, the control design goal is to guarantee that the z_i components of each solution satisfy

$$\lim_{t \to \infty} |z_i(t) - z_r(t)| = 0 \qquad \forall i, r \in \mathcal{V}$$

Since the state of each system is available to its neighbors only at (not necessarily periodic!) isolated time instances, the design of a controller guaranteeing synchronization is challenging.

To illustrate this point, consider the case when the dynamics of the systems are $A = \begin{bmatrix} 0 & 1 \\ -1 & 0 \end{bmatrix}$, $B = \begin{bmatrix} 0 \\ 1 \end{bmatrix}$, state $z_i \in \mathbb{R}^2$, input $u_i \in \mathbb{R}$, and output $y_i = z_i$. These correspond to the dynamics of harmonic oscillators with a scalar input. Suppose the systems are connected over a network through a communication graph that is *completely connected*.[12] Following an emulation design approach, consider the case when the network provides continuous measurements of the z_i's and N is equal to four. With K designed such that the matrix $A + 4BK$ is Hurwitz, the static

---

[12]A directed graph is completely connected if every pair of distinct vertices in the graph is connected by a pair of unique edges, one in each direction.

state-feedback controller assigning u_i as

$$u_i = K \sum_{r \in \mathcal{V} \backslash \{i\}} (z_i - z_r) \qquad \forall i \in \mathcal{V}$$

guarantees synchronization. Figure 1.11(a) shows a simulated solution for $K = \begin{bmatrix} -1 & -2 \end{bmatrix}$ and initial condition for z equal to $((-1, 0), (1, 0), (0.5, 0), (2, 0))$. This figure also shows the norm of $\sum_{r \in \mathcal{V} \backslash \{1\}} (z_1 - z_r)$ along the simulated solution, which captures the synchronization error relative to z_1. The simulated solution shows that synchronization is achieved in the limit, as time goes to infinity. Figure 1.11(b) shows an aperiodic sample-and-hold implementation of this controller. In this implementation, the local control input is updated upon the reception of new state values from the neighbors. The communication events occur no later than 0.65 seconds and no sooner than 0.3 seconds since the last communication event. As the figure shows, synchronization is not achieved. Associated simulation files are online at @BookSite/Simulation/Sync.

Hybrid systems theory can be employed to design discrete-time controllers for continuous-time plants that not only stabilize the origin of the plant in a discrete sense (as a discrete-time controller designed via the direct method), but also quantify the behavior of solutions in between samples, even when they are aperiodic.

Tools for the design of algorithms that control a plant with information that is available aperiodically (or intermittently) are presented in Chapter 5.

1.2.6 Hybrid Feedback Control Improves Performance

The design of feedback controllers involves the selection of the type of control algorithm, associated sensors and actuators, as well as their parameters. Arguably, static continuous-time and discrete-time controllers are among the most popular choices used in practice. Unfortunately, at times, a single feedback controller is not enough to satisfy all of the design specifications. In particular, when the eigenvalues of a linear time-invariant continuous-time plant cannot be pushed deep into the left-half plane through the action of a feedback controller, it might not be possible to satisfy all of the given specifications simultaneously. For instance, the satisfaction of a rise time specification may compromise meeting a desired overshoot percentage – the apparent trade-off between these competing specifications can be confirmed analytically for second-order plants. Such a challenge is due to insisting on employing a single feedback controller to satisfy both specifications simultaneously. On the other hand, it might be possible to design a control algorithm that employs multiple feedback controllers, each of them designed to satisfy particular specifications.

To illustrate this point, consider the problem of designing a feedback controller rendering a set-point z^* stable and, in addition, guaranteeing that the solutions to the plant converge to the set-point rapidly, with small overshoot. Instead of attempting to design a state-feedback controller fulfilling both specifications, which might be difficult, suppose that it is possible to design the following two controllers:

- A feedback controller that, for every initial condition, steers the solutions to the plant to a small neighborhood of z^* with desired performance (e.g., rise time);

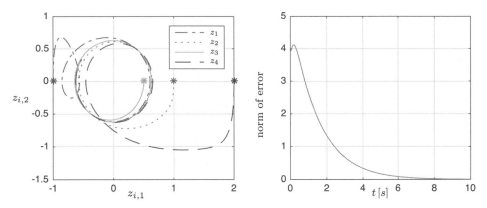

(a) The components of a simulated solution obtained with a static state-feedback law approach each other (left). The synchronization error approaches zero, asymptotically (right).

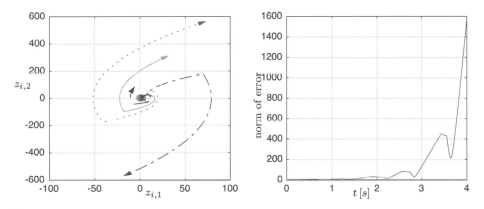

(b) With $0.3 \leq t_{s+1} - t_s \leq 0.65$, the static controller does not guarantee synchronization, as the components of the simulated solution (left) and the synchronization error show (right).

Figure 1.11: A simulated solution to the network of linear time-invariant systems in § 1.2.5 with controller $u_i = K \sum_{r \in \mathcal{V} \setminus \{i\}} (z_i - z_r)$, $K = [-1 \quad -2]$, for the following settings: (a) continuously available information, and (b) information available intermittently.

- Another feedback controller that, for every initial condition in the said small neighborhood of z^*, guarantees that the solutions to the plant do not overshoot outside of the neighborhood and converge to z^*, asymptotically.

These two feedback laws can effectively be combined by designing a hybrid controller that has a logic variable $q \in Q := \{0, 1\}$ as state, denoting the current controller being applied to the plant, and that toggles q according to whether the state of the plant is close to z^* or not. To solve the stabilization problem, such a hybrid controller would have to preserve the properties conferred on the plant by the individual controllers. This requirement suggests that the design of each of the individual controllers can be done mostly independently. Details behind the construction and design of such a hybrid controller are provided in Chapter 4.

1.3   EXERCISES

Exercise 1 (Resettable timer). Outline a mathematical model of a resettable timer with two inputs having the following dynamics:

- When the value of the timer is larger than or equal to the first input, the timer is reset to the value of the second input.

- When the value of the timer is less than the first input, the timer is allowed to grow with ordinary time.

For the first input equal to 1 and the second input equal to -1, sketch the evolution of the system over five seconds of ordinary time and seven jumps, for each of the following initial conditions:

1. Initial timer equal to 0;

2. Initial timer equal to 1;

3. Initial timer equal to -1.

Exercise 2 (Bouncing ball). Consider a (point-mass) ball moving vertically and bouncing on a horizontal platform whose height is zero.

1. Outline a mathematical model capturing the dynamics of the bouncing ball. The model of this system should describe the behavior of the ball's height and vertical velocity only. Assume that potential and kinetic energy of the ball are conserved both at impacts with the ground and in between impacts. Use $\gamma > 0$ to denote gravity.

2. Sketch the trajectories for this system from the following initial conditions:

   a) Initial height equal to 1 and initial velocity equal to 0;
   b) Initial height equal to 1 and initial velocity equal to -1;
   c) Initial height equal to 0 and initial velocity equal to 1;
   d) Initial height equal to 0 and initial velocity equal to -1;
   e) Initial height equal to 0 and initial velocity equal to 0.

Exercise 3 (Sample-and-hold control). Consider the implementation of a static controller

$$\kappa : \mathbb{R}^{n_P} \to \mathbb{R}^{m_P}$$

for the continuous-time plant

$$\dot{z} = F_P(z, u), \qquad z \in \mathbb{R}^{n_P}, u \in \mathbb{R}^{m_P}$$

on a *digital device*, e.g. computer, microcontroller, digital signal processor, etc. This is depicted in Figure 1.12, where the controller is interfaced with sample-and-hold devices. The sample-and-hold device that samples the state z of the plant is referred to as *the sampling device* (or analog-to-digital (A/D) converter), while the sample-and-hold device that stores the output of the controller in between computations is referred to as *hold device* (or digital-to-analog (D/A) converter), which is assumed to be of zero-order type, that is, a zero-order hold (ZOH). Sampling events occur every T_s^* seconds and updates of the control input occur every T_c^* seconds.

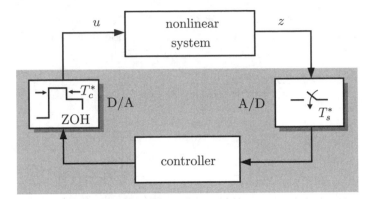

Figure 1.12: Sample-and-hold control of a nonlinear system.

1. Define the state, input, output of the hybrid plant and of the hybrid controller.

2. Outline a mathematical model capturing the evolution of the closed-loop system assuming the following:

   • The computation of the static feedback law takes no time, i.e., is instantaneous.

   • The positive constants T_s^* and T_c^* are not necessarily equal.

Exercise 4 (Walking robot). Consider the walking robot model presented in Example 1.2.

1. Using the Euler-Lagrange method (see [3, Chapter 6]), derive the differential equations capturing the change of θ and ω in between impacts;

2. Follow the steps in [4, Section II.B] to derive the function Γ used in the model of the hybrid plant in Example 1.2;

3. Propose a complete model $\mathcal{H}_P$ capturing the hybrid dynamics of the robot.

Exercise 5 (Stabilization of a two-point set). Show that the construction of the measurement signal $t \mapsto m(t)$ in Section 1.2.4 leads to a unique solution $t \mapsto z(t)$ from $z(0) = \frac{\varepsilon}{2}$ that remains in $\left[\frac{\varepsilon}{4}, \frac{\varepsilon}{2}\right]$ for all time.

Exercise 6 (Hard disk drive control). Reading and writing magnetic heads in hard disk drives require control algorithms that guarantee precise positioning and rapid transitioning between tracks for good performance. Denote by $p \in \mathbb{R}$ the position and by $v \in \mathbb{R}$ the velocity of a magnetic head on a track, and by p^* the position of the location of the data to read/write on the current track. Assuming the second order dynamics given by $\dot{p} = v$, $\dot{v} = u$ and following the example outlined in Section 1.2.6, propose a hybrid control algorithm that assigns the input u using two control laws, one to rapidly steer the magnetic head to a neighborhood of p^* on the current track, and another one to regulate position p to p^* with zero velocity.

Exercise 7 (Boost converter). The circuit of a DC-DC Boost converter is shown in Figure 1.13. This converter consists of an ideal diode d, a capacitor C_a, a DC

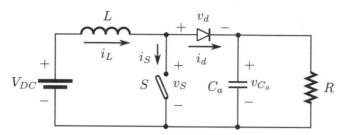

Figure 1.13: Schematic representation of the DC-DC Boost converter.

voltage source V_{DC}, a resistor R, an inductor L, and an ideal switch S. The voltage on the capacitor is denoted v_{C_a} and the current through the inductor is denoted i_L. The voltage on the diode is denoted v_d and the current through it is denoted i_d. The voltage on the switch is denoted v_S. The goal of the circuit is to convert the input DC voltage to a higher output voltage value v_{C_a}. The presence of switching elements (d and S) causes the overall system to be of a switching/hybrid nature. Depending on the (discrete) state of the diode and of the switch, one can distinguish four modes of operation:

$$\text{mode 1: } (S = 0, d = 1), \text{ mode 2: } (S = 1, d = 0)$$

$$\text{mode 3: } (S = 0, d = 0), \text{ mode 4: } (S = 1, d = 1)$$

1. Using the ideal diode model

   $$\text{conducting } (d = 1): \quad i_d \geq 0, v_d = 0; \quad \text{blocking } (d = 0): \quad i_d = 0, v_d \leq 0$$

   and the ideal switch model

   $$\text{conducting } (S = 1): \quad v_S = 0; \quad \text{blocking } (S = 0): \quad i_S = 0$$

   derive the differential equations that govern the evolution of the state variables v_c and i_L for each operating mode;

2. Using the model in item 1 of this exercise, define a hybrid plant $\mathcal{H}_P$ that models the Boost converter.

1.4   NOTES

An in-depth introduction to modeling, stability, and robustness for hybrid dynamical systems modeled as hybrid equations or, more generally, hybrid inclusions is in [1]. Hybrid inclusions without inputs were first introduced in [5], followed by initial contributions on solution properties, asymptotic stability, invariance, and robustness in the journal publications [6, 7, 8, 9, 10] and the *IEEE Control Systems Magazine* special issue in [11]. A self-contained comprehensive version of the robust stability theory for hybrid inclusions without inputs appeared in the book [1]. As the reader familiar with the vast hybrid systems literature may recognize, there are myriad of frameworks to model, analyze, and design hybrid systems. An in-depth discussion of many such frameworks is provided in the next chapter, Chapter 2, which formally introduces the modeling framework employed in this book; see § 2.6. The power of hybrid control has already been highlighted in Springer's *Lecture Notes in Computer Science (LNCS)* volumes and ACM proceedings from the Hybrid Systems: Computation and Control Conference; journal publications [12, 13, 14, 15, 16]; and books [17, 18, 19].

The switched system model of the DC/AC inverter in Example 1.1 is similar to the one presented in [2]; see also [20]. The constructions pertaining to Example 1.4 were proposed therein as well; see also Example 11.12 in Chapter 11, which is a chapter devoted to forward invariance of sets in hybrid closed-loop systems. The model of the walking robot in Example 1.2 appeared in [21], which follows from the one in [4]. The model of the sample-and-hold controller in Example 1.3 is similar to the one in [1, Example 1.4]. A general version of the hybrid control strategy in Example 1.5 for swinging up the pendubot was presented in [22] for general nonlinear systems and illustrated in the pendubot system. This strategy is presented in full detail in Chapter 6. Furthermore, this strategy was validated experimentally in [23].

Necessary conditions for the stabilization of a point by means of continuous feedback discussed in § 1.2.2 appeared in [24, Theorem 1] for continuous-time systems. A necessary condition for the existence of a continuous feedback stabilizing the origin of $\dot{z} = F_P(z, u)$ is that the image of F_P must contain an open neighborhood of the origin. Necessary conditions for robust stabilization of a point by locally bounded feedback are in [25]. Stabilization of the origin of the Brockett integrator via time-varying feedback and via switching control appeared in [26]. The hybrid controller for global and robust stabilization of the origin of Brockett integrator was proposed in [27] (see also [11, pages 87-88]).

The impossibility of globally asymptotically stabilizing a nonlinear system on a compact manifold alluded to in § 1.2.3 is discussed in [28] – in fact, every boundaryless compact manifold does not have the topological property of contractibility, which is a fundamental and necessary property of the basin of attraction for nonlinear systems with locally Lipschitz right-hand side; see [29]The system on a circle illustrating almost global asymptotic stability in § 1.2.3 and how to make it global is revisited in Example 8.1 in the context of a general supervisory control strategy. The hybrid controller for global asymptotic stabilization outlined in § 1.2.3 was first introduced in [30], where extensions to the stabilization of an arbitrary point and for tracking a reference signal in $\mathbb{S}^1$ are also given; see also [31].

A solution similar to the one for the two-point set stabilization problem using a hybrid controller in § 1.2.4 was introduced in [32]. The hybrid controller defined in § 1.2.4 follows closely the construction proposed in [1, Example 4.19]; see also [27] and [31].

The synchronization problem under intermittent information presented in § 1.2.5 was studied in [33]. Chapter 5 covers the design of hybrid control algorithms that exhibit events due to information being available intermittently.

Different strategies have been proposed in the literature solving the problem of uniting local and global controllers, including [34], [35], [36], [37], and [38]. The state-feedback construction in § 1.2.6 follows the one in [38]. Chapter 4 focuses on hybrid control algorithms that unites two feedback controllers.

Chapter Two

Modeling Framework

In this book, hybrid control systems are modeled as *hybrid equations* or, more generally, as *hybrid inclusions*. As depicted in Figure 1.1 a hybrid control system involves models of the system to be controlled, the algorithms used for communication and control, as well as the interfaces needed for operation. Such a model defines a hybrid dynamical system: its solutions can *flow* and *jump*. As outlined in § 1.1, a hybrid control system is partitioned into two main components: *a hybrid plant*, denoted $\mathcal{H}_P$, and *a hybrid controller*, denoted $\mathcal{H}_K$. The hybrid plant model mainly captures the dynamics of the system to be controlled and, if needed, dynamics of other relevant mechanisms; e.g., signal conditioners, sensors, interfaces to algorithms, etc. The hybrid controller model captures the dynamics of the algorithms used for communication and control, as well as dynamics of mechanisms that are needed to define a complete model of the hybrid control system. Figure 2.1 depicts a simplified description of the feedback interconnection between a hybrid plant and a hybrid controller, some which may also incorporate the interfaces needed for the interconnection.

This chapter provides an introduction to the hybrid equations/inclusions modeling framework for control. First, the objects defining the subsystems $\mathcal{H}_P$ and $\mathcal{H}_K$ of a *hybrid closed-loop system* $\mathcal{H}$ are formally defined. A concept of solution to define the evolution of the state and input of a hybrid plant, a hybrid controller, and a hybrid closed-loop system is also formally introduced. This concept requires the definition of hybrid time, hybrid trajectory, and hybrid input. Conditions for the existence of solutions are presented. In addition, a tool for numerical simulation is introduced.

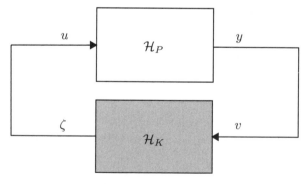

Figure 2.1: A hybrid closed-loop system $\mathcal{H}$ resulting from controlling a hybrid plant $\mathcal{H}_P$ by a hybrid controller $\mathcal{H}_K$.

2.1   OVERVIEW

Independently of the mathematical model used to describe a dynamical system, a notion of state trajectory or, more generally, *solution*, needs to be specified for analysis and design. For autonomous dynamical systems modeled as continuous-time systems of the form

$$\dot{x} = f(x)$$

where f is a given function,[1] a solution consists of a function of ordinary time $t \in \mathbb{R}_{\geq 0}$. Though a formal definition is needed, a solution to this system typically consists of a time function

$$t \mapsto x(t)$$

that is smooth enough so that

$$\frac{d}{dt}x(t) - f(x(t))$$

holds "for all" t in its domain of definition dom $x \subset \mathbb{R}_{\geq 0}$. When an initial condition $x_\circ$ at time $t = 0$ is specified, then the solution has to further satisfy $x(0) = x_\circ$. It could be that there is more than one solution from a given initial condition. For example, the system $\dot{x} = x^{1/3}$ has multiple solutions from $x_\circ = 0$: in particular, the zero function and the function $x(t) = (2t/3)^{3/2}$, both defined on $\mathbb{R}_{\geq 0}$, are solutions to it from zero.

Note that a solution to $\dot{x} = f(x)$ may be defined up to some finite value of ordinary time, and that defining the solution beyond that time may not be possible. This means that a system may have solutions whose *maximal domain of definition* is not $\mathbb{R}_{\geq 0}$. In simple words, this pathology is due to the impossibility of indefinitely continuing a solution, as the examples in the itemized list below indicate. A solution that cannot be further extended is said to be *maximal*. For example, each element in the sequence of functions $x_i(t) = t$ defined for all $t \in [0, i]$, $i \in \mathbb{N}$, is a solution to $\dot{x} = 1$ from $x_\circ = 0$, but none of them is maximal since each function $t \mapsto x_i(t)$ can be further extended, say, to the domain of definition $[0, 2i]$ – in fact, the only maximal solution to this system is the solution $x(t) = t$ defined for each $t \in \mathbb{R}_{\geq 0}$. Maximal solutions that have a domain of definition that is equal to $\mathbb{R}_{\geq 0}$ are said to be *complete*.

Examples of continuous-time systems with maximal solutions that are not complete are abundant.

- *Systems with discontinuous right-hand side:* For the scalar system

$$\dot{x} = -\text{sign}(x)$$

  with sign defined as $\text{sign}(x) = -1$ for each $x \geq 0$ and $\text{sign}(x) = 1$ for each $x < 0$, a solution $t \mapsto x(t)$ as outlined above from $x(0) = 1$ would only be defined on the time interval $[0, 1]$; i.e., dom $x = [0, 1]$. In fact, when $t = 1$, the solution is equal to zero and it cannot be extended beyond one second. Due to the discontinuity of the right-hand side $-\text{sign}(x)$ at $x = 0$, there is no

---

[1] The function f is typically referred to as the *right-hand side* of the system.

extension of the function $t \mapsto x(t)$ such that $\dot{x}(t) = -\text{sign}(x(t))$ for $t > 1$.

- *Systems with finite escape times:* Another (more extreme) example with maximal solutions that are not complete is the system $\dot{x} = -x^2$. For this system, the function $x(t) := \frac{1}{t-1}$ defined for all $t \in [0,1)$ is a maximal solution. But it has the property that $\lim_{t \nearrow 1} |x(t)| = \infty$, so it cannot be further extended for $t \geq 1$ since the value of x as t converges to one is unbounded. A system with such a maximal solution is said to have a *finite escape time*.

For autonomous dynamical systems in discrete time of the form

$$x^+ = g(x)$$

a solution is typically defined to be a function parameterized by discrete time. Denoting discrete time by $k \in \mathbb{N}$, a solution to such a system would consist of a function

$$k \mapsto x(k)$$

satisfying

$$x(k+1) = g(x(k))$$

for each k such that k and $k+1$ are in the domain of the solution x. If an initial condition $x_\circ$ is specified, it is also required that $x(0) = x_\circ$. As long as the right-hand side g of the discrete-time system is defined, solutions can be extended indefinitely. In other words, in such a case, maximal solutions to the system are those defined on $\mathbb{N}$ and, as already introduced in this section, are said to be complete. However, the addition of constraints on the state could prevent solutions from evolving forward in time. For instance, for $x^+ = x/2$ the solution from $x_\circ = 1$ is given by $x(k) = (1/2)^k$ for each $k \in \mathbb{N}$. The addition of the constraint $x \in [1/4, 1]$ only allows the maximal solution from $x_\circ = 1$ to exist for a finite number of discrete time instances while satisfying the constraint. Note that unlike continuous-time systems,[2] maximal solutions to the discrete-time system $x^+ = g(x)$ are unique.

A modeling framework for hybrid dynamical systems with a solution concept that properly handles all of the features outlined above, and that also allows for set-valued dynamics and solutions that have multiple jumps at the same time instant, is presented in § 2.3. A solution to a hybrid system includes "state trajectories" – simply referred to as *solutions* – that flow and jump. Due to the combination of these two behaviors, the state trajectories are given by hybrid arcs, while inputs that may influence the evolution of the state trajectory are given by hybrid inputs. Before the framework is introduced in detail, § 2.2 provides examples of dynamical systems that, though considered to be hybrid in the literature, do not require a hybrid system model with both continuous and discrete dynamics as those that are studied in this book. Tools for systematic analysis of general hybrid systems within the proposed framework that are employed in the remainder of this book are given in Chapter 3. Finally, in § 2.4 a software toolbox for numerical simulation of hybrid dynamical systems is presented. This toolbox is used throughout this book

---

[2]A sufficient condition for maximal solutions to $\dot{x} = f(x)$ to be unique is the right-hand side f is locally Lipschitz. See Definition A.21.

to validate the proposed hybrid control algorithms as well as in the exercises at the end of each chapter.

2.2   ON TRULY HYBRID MODELS

As outlined in Chapter 1, the term "hybrid" is used in this book to denote systems with state variables that may flow – or, equivalently, evolve continuously – and, at times, jump – or, equivalently, evolve discretely, as Figure 1.2 shows. On the other hand, the term hybrid has been used in the hybrid systems literature to denote other features. For instance, one can find models of systems in the literature that are deemed as hybrid due to having a state or an input with real-valued and discrete-valued components. Motivated by the many uses of the term *hybrid*, before diving into the modeling framework and associated analysis and design tools, this section provides examples of systems that are not hybrid according to the definition employed in this book. The intention is not to argue that the term *hybrid* has been used inappropriately or to advocate for a particular modeling framework, but rather to help avoid the use (and potential misuse) of tools presented in this book when simpler methods available in the literature can be applied.

Due to its impact on the overall research effort, the very basic task of developing a mathematical model of a system that is presumed to have hybrid dynamics should be done with utmost care. While continuous-valued and discrete-valued variables, nonsmothness of the system behavior, and different modes of operation may suggest the presence of hybrid dynamics in a system being studied, one needs to carefully identify if a model given in terms of a hybrid dynamical system is really necessary before getting too invested in applying hybrid systems tools.

The following examples provide concrete cases of systems exhibiting behavior that, at first sight, could be considered to be hybrid, but that after further review one realizes that they can be represented by a continuous-time model or by a discrete-time model.

Example 2.1 (Continuous-time systems with impulse-free inputs). *Consider a continuous-time system of the form*

$$\dot{x} = f(x, u)$$

where x is the state, u is the input, and f is the right-hand side of the system, that is, a function that defines the continuous change of x in terms of x and u. When the input signal $t \mapsto u(t)$ applied to it is impulse free, then nonsmooth behavior is only possible on the derivative of a solution $t \mapsto x(t)$. For instance, when $t \mapsto u(t)$ is a piecewise constant signal, the derivative of the solution might exhibit jumps, but not the solution itself. This system is not a hybrid system as defined in this book.

To illustrate this point, consider the so-called Dubins model of a vehicle on the plane given by

$$u_1 \in [0, v_{\max}], \ |u_2| \leq \omega_{\max}, \qquad \begin{cases} \dot{p}_1 = u_1 \sin\theta \\ \dot{p}_2 = u_1 \cos\theta \\ \dot{\theta} = u_2 \end{cases} \qquad (2.1)$$

where (p_1, p_2) denotes the position of the vehicle, θ is its orientation, which is defined as the angle between the vehicle and the vertical axis on the plane, u_1 is its forward velocity input, and u_2 is its angular velocity input. The positive constants $v_{\max}$ and $\omega_{\max}$ denote the bounds on forward and angular velocity inputs, respectively. A particular family of inputs of interest is one that steers the vehicle between two points (in position and orientation space) along a path of minimum time. It can be shown that such optimal inputs are given by piecewise continuous functions of time $t \mapsto (u_1(t), u_2(t))$ that satisfy $u_1(t) \in \{0, v_{\max}\}$ and $u_2(t) \in \{-\omega_{\max}, 0, \omega_{\max}\}$ for all $t \in [0, T]$, where $T \geq 0$ is the time to reach the final point. In particular, when the angles of the initial and final points are different, the input u_2 needed to steer the vehicle between them has to assume different values from the set $\{-\omega_{\max}, 0, \omega_{\max}\}$ over the interval $[0, T]$. Due to such a change in the value of the input, the angle component of the solution is a continuous function of time that is not differentiable at each point in the interval $[0, T]$. It only has "jumps" on its derivative.

For instance, to steer the vehicle along a straight path, $\dot{\theta}$ needs to be zero, but to generate a turn, $\dot{\theta}$ needs to be equal to $\omega_{\max}$ or $-\omega_{\max}$. Suppose that the angular velocity u_2 is zero up to time $t = t_1 \in (0, T)$ and that, after that time, u_2 is equal to $\omega_{\max}$. Then, the minimum time solution $t \mapsto x(t) = (p_1(t), p_2(t), \theta(t))$ satisfies[3]

$$\left. \begin{array}{l} \dot{p}_1(t) = v_{\max} \sin\theta(t) \\ \dot{p}_2(t) = v_{\max} \cos\theta(t) \\ \dot{\theta}(t) = u_2(t) = 0 \end{array} \right\} \qquad \forall t \in [0, t_1)$$

and

$$\left. \begin{array}{l} \dot{p}_1(t) = v_{\max} \sin\theta(t) \\ \dot{p}_2(t) = v_{\max} \cos\theta(t) \\ \dot{\theta}(t) = u_2(t) = \omega_{\max} \end{array} \right\} \qquad \forall t \in [t_1, T]$$

Note that since $t \mapsto u_2(t)$ changes from zero to $\omega_{\max}$ at $t = t_1$, the derivative of $\theta(t)$ has a "jump" in its value at that time, but the solution $t \mapsto x(t)$ is continuous (as a function) over the interval $[0, T]$. Figure 2.2 shows the solution for this input, with $T = 10 \ s$, $t_1 = 5 \ s$, $v_{\max} = 1 \ m/s$, and $\omega_{\max} = 0.5 \ rad/s$. As expected, the derivative of θ is discontinuous – it has a jump at $t = t_1$. Associated simulation files are at @BookSite/Simulation/Dubins.

The system in (2.1) is not considered to be hybrid in this book since its state does not exhibit jumps. The "jump" is in its input – and, consequently, in the derivative of the state. Though the model of a hybrid plant $\mathcal{H}_P$ used in this book is rich enough to model continuous-time systems, considering the system in (2.1) as a hybrid system

---

[3]Without loss of generality, the input is assumed to be a right-continuous function of time.

would imply that, in particular, linear time-invariant systems of the form $\dot{x} = Ax + Bu$ with nonsmooth inputs are hybrid systems.

Example 2.2 (Switched systems). *Consider a continuous-time system of the form*

$$\dot{z} = f_{\sigma(t)}(z) \qquad z \in \mathbb{R}^{n_P} \tag{2.2}$$

where $t \mapsto \sigma(t) \in \Sigma := \{0, 1, \ldots, \sigma_{\max}\}$ is a piecewise constant function called the switching signal. It selects the vector field f_i to be used at each time instant. Given a switching signal, the resulting solution $t \mapsto z(t)$ to the system may not be continuously differentiable at the times the switches occur. Still, even under such nonsmooth behavior, the system can be modeled as a continuous-time system with an input $u = (u_0, u_1, \ldots, u_{\sigma_{\max}}) \in \mathbb{R}^{\sigma_{\max}+1}$ taking the form

$$\dot{z} = \sum_{i=0}^{\sigma_{\max}} u_i(t) f_i(z) \qquad z \in \mathbb{R}^{n_P}$$

where, for each $i \in \Sigma$, $u_i(t) = 1$ if and only if $\sigma(t) = i$; otherwise, u_i is equal to zero. As a result, the resulting system is a continuous-time nonlinear system with input u.

The system in (2.2) can be written as a standard control nonlinear continuous-time system of the form $\dot{z} = f(z, u)$. Hence, in this book, it is not considered to be a hybrid system. The model of the inverter in Example 1.1 fits this type of model and, in this book, is treated as a switched system defining a continuous-time plant. As pointed out in §1.1, it can be modeled as a hybrid plant $\mathcal{H}_P$ without jumps.

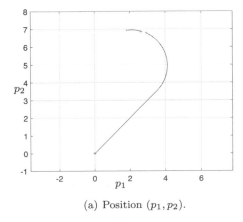

(a) Position (p_1, p_2).

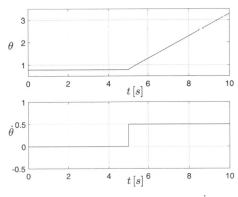

(b) Orientation θ and angular velocity $\dot{\theta}$.

Figure 2.2: Solution and angular velocity for the vehicle model in Example 2.1 to navigate between two points. The state (p, θ) evolves continuously, without jumps.

Example 2.3 (Piecewise-affine systems). *Consider a continuous-time system of the form*[4]

$$\dot{z} = A_{\sigma(z)} z + B_{\sigma(z)} u + f_{\sigma(z)} \qquad (z, u) \in \mathcal{P}_{\sigma(z)}$$

where $\Sigma := \{0, 1, \ldots, \sigma_{\max}\}$ *is a finite discrete set,* $z \mapsto \sigma(z) \in \Sigma$, *and* $\{\mathcal{P}_i\}_{i \in \Sigma}$ *is a collection of disjoint sets. According to the partition to which the state and the input belong to, the function* σ *selects the matrices* (A_i, B_i, f_i) *to be used. Due to the state and input partition, the evolution of the state* z *is governed by different dynamics and under the effect of different inputs on each set* $\mathcal{P}_i$.

It is very tempting to add a logic variable $q \in Q := \Sigma$ *to the model of the system so as to determine the partition and matrices being used. One such model is*

$$\begin{aligned}
\dot{z} &= A_q z + B_q u + f_q, & \dot{q} &= 0 & (z, u) \in \mathcal{P}_q, \quad q \in Q \\
z^+ &= z, & q^+ &= \sigma(z) & (z, u) \notin \mathcal{P}_q, \quad q \in Q
\end{aligned} \qquad (2.3)$$

However, note that the following continuous-time model with discontinuous right-hand side already captures the dynamics of the system of interest:

$$\dot{z} = F_P(z, u) := \begin{cases}
A_0 z + B_0 u + f_0 & \text{if } \sigma(z) = 0, (z, u) \in \mathcal{P}_0 \\
A_1 z + B_1 u + f_1 & \text{if } \sigma(z) = 1, (z, u) \in \mathcal{P}_1 \\
\quad \vdots & \qquad \vdots \\
A_{\sigma_{\max}} z + B_{\sigma_{\max}} u + f_{\sigma_{\max}} & \text{if } \sigma(z) = \sigma_{\max}, (z, u) \in \mathcal{P}_{\sigma_{\max}}
\end{cases} \qquad (2.4)$$

When some of the (A_i, B_i, f_i)*'s are not the same, the right-hand side* F_P *of (2.4) is discontinuous and the system is not smooth. However, such a system is not hybrid as defined in this book. A mathematical model describing its behavior does not require incorporating an additional logic variable. Such a system is a nonlinear continuous-time system with a discontinuous right-hand side.*

For instance, consider the case of a scalar state z *with the state space* $\mathbb{R}$ *partitioned in two (i.e.,* $\Sigma = \{0, 1\}$*), and no inputs, along with the data*

$$A_0 = -A_1 = -1, \quad B_0 = B_1 = 0, \quad f_0 = f_1 = 0, \quad \mathcal{P}_0 := [0, \infty), \quad \mathcal{P}_1 := (-\infty, 0)$$

Then, the model in (2.4) is simply given by

$$\dot{z} = -\mathrm{sign}(z)$$

with $s \mapsto \mathrm{sign}(s)$ *being the (single-valued) sign map defined to be* -1 *at zero. This type of system has been well studied in the literature of sliding mode control. It can be interpreted as the closed loop resulting from controlling* $\dot{z} = u$ *with the discontinuous static state-feedback law* $\kappa(z) = -\mathrm{sign}(z)$. *Note that from the origin, the only solution is the solution that does not flow – namely, the trivial solution. Furthermore, adding a logic variable as suggested in (2.3) has no effect on the solutions to the system.*

A way to have solutions that can flow at the origin is to construct the Filippov

---

[4]An output of the form $y = H_{\sigma(z)} z + E_{\sigma(z)} u$ for each $(z, u) \in \mathcal{P}_{\sigma(z)}$ is typically defined in piecewise-affine system models.

system associated with the discontinuous system $\dot{z} = -\text{sign}(z)$. This system is given by the differential inclusion

$$\dot{z} \in -\overline{\text{sign}}(z)$$

where $s \mapsto \overline{\text{sign}}(s)$ is the set-valued sign map defined at zero as $\overline{\text{sign}}(0) = [-1, 1]$. The Filippov system is obtained by convexifying the sign function (single valued) at the origin and adding the origin to $\mathcal{P}_1$ – namely, using its closure $(-\infty, 0]$. In particular, the Filippov system admits as a solution the function $t \mapsto x(t)$ that, regardless of its initial condition, reaches the origin in finite time and stays flowing there forever. Similarly, the Filippov system associated to the hybrid model in (2.3) has $\mathcal{P}_1$ replaced by its closure. Though the change is minor – just one point is added to $\mathcal{P}_1$ – it leads to the existence of solutions from the origin that are not trivial.

Example 2.4 (Mixed logical dynamical systems). *Consider the discrete-time system involving continuous-valued and discrete-valued states, inputs, and outputs, as well as constraints depending on the states, the inputs, and the outputs, given by*

$$\begin{aligned}
z^+ &= Az + B_1\tilde{u} + B_2\delta + B_3\mu + B_4 \\
y &= Mz + E_1\tilde{u} + E_2\delta + E_3\mu + E_4 \\
&\quad\text{subject to } F_2\delta + F_3\mu \leq F_1\tilde{u} + F_4z + F_5
\end{aligned} \tag{2.5}$$

where the state vector z is partitioned as (z_c, z_d), where $z_c \in \mathbb{R}^{n_{Pc}}$ are the continuous-valued components and $z_d \in \{0,1\}^{n_{Pd}}$ are the discrete-valued components of z. Similarly, the input $\tilde{u}$ is partitioned as $(\tilde{u}_c, \tilde{u}_d) \in \mathbb{R}^{m_{Pc}} \times \{0,1\}^{m_{Pd}}$ and the output y as $(y_c, y_d) \in \mathbb{R}^{r_{Pc}} \times \{0,1\}^{r_{Pd}}$. The continuous-valued auxiliary variables $\mu \in \mathbb{R}^{m_{Pc}^a}$ and the discrete-valued auxiliary variables $\delta \in \{0,1\}^{m_{Pd}^a}$ are added to capture constraints, logic statements, and other conditions. The matrices A, $\{B_i\}_{i=1}^4$, M, $\{E_i\}_{i=1}^4$, and $\{F_i\}_{i=1}^5$ have suitable dimensions.

Though the combination of continuous-valued and discrete-valued states, inputs, and outputs along with the constraints may likely lead to nonsmooth evolution, the system in (2.5) can be modeled by a constrained discrete-time system.

In fact, the system in (2.5) can be written as

$$\left.\begin{aligned}
z^+ &= Az + \begin{bmatrix} B_1 & B_2 & B_3 \end{bmatrix} u + B_4 \\
y &= Mz + \begin{bmatrix} E_1 & E_2 & E_3 \end{bmatrix} u + E_4
\end{aligned}\right\} \ (z, u) \ : \ F_2\delta + F_3\mu \leq F_1\tilde{u} + F_4z + F_5$$

where the input u is the vector with components $\tilde{u}$, δ, and μ. The control input is still $\tilde{u}$, while the time evolution of the auxiliary variables δ and μ is to be defined so as to satisfy the constraints at each discrete time instant.

2.3 MODELING

As depicted in Figure 1.1 and outlined in § 1.1, each of the subsystems in a hybrid control system might have hybrid dynamics. To accommodate for these, hybrid models of the plant and of the controller – and of the resulting hybrid closed-loop system – are proposed.

The continuous and discrete dynamics of the interfaces interconnecting the plant and the controller are included in the model of the plant or in the model of the controller.

This simplification streamlines the presentation of the mathematical models and, in particular, avoids the definition of interconnections of more than two models.

2.3.1   From Plants and Controllers to Closed-Loop Systems

The control problems outlined in Chapter 1 specify the system to control, or equivalently, the hybrid plant $\mathcal{H}_P$, and a desired goal for which a hybrid control algorithm $\mathcal{H}_K$ is to be designed. For instance, for the DC/AC inverter in Example 1.1, the plant is given by a switched system with three modes of operation, uniquely defined by the value of the input u – see the model in terms of $\mathcal{H}_P$ therein. The desired goal associated with this control problem is to render a particular set of points forward invariant. In the sample-and-hold control problem in Example 1.3, the closed-loop system is hybrid (as defined in this book) since the memory state in the controller is reset instantaneously upon the expiration of a timer that flows toward a constant threshold T^* defining the events rate. The desired goal in such a problem is to asymptotically stabilize the plant and the memory state to zero, while keeping the timer within the range $[0, T^*]$. The pendubot system in Example 1.5 can also be modeled as a plant $\mathcal{H}_P$ with only continuous-time dynamics, namely, as $\dot{z} = F_P(z, u)$, but for this problem the desired goal is to render the upright equilibrium – namely, both links of the pendubot system shown in Figure 1.6 "up" – globally asymptotically stable, with robustness to small disturbances. To capture such variety of plant and controller models, this section introduces general enough models that are suitable to fit the needs of several hybrid control design problems.

Following Chapter 1, a hybrid plant $\mathcal{H}_P$ to be controlled is modeled by the hybrid equation

$$\mathcal{H}_P : \begin{cases} (z, u) \in C_P & \dot{z} = F_P(z, u) \\ (z, u) \in D_P & z^+ = G_P(z, u) \\ & y = h(z) \end{cases} \tag{2.6}$$

or, more generally, by the hybrid inclusion

$$\mathcal{H}_P : \begin{cases} (z, u) \in C_P & \dot{z} \in F_P(z, u) \\ (z, u) \in D_P & z^+ \in G_P(z, u) \\ & y = h(z) \end{cases} \tag{2.7}$$

The collection (C_P, F_P, D_P, G_P, h) is the data of $\mathcal{H}_P$. A hybrid plant $\mathcal{H}_P$ is represented by the convenient notation

$$\mathcal{H}_P = (C_P, F_P, D_P, G_P, h)$$

The data (C_P, F_P, D_P, G_P, h) of $\mathcal{H}_P$ is formally defined as follows.

Definition 2.5 (Data of a hybrid plant). *The data of a hybrid plant $\mathcal{H}_P$ with state $z \in \mathbb{R}^{n_P}$, input $u \in \mathbb{R}^{m_P}$, and output $y \in \mathbb{R}^{r_P}$ consists of five elements:*[5]

- *A set $C_P \subset \mathbb{R}^{n_P} \times \mathbb{R}^{m_P}$, called the flow set of the plant;*

- *A set-valued map $F_P : \mathbb{R}^{n_P} \times \mathbb{R}^{m_P} \rightrightarrows \mathbb{R}^{n_P}$ with $C_P \subset \operatorname{dom} F_P$, called the flow map of the plant;*

- *A set $D_P \subset \mathbb{R}^{n_P} \times \mathbb{R}^{m_P}$, called the jump set of the plant;*

- *A set-valued map $G_P : \mathbb{R}^{n_P} \times \mathbb{R}^{m_P} \rightrightarrows \mathbb{R}^{n_P}$ with $D_P \subset \operatorname{dom} G_P$, called the jump map of the plant;*

- *A single-valued map $h : \mathbb{R}^{n_P} \to \mathbb{R}^{r_P}$ called the output map of the plant.*[6]

As outlined in § 1.1, flows of $\mathcal{H}_P$ are governed by

$$(z, u) \in C_P \qquad \dot{z} \in F_P(z, u)$$

where the set C_P captures the values of z and u at which flows of $\mathcal{H}_P$ are allowed. In the model proposed in Example 1.1, the flow set includes any possible value for the voltage, current, and the (discrete-valued) input of the circuit therein. So, as long as the input assumes valid discrete values, flows are possible. On the other hand, in the model of sample-and-hold control algorithm in Example 1.3, the flow set C_K imposes the constraint $\tau \subset [0, T^*]$. Due to this constraint, when the timer reaches the threshold T^* flows are no longer possible since F_K would force τ to reach a value that is larger than T^*. In this way, jumps from $\tau = T^*$ are forced. This suggests that, in general, there might be points in the boundary of C_P from where flows are not possible due to the values that F_P assumes – a more elementary case of such behavior is in the system in Exercise 8.

The informal introduction to $\mathcal{H}_P$ in § 1.1 suggests that jumps of $\mathcal{H}_P$ are governed by

$$(z, u) \in D_P \qquad z^+ \in G_P(z, u)$$

where the set D_P captures the values of z and u at which jumps are permitted. Unlike flows, when both the state and the input are in D_P, a jump is always possible, since by Definition 2.5, G_P is defined on D_P. Conveniently, the jump set of $\mathcal{H}_K$ in Example 1.3 includes the points in the boundary of C_K from which no flows are possible. In this way, a solution can only continue evolving by jumping, when reaching the boundary of that set.

Remark 2.6 (On enabling semantics). As formally introduced later in this chapter, in § 2.3.3, *enabling semantics* are employed for the definition of solutions to the hybrid models used in this book. The reason for using such semantics is the potential overlap between the z component of the flow and jump sets. In fact, from such overlap it might be that both flows and jumps are possible – indeed, jumps are always

---

[5]Hybrid inclusions have right-hand sides given by set-valued maps. Unlike functions or *single-valued maps*, set-valued maps may return a set when evaluated at a point. For instance, at points in C_P, the set-valued flow map F_P of the hybrid plant $\mathcal{H}_P$ might return more than one value, allowing for different values of the derivative of z. For more background about these concepts, see Appendix A.

[6]The case when h depends on u can easily be accommodated, but to simplify the definition of closed-loop systems, h is assumed to only depend on z.

possible from such points. Hence, enabling semantics permit capturing all possible solutions and, very importantly, solving the hybrid control problem formulated in page 14 for all solutions to the system. If *forcing semantics* were to be used, then jumps from the said overlap would be always forced. As a consequence, a control algorithm solving the hybrid control problem would not assure that solutions flowing from such overlap are well behaved. Small perturbations on the state can lead to solutions that flow from such overlap and, for instance, do not converge to the desired set-point, even when the size of the noise is arbitrarily small.   △

Remark 2.7 (On specific state, input, and output values of $\mathcal{H}_P$). At times, for convenience, the sets X_P, U, and Y are introduced to denote the allowed values for the state z, the input u, and the output y, respectively. In some chapters, defining X_P, U, and Y first, followed by a definition of C_P and D_P that involves these sets, eases the presentation.   △

When the plant $\mathcal{H}_P$ only has continuous dynamics, the jump set is empty so that jumps are not possible. In such a case, the jump map can be arbitrary and the hybrid plant reduces to a continuous-time system. Such a plant is written as the simplified hybrid equation

$$\mathcal{H}_P \; : \; \begin{cases} (z, u) \in C_P & \dot{z} = F_P(z, u) \\ & y = h(z) \end{cases} \tag{2.8}$$

or, more generally, as the simplified hybrid inclusion

$$\mathcal{H}_P \; : \; \begin{cases} (z, u) \in C_P & \dot{z} \in F_P(z, u) \\ & y = h(z) \end{cases} \tag{2.9}$$

In such cases, the data of $\mathcal{H}_P$ can be denoted $(C_P, F_P, \emptyset, \star, h)$, where the symbol $\star$ indicates that the jump map is arbitrary. With $\emptyset$ denoting the empty set, in these cases, the jump set of $\mathcal{H}_P$ is empty. When the output of the plant is the state, the output function h is given by the identity function Id, leading to $\mathcal{H}_P = (C_P, F_P, \emptyset, \star, \mathrm{Id})$.

Controllers for the hybrid plant $\mathcal{H}_P$ are also modeled as hybrid systems. The data of a generic hybrid controller is defined as follows.

Definition 2.8 (Data of a hybrid controller). *The data of a hybrid controller $\mathcal{H}_K$ with state $\eta \in \mathbb{R}^{n_K}$, input $v \in \mathbb{R}^{m_K}$, and output $\zeta \in \mathbb{R}^{r_K}$ consists of five elements:*

- *A set $C_K \subset \mathbb{R}^{m_K} \times \mathbb{R}^{n_K}$, called the flow set of the controller;*

- *A set-valued map $F_K : \mathbb{R}^{m_K} \times \mathbb{R}^{n_K} \rightrightarrows \mathbb{R}^{n_K}$ with $C_K \subset \mathrm{dom}\, F_K$, called the flow map of the controller;*

- *A set $D_K \subset \mathbb{R}^{m_K} \times \mathbb{R}^{n_K}$, called the jump set of the controller;*

- *A set-valued map $G_K : \mathbb{R}^{m_K} \times \mathbb{R}^{n_K} \rightrightarrows \mathbb{R}^{n_K}$ with $D_K \subset \mathrm{dom}\, G_K$, called the jump map of the controller;*

- *A single-valued map $\kappa : \mathbb{R}^{m_K} \times \mathbb{R}^{n_K} \to \mathbb{R}^{r_K}$ called the output map of the controller.*

Remark 2.9 (On specific state and input values of $\mathcal{H}_K$). Similar to the hybrid plant, the sets X_K, U_K, and Y_K can be implicitly included in the definitions of C_K and D_K. At times, they are defined first and then used in the definitions of C_K and D_K. △

To capture a wide range of hybrid controllers, like the ones presented in this chapter and in the forthcoming chapters – see also those in § 1.2 – the hybrid controller with data as in Definition 2.8 is represented as

$$\mathcal{H}_K = (C_K, F_K, D_K, G_K, \kappa)$$

and, at times, written as

$$\mathcal{H}_K \ : \ \begin{cases} (v, \eta) \in C_K & \dot{\eta} \ \in \ F_K(v, \eta) \\ (v, \eta) \in D_K & \eta^+ \in \ G_K(v, \eta) \\ & \zeta \ = \kappa(v, \eta) \end{cases} \tag{2.10}$$

When the controller $\mathcal{H}_K$ only has continuous dynamics, its jump set is empty and its jump map is arbitrary. In such a case, the controller is written as

$$\mathcal{H}_K \ : \ \begin{cases} (v, \eta) \in C_K & \dot{\eta} = F_K(v, \eta) \\ & \zeta = \kappa(v, \eta) \end{cases} \tag{2.11}$$

or, more generally, as the simplified hybrid inclusion

$$\mathcal{H}_K \ : \ \begin{cases} (v, \eta) \in C_K & \dot{\eta} \in F_K(v, \eta) \\ & \zeta = \kappa(v, \eta) \end{cases} \tag{2.12}$$

The model for $\mathcal{H}_K$ also allows to capture feedback controllers defined by a static (potentially nonlinear) function of the state or of the output of the plant, depending on how v is assigned. Such controllers include proportional control, neural networks, look-up tables, etc. It reduces to

$$\mathcal{H}_K \ : \ \ \ \zeta = \kappa(v) \tag{2.13}$$

where $v = z$ or $v = y$. This controller itself is simply referred to as κ.

The first hybrid controller modeled in this book using the formulation above is given in Example 1.3. This controller triggers events when a timer reaches a threshold. As indicated therein, its flow and jump sets can be defined to trigger events aperiodically. The hybrid controller given next generalizes the condition triggering the events.

Example 2.10 (Event-triggered sample-and-hold control). *When the events in the sample-and-hold control strategy in Example 1.3 are triggered when a function*

γ *becomes positive (or nonnegative), the data of the controller is changed into*

$$F_K(\eta) := 0 \qquad \forall \eta \in C_K := \{(z, \eta) : \gamma(y, \eta) \leq 0, y = h(z)\}$$
$$G_K(z, \eta) := \kappa_c(h(z)) \qquad \forall (z, \eta) \in D_K := \{(z, \eta) : \gamma(y, \eta) \geq 0, y = h(z)\}$$
$$\kappa(z, \eta) := \ell_v$$

where $\eta = (\ell_v, \tau)$. *In this construction, it is typically desired to only allow jumps when* γ *becomes zero, in this way, triggering events when* γ *transitions from negative to positive. Such a control strategy is known as event-triggered control and will be treated in detail in Chapter 5.*

To control the hybrid plant $\mathcal{H}_P$ with a hybrid controller $\mathcal{H}_K$, the dimension of their inputs and outputs should match so that they can be interconnected. In other words, it is required that

$$m_P = r_K, \qquad m_K = r_P$$

When this condition holds, a hybrid plant controlled by a hybrid controller leads to a hybrid closed-loop system, which is denoted $\mathcal{H}$. As depicted in Figure 2.1, the interconnection assignment

$$u = \kappa(v, \eta), \qquad v = h(z) \tag{2.14}$$

leads to a hybrid system with data defined as follows.[7]

Definition 2.11 (Data of a hybrid closed-loop system). *Given a hybrid plant* $\mathcal{H}_P$ *and a hybrid controller* $\mathcal{H}_K$, *the resulting hybrid closed-loop system has state* $x = (z, \eta) \in \mathbb{R}^{n_P} \times \mathbb{R}^{n_K}$ *and its data consists of four elements:*

- *A set* $C \subset \mathbb{R}^{n_P} \times \mathbb{R}^{n_K}$, *called the flow set of the closed loop, given by*

$$C = \{x = (z, \eta) : (z, u) \in C_P, (v, \eta) \in C_K, u = \kappa(v, \eta), v = h(z)\} \tag{2.15}$$

- *A set-valued map* $F : \mathbb{R}^{n_P} \times \mathbb{R}^{n_K} \rightrightarrows \mathbb{R}^{n_P + n_K}$, *called the flow map of the closed loop, given by*

$$F(x) = \begin{bmatrix} F_P(z, \kappa(h(z), \eta)) \\ F_K(h(z), \eta) \end{bmatrix} \qquad \forall x \in C \tag{2.16}$$

- *A set* $D \subset \mathbb{R}^{n_P} \times \mathbb{R}^{n_K}$, *called the jump set of the closed loop, given by*

$$D = D_1 \cup D_2 \tag{2.17}$$

---

[7] An alternative way to define the jump map of $\mathcal{H}$ is considered in Exercise 19.

where

$$D_1 = \{x = (z, \eta) : (z, u) \in D_P, u = \kappa(v, \eta), v = h(z)\}$$
$$D_2 = \{x = (z, \eta) : (v, \eta) \in D_K, u = \kappa(v, \eta), v = h(z)\}$$

- *A set-valued map $G : \mathbb{R}^{n_P} \times \mathbb{R}^{n_K} \rightrightarrows \mathbb{R}^{n_P + n_K}$, called the jump map of the hybrid closed-loop system, given by*

$$G(x) = \begin{cases} \begin{bmatrix} G_P(z, \kappa(h(z), \eta)) \\ \eta \end{bmatrix} & \text{if } x \in D_1 \setminus D_2 \\ \left\{ \begin{bmatrix} G_P(z, \kappa(h(z), \eta)) \\ \eta \end{bmatrix}, \begin{bmatrix} z \\ G_K(h(z), \eta) \end{bmatrix} \right\} & \text{if } x \in D_1 \cap D_2 \quad \forall x \in D \\ \begin{bmatrix} z \\ G_K(h(z), \eta) \end{bmatrix} & \text{if } x \in D_2 \setminus D_1 \end{cases}$$

$$(2.18)$$

Remark 2.12 (On the composition of flows and jumps). As the construction of the data of the hybrid closed-loop system $\mathcal{H}$ in Definition 2.11 suggests, flows of $\mathcal{H}$ are only allowed when the closed-loop system state x is such that its components satisfy the conditions imposed by the flow set C_P of the plant *and* by the flow set C_K of the controller. On the other hand, jumps of $\mathcal{H}$ are allowed when *either* the appropriate state variables in x satisfy the conditions imposed by the jump set D_P of $\mathcal{H}_P$ or by the jump set D_K of $\mathcal{H}_K$. The reason for defining C as the "intersection" of the flow sets is that if flows were to be allowed, say, when only the conditions imposed by C_P were satisfied, then the conditions imposed by C_K might be violated. Alternatively, one could alter the flow of the state components that should not flow, or even reset them to appropriate values to avoid this issue. But such an approach requires modifying the dynamics of the system, which is a delicate matter. The reason for defining D as the "union" of the jump sets is that if jumps are possible for either the plant or the controller, then they should be allowed. $\triangle$

Remark 2.13 (On the state space of the closed loop). At times, it is convenient to specify the set where x takes values from. This set is denoted X, and is implicitly defined by the data of $\mathcal{H}$. In fact, as it will be clear when a formal notion of solution is introduced in Definition 2.29, the allowed values for x are those in $\overline{C} \cup D \cup G(D)$. $\triangle$

Similar to the hybrid plant and hybrid controller, the hybrid closed-loop system is represented as $\mathcal{H} = (C, F, D, G)$ and, at times, written as

$$\mathcal{H} : \begin{cases} x \in C & \dot{x} \in F(x) \\ x \in D & x^+ \in G(x) \end{cases} \qquad (2.19)$$

Note that this model of the closed-loop system is under nominal operating conditions, namely, without disturbances.

Example 2.14 (Sample-and-hold control, revisited). *Consider the sample-and-hold controller in Example 1.3. For a general hybrid plant with data (C_P, F_P, D_P, G_P, h), along with a static controller κ_c as in Example 1.3, the resulting hybrid closed-loop system $\mathcal{H}$ has state $x = (z, \eta) = (z, \ell_v, \tau) \in \mathbb{R}^{n_P} \times \mathbb{R}^{m_P} \times \mathbb{R}_{\geq 0} =: X$ and data given as in Definition 2.11. This construction leads to the following closed loop:*

$$C = \{x \in X \ : \ (z, \ell_v) \in C_P, \tau \in [0, T^*]\},$$

$$F(x) = \begin{bmatrix} F_P(z, \ell_v) \\ 0 \\ 1 \end{bmatrix} \qquad \forall x \in C$$

$$D = D_1 \cup D_2$$

$$D_1 = \{x \in X \ : \ (z, \ell_v) \in D_P\}, \quad D_2 = \{x \in X \ : \ \tau \geq T^*\}$$

$$G(x) = \begin{cases} \begin{bmatrix} G_P(z, \ell_v) \\ \ell_v \\ \tau \end{bmatrix} & \text{if } x \in D_1 \setminus D_2 \\[2em] \left\{ \begin{bmatrix} G_P(z, \ell_v) \\ \ell_v \\ \tau \end{bmatrix}, \begin{bmatrix} z \\ \kappa_c(h(z)) \\ 0 \end{bmatrix} \right\} & \text{if } x \in D_1 \cap D_2 \qquad \forall x \in D \\[2em] \begin{bmatrix} z \\ \kappa_c(h(z)) \\ 0 \end{bmatrix} & \text{if } x \in D_2 \setminus D_1 \end{cases}$$

These sets indicate that flows are only possible when the conditions for flows of the plant and of the controller hold simultaneously. On the other hand, jumps of the closed loop may occur when either the jump conditions of the plant or of the controller hold.

For the particular case that $C_P = \mathbb{R}^{n_P} \times \mathbb{R}^{m_P}$ and $D_P = \emptyset$, i.e., the plant is a purely continuous-time system, namely, $\dot{z} \in F_P(z, u)$, jumps of $\mathcal{H}$ are solely triggered by the hybrid controller $\mathcal{H}_K$. This case is treated in Example 2.22 and Example 3.15.

The model of the hybrid controller proposed in Example 1.3 can be extended to model a skewed timer using a set-valued flow map F_K. A timer τ with a rate of change that is not equal to one but, instead, counts time slower or faster is captured by the differential inclusion

$$\dot{\tau} \in [1 - w_s, 1 + w_s]$$

governing the flow of τ, where w_s is a constant that characterizes the minimum and maximum skew of the timer. The model can also be extended to model the situation in which τ is not necessarily reset to zero, but rather, to a (positive) value nearby zero. Such a feature can be captured by a set-valued jump map G_K. In fact, when jumps of the timer τ are governed by the difference inclusion

$$\tau^+ \in [0, w_r]$$

then τ is reset to any point in the interval $[0, w_r]$ after jumps, where w_r is a constant that defines the maximum error possible incurred in resetting the timer to zero. Given $w_s \in [0, 1)$ and $w_r \in [0, T^*)$, these two features lead to a hybrid closed-loop *system with flow and jump maps given by*

$$F(x) = \begin{bmatrix} F_P(z, \ell_v) \\ 0 \\ [1 - w_s, 1 + w_s] \end{bmatrix} \qquad \forall x \in C$$

$$G(x) = \begin{cases} \begin{bmatrix} G_P(z, \ell_v) \\ \ell_v \\ \tau \end{bmatrix} & \text{if } x \in D_1 \setminus D_2 \\ \left\{ \begin{bmatrix} G_P(z, \ell_v) \\ \ell_v \\ \tau \end{bmatrix}, \begin{bmatrix} z \\ \kappa_c(h(z)) \\ [0, w_r] \end{bmatrix} \right\} & \text{if } x \in D_1 \cap D_2 \qquad \forall x \in D \\ \begin{bmatrix} z \\ \kappa_c(h(z)) \\ [0, w_r] \end{bmatrix} & \text{if } x \in D_2 \setminus D_1 \end{cases}$$

When the hybrid controller $\mathcal{H}_K = (C_K, F_K, D_K, G_K, \kappa)$ is a static output-feedback law κ given as in (2.13), the data of the hybrid closed-loop system simplifies as stated next. This construction follows directly from Definition 2.11.

Definition 2.15 (Data of a hybrid closed-loop system with static feedback). *Given a hybrid plant $\mathcal{H}_P$ and a static output-feedback controller κ, the resulting hybrid closed-loop system has state $x = z \in \mathbb{R}^{n_P}$ and its data consists of four elements:*

- *A set $C \subset \mathbb{R}^{n_P}$, called the flow set of the closed loop, given by*

$$C = \{z \; : \; (z, h(z)) \in C_P\} \tag{2.20}$$

- *A set-valued map $F : \mathbb{R}^{n_P} \rightrightarrows \mathbb{R}^{n_P}$, called the flow map of the closed loop, given by*

$$F(z) = F_P(z, \kappa(h(z))) \qquad \forall z \in C \tag{2.21}$$

- *A set $D \subset \mathbb{R}^{n_P}$, called the jump set of the closed loop, given by*

$$D = D_1 = \{z \; : \; (z, h(z)) \in D_P\} \tag{2.22}$$

- *A set-valued map $G : \mathbb{R}^{n_P} \rightrightarrows \mathbb{R}^{n_P}$, called the jump map of the closed loop, given by*

$$G(z) = G_P(z, \kappa(h(z))) \qquad \forall z \in D \tag{2.23}$$

Remark 2.16 (*On flow and jump inputs*). At times, it is convenient to define inputs $u_c \in \mathbb{R}^{m_{P,c}}$ and $u_d \in \mathbb{R}^{m_{P,d}}$ collecting every component of the input u of the plant that affect flows and that affect jumps, respectively. Some of the components of u can be used to define both u_c and u_d, that is, there could be inputs that affect both flows and jumps. Similarly, one can define $y_c \in \mathbb{R}^{r_{P,c}}$ and $y_d \in \mathbb{R}^{r_{P,d}}$ as the components of y that are measured during flows and jumps, respectively, in which case output functions $z \mapsto h_c(z)$ and $z \mapsto h_d(z)$ would be used to define y_c and y_d. The same grouping can be done for the inputs and outputs of the hybrid controller, leading to $v_c \in \mathbb{R}^{m_{K,c}}$ and $(z, v_c) \mapsto \kappa_c(h(z), v_c)$ for flows, and $v_d \in \mathbb{R}^{m_{K,d}}$ and $(z, v_d) \mapsto \kappa_d(h(z), v_d)$ for jumps. The following example illustrates this point. $\triangle$

Example 2.17 (Hybrid closed-loop system with static state-feedback law). *When the controller $\mathcal{H}_K$ is given by the static state-feedback pair as in (2.13) that is partitioned as $\kappa = (\kappa_c, \kappa_d)$, the hybrid closed-loop system resulting from controlling $\mathcal{H}_P$ by the feedback pair is given by*

$$\mathcal{H} \; : \; \begin{cases} (z, \kappa_c(z)) \in C_P & \dot{z} \;\in F_P(z, \kappa_c(z)) \\ (z, \kappa_d(z)) \in D_P & z^+ \in G_P(z, \kappa_d(z)) \end{cases} \tag{2.24}$$

which can be rewritten as (2.19) by defining

$$\begin{aligned} C &:= \{z \in \mathbb{R}^{n_P} : (z, \kappa_c(z)) \in C_P\}, & F(z) := F_P(z, \kappa_c(z)) & \quad \forall z \in C \\ D &:= \{z \in \mathbb{R}^{n_P} : (z, \kappa_d(z)) \in D_P\}, & G(z) := G_P(z, \kappa_d(z)) & \quad \forall z \in D \end{aligned} \tag{2.25}$$

with $x = z$ and $n = n_P$.

2.3.2 Hybrid Basic Conditions

The examples in the previous section illustrate how to model feedback controllers as hybrid controllers $\mathcal{H}_K$ as in (2.10). The hybrid closed-loop system model $\mathcal{H}$ combines the hybrid plant $\mathcal{H}_P$ and the hybrid controller $\mathcal{H}_K$ to define a hybrid control system. As suggested by the examples, the dynamics of $\mathcal{H}$ depend on the data of the individual subsystems, namely, (C_P, F_P, D_P, G_P, h) and $(C_K, F_K, D_K, G_K, \kappa)$. As shown in the next chapter, this data plays a unique role in establishing asymptotic stability, invariance, and, very importantly, robustness of the closed loop. With the objective of presenting hybrid control design techniques that yield closed-loop systems with some robustness to disturbances, the following mild conditions on the data of the hybrid closed-loop system $\mathcal{H}$ are required. Due to the desired properties guaranteed for its set of solutions, $\mathcal{H}$ is said to be <u>well-posed</u> when its data satisfies the hybrid basic conditions.

Definition 2.18 (Hybrid basic conditions – first pass). *Given a hybrid plant $\mathcal{H}_P$ and a hybrid controller $\mathcal{H}_K$, the hybrid closed-loop system $\mathcal{H}$ as in Definition 2.11 with single-valued flow and jump maps, namely,*

$$\mathcal{H} \; : \; \begin{cases} x \in C & \dot{x} \;= F(x) \\ x \in D & x^+ = G(x) \end{cases} \tag{2.26}$$

is said to satisfy the <u>hybrid basic conditions</u> if its data (C, F, D, G) satisfies the following properties:

(A1) C and D are closed subsets of $\mathbb{R}^{n_P} \times \mathbb{R}^{n_K}$;

(A2) F is a single-valued map that is defined on C and continuous; and

(A3) G is a single-valued map that is defined on D and continuous.

Remark 2.19 (*On the mildness of the hybrid basic conditions*). The conditions in Definition 2.20 can be easily checked without computing solutions to $\mathcal{H}$. Item (A1) requires the flow and jump sets to be closed; see Definition A.5. Recall that a set $C \subset \mathbb{R}^n$ is closed if for each convergent sequence of points $x_i \in C$, the limit $\lim_{i \to \infty} x_i$ is in C. Items (A2) and (A3) require F and G to have a domain of definition that contains C and D, respectively; namely, $C \subset \operatorname{dom} F$ and $D \subset \operatorname{dom} G$. In addition, items (A2) and (A3) require F and G to be continuous functions; see Definition A.13. Recall that a function F is continuous if for each $x \in \operatorname{dom} F$, every sequence of points $x_i \in \operatorname{dom} F$ converging to x is such that $\lim_{i \to \infty} F(x_i) = F(x)$. See Appendix A for more details. $\triangle$

In many cases, as seen in Example 1.3, Example 2.24 and Example 2.23 below, and other hybrid systems treated in this book, the maps F or G might be set valued. Items (A2) and (A3) can be reformulated for such a case, with the continuity property therein replaced by *outer semicontinuity*. A set-valued map F is outer semicontinuous if for each $x \in \operatorname{dom} F$, every sequence of points $x_i \in \operatorname{dom} F$ converging to x is such that every sequence $y_i \in F(x_i)$ converging to y satisfies $y \in F(x)$. It turns out that a set-valued map is outer semicontinuous if and only if its graph is closed. Certainly, a continuous function is outer semicontinuous. In addition to outer semicontinuity for both F and G, it is required that the values of F are convex – that is, each set $F(x)$ is convex. These notions and properties are presented in detail in Appendix A.

Definition 2.20 (Hybrid basic conditions – second pass). *Given a hybrid plant $\mathcal{H}_P$ and a hybrid controller $\mathcal{H}_K$, the hybrid closed-loop system $\mathcal{H}$ as in Definition 2.11 is said to satisfy the <u>hybrid basic conditions</u> if its data (C, F, D, G) satisfies the following properties:*

(A1) C and D are closed subsets of $\mathbb{R}^{n_P} \times \mathbb{R}^{n_K}$;

(A2) $F : \mathbb{R}^{n_P} \times \mathbb{R}^{n_K} \rightrightarrows \mathbb{R}^{n_P + n_K}$ is outer semicontinuous and locally bounded relative to C, $C \subset \operatorname{dom} F$, and $F(x)$ is convex for each $x \in C$;

(A3) $G : \mathbb{R}^{n_P} \times \mathbb{R}^{n_K} \rightrightarrows \mathbb{R}^{n_P + n_K}$ is outer semicontinuous and locally bounded relative to D, and $D \subset \operatorname{dom} G$.

Conditions on the data of $\mathcal{H}_P$ and of $\mathcal{H}_K$ can be imposed so that the hybrid basic conditions hold for $\mathcal{H}$. The following result establishes one set of conditions guaranteeing that hybrid basic conditions hold for $\mathcal{H}$. Its proof is left as an exercise to the reader; see Exercise 19.

Lemma 2.21 (Well-posedness of $\mathcal{H}$). *Suppose that the data (C_P, F_P, D_P, G_P, h) of the hybrid plant $\mathcal{H}_P$ satisfies*

(A1$_P$) C_P and D_P are closed subsets of $\mathbb{R}^{n_P} \times \mathbb{R}^{m_P}$;

(A2$_P$) $F_P : \mathbb{R}^{n_P} \times \mathbb{R}^{m_P} \rightrightarrows \mathbb{R}^{n_P}$ is outer semicontinuous and locally bounded relative to C_P, $C_P \subset \operatorname{dom} F_P$, and $F_P(z, u)$ is convex for each $(z, u) \in C_P$;

(A3$_P$) $G_P : \mathbb{R}^{n_P} \times \mathbb{R}^{m_P} \rightrightarrows \mathbb{R}^{n_P}$ is outer semicontinuous and locally bounded relative to D_P, and $D_P \subset \operatorname{dom} G_P$;

(A4$_P$) $h : \mathbb{R}^{n_P} \to \mathbb{R}^{r_P}$ is continuous;

and that the data $(C_K, F_K, D_K, G_K, \kappa)$ of the hybrid controller $\mathcal{H}_K$ satisfies

(A1$_K$) C_K and D_K are closed subsets of $\mathbb{R}^{m_K} \times \mathbb{R}^{n_K}$;

(A2$_K$) $F_K : \mathbb{R}^{m_K} \times \mathbb{R}^{n_K} \rightrightarrows \mathbb{R}^{n_K}$ is outer semicontinuous and locally bounded relative to C_K, $C_K \subset \operatorname{dom} F_K$, and $F_K(v, \eta)$ is convex for each $(v, \eta) \in C_K$;

(A3$_K$) $G_K : \mathbb{R}^{m_K} \times \mathbb{R}^{n_K} \rightrightarrows \mathbb{R}^{n_K}$ is outer semicontinuous and locally bounded relative to D_K, and $D_K \subset \operatorname{dom} G_K$;

(A4$_K$) $\kappa : \mathbb{R}^{m_K} \times \mathbb{R}^{n_K} \to \mathbb{R}^{r_K}$ is continuous.

Then, the resulting hybrid closed-loop system $\mathcal{H}$ as in Definition 2.11 satisfies the hybrid basic conditions.

Example 2.22 (Sample-and-hold control, revisited). *The definition of the hybrid controller in Example 1.3 is such that C_K and D_K are closed and that F_K is a continuous single-valued map. Moreover, G_K is continuous when the given static output-feedback κ_c is continuous. Under these conditions, (A1$_K$)-(A4$_K$) in Lemma 2.21 hold. Then, according to Lemma 2.21, the resulting hybrid closed-loop system $\mathcal{H}$ satisfies the hybrid basic conditions when the hybrid plant satisfies (A1$_P$)-(A4$_P$) therein.*

Example 2.23 (Synchronization over a network). *Consider the synchronization problem introduced in § 1.2.5. Due to the impulsive nature of the communication structure outlined therein, a resettable timer is used to trigger the communication events. For each $i \in \mathcal{V}$, let $\tau_i \in [0, T_{2,i}^*]$ be a timer state that decreases with respect to continuous time. When the timer reaches zero, it is reset to a point in the interval $[T_{1,i}^*, T_{2,i}^*]$ and, at the same time, system i receives the value of the output of each of its connected system. Namely, τ_i has the following hybrid dynamics:*

$$
\begin{aligned}
\dot{\tau}_i &= -1 &\qquad \tau_i &\in [0, T_{2,i}^*], \\
\tau_i^+ &\in [T_{1,i}^*, T_{2,i}^*] &\qquad \tau_i &= 0.
\end{aligned}
\tag{2.27}
$$

This hybrid model has the nice feature that it generates any possible sequence of time instances $\{t_s^i\}_{s=1}^\infty$ at which events occur and satisfy

$$
T_{1,i}^* \le t_{s+1}^i - t_s^i \le T_{2,i}^* \qquad \forall s \in \mathbb{N} \setminus \{0\}, \qquad \forall i \in \mathcal{V}
\tag{2.28}
$$

Indeed, except the first jump, every jump occurs no later than $T_{2,i}^*$ seconds and no sooner than $T_{1,i}^*$ since the last jump.

  A memory state is used to store the information received by each of the systems. For each $i \in \mathcal{V}$, let ℓ_i denote such a memory state for the i-th system. A particular choice of the dynamics for this memory state is as follows:

  * *At each event triggered by the hybrid model in (2.27), reset ℓ_i via*

$$
\ell_i^+ = K \sum_{k \in \mathcal{N}_i} (y_i - y_k) = KM \sum_{k \in \mathcal{N}_i} (z_i - z_k)
\tag{2.29}
$$

  where the fact that the output is given by $y_i = M z_i$ was used. In this way, ℓ_i

stores the current synchronization (output) error.

- *In between events, the memory state ℓ_i is updated continuously according to*

$$\dot{\ell}_i = R\ell_i \tag{2.30}$$

The constant matrices R and K define the tuning parameters for the control algorithm. For the design of these parameters, the following coordinates are employed:

$$\theta_i = KM \sum_{k \in \mathcal{N}_i} (z_i - z_k) - \ell_i \tag{2.31}$$

which leads to

$$\theta = (\mathcal{L} \otimes KM)z - \ell \tag{2.32}$$

with $z = (z_1, z_2, \ldots, z_N)$, $\theta = (\theta_1, \theta_2, \ldots, \theta_N)$, $\ell = (\ell_1, \ell_2, \ldots, \ell_N)$, and $\mathcal{L}$ being the Laplacian matrix given by the directed graph Γ of the network. Let $x = (z, \theta, \tau) \in \mathbb{R}^{n_p N} \times \mathbb{R}^{rN} \times \mathcal{T} =: X$, where $\tau = (\tau_1, \tau_2, \ldots, \tau_N)$, $\mathcal{T} = [0, T^*_{2,1}] \times [0, T^*_{2,2}] \times \cdots \times [0, T^*_{2,N}]$, and r is the dimension of the outputs to each system. Then, a hybrid closed-loop system $\mathcal{H}$ is defined as the collection of all systems with their dynamics and the dynamics of the additional states in (2.29) and (2.30). The data (C, F, D, G) of this system is given as follows.

- *The flow set is the entire set of allowed values for x, namely, $C := X$.*

- *The flow map follows directly from the continuous dynamics of the state x, which is given by*

$$\dot{x} = \begin{bmatrix} A_f \begin{bmatrix} z \\ \theta \end{bmatrix} \\ -1_N \end{bmatrix} =: F(x) \tag{2.33}$$

where A_f is given by

$$A_f = \begin{bmatrix} A_1 & -\widetilde{B} \\ \widetilde{K}A_1 - \widetilde{R}\widetilde{K} & \widetilde{R} - \widetilde{K}\widetilde{B} \end{bmatrix}$$

with $A_1 = I \otimes A + \widetilde{B}\widetilde{K}$, $\widetilde{B} = I \otimes B$, $\widetilde{K} = \mathcal{L} \otimes KM$, and $\widetilde{R} = I \otimes R$.

- *The jump set is given by $D := \cup_{i \in \mathcal{V}} D_i$, where $D_i := \{x \in X : \tau_i = 0\}$.*

- *When $\tau_i = 0$ for some $i \in \mathcal{V}$, a jump of the i-th agent occurs. At such events, the i-th components of θ and τ are reset via*

$$\theta_i^+ = 0$$

and

$$\tau_i^+ \in [T^*_{1,i}, T^*_{2,i}]$$

The state component z_i remains constant. Moreover, for each $k \in \mathcal{V} \setminus \{i\}$ the state components z_k, ℓ_k, and τ_k are held constant when only τ_i is the *component of τ that is zero. Combining these constructions, for each $x \in D$, x is reset via*

$$x^+ \in G(x) := \{G_i(x) \, : \, x \in D_i, i \in \mathcal{V}\} \qquad (2.34)$$

where

$$G_i(x) = \begin{bmatrix} z \\ (\theta_1, \theta_2, \ldots, \theta_{i-1}, 0, \theta_{i+1}, \ldots, \theta_N) \\ (\tau_1, \tau_2, \ldots, \tau_{i-1}, [T_{1,i}^*, T_{2,i}^*], \tau_{i+1}, \ldots, \tau_N) \end{bmatrix}$$

This hybrid closed-loop system satisfies the hybrid basic conditions. In fact, by construction, the sets C and D are closed. The flow map F is continuous. The jump map G is outer semicontinuous since its graph is closed; moreover, it is locally bounded on D.

Example 2.24 (Network control). *The control of a physical system via an algorithm that is running at a remote location requires coping with information available at isolated time instances, at which communication events occur. In fact, a digital communication network triggers events in the entire system when transmission and reception of information occur. Similar to the model in Example 2.23, suppose that the output of the plant and the output of the algorithm are digitally transmitted through a network that is assumed to have zero transmission delay, but with communication events triggered at times $\{t_s\}_{s=1}^{\infty}$ satisfying*

$$T_1^* \leq t_{s+1} - t_s \leq T_2^* \qquad \forall s \in \mathbb{N} \setminus \{0\} \qquad (2.35)$$

where T_1^* and T_2^* are constants that define the minimum and maximum rate of communication events, respectively. At such events, the value of the output of the plant is transmitted. A hybrid model that captures such a mechanism incorporates the following additional variables:

- *A timer state, denoted τ_P, that triggers the communication events at times $\{t_s\}_{s=1}^{\infty}$ satisfying the minimum and maximum communication rate specified by T_1^* and T_2^*, respectively;*

- *A memory state, denoted ℓ_P, to store the output of the plant (or a function of it) at every communication event.*

Suppose the physical system is modeled by a differential equation with inputs

$$\dot{z}_1 = F_{P1}(z_1, u), \qquad \widetilde{y} = \widetilde{h}(z_1, u), \qquad z \in \mathbb{R}^{n_{P,1}}, u \in \mathbb{R}^{m_P}, \widetilde{y} \in \mathbb{R}^{\widetilde{r}_P} \qquad (2.36)$$

where z_1 is the state, u the input, and $\widetilde{y}$ is the measured output. These dynamics along with the network mechanism outlined above are captured in a hybrid plant model $\mathcal{H}_P$. This model has state $z = (z_1, z_2)$ with $z_2 = (\tau_P, \ell_P)$. The set of allowed values for z is $X_P := \mathbb{R}^{n_{P,1}} \times [0, T_2^*] \times \mathbb{R}^{\widetilde{r}_P}$. The flow of the state z is given by

$$\dot{z}_1 \;=\; F_{P1}(z_1, u), \quad \dot{z}_2 \;=\; \begin{bmatrix} \dot{\tau}_P \\ \dot{\ell}_P \end{bmatrix} = \begin{bmatrix} -1 \\ 0 \end{bmatrix}$$

during flow, and by

$$z_1^+ = z_1, \qquad z_2^+ = \begin{bmatrix} \tau_P^+ \\ \ell_P^+ \end{bmatrix} \in \begin{bmatrix} [T_1^*, T_2^*] \\ \widetilde{y} \end{bmatrix}$$

when $\tau_P = 0$. The model for τ_P is the same as the one in Example 2.23. Its output is $y = h(z) = \ell_P$. The resulting data of $\mathcal{H}_P$ is

$$C_P = \{(z, u) \in \mathbb{R}^{n_{P,1}} \times \mathbb{R}^{m_P} : \tau_P \in [0, T_2^*]\}$$

$$F_P(z, u) = \begin{bmatrix} F_{P1}(z_1, u) \\ -1 \\ 0 \end{bmatrix} \qquad \forall (z, u) \in C_P$$

$$D_P = \{(z, u) \in \mathbb{R}^{n_{P,1}} \times \mathbb{R}^{m_P} : \tau_P = 0\}$$

$$G_P(z, u) = \begin{bmatrix} z_1 \\ [T_1^*, T_2^*] \\ \widetilde{h}(z_1, u) \end{bmatrix} \qquad \forall (z, u) \in D_P$$

$$h(z) = \ell_P$$

In most network control settings, the control problem consists of asymptotically stabilizing the state z_1 of the plant to the origin, with robustness to small disturbances. A general dynamic output-feedback controller for this plant is given by

$$\dot{\eta}_1 = F_{K1}(\eta_1, \widetilde{y}), \qquad \widetilde{\zeta} = \widetilde{\kappa}(\eta_1, \widetilde{y}), \qquad \eta \in \mathbb{R}^{n_{K,1}}, \widetilde{\zeta} \in \mathbb{R}^{\widetilde{r}_K} \qquad (2.37)$$

with output transmitted via a communication network. Similar to the model of the plant, such a control algorithm can be modeled by a hybrid controller $\mathcal{H}_K$ with state $\eta = (\eta_1, \eta_2) \in X_K := \mathbb{R}^{n_{K,1}} \times [0, T_2^*] \times \mathbb{R}^{\widetilde{r}_K}$, $\eta_2 = (\tau_K, \ell_K)$, and the following data:

$$C_K = \left\{ (\ell_P, \eta) \in \mathbb{R}^{\widetilde{r}_P} \times X_K : \tau_K \in [0, T_2^*] \right\}$$

$$F_K(\ell_P, \eta) = \begin{bmatrix} F_{K1}(\eta_1, \ell_P) \\ -1 \\ 0 \end{bmatrix} \qquad \forall (\ell_P, \eta) \in C_K$$

$$D_K = \left\{ (\ell_P, \eta) \subset \mathbb{R}^{\widetilde{r}_P} \times X_K : \tau_K - 0 \right\}$$

$$G_K(\ell_P, \eta) = \begin{bmatrix} \eta_1 \\ [T_1^*, T_2^*] \\ \widetilde{\kappa}(\eta_1, \ell_P) \end{bmatrix} \qquad \forall (\ell_P, \eta) \in D_K$$

$$\kappa(\ell_P, \eta) = \ell_K$$

where the controller output ζ is equal to $\kappa(\ell_P, \eta)$. The input v of the controller has been assigned to the memorized plant output, ℓ_P, since $\widetilde{y}$ cannot be measured continuously.

Assigning the input u of the plant to the output ζ of the controller defines a hybrid closed-loop system $\mathcal{H}$. The sets C_P and D_P are closed by definition. Also, by definition, h is continuous. Then, when F_{P1} and $\widetilde{h}$ are continuous, items $(A1_P)$-$(A4_P)$ in Lemma 2.21 hold. Also, by definition, the sets C_K and D_K are closed and the map κ is continuous. Then, when F_{K1} and $\widetilde{\kappa}$ are continuous, items $(A1_K)$-$(A4_K)$ in Lemma 2.21 also hold. This lemma implies that the resulting hybrid closed-loop system $\mathcal{H}$ as in Definition 2.11 satisfies the hybrid basic conditions.

The examples above and Lemma 2.21 motivate the following definition of hybrid basic conditions for a hybrid plant.

Definition 2.25 (Hybrid basic conditions – third pass). *A hybrid plant $\mathcal{H}_P$ is said to satisfy the* hybrid basic conditions *if its data (C_P, F_P, D_P, G_P, h) satisfies $(A1_P)$-$(A4_P)$ in Lemma 2.21.*

Since the structure of models of the hybrid plant and of the hybrid controller coincide,[8] a similar definition of the hybrid basic conditions can be formulated for a hybrid controller. These conditions are precisely those in $(A1_K)$-$(A4_K)$.

2.3.3 Solution Concept

As outlined in § 1.1 and depicted in Figure 1.2, the combination of continuous and discrete dynamics in hybrid control systems gives rise to solutions with intervals of flow and also instants at which the solutions jump. For instance, the angular velocity of the legs of the walking robot in Example 1.2 evolves continuously in between impacts of the legs with the ground, and at instants where impacts occur, their angles and angular velocities jump to new values – in particular, the values of the states modeling the angles of the legs are swapped. Similarly, in the model of a closed-loop system resulting from sample-and-hold control in Example 1.3, the timer variable τ grows linearly and continuously at unitary rate and, when it reaches T^*, jumps to zero. These suggest that ordinary time t is a natural choice of a parameter to determine the amount of flow, as it is typically used to parameterize the evolution of solutions to continuous-time systems; see § 2.1. In principle, the instants at which jumps occur can be identified by simply partitioning the ordinary time semiaxis $\mathbb{R}_{\geq 0}$ into intervals with ending point being the jump time.

As illustrated in Example 1.2, hybrid control systems may have solutions with accumulations of jumps on a finite amount of flow time or even with consecutive jumps without flow in between; see Figure 1.2. In such cases, keeping track of the amount of flow and of the number of jumps using a partitioned ordinary time line is not adequate.

An alternative approach is to incorporate a discrete parameter $j \in \mathbb{N}$ that counts the number of jumps in the solution so as to parameterize the jumps – j plays a role similar to the role played by the discrete parameter, typically denoted as k, used in the definition of trajectories to discrete-time systems; see § 2.1. The combination of the parameters t and j gives rise to the notion of hybrid time and hybrid time domain defined next. This hybrid notion of time is used in this book, not only due to making the parameterization of flow intervals and jump instants more convenient, but also due to leading to a suitable notion of distance between solutions and robustness of asymptotic stability, as stated in Chapter 3.

---

[8]Except for the fact that, for convenience, the controller input is listed first in $\mathcal{H}_K$.

Definition 2.26 (Hybrid time and hybrid time domain). *A set $E \subset \mathbb{R}_{\geq 0} \times \mathbb{N}$ is a hybrid time domain if, for each $(T, J) \in E$, the set*

$$E \cap ([0, T] \times \{0, 1, \ldots, J\})$$

can be written in the form

$$\bigcup_{j=0}^{J} ([t_j, t_{j+1}] \times \{j\})$$

for some finite sequence of times $\{t_j\}_{j=0}^{J+1}$ satisfying $0 = t_0 \leq t_1 \leq t_2 \leq \ldots \leq t_J \leq t_{J+1}$. Each element $(t, j) \in E$ denotes the elapsed hybrid time, which indicates that t seconds of flow time and j jumps have occurred, or, equivalently, that $t + j$ "units" of hybrid time have elapsed. A hybrid time domain E induces a natural ordering among its elements: for each $(t, j), (t', j') \in E$ such that $t + j \leq t' + j'$, the ordering is denoted as $(t, j) \preceq (t', j')$.

The input, the state trajectory, and the output to a hybrid system are given by functions defined on hybrid time domains. For a hybrid plant $\mathcal{H}_P$, its input is given by functions of the form $(t, j) \mapsto u(t, j)$ with domain denoted as dom u and its state trajectory by $(t, j) \mapsto z(t, j)$ with domain dom z. Both dom u and dom z are hybrid time domains. Similar signals are used to define inputs and state trajectories to a hybrid controller $\mathcal{H}_K$ and to a hybrid closed-loop system $\mathcal{H}$. These functions are given by *hybrid inputs* and *hybrid arcs*, defined below.

A few notions are recalled first. To allow for general inputs u, the function u over intervals of flow is only required to be *Lebesgue measurable*. A set $S \subset \mathbb{R}$ is said to be *Lebesgue measurable* if it has positive Lebesgue measure $\mu(S)$, where $\mu(S) = \sum_k (b_k - a_k)$ and the (a_k, b_k)'s are all of the disjoint open intervals in S. Then, $s : S \to \mathbb{R}^{n_s}$ is said to be a *Lebesgue measurable function* if for every open set $\mathcal{U} \subset \mathbb{R}^{n_s}$ the set $\{r \in S : s(r) \in \mathcal{U}\}$ is Lebesgue measurable. Certainly, every continuous function s is also Lebesgue measurable.[9] A property is said to hold *almost everywhere* or *for almost all points* in a set S if the set of elements of S at which the property does not hold has zero Lebesgue measure. The function $s : S \to \mathbb{R}^{n_s}$ is said to be *locally essentially bounded* if for any $r \in S$ there exists an open neighborhood $\mathcal{U}$ of r such that s is bounded almost everywhere on $\mathcal{U}$; i.e., there exists $c \geq 0$ such that $|s(r)| \leq c$ for almost all $r \in \mathcal{U} \cap S$.

Definition 2.27 (Hybrid input). *A function $u : \text{dom} \, u \to \mathbb{R}^{m_P}$ is a hybrid input if dom u is a hybrid time domain and if, for each $j \in \mathbb{N}$, the function $t \mapsto u(t, j)$ is Lebesgue measurable and locally essentially bounded on the interval $I_u^j := \{t : (t, j) \in \text{dom} \, u\}$.*

The following definition introduces the notion of a hybrid arc. As suggested at the beginning of this chapter, a hybrid arc describes the state trajectory of the system. To guarantee that the state trajectory has a derivative with respect to ordinary t during flows – at least, for almost all t's on each interval of flow – state

---

[9]Informally, almost every function of practical interest is measurable.

trajectories $(t, j) \mapsto z(t, j)$ are assumed to be *locally absolutely continuous* for each fixed j; see Definition A.20 and the discussion following it.

Definition 2.28 (Hybrid arc). *A function $x : \operatorname{dom} x \to \mathbb{R}^{m_x}$ is a hybrid arc if $\operatorname{dom} x$ is a hybrid time domain and if, for each $j \in \mathbb{N}$, the function $t \mapsto x(t, j)$ is locally absolutely continuous on the interval $I_x^j = \{t : (t, j) \in \operatorname{dom} x\}$.*

A hybrid input and a hybrid arc define a solution to a hybrid system if, for the given input, the hybrid arc satisfies the dynamics of the hybrid system. This concept is formalized in the following definition, which is stated for a hybrid plant $\mathcal{H}_P$.

Definition 2.29 (Solution to a hybrid plant $\mathcal{H}_P$). *A hybrid input u and a hybrid arc z define a solution (z, u) to the hybrid plant $\mathcal{H}_P = (C_P, F_P, D_P, G_P, h)$ if*

(S0) $(z(0,0), u(0,0)) \in \overline{C}_P$ or $(z(0,0), u(0,0)) \in D_P$, and $\operatorname{dom} z = \operatorname{dom} u$ ($= \operatorname{dom}(z, u)$);[10]

(S1) For each $j \in \mathbb{N}$ such that I_z^j has a nonempty interior $\operatorname{int}(I_z^j)$, (z, u) satisfies

$$(z(t,j), u(t,j)) \in C_P \qquad \text{for all } t \in \operatorname{int}(I_z^j)$$

and

$$\frac{d}{dt} z(t,j) \in F_P(z(t,j), u(t,j)) \qquad \text{for almost all } t \in I_z^j$$

(S2) For each $(t, j) \in \operatorname{dom}(z, u)$ such that $(t, j+1) \in \operatorname{dom}(z, u)$, (z, u) satisfies

$$(z(t,j), u(t,j)) \in D_P$$

and

$$z(t, j+1) \in G_P(z(t,j), u(t,j))$$

Remark 2.30 (*About solution concept*). The solution concept in Definition 2.29 basically requires the state trajectory and input pair (z, u) to satisfy the dynamics of $\mathcal{H}_P$ in (2.7). Item (S0) requires that the initial condition for the state, $z(0,0)$, and the initial value of the input, $u(0,0)$, belong to either the jump set, so that a jump is possible, or to the closure of the flow set. When C_P is closed, then $\overline{C}_P$ coincides with C_P. A reason to put the closure on the flow set is to allow for solutions that "flow into" C_P. Item (S1) enforces that (z, u) satisfies the flow dynamics

$$(z, u) \in C_P \quad \dot{z} \in F_P(z, u)$$

during intervals of flow. In fact, the intervals of flow of (z, u) are the I_z^j's with nonempty interior, or, equivalently, with nonzero Lebesgue measure – such intervals are those that can be written as $[t_j, t_{j+1}]$ or $[t_j, t_{j+1})$ with $t_j < t_{j+1}$. Since over each

---

[10]When an input $u : \operatorname{dom} u \to \mathbb{R}^{m_P}$ is given and the hybrid arc $z : \operatorname{dom} z \to \mathbb{R}^n$ has a domain that is a truncation of $\operatorname{dom} u$ then, with some abuse of notation, $\operatorname{dom}(z, u)$ represents $\operatorname{dom} z \cap \operatorname{dom} u$.

such interval $t \mapsto z(t,j)$ is locally absolutely continuous, the time derivative of the state trajectory exists almost everywhere; see Definition A.20 and the discussion below it. Note also that the flow condition $(z,u) \in C_P$ is only required to hold on the interior of the interval of flow. This means that (z,u) has to belong to C_P for all times in each flow interval, except potentially at the initial and at the final time of each such interval. Item (S2) guarantees that jumps of z only occur from values of (z,u) that are in D_P, and that the value of z after the jump is assigned via the jump map G_P. $\triangle$

Solutions can be classified, in particular, according to the properties of their hybrid time domains. The type of solutions that are considered in this book are listed below. For simplicity, they are stated for $\mathcal{H}_P$, but the same definitions apply for solutions to $\mathcal{H}_K$ and to $\mathcal{H}$.

Definition 2.31 (Type of solutions). *A solution (z,u) to $\mathcal{H}_P$ is said to be*

- *nontrivial if* $\mathrm{dom}(z,u)$ *contains at least two points;*

- *complete if* $\mathrm{dom}(z,u)$ *is unbounded;*

- *bounded if* $\mathrm{rge}\, z = \{z(t,j) : (t,j) \in \mathrm{dom}\, z\}$ *is bounded;*

- *precompact if complete and* $\mathrm{rge}\, z$ *is bounded;*

- *Zeno if it is complete and the projection of* $\mathrm{dom}(z,u)$ *onto* $\mathbb{R}_{\geq 0}$ *is bounded;*

- *discrete if nontrivial and* $\mathrm{dom}(z,u) \subset \{0\} \times \mathbb{N}$;

- *continuous if nontrivial and* $\mathrm{dom}(z,u) \subset \mathbb{R}_{\geq 0} \times \{0\}$;

- *maximal if there does not exist another solution (z',u') such that (z,u) is a truncation of (z',u') to some proper subset of* $\mathrm{dom}(z,u)'$.

The following example illustrates the notion of solution in Definition 2.29 as well as some of the possible types of solutions introduced in Definition 2.31.

Example 2.32 (A one-degree-of-freedom juggling system). *Consider the juggling system shown in Figure 2.3. It consists of a ball of mass m_b moving along a vertical rod and a force actuated piston of mass m_p at the lower end. Assume that there is no friction between the rod and the ball, and that the rod is infinitely long. Furthermore, suppose that the ball and the piston have their mass concentrated at a point and that gravity is constant and given by $\gamma > 0$. To write the equations of motion, the state variables and input playing a role in the dynamics of the system have to be defined. The state vector z is defined as $z = (z_1, z_2, z_3, z_4)$, where, as Figure 2.3 indicates, the first component, z_1, represents the position of the ball and the second component, z_2, denotes the velocity of the ball. Similarly, z_3 represents the position of the piston and z_4 its velocity. By convention, z_2 is negative when the ball falls and z_4 is positive when the piston moves upwards. The force input applied to the piston is denoted by $u \in U := [\underline{u}, \overline{u}] \subset \mathbb{R}$, where $\underline{u} < \overline{u}$ are constants defining the boundary values of u, respectively. As shown below, this constraint on the input u can be conveniently captured by C_P.*

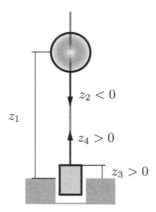

z_1

$z_2 < 0$

$z_4 > 0$

$z_3 > 0$

Figure 2.3: One-degree-of-freedom juggling system: a ball controlled by an actuated piston. Their positions are denoted by z_1, z_3 and their velocities by z_2, z_4, respectively.

Using Newton's second law of motion, the equation of motion of the ball in free flight can be written as

$$m_b \ddot{z}_1 = -m_b \gamma$$

A state model for the ball can be obtained using the fact that $z_2 = \dot{z}_1$, which leads to

$$\begin{bmatrix} \dot{z}_1 \\ \dot{z}_2 \end{bmatrix} = \begin{bmatrix} z_2 \\ -\gamma \end{bmatrix}$$

Under the same assumptions, the equation of motion for the piston in between impacts is given as

$$\begin{bmatrix} \dot{z}_3 \\ \dot{z}_4 \end{bmatrix} = \begin{bmatrix} z_4 \\ \frac{u}{m_p} \end{bmatrix}$$

The impacts between the ball and the piston occur when their positions coincide and they are moving towards each other. In terms of the state z, this condition is given by

$$z_1 = z_3, \qquad z_2 \leq z_4$$

From such points, the state is instantaneously updated via the following impact rule with conservation of momentum:

$$z_2^+ - z_4^+ = -\lambda(z_2 - z_4)$$
$$m_b z_2^+ + m_p z_4^+ = m_b z_2 + m_p z_4$$

where $\lambda \in [0,1]$ is the restitution coefficient. Let $\overline{m} = \frac{m_b}{m_b + m_p}$. The resulting update law at impacts for z_2 and z_4 is given by

$$\begin{bmatrix} z_2^+ \\ z_4^+ \end{bmatrix} = \begin{bmatrix} \overline{m} - (1 - \overline{m})\lambda & (1 - \overline{m})(1 + \lambda) \\ \overline{m}(1 + \lambda) & 1 - \overline{m} - \overline{m}\lambda \end{bmatrix} \begin{bmatrix} z_2 \\ z_4 \end{bmatrix} =: \Gamma_0 \begin{bmatrix} z_2 \\ z_4 \end{bmatrix}$$

where Γ_0 is uniquely defined by the constants λ and $\overline{m}$. The update law for z_1 and z_3 is given by

$$\begin{bmatrix} z_1^+ \\ z_3^+ \end{bmatrix} = \begin{bmatrix} z_1 \\ z_3 \end{bmatrix}$$

Then, the one-degree-of-freedom juggling system in Figure 2.3 can be described by the hybrid plant $\mathcal{H}_P$ as follows:

$$\mathcal{H}_P : \begin{cases} (z, u) \in C_P & \dot{z} = \begin{bmatrix} z_2 \\ -\gamma \\ z_4 \\ \frac{u}{m_p} \end{bmatrix} =: F_P(z, u) \\ \\ z \in D_P & z^+ = \begin{bmatrix} z_1 \\ [1\ 0]\,\Gamma_0 \begin{bmatrix} z_2 \\ z_4 \end{bmatrix} \\ z_3 \\ [0\ 1]\,\Gamma_0 \begin{bmatrix} z_2 \\ z_4 \end{bmatrix} \end{bmatrix} =: G_P(z) \\ \\ y \quad - z =: h(z) \end{cases} \tag{2.38}$$

where

$$C_P := \left\{ (z, u) \in \mathbb{R}^4 \times \mathbb{R} \ : \ z_1 \geq z_3, u \in [\underline{u}, \overline{u}] \right\}$$
$$D_P := \left\{ z \in \mathbb{R}^4 \ : \ z_1 = z_3, z_2 \leq z_4 \right\}$$

Note that since the piston is controlled through a physical actuator, the input u only has an effect on the flow map of the hybrid plant.

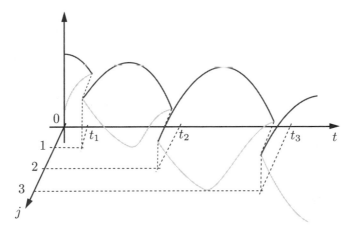

Figure 2.4: Position components of a solution to $\mathcal{H}_P$ on hybrid time domains: height of the ball (in dark trace) and height of the piston (in light trace).

Figure 2.4 depicts the z_1 and z_3 components of a solution to $\mathcal{H}_P$ for a particular choice of a constant control input u. At hybrid times $(t_1, 0), (t_2, 1), (t_3, 2)$ the ball *and the piston impact. At such times, the jump map G_P determines the new value of the state. The continuous evolution of the solution is governed by the differential equations describing the motion of the ball and of the piston. The following pairs also define solutions to $\mathcal{H}_P$ from different initial conditions and under the effect of other inputs:*

a) *With the parameter choices $m_p = m_b = \lambda = 1$ and $\gamma \in [\underline{u}, \overline{u}]$, given $z_1(0,0) > 0$, the pair (z, u) defined as*

$$z(t,j) = \begin{bmatrix} -\gamma\frac{(t-j\overline{t})^2}{2} + z_1(j\overline{t}, j) \\ -\gamma(t - j\overline{t}) \\ \gamma\frac{(t-j\overline{t})^2}{2} - z_1(j\overline{t}, j) \\ \gamma(t - j\overline{t}) \end{bmatrix}, \quad u(t,j) = \gamma$$

for all $t \in [0, \overline{t}]$ and $j = 0$, and for all $t \in [(2j-1)\overline{t}, (2j+1)\overline{t}]$ and $j \in \mathbb{N} \setminus \{0\}$, where $\overline{t} := \sqrt{\frac{2z_1(0,0)}{\gamma}}$, is a solution to $\mathcal{H}_P$ with hybrid time domain

$$\text{dom}(z, u) = ([0, \overline{t}] \times \{0\}) \ \cup \bigcup_{j \in \mathbb{N} \setminus \{0\}} ([[(2j-1)\overline{t}, (2j+1)\overline{t}] \times \{j\})$$

In fact, condition (S0) in Definition 2.29 holds since $(z(0,0), u(0,0)) \in C_P$ due to the fact that $z(0,0) = (z_1(0,0), 0, -z_1(0,0), 0)$, $z_1(0,0) > 0$, and $\gamma \in [\underline{u}, \overline{u}]$. Note that for every $j \in \mathbb{N}$, I_z^j has a nonempty interior and that for every $t \in \text{int}(I_z^j)$ we have

$$\frac{d}{dt}z(t,j) = \begin{bmatrix} -\gamma(t - j\overline{t}) \\ -\gamma \\ \gamma(t - j\overline{t}) \\ \gamma \end{bmatrix} = F_P(z(t,j), u(t,j))$$

and

$$(z(t,j), u(t,j)) \in C_P$$

Hence, condition (S1) holds. To check that condition (S2) holds, note that each $(t,j) \in \text{dom}(z,u)$ such that $(t, j+1) \in \text{dom}(z, u)$ belongs to the set of points

$$\{(\overline{t}, 0)\} \cup \bigcup_{j \in \mathbb{N}, j > 0} \{((2j+1)\overline{t}, j)\}$$

For $(\overline{t}, 0)$ and the given choices of the constants, the solution satisfies

$$z(\overline{t}, 0) = \begin{bmatrix} -\gamma\frac{\overline{t}^2}{2} + z_1(0,0) \\ -\gamma\overline{t} \\ \gamma\frac{\overline{t}^2}{2} - z_1(0,0) \\ \gamma\overline{t} \end{bmatrix} = \begin{bmatrix} 0 \\ -\sqrt{2\gamma z_1(0,0)} \\ 0 \\ \sqrt{2\gamma z_1(0,0)} \end{bmatrix} \in D_P$$

and

$$z(\overline{t}, j+1) = \begin{bmatrix} 0 \\ \sqrt{2\gamma z_1(0,0)} \\ 0 \\ -\sqrt{2\gamma z_1(0,0)} \end{bmatrix} = G_P(z(\overline{t}, j))$$

Similar steps can be used to show that (S2) holds at the other jump times. Hence, (z, u) is a solution to the hybrid plant. Furthermore, this solution is nontrivial since $\mathrm{dom}(z, u)$ has more than two points – this property already follows from $\bar{t}$ being larger than zero. This solution is actually complete since $\mathrm{dom}(z, u)$ is unbounded: in particular, each $j \in \mathbb{N}$ is such that $(t, j) \in \mathrm{dom}(z, u)$ for some $t \geq 0$. Since it is complete, it cannot be further extended; hence, it is maximal. In addition, it is precompact since z is bounded: it is easy to show that, for each $(t, j) \in \mathrm{dom}(z, u)$, the Euclidean norm of $(t, j) \mapsto z(t, j)$ is upper bounded by some constant $c > 0$. The state trajectory z exhibits periodic behavior with period equal to $2\bar{t}$ after the first jump at $(\bar{t}, 0)$.

b) *With the parameter choices $m_p = m_b$, $\lambda \in (0, 1)$, and $\gamma \in [\underline{u}, \overline{u}]$, given $z_1(0, 0) > 0$, the pair (z, u) defined as*

$$z(t, j) = \begin{bmatrix} -\gamma \frac{(t - t_j)^2}{2} + z_2(t_j, j)(t - t_j) + \overline{z}_1 \\ -\gamma(t - t_j) + z_2(t_j, j) \\ \gamma \frac{(t - t_j)^2}{2} - z_2(t_j, j)(t - t_j) - \overline{z}_1 \\ \gamma(t - t_j) - z_2(t_j, j) \end{bmatrix}, \quad u(t, j) = \gamma$$

for each $(t, j) \in \mathrm{dom}(z, u)$ is a solution to $\mathcal{H}_P$, where

$$\overline{z}_1 = -\frac{\gamma}{2} t_1^2 + z_2(0, 0) t_1 + z_1(0, 0)$$

$$t_1 = \frac{z_2(0, 0) + \sqrt{z_2(0, 0)^2 + 2\gamma z_1(0, 0)}}{\gamma}$$

and, for each $j \in \mathbb{N} \setminus \{0\}$,

$$z_2(t_j, j) = -\lambda z_2(t_j, j - 1)$$

and

$$t_{j+1} = t_j + \frac{2 z_2(t_j, j)}{\gamma}$$

which defines the hybrid time domain of (z, u) as

$$\mathrm{dom}(z, u) = \bigcup_{j \in \mathbb{N}} ([t_j, t_{j+1}] \times \{j\})$$

Moreover, $\mathrm{dom}(z, u)$ satisfies the property

$$\sup_t \mathrm{dom}(z, u) = \frac{z_2(0, 0)}{\gamma} + \frac{1 + \overline{m}}{\gamma(1 - \overline{m})} \sqrt{z_2(0, 0)^2 + 2\gamma z_1(0, 0)}$$

which is finite. Following similar steps as those in item 1 of this example, it can be shown that this solution is nontrivial, maximal, and complete. Furthermore, since $\sup_t \mathrm{dom}(z, u)$ is bounded, the solution is also Zeno. This solution represents a trajectory that converges to rest after a finite amount of flow time and infinitely many impacts.

c) *With the parameter choices $\gamma_0 \in [\underline{u}, \overline{u}]$ and $\gamma_0 > \gamma$, given $z_1(0, 0) > z_3(0, 0)$,* the pair (z, u) defined as

$$z(t, 0) = \begin{bmatrix} -\gamma \frac{t^2}{2} + z_1(0,0) \\ -\gamma t \\ -\gamma_0 \frac{t^2}{2} + z_3(0,0) \\ -\gamma_0 t \end{bmatrix}, \quad u(t, 0) = -\gamma_0$$

for all $t \in \mathbb{R}_{\geq 0}$ is a solution to $\mathcal{H}_P$ with hybrid time domain

$$\mathrm{dom}(z, u) = \mathbb{R}_{\geq 0} \times \{0\}$$

It is immediate to check that $t \mapsto z(t, 0)$ is a solution to $\dot{z} = F_P(z)$. Since $z_1(0, 0) > z_3(0, 0)$ and $\gamma_0 > \gamma > 0$, $z_1(t, 0) > z_3(t, 0)$ and $(z(t, 0), u(t, 0)) \in C_P$ for all $t \in \mathrm{dom}(z, u)$. This implies that (z, u) is a nontrivial, complete, pre-compact, continuous, and maximal solution to the hybrid plant. This solution represents the ball and the actuated piston moving down indefinitely, without any impacts.

d) Given $z_1(0, 0) = z_3(0, 0)$ and $z_2(0, 0) = z_4(0, 0)$, the pair (z, u) defined as

$$z(0, j) = \begin{bmatrix} z_1(0, 0) \\ z_2(0, 0) \\ z_3(0, 0) \\ z_4(0, 0) \end{bmatrix}, \quad u(0, j) = \star$$

for all $j \in \mathbb{N}$, where $\star$ denotes an arbitrary choice of the input u, is a solution to $\mathcal{H}_P$ with hybrid time domain

$$\mathrm{dom}(z, u) = \{0\} \times \mathbb{N}$$

In fact, $z(0, j + 1) = G_P(z(0, j))$ and $z(0, j) \in D_P$ for each $j \in \mathbb{N}$. This solution is nontrivial, complete, precompact, and discrete. This solution is also Zeno and represents the ball and the actuated piston at the same height and with zero velocity, which according to the model in Example 2.32, corresponds to a jump condition.

Certainly, there exist pairs (z, u) that are not solutions or that are trivial solutions to the hybrid plant $\mathcal{H}_P$ in (2.38). For instance, the pair (z, u) with $\mathrm{dom}(z, u)$ having more than one point and with $z(0, 0)$ such that

$$z_1(0, 0) < z_3(0, 0)$$

is not a solution since $z(0, 0) \notin \overline{C}_P \cup D_P$. Another example is a pair (z, u) with $u(t, 0) = \overline{u} + t$ for all $t \in \mathbb{R}_{\geq 0}$, which is not a solution since $u(t, 0) \notin [\underline{u}, \overline{u}]$ for all $t \in \mathrm{dom}(z, u)$, $t > 0$. Note that the pair (z, u) with $\mathrm{dom}(z, u) = \{(0, 0)\}$, $u(0, 0) = \overline{u}$, and $z(0, 0) \in D_P$ is a trivial solution to $\mathcal{H}_P$.

Solutions to $\mathcal{H}_K$ are defined in the same way as for $\mathcal{H}_P$ in Definition 2.29: to define a solution to $\mathcal{H}_K$ replace (C_P, F_P, D_P, G_P), z, and u in Definition 2.29 by (C_K, F_K, D_K, G_K), η, and v, respectively. Similarly, to define a solution to a hybrid closed-loop system $\mathcal{H}$ replace in Definition 2.29 the data (C_P, F_P, D_P, G_P) and the pair (z, u) by (C, F, D, G) and x, respectively, and remove the dependence on the input u. For completeness, the definition of solution to $\mathcal{H}$ is provided next.

Definition 2.33 (Solution to a hybrid closed-loop system $\mathcal{H}$). *A hybrid arc x defines a solution to the hybrid closed-loop system $\mathcal{H} = (C, F, D, G)$ if*

(S0) $x(0,0) \in \overline{C}$ *or* $x(0,0) \in D$;

(S1) *For each $j \in \mathbb{N}$ such that I_x^j has a nonempty interior $int(I_x^j)$, x satisfies*

$$x(t,j) \in C \qquad \text{for all } t \in int(I_x^j)$$

and

$$\frac{d}{dt}x(t,j) \in F(x(t,j)) \qquad \text{for almost all } t \in I_x^j$$

(S2) *For each $(t,j) \in \operatorname{dom} x$ such that $(t, j+1) \in \operatorname{dom} x$, x satisfies*

$$x(t,j) \in D$$

and

$$x(t, j+1) \in G(x(t,j))$$

The reader may have noted that x is used to denote both the state and a solution to $\mathcal{H}$. The same comment applies for the state and a solution to $\mathcal{H}_P$ and $\mathcal{H}_K$. While a different label to denote a solution could be used, making such a distinction increases the notational burden.

In this book, the same symbols are used to denote the state and a solution to the systems. Either context or the writing itself, in particular, as "the state x" or "the solution x," clarify which one of the two the symbol x represents.

2.3.4 Existence of Solutions to Closed-Loop Systems

The following result provides checkable conditions that guarantee that solutions to a hybrid closed-loop system $\mathcal{H}$ exist. Furthermore, it provides conditions on the data of $\mathcal{H}$ guaranteeing that its solutions satisfy the properties in Definition 2.31 defining different types of solutions. Below, $T_C(x)$ denotes the tangent cone[11] of C at x.

Proposition 2.34 (Existence of solutions to $\mathcal{H}$). *Consider the closed-loop system $\mathcal{H}$ and suppose it satisfies the hybrid basic conditions in Definition 2.20. Let $x_\circ \in C \cup D$ be arbitrary. If $x_\circ \in D$ or*

---

[11]See Definition A.8.

(VC) there exists a neighborhood $\mathcal{U}$ of $x_\circ$ such that for every $x \in \mathcal{U} \cap C$,

$$F(x) \cap T_C(x) \neq \emptyset,$$

then there exists a nontrivial solution x to $\mathcal{H}$ with $x(0,0) = x_\circ$. If (VC) holds for every $x_\circ \in C \setminus D$, then there exists a nontrivial solution to $\mathcal{H}$ from every initial point in $C \cup D$, and every maximal solution x to $\mathcal{H}$ satisfies exactly one of the following conditions:

  a) x is complete;

  b) $\operatorname{dom} x$ is bounded and the interval I_x^J, where $J = \sup_j \operatorname{dom} x$, has nonempty interior and $t \mapsto x(t, J)$ is a maximal solution to $\dot{x} \in F(x)$, $x \in C$ satisfying $\lim_{t \nearrow T} |x(t, J)| = \infty$, where $T = \sup_t \operatorname{dom} x$;[12]

  c) $x(T, J) \notin C \cup D$, where $(T, J) = \sup \operatorname{dom} x$.

Furthermore, the following hold:

  1. If $G(D) \cap D = \emptyset$, then

    i) every maximal solution x to $\mathcal{H}$ with $\operatorname{dom} x$ having more than two points is not Zeno; and

    ii) every bounded solution x to $\mathcal{H}$ has jump times that are uniformly lower bounded by a positive constant, i.e., for each bounded solution x to $\mathcal{H}$ there exists $\gamma > 0$ such that $t_{j+1} - t_j \geq \gamma$ for all $j \geq 1$, where $I_x^j = [t_j, t_{j+1}]$.

  2. If C is compact, F has linear growth on C, or F is single valued and locally Lipschitz on C, then item b does not occur;

  3. If $G(D) \subset C \cup D$, then item c above does not occur.

In most cases, a hybrid controller should assure that every maximal solution to the closed loop is complete. In this way, the solutions to the hybrid plant can evolve for arbitrarily long hybrid time. However, solutions may not exist from every point in the state space of the hybrid closed-loop systems – they would exist only from the set $\overline{C} \cup D$, which involves the flow and jump sets of the controller. Actually, a hybrid controller constrains – or "shapes" – the behavior of a hybrid plant through the definition of its data.

The following examples revisit hybrid systems and controllers introduced earlier, to illustrate Proposition 2.34.

Example 2.35 (Juggling systems, revisited). *Consider the juggling system in Example 2.32 with the parameters and inputs assigned in item b) therein. The hybrid closed-loop system $\mathcal{H}$ has state $x = z \in \mathbb{R}^4 =: X$, input $u = \gamma$, and dynamics*

---

[12]In this case, the maximal solution x is unbounded.

as those of $\mathcal{H}_P$ in (2.38) but with the input u assigned: $C = \{x \in X : z_1 \geq z_3\}$, $F(x) = F_P(z, \gamma)$ for each $x \in C$, $D = \{x \in X : z_1 = z_3, z_2 \leq z_4\}$ and $G(x) = G_P(z)$ for each $x \in D$, where z is used in the definitions for clarity. Existence of (nontrivial) solutions from every point in $C \cup D$ is shown via Proposition 2.34. To that end, note that the construction of C and D lead to

$$C \setminus D = \{x \in X : z_1 > z_3\} \cup \{x \in X : z_1 = z_3, z_2 > z_4\}$$

Now, pick $x_\circ \in C \cup D$.

- *If $x_\circ \in D$, then a jump is possible and a nontrivial solution exists from $x_\circ$.*

- *If $x_\circ \in C \setminus D$, the following two cases are possible based on the expression of $C \setminus D$ above:*

  - *If $x_\circ \in \{x \in X : z_1 > z_3\}$, $T_C(x_\circ) = \mathbb{R}^4$ due to $x_\circ$ being in the interior of C. This implies that $F(x_\circ) \cap T_C(x_\circ) \neq \emptyset$ holds trivially.*

  - *If $x_\circ \in \{x \in X : z_1 = z_3, z_2 > z_4\}$, then $x_{\circ,2} > x_{\circ,4}$ and $T_C(x_\circ) = C$. Let $\chi := F(x_\circ)$. Since $F(x_\circ) = F_P(x_\circ, \gamma)$, then $\chi_1 = x_{\circ,2}$ and $\chi_3 = x_{\circ,4}$, and, consequently, $\chi_1 > \chi_3$ due to the choice of $x_\circ$. This implies that $\chi \in C \setminus D$. Hence, $F(x_\circ) \subset T_C(x_\circ)$,*

  Then, (VC) holds for each point $x_\circ$ in $C \setminus D$.

By Proposition 2.34, for each $x_\circ \in C \cup D$ there exists a nontrivial solution x to $\mathcal{H}$ with $x(0,0) = x_\circ$.

To establish that each maximal solution to $\mathcal{H}$ is complete, it is enough to show that cases b and c in Proposition 2.34 cannot occur. This is shown next.

- *The flow map F has linear growth on C. In fact, $|F(x)| \leq k(1 + |x|)$ for each $x \in C$ with $k = \max\left\{1, \gamma, \frac{\gamma}{m_P}\right\}$. Then, by item 2 of Proposition 2.34, case b does not hold.*

- *Note that for each $x \in D$, the new value of the state after a jump, denoted x^+, is such that $x_1^+ = x_1$ and $x_3^+ = x_3$. Since $x \in D$ implies that $x_1 = x_3$, then $x_1^+ = x_3^+$. From the definition of C, it follows that $x^+ \in C$. By item 3 of Proposition 2.34, case c does not hold.*

Hence, since only case a holds, every maximal solution to $\mathcal{H}$ is complete.

Example 2.36 (Sample-and-hold control, revisited). *Consider the hybrid closed-loop system given in Example 2.14 resulting from controlling a hybrid plant via sample-and-hold control. When the plant $\mathcal{H}_P$ is a purely continuous-time system without constraints, the sample-and-hold strategy leads to a hybrid closed-loop system with flow set $C = \{x \in \mathbb{R}^n : \tau \in [0, T^*]\}$, same flow map as in Example 2.14, namely, $F(x) = (F_P(z, \ell_v), 0, 1)$, jump set $D = D_2$, and jump map G given as $G(x) = (z, \kappa_c(h(z)), 0)$. To apply Proposition 2.34, F_P and κ_c have to be continuous for $\mathcal{H}$ to satisfy the hybrid basic conditions. It follows that condition (VC) in Proposition 2.34 holds at points $x \in C \setminus D$. Then, invoking Proposition 2.34, for every point in $C \cup D$ there exists a nontrivial solution to $\mathcal{H}$. If $\dot{z} = F_P(z, \ell_v)$ does not exhibit finite escape times for each possible value of ℓ_v, then case b in Proposition 2.34 cannot happen. Furthermore, since for this system $G(D) \subset C \cup D$ holds,* *then, by item 3 of Proposition 2.34, case c cannot hold either. Then, only case a is possible, implying that every maximal solution to $\mathcal{H}$ is complete. Finally, by item 1 of Proposition 2.34, each maximal solution is not Zeno, and each bounded solution has jump times that are separated by a uniform (for the particular solution) positive lower bound. The case when the plant is constrained requires the viability condition (VC) to hold at points in $C \setminus D$.*

Example 2.37 (DC/AC converter, revisited). *Consider the controlled single-phase DC/AC inverter shown in Figure 1.3 and introduced in Example 1.1; see also Example 1.4. The dynamics of the combined inverter and filter are given by a switched system, which is modeled as a hybrid plant $\mathcal{H}_P$. Its data is defined in Example 1.1. A hybrid controller is designed to generate a sinusoidal-like output v_{C_a} approximating a given reference voltage. As explained in Example 1.4, a reference signal – in current-voltage space – is given by a function of time making the function V therein constant. Denoting the reference signal as $t \mapsto (i_L^r(t), v_{C_a}^r(t))$ and choosing a unitary constant c, the voltage reference is given by $t \mapsto v_{C_a}^r(t) = b \sin(\omega t + \theta)$, where b is the amplitude, ω is the frequency, and θ is the initial phase – recall that current and voltage in the circuit are related via $i_L = C_a \dot{v}_{C_a}$. and that $a = C_a \omega b$, where the constants a and b are defined in Example 1.4. Then, the problem to solve consists of keeping the state of the plant within the tracking band, which, as defined in Example 1.4, is given by $\{z = (i_L, v_{C_a}) \in \mathbb{R}^2 : c_i \leq V(z) \leq c_o\}$, where $0 < c_i < 1 < c_o$ are parameters of the controller.*

To solve the stated control problem, a logic-based hybrid controller $\mathcal{H}_K$ with state $\eta = q \in Q := \{-1, 0, 1\}$ is designed. The input v of the controller is assigned to z, and its dynamics are given by

$$\mathcal{H}_K : \quad \begin{cases} (z, q) \in C_K & \dot{q} = F_K(z, q) := 0 \\ (z, q) \in D_K & q^+ \in G_K(z, q) \\ & \zeta = \kappa(q) := q \end{cases} \tag{2.39}$$

The set-valued map G_K provides the values of the logic state q that guarantee forward invariance of the set

$$K = \{z \in \mathbb{R}^2 : c_i \leq V(z) \leq c_o\} \times Q \tag{2.40}$$

for the resulting hybrid closed-loop system. In this way, the output of the hybrid controller determines the position of the switches of the inverter.

To complete the construction of the controller, the flow set, the jump set, and the jump map of the controller are designed as follows. For simplicity, their expressions are given for $v = z$, in terms of $(z, q) \in X := \mathbb{R}^2 \times Q$ and of the sets S_i and S_o defined in Example 1.4.

- *The jump set D_K is defined using the logic depicted in Figure 2.5. As shown therein, when $q = -1$ and z reaches a point in S_i with $i_L \geq 0$, then the controller triggers a jump and updates q to one. This selection is due to the fact that, from such points, $q = 1$ forces the state z to stay within the tracking band, by flowing. This property, along with the properties leading to the other parts of the logic shown in Figure 2.5, is established in Lemma 2.38 below. In* *this way, following Figure 2.5, the jump set D_K is given as*

$$D_K := \{(z,q) \in X : V(z) = c_i, i_L q \leq 0, q \in \{-1,1\}\}$$
$$\bigcup \{(z,q) \in X : V(z) = c_o, i_L q \geq 0, q \in \{-1,1\}\}$$
$$\bigcup \{(z,q) \in X : V(z) = c_i, q = 0\}$$

- *The jump map G_K implements the logic depicted in Figure 2.5. More precisely, using measurements of $z = (i_L, v_{C_a})$ and the current value of q, the hybrid controller updates q according to the following logic: with $\epsilon > 0$ a (small) positive parameter, and sets*

$$M_1 = \{z \in \mathbb{R}^2 : z \in S_o, 0 \leq i_L \leq \epsilon, v_{C_a} \leq 0\}$$
$$M_2 = \{z \in \mathbb{R}^2 : z \in S_o, -\epsilon \leq i_L \leq 0, v_{C_a} \geq 0\}$$

  a) *If $z \in S_o \backslash M_1$, $i_L \geq 0$, and $q \in \{0,1\}$, then reset q to -1 to steer the state z to S_i. Define $D_a := \{(z,q) \in X : z \in S_o \backslash M_1, i_L \geq 0, q \in \{0,1\}\}$.*

  b) *If $z \in S_o \backslash M_2$, $i_L \leq 0$, and $q \in \{-1,0\}$, then reset q to 1 to steer the state z to S_i. Define $D_b := \{(z,q) \in X : z \in S_o \backslash M_2, i_L \leq 0, q \in \{-1,0\}\}$.*

  c) *If $z \in S_i$, $i_L \geq 0$, and $q \in \{-1,0\}$, then reset q to 1 to steer the state z to S_o. Define $D_c := \{(z,q) \in X : z \in S_i, i_L \geq 0, q \in \{-1,0\}\}$.*

  d) *If $z \in S_i$, $i_L \leq 0$, and $q \in \{0,1\}$, then reset q to -1 to steer the state z to S_o. Define $D_d := \{(z,q) \in X : z \in S_i, i_L \leq 0, q \in \{0,1\}\}$.*

  e) *If $z \in M_1$ and $q = 1$, then reset q to 0 to steer the state z to the right-hand side of the z plane. Define $D_e := \{(z,q) \in X : z \in M_1, q = 1\}$.*

  f) *If $z \in M_2$ and $q = -1$, then reset q to 0 to steer the state z to the left-hand side of the z plane. Define $D_f := \{(z,q) \in X : z \in M_2, q = -1\}$.*

The sets M_1 and M_2 of the jump set lead to jumps to $q = 0$. This mechanism is included to rule out the existence of Zeno solutions from points in $\{z \subset \mathbb{R}^2 : V(z) = c_o, i_L = 0\}$. Indeed, from such points, solutions would flow "horizontally" – namely, to the left or to the right – on the (i_L, v_{C_a}) plane when $q \in \{-1,1\}$. Using this logic, the jump map G_K is given by

$$G_K(z,q) := \begin{cases} -1 & \text{if } (z,q) \in D_a; \text{ or } (z,q) \in D_d, q \neq 0, i_L \neq 0 \\ 1 & \text{if } (z,q) \in D_b; \text{ or } (z,q) \in D_c, q \neq 0, i_L \neq 0 \\ \{-1,1\} & \text{if } (z,q) \in D_c \cap D_d, q = 0, i_L = 0 \\ 0 & \text{if } (z,q) \in D_e \cup D_f, |i_L| \neq \epsilon \\ \{0,1\} & \text{if } z \in M_2, i_L = -\epsilon \\ \{-1,0\} & \text{if } z \in M_1, i_L = \epsilon \end{cases}$$

The second and the last two pieces in the definition of G_K are set valued so as to guarantee outer semicontinuity at points where S_o intersects with M_1 and M_2.

- *The flow set C_K is chosen as the closed complement of the jump set, which is equal to K as in (2.40). This choice guarantees that flows converge to and* *remain in the tracking band, for each possible value of q.*

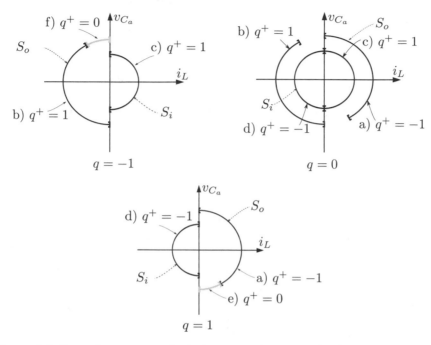

Figure 2.5: Sets of points at which the control logic in a)-f) – see the definition of the jump map – triggers an update of the logic state q. *Source: Chai and Sanfelice, 2014 [2]. Reproduced by permission of IEEE.*

The key property that makes the proposed logic work is the following fact about the inner product between the gradient of V and the map F_P defined in Example 1.1.

Lemma 2.38 (Inner product properties). *Given positive system constants R, L, C_a, ω, V_{DC} such that $LC_a\omega^2 \geq 1$, and with $\alpha = \frac{2}{a^2 L}$, $\beta = \frac{2}{b^2 C_a}$, and*

$$\Gamma := \left\{ z \in \mathbb{R}^2 : -\alpha V_{DC} \leq -\alpha R i_L + (\beta - \alpha)v_{C_a} \leq \alpha v_{DC} \right\}$$

the following hold:

a) $\langle \nabla V(z), F_P(z,q) \rangle \leq 0$ for all $(z,q) \in X$ such that $z \in \Gamma$ and, ($i_L \leq 0, q = 1$) or ($i_L \geq 0, q = -1$).

b) $\langle \nabla V(z), F_P(z,q) \rangle \geq 0$ for all $(z,q) \in X$ such that $z \in \Gamma$ and, ($i_L \leq 0, q = -1$) or ($i_L \geq 0, q = 1$).

c) $\langle \nabla V(z), F_P(z,q) \rangle \leq 0$ for all $(z,q) \in X$ such that $z \in M_1 \cup M_2$ and $q = 0$.

When $\mathcal{H}_K$ is used to control the plant in Example 1.1, the resulting hybrid closed-loop system $\mathcal{H} = (C,F,D,G)$ has state variable $x = (z,q) \in X$ and data *given by*

$$F(x) := \begin{bmatrix} \frac{v_{DC}}{L}q - \frac{R}{L}i_L - \frac{1}{L}v_{C_a} \\ \frac{1}{C_a}i_L \\ 0 \end{bmatrix} \qquad \forall x \in C := C_K \qquad (2.41)$$

$$G(x) := \begin{bmatrix} i_L \\ v_{C_a} \\ G_K(z,q) \end{bmatrix} \qquad \forall x \in D := D_K \qquad (2.42)$$

Furthermore, it satisfies the hybrid basic conditions. Showing this property is left as an exercise. Now, Proposition 2.34 is employed to show that maximal solutions from every point in $C \cup D = K$ are bounded, complete, and have jump times that are uniformly lower bounded by a positive constant. According to the viability condition (VC) in Proposition 2.34, to verify the sufficient conditions for the existence of a solution to $\mathcal{H}$ from $x_\circ \in C \cup D$, it is enough to show that there is a solution that flows from $x_\circ \in C \setminus D$ for some time. This property is satisfied since the function V defined in Example 1.1 satisfies the properties stated in Lemma 2.38. In fact, due to $\dot{V}$ being nonpositive at points in S_o and nonnegative at points in S_i, flow from S_o and S_i is possible to points inside the tracking band.

2.3.5 Hybrid System Models with Disturbances

As it is the case for continuous-time and discrete-time systems, a model of a hybrid plant may not capture all of the dynamics of the actual system being modeled. Modeling error, measurement error, and actuator noise are unavoidable due to lack of full information of the system being modeled. Typical sources of modeling error include uncertainty on the parameter values, on the exact form of the nonlinearities, and on the conditions triggering the events. This modeling limitation motivates the study of the inherent robustness properties of asymptotic stability of a hybrid closed-loop system, in particular, in light of *unknown disturbances* leading to solutions that jump at time instances that are almost surely different from those of the nominal solutions. Without robustness guarantees, the behavior of a hybrid system under disturbances might be totally different from the nominal behavior, even if the magnitude of the disturbances is arbitrarily small.

Given a (nominal) model $\mathcal{H}_P$ of a hybrid plant, a model with added disturbances that has practical interest is given by

$$\mathcal{H}_{P,w} : \begin{cases} (z, u + w_u) \in C_P & \dot{z} \in F_P(z, u + w_u) + w_{F_P} \\ (z, u + w_u) \in D_P & z^+ \in G_P(z, u + w_u) + w_{G_P} \\ & y = h(z) + w_y \end{cases} \qquad (2.43)$$

where w_u, w_{F_P}, w_{G_P}, and w_y represent disturbances. For example, w_u models actuator noise; w_{F_P} and w_{G_P} capture modeling error and disturbances on the continuous and on the discrete dynamics, respectively; and w_y represents disturbances on the model of the output or measurement noise. These disturbances may be passed through functions so that they affect only particular components of the state, input, and maps. They may also depend on the state, the input, or even on hybrid time (t, j).

These type of disturbances arise in most practical situations. For instance, the model of the pendulum system in Example 1.5 may not completely capture the dynamics of a real-world pendulum system, leading to unmodeled plant dynamics. In the network control problem in Example 2.24, the measurements obtained from a network are unavoidably perturbed by measurement noise. Robustness to these disturbances is addressed in each of the forthcoming chapters for different hybrid feedback controllers. In addition, in many hybrid control problems, the variables that are used to trigger the jumps are affected by disturbances. Such situation can be captured with an "inflated" version of the nominal jump set. In simple words, the jump condition $(z, u + w_u) \in D_P$ can be replaced by a "worst case" set condition obtained from the nominal jump set D_P and the allowed disturbances w_u. If w_u is such that $|w_u| \le \rho_u$ for some $\rho_u > 0$, then a construction incorporating the effect of w_u is given by

$$\{(z, u) \,:\, (z, u + w_u) \in D_P, |w_u| \le \rho_u\}$$

A similar reformulation of the flow set is also possible.[13] This modeling approach becomes particularly relevant when deriving a model of a hybrid controller $\mathcal{H}_K$ with added uncertainty, as it would permit capturing the situation where, due to imperfect implementation of the controller, jumps are triggered at slightly different conditions than those imposed by the model.

In general, when either the hybrid plant or the hybrid controller are subject to uncertainty, a model of the hybrid closed-loop system $\mathcal{H}$ that captures the effects of disturbances is given by the following ρ-perturbation of $\mathcal{H}$:

Denoting by $\rho : \mathbb{R}^n \to \mathbb{R}_{\ge 0}$ the state-dependent perturbation function and with (C, F, D, G) being the nominal data of $\mathcal{H}$,

- The ρ-perturbation of C is given by all points x such that the ball of size $\rho(x)$ around x has nonempty intersection with C. Such a set is denoted C^ρ and is defined as

$$C^\rho := \{x \in \mathbb{R}^n \,:\, (x + \rho(x)\mathbb{B}) \cap C \ne \emptyset\} \qquad (2.44)$$

  The definition of C^ρ allows to model perturbations on the state and on the nominal set C. For instance, any signal w_x added to the state as $x + w_x$ and being such that $x + w_x$ belongs to both C and to $x + \rho(x)\mathbb{B}$ is modeled by C^ρ. Also, any perturbation of C that is contained in C^ρ is also captured by C^ρ. Note that C^ρ is essentially an inflation of C modulated by ρ.

- The ρ-perturbation of the flow map F is given by the directions of flow resulting from perturbing both the argument of F – namely, the state x is replaced by $x + w_x$ for any disturbance w_x such that $x + w_x$ belongs to $x + \rho(x)\mathbb{B}$ – and from perturbing the actual values of F directly. This perturbation is denoted F^ρ and is defined as

$$F^\rho(x) = \overline{\mathrm{con}}F(x + \rho(x)\mathbb{B}) + \rho(x)\mathbb{B} \qquad \forall x \in C^\rho \qquad (2.45)$$

  In particular, this perturbation allows to capture state noise and unmodeled dynamics. For instance, denoting by w_x the noise in the state, and by w_F the unmodeled continuous dynamics, F^ρ allows to capture the perturbed flow

---

[13]When the sets are inflated, appropriate definitions of the flow and jump maps over the new sets are required.

$\dot{x} \in F(x + w_x) + w_F$, $x + w_x \in C$ when w_F belongs to $\rho(x)\mathbb{B}$, and $x + w_x$ belongs to $x + \rho(x)\mathbb{B}$. Example 2.14 presents an inflated flow map capturing a disturbance on the continuous evolution of a timer.

• Similarly to C^ρ, the ρ-perturbation of D is given by all points x that, when a ball of size $\rho(x)$ is put around x, $x + \rho(x)\mathbb{B}$ intersects D. In this way, the set D^ρ is given by

$$D^\rho = \{x \in \mathbb{R}^n \ : \ (x + \rho(x)\mathbb{B}) \cap D \neq \emptyset\} \tag{2.46}$$

Similar to C^ρ, the definition of D^ρ captures perturbations on the state and on the nominal jump set D.

• Similar to the construction of F^ρ, the ρ-perturbation of the jump map G is given by the values of the nominal jump map G resulting from perturbing both the argument of G and the actual values of G directly. More precisely, the argument x of G is perturbed by $\rho(x)$ and each value χ from G is further perturbed by $\rho(x)$. This perturbation is denoted G^ρ and is defined as

$$G^\rho(x) = \{x' \ : \ x' \in \chi + \rho(\chi)\mathbb{B}, \ \chi \in G(x + \rho(x)\mathbb{B})\} \qquad \forall x \in D^\rho \tag{2.47}$$

In particular, this perturbation allows to capture state noise and unmodeled dynamics at jumps. As for the case during flows, denoting by w_x the noise in the state, and by w_G the unmodeled discrete dynamics, G^ρ allows to capture the perturbed discrete dynamics $x^+ \in G(x + w_x) + w_G$, $x + w_x \in D$ when w_G belongs to $\rho(x)\mathbb{B}$, and $x + w_x$ belongs to both D and $x + \rho(x)\mathbb{B}$. Example 2.14 features an inflated jump map capturing a disturbance on the value after jumps of a timer.

The resulting ρ-perturbation of $\mathcal{H}$ is denoted $\mathcal{H}^\rho = (C^\rho, F^\rho, D^\rho, G^\rho)$, and is given by

$$\mathcal{H}^\rho \ : \ \begin{cases} x \in C^\rho & \dot{x} \in F^\rho(x) \\ x \in D^\rho & x^+ \in G^\rho(x) \end{cases} \tag{2.48}$$

2.4 NUMERICAL SIMULATION

Most of the models of the plants, controllers, and closed-loop systems studied in this book can be simulated in the Hybrid Equations (HyEQ) Toolbox.[14] This toolbox consists of open source computer code developed to numerically approximate solutions to hybrid systems given in terms of hybrid equations.

The HyEQ Toolbox includes the following components and features:

• Computer code for simulation of hybrid equations;

---

[14] Available in Python, Java, Octave, and MATLAB at @BookSite/Simulation.

- Computation of state trajectories that are Zeno and that have multiple jumps at the same instant;

- Basic event detection and capability to implement advanced crossing detection algorithms.

The toolbox includes functions to plot hybrid arcs, on hybrid time domains, as a function of ordinary time t, and as a function of jump time j.

The latest version of this toolbox and instructions on how to use it are available at @BookSite/Simulation. In simple words, and for the case of a hybrid equation describing the hybrid closed-loop system $\mathcal{H}$, the HyEQ Toolbox computes an approximation of the solution x by evaluating the flow condition $x \in C$ and the jump condition $x \in D$. According to the result of this evaluation, it appropriately discretizes the differential equation defined by the (single-valued) flow map F of $\mathcal{H}$ or computing the new value of the state after jumps according to the jump map G. This computation occurs over a pre-specified finite amount of flow and finite number of jumps. In this way, the HyEQ Toolbox returns a discrete version of x and of its hybrid time domain dom x. More precisely, the computed version of the associated solution x, when it exists, is denoted

$$x_s : \operatorname{dom} x_s \to \mathbb{R}^n$$

which is referred to as *a simulated solution* to $\mathcal{H}$, and satisfies

$$x_s^+ = F^s(x_s) \qquad x_s \in C \tag{2.49}$$

over the intervals of flow and, at jumps, satisfies the discrete dynamics

$$x_s^+ = G(x_s) \qquad x_s \in D \tag{2.50}$$

The function F^s is the resulting discretized flow map F, obtained when employing an integration scheme for the differential equation $\dot{x} = F(x)$. For instance, when the integration scheme is given by the forward Euler integration scheme, $F_s(x_s) = x_s + sF(x_s)$ with $s > 0$ denoting the step size for integration. The flow set, jump set, and jump map could also be affected by discretization occurring in the code implementation stage, leading to C_s, D_s, and G_s, respectively. The simulation of a hybrid plant and of a hybrid controller that are part of the hybrid closed-loop system can be performed similarly.

The HyEQ Toolbox includes computer code written in a variety of programming languages for the simulation of a wide range of hybrid systems. The simplest way to simulate a hybrid closed-loop system $\mathcal{H}$ is to use the "lite" code included in the toolbox, which requires the creation of the following files:[15]

i) C.sim, a function defining the flow set. The input to this function is a vector with components defining the state of the system x_s. Its output is equal to 1 if the state belongs to the set C or equal to 0 otherwise.

---

[15]The extension sim is to be replaced appropriately based on the code used (e.g., m for MAT-LAB, py for Python, etc.).

ii) `f.sim`, a function defining a single-valued flow map. The input to this function is a vector with components defining the state of the closed-loop system x_s. Its output is the value of the flow map F evaluated at x_s.

iii) `D.sim`, a function defining the jump set. Its input is a vector with components defining the state of the system x_s. Its output is equal to 1 if the state belongs to D or equal to 0 otherwise.

iv) `g.sim`, a function defining a single-valued jump map. Its input is a vector with components defining the state of the system x_s. Its output is the value of the jump map G evaluated at x_s.

The function `HyEQsolver` implements the simulation of a hybrid equation $\mathcal{H}$ defined by functions `C.sim`, `f.sim`, `D.sim`, and `g.sim`, encoding its data (C, F, D, G). These functions are used by `HyEQsolver` to integrate the differential equation during flows, to trigger jumps, and to reset the state. To determine the appropriate behavior (flow or jump), the algorithms in `HyEQsolver` check at each integration step if x_s is in the set C or D. Depending on which set x_s is in, the state is updated accordingly following the dynamics given in F and G. If the state is not in either set, then the simulation is stopped. The function `HyEQsolver` requires setting the following simulation parameters:

- The $n \times 1$ vector `x0` defines the initial condition.

- The 2×1 parameter `TSPAN = [TSTART TFINAL]` defines the initial and final values of the flow variable t, i.e., the *continuous horizon*.

- The 2×1 parameter `JSPAN = [JSTART JFINAL]` defines the initial and final values of the jump index j, i.e., the *discrete horizon*.

- The scalar parameter `rule` defines whether the simulator gives priority to jumps (`rule=1`), priority to flows (`rule=2`), or no priority (`rule=3`) when both $x_s \in C$ and $x_s \in D$ hold. When no priority is selected, then the simulator selects flowing or jumping randomly.

- The parameter `options` configures the relative tolerance, maximum integration step allowed, and other knobs of the integration scheme used to approximate the flows.

The simulation stops when either when t reaches `TFINAL` or j reaches `JFINAL`. The `HyEQsolver` function returns the computed state x_s along with the (discretized) hybrid time domain dom x_s.

The `run.sim` function is provided to initialize these parameters and to run the simulation. This function is also used to plot the computed trajectories after the simulation is complete.

Example 2.39 (Juggling system, revisited). *This example illustrates how to use the lite code provided by the simulator to approximate the solutions to the hybrid closed-loop system resulting from controlling the hybrid system in Example 2.32 with a constant input; see also Example 2.35. Following the four cases in Example 2.32, the input u is taken to be a constant given by $u = \gamma$, and $\underline{u} = \overline{u} = \gamma$. The MATLAB code implementing the data of this hybrid system is given next. To match the construction in Example 2.32, the data of the hybrid plant therein with constant input is coded using MATLAB language.*

i) Cp.m: *This function returns* 1 *when* $z \in C_P$, *with* C_P *as in Example 2.32. Since* u *is a constant designed to satisfy the conditions in* C_P, *then the the function does not parse* u. *This function is given as follows:*

```matlab
function v  = Cp(z)
%-------------------------------------------------------------
% Project: Hybrid Feedback Control book
% Description: 1-DOF juggling system
% https://hybrid.soe.ucsc.edu/software
% http://hybridsimulator.wordpress.com/
% @BookSite/Simulation
% Filename: Cp.m
%-------------------------------------------------------------
% Return 0 if outside of Cp, and 1 if inside Cp
%-------------------------------------------------------------

% Definition of state
z1 = z(1);
z2 = z(2);
z3 = z(3);
z4 = z(4);

% Check of flow condition
if (z1 >= z3)
   v = 1;  % report flow
else
   v = 0;  % do not report flow
end

end
```

Note that the constraint on the input u *is not included in the code for* C_P *due to that constraint being satisfied by the choice of* u.

ii) Fp.m: *This function returns the value of the single-valued flow map* F_P *in Example 2.32. As in* Cp.m, *since* u *is a constant, then the function is just a function of* z. *For simplicity, the gravity constant is defined within the function, but could be defined as a function argument. This function is given as follows:*

```matlab
function zdot = Fp(z)
%-------------------------------------------------------------
% Project: Hybrid Feedback Control book
% Description: 1-DOF juggling system
% https://hybrid.soe.ucsc.edu/software
% http://hybridsimulator.wordpress.com/
% @BookSite/Simulation
% Filename: Fp.m
%-------------------------------------------------------------
```

```
% Definition of state
z1 = z(1);
z2 = z(2);
z3 = z(3);
z4 = z(4);

% Definition of constants
gamma = 9.81;
u = gamma;

% Definition of zdot, with constant input
zdot = [z2; -gamma; z4; u];

end
```

iii) **Dp.m**: *Similar to* **Cp.m**, *this function returns* 1 *when* $z \in D_P$, *with* D_P *as in Example 2.32. The condition* $z_1 = z_3$ *is implemented as* $z_1 \le z_3$ *so that the simulator can detect when a simulated solution crosses the condition* $z_1 = z_3$. *This function is given as follows:*

```
function v = Dp(z)
%--------------------------------------------------------------
% Project: Hybrid Feedback Control book
% Description: 1-DOF juggling system
% https://hybrid.soe.ucsc.edu/software
% http://hybridsimulator.wordpress.com/
% @BookSite/Simulation
% Filename: Dp.m
%--------------------------------------------------------------
% Return 0 if outside of Dp, and 1 if inside Dp
%--------------------------------------------------------------

% Definition of state
z1 = z(1);
z2 = z(2);
z3 = z(3);
z4 = z(4);

% Definition of constants
gamma = 9.81;
u = gamma;

% Check of jump condition
if (z1 <= z3 && z2 <= z4)
    v = 1;
else
    v = 0;
end

end
```

iv) Gp.m: *This function returns the value of the single-valued jump map* G_P *in Example 2.32. The third entry of* G_P *is implemented as* z_1 *so that, after a jump, the simulated solution is in* C_P *and the simulator can continue computing forward in time. The variable* ztemp *is an auxiliary variable storing intermediate computations. In MATLAB language, this function is given as follows:*

```
function zplus = Gp(z)
%-----------------------------------------------------------------
% Project: Hybrid Feedback Control book
% Description: 1-DOF juggling system
% https://hybrid.soe.ucsc.edu/software
% http://hybridsimulator.wordpress.com/
% @BookSite/Simulation
% Filename: Gp.m
%-----------------------------------------------------------------

% Definition of state
z1 = z(1);
z2 = z(2);
z3 = z(3);
z4 = z(4);

% Definition of constants
lambda = 1;
mbar = 0.5;
Gamma0 = [mbar - (1-mbar)*lambda, (1-mbar)*(1+lambda);...
                        mbar*(1+lambda), 1-mbar-mbar*lambda];

% Definition of zplus
ztemp = Gamma0*[z2; z4];
zplus = [z1; ztemp(1); z3; ztemp(2)];

end
```

Even though it consists of a set-valued jump map, a model with uncertainty in the restitution coefficient can be easily implemented by adding a line of code that selects lambda *randomly from a set of possible values for the restitution coefficient.*

A simulated solution to the juggling system in Example 2.39 from the initial condition $z(0,0) = [1 \ 0 \ -1 \ 0]^\top$ and with TSPAN= $[0 \ 3]$, JSPAN= $[0 \ 20]$, rule= 1, $\gamma = 9.81$, $\overline{m} = 0.5$, and $\lambda = 1$ is depicted in Figure 2.39. The projection onto t of the components of $(t, j) \mapsto z(t, j)$ is shown. In MATLAB language, the code in run.m used to run this simulation is as follows:

```
% initial conditions
z1_0 = 1;
z2_0 = 0;
z3_0 = -1;
```

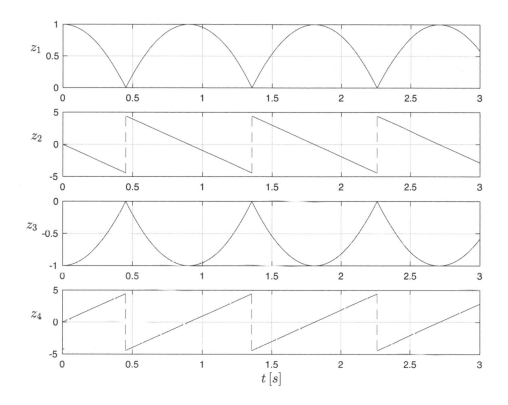

Figure 2.6: Simulated solution to the hybrid system in Example 2.39.

```
z4_0 = 0;
z0 = [z1_0;z2_0;z3_0;z4_0];

% simulation horizon
TSPAN=[0 3];
JSPAN = [0 20];

% rule for jumps
% rule = 1 -> priority for jumps
% rule = 2 -> priority for flows
rule = 1;

options = odeset('RelTol',1e-6,'MaxStep',.001);

% simulate
[t,j,z] = HyEQsolver(@Fp,@Gp,@Cp,@Dp,...
    z0,TSPAN,JSPAN,rule,options,'ode23t');

% plot solution
```

```
figure(1), clf
subplot(4,1,1), plotflows(t,j,z(:,1));
grid on
ylabel('z1')
subplot(4,1,2), plotflows(t,j,z(:,2));
grid on
ylabel('z2')
subplot(4,1,3), plotflows(t,j,z(:,3));
grid on
ylabel('z3')
subplot(4,1,4), plotflows(t,j,z(:,4));
grid on
ylabel('z4')
```

Simulation files associated to this example are available at @BookSite/Simulation/Juggling.

2.5 EXERCISES

Exercise 8 (Model an unknown hybrid system). Open the simulation files associated with this exercise, which are available at @BookSite/Exercises, and determine the following about the hybrid system it simulates:

1. Determine its state, input, and output;

2. Define the flow set, flow map, jump set, jump map, and output map;

3. Simulate the hybrid system from the three sets of initial conditions listed in the simulation files;

4. Use your model to justify the trajectories obtained in the simulations.

Exercise 9 (Model another unknown hybrid system). Open the simulation files associated with this exercise, which are available at @BookSite/Exercises, and determine the following about the hybrid system it simulates:

1. Determine its state, input, and output;

2. Define the flow set, flow map, jump set, jump map, and output map;

3. Simulate the hybrid system from the three sets of initial conditions listed in the simulation files;

4. Use your model to justify the trajectories obtained in the simulations.

Exercise 10 (Resettable timer). Derive a hybrid model of a resettable timer with two inputs having the dynamics in Exercise 1. Set the first input equal to 1 and the second input equal to -1.

1. Simulate the resulting hybrid closed-loop system for the following initial conditions:

   a) initial timer equal to zero;
   b) initial timer equal to 1;
   c) initial timer equal to −1;

   and plot each of the resulting trajectories over five seconds of ordinary time and seven jumps.

2. Show that every maximal solution to the hybrid closed-loop system is complete and bounded (hence, precompact).

Exercise 11 (bouncing ball with input). Consider the point-mass bouncing ball system in Exercise 2.

1. Derive a hybrid system model $\mathcal{H}_P$ capturing the dynamics of the bouncing ball with an input that determines the height of the ground. The model of this hybrid plant should describe the behavior of the height and vertical velocity of the ball only. Assume that potential and kinetic energy of the ball are conserved both at impacts with the ground and in between impacts. Let $\gamma > 0$ denote the gravity.

2. For inputs that equal a constant, simulate the system from

   a) unitary initial height, zero initial vertical velocity, and zero constant input;
   b) zero initial height, unitary initial vertical velocity, and zero constant input;
   c) zero initial height, negative unitary initial vertical velocity, and zero constant input;
   d) zero initial height, zero initial vertical velocity, and zero constant input;
   e) negative unitary initial height, zero initial vertical velocity, and zero constant input;

   and plot each of the resulting trajectories over ten seconds of ordinary time and ten jumps. Use $\gamma = 9.8$ m/s^2.

3. Modify your model to allow for dissipation of kinetic energy at the impacts due to a constant restitution coefficient.

4. Repeat the simulations in item 2 for the revised model in item 3 with restitution coefficient equal to 0.5.

5. For the modified model allowing for dissipation, show that no matter what the constant value of its input is, every maximal solution is Zeno.

Exercise 12 (Two impacting pendulums). Consider the pair of pendulums with mass m_L, m_R as shown in Figure 2.7, where the pendulum on the right includes a torque input for control. Let the parameter $\gamma > 0$ denote gravity.

1. Determine the state z of the hybrid plant.

2. Define each element of $\mathcal{H}_P$ assuming the following:

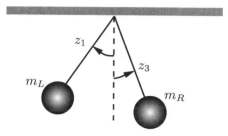

Figure 2.7: Two pendulums exhibiting impacts. The torque of the pendulum on the right is controllable.

- Viscous friction for circular motion.

- Conservation of momentum at impacts.

- Dissipation of energy at impacts.

3. For torque input equal to zero, perform the following simulations of the resulting hybrid system for parameters $m_L = 1, m_R = 2$, unitary viscous friction, a restitution law with parameters such that there is dissipation at impacts, and a sinusoidal input of unitary angular frequency:

   a) Plot trajectories as a function of t for pendulums starting at $z_1 = -\pi/4$ and $z_3 = \pi/4$ with zero velocity.

   b) Plot trajectories as a function of t for pendulums starting at $z_1 = 0$ and $z_3 = \pi/4$ with zero velocity.

   c) Plot trajectories as a function of t for pendulums starting at $z_1 = \pi/4$ and $z_3 = \pi/2$ with zero velocity.

4. For torque input equal to zero, show that every maximal solution to the system is complete.

5. For torque input equal to zero, show that the system admits a discrete solution that is complete.

Exercise 13 (Juggling system, revisited). Consider the juggling system in Example 2.32, revisited in Example 2.35.

1. Show that the solution in item 1 therein is nontrivial, complete, maximal, and precompact.

2. Show that the solution in item 2 therein is Zeno.

Exercise 14 (Sample-and-hold control). Consider the implementation of a static controller using a sample-and-hold strategy as in Exercise 3.

1. Formulate a mathematical model for the hybrid plant, the hybrid controller, and the resulting closed-loop system assuming the following:

- The computation of the static feedback law takes no time, i.e., is instantaneous.

- The positive constants T_s and T_c are not necessarily equal.

2. Show that for every point in the union of the flow set and jump set of your model, there exists a nontrivial solution. If your current model does not guarantee that, revise it appropriately.

3. Suppose that the computation of the control law takes $\delta > 0$ units of time. Modify your model in item 1 to account for it.

Exercise 15 (Stabilization of a point on the unit circle). An implementation of the strategy outlined in Section 1.2.3 is given by a hybrid controller $\mathcal{H}_K$ with state $\eta = q \in Q := \{0, 1\}$, input $v = z \in \mathbb{R}^2$, output $\zeta \in \mathbb{R}$, and data as follows:

$$C_K = \bigcup_{q \in Q} (C_{K,q} \times \{q\}), \quad \begin{cases} C_{K,0} = \{z \in \mathbb{S}^1 : z_1 \geq c_0\} \\ C_{K,1} = \{z \in \mathbb{S}^1 : z_1 \leq c_1\} \end{cases}$$

$$F_K(z, q) = 0 \quad \forall (z, q) \in C_K$$

$$D_K = \bigcup_{q \in Q} (D_{K,q} \times \{q\}), \quad \begin{cases} D_{K,0} = \{z \in \mathbb{S}^1 : z_1 \leq c_0\} \\ D_{K,1} = \{z \in \mathbb{S}^1 : z_1 \geq c_1\} \end{cases}$$

$$G_K(z, q) = 1 - q \quad \forall (z, q) \in D_K$$

$$\kappa(z, q) = \kappa_q(z), \quad \begin{cases} \kappa_0(z) = -z_2 \ \forall z \in C_{K,0} \\ \kappa_1(z) = -z_1 \ \forall z \in C_{K,1} \end{cases}$$

The output ζ of the controller is assigned to the input u to the plant. The constants c_0 and c_1 satisfy $-1 < c_0 < c_1 < 0$.

1. Define the data of the closed-loop system resulting from interconnecting the plant and the controller given above.

2. Show that every maximal solution to the resulting closed-loop system is complete.

3. For $c_0 = -\frac{2}{3}$ and $c_1 = -\frac{1}{3}$, simulate the closed-loop system from the following initial conditions:

   a) initial z equal to $(c_0, \sqrt{1 - c_0^2})$ and initial q equal to 0;

   b) initial z equal to $(c_0, \sqrt{1 - c_0^2})$ and initial q equal to 1;

   c) initial z equal to $(c_0, -\sqrt{1 - c_0^2})$ and initial q equal to 0;

   d) initial z equal to $(c_0, -\sqrt{1 - c_0^2})$ and initial q equal to 1;

   e) initial z equal to $(c_1, \sqrt{1 - c_1^2})$ and initial q equal to 0;

   f) initial z equal to $(c_1, \sqrt{1 - c_1^2})$ and initial q equal to 1;

   g) initial z equal to $(c_1, -\sqrt{1 - c_1^2})$ and initial q equal to 0.

   h) initial z equal to $(c_1, -\sqrt{1 - c_1^2})$ and initial q equal to 1.

4. Explain the differences between the solutions obtained and point out reasons.

Exercise 16 (DC/AC inverter). Consider the DC/AC inverter and hybrid controller proposed in Example 2.37.

1. Show that the hybrid closed-loop system $\mathcal{H}$ with data given in (2.41)-(2.42) satisfies the hybrid basic conditions.

2. Show Lemma 2.38.

Exercise 17 (Discretized event-triggered control). A discretized version of an event-triggered implementation of the following proportional-integral controller in Laplace domain

$$U(s) = K\left(\beta Y_{sp}(s) - Y(s) + \frac{1}{sT_i}\left(Y_{sp}(s) - Y(s)\right)\right)$$

for the linear, time-invariant system

$$\dot{z} = Az + Bu, \qquad y = Mz$$

is given by

```
(*Pre-calculated parameter*)
bi := K / Ti;
hact := 0;
WHILE(1)
      (*Event detection*)
      ysp := ADIn(ch1);
      y   := ADIn(ch2);
      e   := ysp - y;
      hact := hact + hnom;
      IF (abs(e - es) > elim) OR (hact >= hmax) THEN
            es := e;
            (*Calculate control signal*)
            up := K*(beta * ysp - y);
            u  := up + ui;
            DAOut(u,ch3);
            (*Update states*)
            ui := ui + bi*hact*(ysp - y);
            hact := 0;
      ENDIF;
ENDWHILE;
```

where U is the control signal, Y_{sp} is the set-point, and Y is the measurement signal, all of them in the Laplace domain. In the (discrete) time domain, these signals are represented as u, ysp, and y, respectively. The function call `ADIn(chx)` measures a signal in channel x, and the function call `DAOut(y,chx)` outputs signal y to channel x. The constants K, T_i are tuning parameters and β is a set-point weighting term. The positive constants `hnom` and `elim` are tunable parameters associated to the event-triggered implementation of the controller. The constants `bi` and `hact` are defined/initialized at the beginning of the implementation.

1. Derive a model of the hybrid controller implementing the event-triggered algorithm given above so that the resulting hybrid closed-loop system satisfies the hybrid basic conditions. Write down the hybrid closed-loop system and show the latter property.

2. Show that every maximal solution to the hybrid closed-loop system is complete.

3. For the plant given by a scalar system with $A = B = M = 1$, determine numerically parameters of the hybrid controller to guarantee that solutions to the closed-loop system are such that the z component converges to zero.

Exercise 18 (Stabilization of the inverted configuration of a pendulum). The mathematical model of the angle and the angular velocity of a pendulum on a cart is given by

$$\begin{bmatrix} \dot{\xi} \\ \dot{\omega} \end{bmatrix} = \begin{bmatrix} \omega R(-\pi/2)\xi \\ \xi_1 + \xi_2\alpha \end{bmatrix} \quad (\xi, \omega) \in S^1 \times \mathbb{R},$$

where $\xi = (\xi_1, \xi_2)$ is a unit vector whose angle denotes the angle of the pendulum itself, ω corresponds to the angular velocity, with positive velocity in the clockwise direction, and α the angular acceleration, which is treated as an input. The point $\xi = \mathbf{1}$ corresponds to the upright position while $\xi = -\mathbf{1}$ corresponds to the down position of the pendulum. This model was obtained after an input transformation from force to acceleration α and with ratio between the gravitational constant and the pendulum length equal to one. The cart dynamics are ignored to simplify the exercise – the derivation of a model including the position and velocity of the cart is in Exercise 46. The control objective is to design α such that the upright configuration of the pendulum is globally stabilized by relying on measurements of ξ and ω.

1. Derive a model of the plant and formulate the control problem to solve.

2. Using linear feedback control design tools, propose a feedback controller that accomplishes the task locally. Call this feedback the local controller. Characterize the basin of attraction of that controller, analytically or numerically.

3. Design an energy pumping control algorithm as follows:

   a) Write down the total energy of the system so that it is zero when the pendulum is at rest, at the downward configuration. Call it E.
   b) Compute its time derivative and express it in terms of infinitesimal quantities.
   c) Pick the control input as a state-feedback law such that the time derivative is strictly increasing away from the equilibrium points.

   Call this feedback the global controller.

4. Design control logic that assures that the local controller is used when the state of the system is in the basin of attraction of the local controller, while otherwise, the global controller is used.

5. Model the complete algorithm outlined above as a hybrid controller.

6. Model the hybrid closed-loop system and characterize its basin of attraction.

7. Validate your results numerically.

Exercise 19 (Well-posedness of $\mathcal{H}$). Show Lemma 2.21 – solving Exercise 95, Exercise 97, and Exercise 100 first is recommended. Then, consider the following alternative definition of the jump map of $\mathcal{H}$:

$$
G(x) = \begin{cases}
\begin{bmatrix} G_P(z, \kappa(\eta, h(z))) \\ \eta \end{bmatrix} & \text{if } x \in D_1 \setminus D_2 \\[2ex]
\begin{bmatrix} G_P(z, \kappa(\eta, h(z))) \\ G_K(\eta, h(z)) \end{bmatrix} & \text{if } x \in D_1 \cap D_2 \\[2ex]
\begin{bmatrix} z \\ G_K(\eta, h(z)) \end{bmatrix} & \text{if } x \in D_2 \setminus D_1
\end{cases} \tag{2.51}
$$

Suppose $(A1_P)$-$(A4_P)$ and $(A1_K)$-$(A4_K)$ hold and argue – via a counterexample – that the claim in Lemma 2.21 does not hold when G is given as in (2.51).

2.6 NOTES

The framework to model, analyze, and design hybrid systems used in this book is *hybrid equations* – as in (2.6) – or, more generally, *hybrid inclusions* – as in (2.7). The interested reader is referred to [5, 6, 7, 8, 9, 10, 1] and the resources listed in Preface for more details about this framework. As the hybrid systems literature indicates, there are other powerful frameworks for modeling, analysis, and design. Next, the two main frameworks that are most related to hybrid inclusions are summarized and recast into the hybrid inclusions framework.

- *Impulsive systems framework:* In this framework, the state changes continuously according to an ordinary differential equation over pre-specified time intervals of nonzero length. The state is updated instantaneously at the end time of those time intervals, when finite. More precisely, given a sequence of positive impulse times t_i satisfying $t_1 < t_2 < t_3 \ldots$, a solution $t \mapsto z(t)$ given by a right-continuous function satisfies

$$
\dot{z}(t) = \widetilde{f}(z(t))
$$

except when $t = t_i$, at which times the new value of z is given by

$$
\lim_{t \nearrow t_i} z(t) + \widetilde{g}\left(\lim_{t \nearrow t_i} z(t)\right)
$$

where, for each i, $\widetilde{g}$ is a function of the state capturing its variation at the impulse times. The impulse times are often fixed a priori for each particular solution, which are typically given by piecewise differentiable or piecewise absolutely continuous functions parameterized by ordinary time t only. For $t \mapsto z(t)$ to be uniquely defined, it is typically defined as a right-continuous function, namely, for each t in dom z, the value of the function is given by its right limit and its limit from the left always exists (left-continuous functions can also be used).[16] More details about the impulsive systems framework can be found in [39], [40], [41], [42], and [19], to just list a few.

Impulsive systems given as above can be rewritten as hybrid inclusions $\mathcal{H}$. When the impulse times are determined by a condition of the form $x(t) \in D$, then a hybrid inclusion model follows immediately More generally, the impulsive system defined by $\widetilde{f}$, $\widetilde{g}$, and an unbounded increasing sequence t_i can be modeled as the following hybrid inclusion with state $x = (z, \tau, \ell)$:

$$\mathcal{H} \ : \ \begin{cases} (z, \tau, \ell) \in C & \dot{x} = \begin{bmatrix} \widetilde{f}(z) \\ 1 \\ 0 \end{bmatrix} \\ (z, \tau, \ell) \in D & x^+ = \begin{bmatrix} z + \widetilde{g}(z) \\ 0 \\ \ell + 1 \end{bmatrix} \end{cases} \tag{2.52}$$

with

$$C := \{x \ : \ \tau \in [0, t_{\ell+1} - t_\ell]\}$$

and

$$D := \{x \ : \ \tau = t_{\ell+1} - t_\ell\}$$

When τ and ℓ are initialized at zero, the jumps occur at the impulse times t_i. Impulsive systems with finite sequences t_i as well as the case when $\widetilde{g}$ varies with i can also be captured by a hybrid inclusion. Also, the natural generalization of impulsive systems in which $\widetilde{f}$ and $\widetilde{g}$ are set valued – known as *impulsive differential inclusions* – can also be modeled as hybrid inclusions.

- *Hybrid automata framework (and other frameworks with explicit logic variables)*: Several hybrid systems frameworks explicitly partition the state of the system into a continuous state and a discrete state (or logic variable) – as in § 1.2.1, the logic variable may denote the mode of operation of the system. These frameworks involve the following objects:

  - A continuous state $x_c \in \mathbb{R}^{n_c}$;

  - A discrete state $q \subset Q$, where Q is a finite discrete set;

  - A map Dom $: Q \rightrightarrows \mathbb{R}^{n_c}$ defining, for each $q \in Q$, the domain Dom(q) in which the continuous state x_c evolves;

  - A vector field $\widetilde{f} : \mathbb{R}^{n_c} \times Q \to \mathbb{R}^{n_c}$ defining, for each $q \in Q$, the right-hand side of an ordinary differential equation governing the continuous evolution of x_c for mode q;

  - A set of edges E $\subset Q \times Q$ consisting of pairs (q, q') for which a transition from q to q' is possible;

  - A guard map Guard $:$ E $\rightrightarrows \mathbb{R}^{n_c}$ defining, for each edge in E, the set of points from which mode transitions can occur: for each pair $(q, q') \in$ E, transitions from q to q' can occur when $x_c \in$ Guard(q, q');

---

[16]Right-continuous and left-continuous functions cannot model multiple jumps at a single time instant in a solution.

– A reset map Reset : $E \times \mathbb{R}^{n_c} \to \mathbb{R}^{n_c}$ defines, for each edge $(q, q') \in E$, the values to which the continuous state can be reset during a transition from q to q'.

Frameworks employing such data to define a hybrid system include differential automata in [43], hybrid automata in [44],[45], and the ones in [46], [12], [17], [47], and [48].

Hybrid systems given in the hybrid automata framework, or in similar frameworks with an explicit logic state, can be modeled as a hybrid inclusion. To that end, for each $q \in Q$, define

$$C_q := \mathrm{Dom}(q), \qquad F_q(x_c) := \widetilde{f}(x_c, q) \qquad \forall x_c \in C_q,$$
$$D_q := \bigcup_{(q,q') \in E} \mathrm{Guard}(q, q'),$$
$$G_q(x_c) := \bigcup_{\{q' : x_c \in \mathrm{Guard}(q,q')\}} (\mathrm{Reset}(q, q', x_c), q'), \qquad \forall x_c \in D_q.$$

This construction allows for overlapping guards, namely, for x_c to belong to the map Guard evaluated at two different edges in E. In turn, this leads to G_q being set valued even if Reset is not. With these definitions, the hybrid inclusion $\mathcal{H}$ with data (C, F, D, G) associated to the hybrid automaton has state $x = (x_c, q) \in \mathbb{R}^{n_c} \times Q$ and dynamics

$$x \in C := \bigcup_{q \in Q} (C_q \times \{q\}) \qquad \dot{x} = \begin{bmatrix} \dot{x}_c \\ \dot{q} \end{bmatrix} = \begin{bmatrix} F_q(x_c) \\ 0 \end{bmatrix} =: F(x)$$
$$x \in D := \bigcup_{q \in Q} (D_q \times \{q\}) \qquad x^+ = \begin{bmatrix} x_c^+ \\ q^+ \end{bmatrix} \in G_q(x_c) =: G(x)$$

Among the many other frameworks in the hybrid systems literature, the measure-driven differential equations (or inclusions) framework is widely used for optimal control and for the study of systems with unilateral constraints, which naturally emerge in mechanical systems with impacts and friction. In this framework, a single ordinary differential equation (or inclusion) with impulses is employed to capture both the continuous and the discrete dynamics of a hybrid system. The model exploits the fact that a solution $t \mapsto z(t)$ to $\dot{z}(t) = \widetilde{f}(z(t))$ can be rewritten as $dz(t) = \widetilde{f}(z(t))dt$ and that jumps in the solution can be induced by incorporating impulses in the right-hand side $\widetilde{f}(z(t))dt$. More precisely, given a nonnegative scalar Borel measure μ, a measure-driven differential equation consists of

$$dz(t) = \widetilde{f}(z(t))dt + \widetilde{g}(z(t))\mu(dt)$$

(An extension of this model to the inclusion case is immediate.) A solution to this model is typically defined as a function $t \mapsto z(t)$ of bounded variation.[17] Clearly, the

---

[17]A function $s \mapsto z(s)$ is said to be of bounded variation if the supremum of the sum of $|z(s_{i+1}) - z(s_i)|$ over all partitions $\{s_1, s_2, \ldots\}$ of dom t is bounded. The said supremum is called the total variation of the function.

flows of the solution are governed by $\widetilde{f}$. The time instants t_i at which the solution jumps are defined by the so-called atoms of the measure μ. At such instants, the solution exhibits an impulse of value determined by $\widetilde{g}$, meaning that, as in the impulsive systems framework, z is reset to $\lim_{t \nearrow t_i} z(t) + \widetilde{g}\left(\lim_{t \nearrow t_i} z(t)\right)$. A salient feature of the measure-driven differential equations (or inclusions) framework is that the impulse times t_i can accumulate over a bounded subset of $\mathbb{R}_{\geq 0}$ – meaning that Zeno solutions are allowed – and also be defined beyond an infinite number of them over a bounded subset of $\mathbb{R}_{\geq 0}$ – namely, solutions can be defined past Zeno times. For more details about the measure-driven differential equations (or inclusions) framework, also known as systems with unilateral constraints, see [49], [50], [51], [52], and [53].

A framework for unifying the classical theories of differential and difference equations is that of dynamical systems on time scales [54]. Given a time scale $\mathbb{T}$, which is a nonempty closed subset of $\mathbb{R}$, a generalized derivative of a function $\phi : \mathbb{T} \to \mathbb{R}$ relative to $\mathbb{T}$ can be defined. This generalized derivative reduces to the standard derivative when $\mathbb{T} = \mathbb{R}$, and to the difference $\phi(n+1) - \phi(n)$ when evaluated at n and for $\mathbb{T} = \mathbb{N}$. One advantage of the framework of dynamical systems on time scales is the generality of the concept of a time scale. A potential drawback, especially for control engineering purposes, is that a time scale is chosen a priori, and all solutions to a system are defined on the same time scale.

As discussed in § 2.2, there are other frameworks in the literature with some degree of nonsmoothness or featuring some hybrid-like properties. These include continuous-time systems with discontinuous right-hand side [55, 56, 57, 58, 59], complementarity systems [60], piecewise affine systems [61, 62, 63], mixed logical dynamical systems [64], switched systems [65, 66, 67, 68, 69, 70, 71], and discrete-event systems [72, 73, 74, 75]. This broad and important class of systems can be modeled as a hybrid plant, hybrid controller, or hybrid closed-loop system, but, as shown in the examples in § 2.2, may not necessarily have solutions exhibiting both flows and jumps.

A similar model of hybrid systems with inputs to the one in (2.7) with data as in Definition 2.5 was introduced in [76] and [77] for the study of interconnections of hybrid systems, and in [78] for the study of passivity as well as for the design of passivity-based feedback controllers; see also [79]. Special cases of the hybrid controller model in Definition 2.8 and (2.10) were used in [30] and in [11] (see sections "Hybrid controllers for nonlinear systems" and "Supervising a family of hybrid controllers" in the latter reference); see also [31]. The model of the hybrid closed-loop system in (2.19) with generic data (C, F, D, G) was first proposed in [5] and comprehensively studied in [1]; see also [6] and [11].

The hybrid controller model for event-triggered sample-and-hold control in Example 2.10 appeared in [80]; see also the references therein.

Hybrid basic conditions in Definition 2.20 imply that the hybrid closed-loop system $\mathcal{H}$ is well-posed as defined in [1, Definition 6.29]. [1, Definition 6.2], under the name *nominally well-posed*, introduces the same notion but without involving vanishing disturbances. In simple words, $\mathcal{H}$ is nominally well-posed if, in particular, for every (graphically) convergent sequence of solutions that is locally bounded, its (graphical) limit is also a solution. As mentioned at the beginning of this chapter, Solutions to nominally well-posed hybrid systems depend in an upper semicontinuous manner with respect to initial conditions; see [1, Chapter 6]. The conditions in Definition 2.20 imply both nominal well-posedness and well-posedness of $\mathcal{H}$; see [1,

Theorem 6.8 and Theorem 6.30]. Note that according to [1, Lemma 6.9] only the conditions on D and G are necessary.

Hybrid time domains and arcs were first introduced in [5]. They have similarities with hybrid time sets and trajectories in [81] and with the notion of time and hybrid time trajectories in [45]. For a more in-depth discussion, see sections "Generalized time domains" and "Solutions" in [11]. These notions of solution allow for one Zeno event per solution. A notion of solution that allows multiple Zeno events in a single solution with well-posedness appeared in [82].

An in-depth introduction to measurability, in particular, measurable functions and Lebesgue measurable functions is in [83] (see Definition 1.3 on page 8 therein). A well-written introduction to measure theory in an engineering context for the analysis of signals is in [84] (see Section 3.4.1 on page 226 therein). A reader may ask "When is a function not Lebesgue measurable"? One such function can be constructed from a set that is not Lebesgue measurable. In fact, for any $M \subset \mathbb{R}_{\geq 0}$ that is not Lebesgue measurable, the function $s : S \to \mathbb{R}$, $M \subset S \subset \mathbb{R}_{\geq 0}$, defined as $s(r) = r$ if $r \in M$ and $s(r) = -r$ if $r \in S \setminus M$, is not a Lebesgue measurable function: for the open set $\mathcal{U} = (0, \infty)$,

$$\{r \in S \ : \ s(r) \in \mathcal{U}\} = \{r \in S \ : \ s(r) > 0\} = M$$

which is not Lebesgue measurable. Concrete examples of sets that are not Lebesgue measurable include the Cantor set and Vitali set.

Absolutely continuous functions are commonly used as solutions to differential inclusions. A rather complete treatment of such functions in this context is given in [85] (see Section 2 of Chapter 0 on page 12 therein). The key result outlined above Definition 2.28, about almost everywhere differentiability of an absolutely continuous function, is given in [86, Chapter 9] on page 246 therein – see "Corollary" right above Theorem 2.

The notion of hybrid input in Definition 2.27 was given in [87]. The notion of solution to hybrid systems with inputs was also introduced therein and in [76, 77]. The result on existence of solutions for hybrid closed-loop systems satisfying the hybrid basic conditions in Proposition 2.34 is a combination of [1, Proposition 6.10] and [9, Lemma 2.7]. A result on existence of solutions for hybrid systems with inputs appeared in [88].

The hybrid model of the one-degree-of-freedom juggling system in Example 2.32 was inspired by the impulsive models in [89] and [90]. This model reduces to the well-known bouncing ball system; see Exercise 2 and Exercise 11. Further details on the model and a hybrid controller for tracking can be found in [91]. Experimental results for this system appeared in [92].

For more details on the properties of the DC/AC converter in Example 2.37, the reader is referred to [2]. See also Example 1.1 and Example 1.4.

The hybrid system model including disturbances in § 2.3.5 was introduced in [6] and [1]. The construction provided in this chapter is as in [6], the main difference compared to the one in [1] being the intersection by C and by D within the argument of F^ρ and G^ρ, respectively. In this way, the points collected by F^ρ and G^ρ right outside C and D can be specified through the definition of F and G, respectively.

Formal definitions of simulated trajectories (or solutions) and dynamical properties of the discretization (2.49)-(2.50) of $\mathcal{H}$ can be found in [93]. In [94], the Hybrid Equations (HyEQ) Toolbox for MATLAB©/Simulink© is introduced in detail.

Chapter Three

Notions and Analysis Tools ·

With the versatile modeling and simulation techniques introduced in the previous chapters, this chapter introduces the reader to mathematical analysis of hybrid dynamical systems. The goal of this chapter is to provide a concise presentation of the main notions and analysis tools available in the literature that are employed in the upcoming chapters. The presentation of these tools is done in a unified, self-contained manner. In the first part of this chapter, in § 3.2, the main dynamical notions employed in this book are introduced. The property of a set being *stable* is introduced as the property that solutions that start close to the set stay close to the set for all hybrid time. Such a property is known as *Lyapunov stability*. The notion of solutions converging to a set is introduced as well, under the term *(pre-)attractivity*. When a set enjoys these two properties, the set is said to be *(pre-)asymptotically stable* for the hybrid system. This notion is used extensively in the forthcoming chapters. A notion capturing robustness of this property is also introduced towards the end of this chapter. *Forward invariance* of a set is defined as the property of solutions that, if they start in the set, they stay in the set for all time. These notions are illustrated in Figure 3.1.

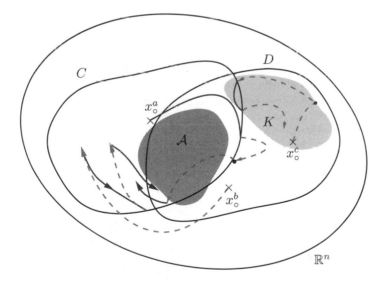

Figure 3.1: Illustration of some of the notions introduced in this chapter. The solution from $x_\circ^a$ starts close and stays close to the set $\mathcal{A}$, in this way, illustrating stability. The solution from $x_\circ^b$ illustrates attractivity of $\mathcal{A}$ as it "converges" to $\mathcal{A}$. The solution from $x_\circ^c$ starts and stays in the set K, pictorially showing invariance of K.

Given a hybrid system and a set, the second part of this chapter, §3.3, presents tools to certify these notions. The concept of Lyapunov function is introduced for hybrid inclusions modeling hybrid closed-loop systems. As in the classical setting of continuous-time systems and discrete-time systems, a Lyapunov function that is a candidate for asymptotic stability of a set is a function with nonnegative values that is zero only at the set of points to render asymptotically stable. A Lyapunov theorem for hybrid closed-loop systems providing sufficient conditions for (pre-)asymptotic stability is introduced. These conditions are under different assumptions, most of which arise in the forthcoming chapters. To certify (pre-)attractivity of a set, an invariance principle for hybrid closed-loop systems is presented. By exploiting invariance, with this tool it is possible to certify (pre-)asymptotic stability with a Lyapunov function that is weak, in the sense that it does not strictly decrease during both flows and jumps. After introducing perturbed hybrid closed-loop systems, robustness of asymptotic stability is introduced and results are formulated using $\mathcal{KL}$ functions.

3.1 OVERVIEW

Stability and attractivity of a point – or, more generally, of a set – for a dynamical system is a property of its solutions. More precisely, for the autonomous dynamical system $\dot{x} = f(x)$, the compact set $\mathcal{A}$ is said to be *stable* in the sense of Lyapunov if for each level of closeness to the set $\mathcal{A}$, denoted $\varepsilon > 0$, there exists $\delta > 0$ defining a neighborhood of $\mathcal{A}$ such that for each solution $t \mapsto x(t)$ the following holds:[1]

$$|x(0)|_{\mathcal{A}} \leq \delta \qquad \Rightarrow \qquad |x(t)|_{\mathcal{A}} \leq \varepsilon \quad \forall t \in \operatorname{dom} x$$

Namely, if a solution is such that its initial distance to $\mathcal{A}$ is less than or equal to δ, then the solution needs to remain within (the given) ε distance to $\mathcal{A}$. When maximal solutions are complete,[2] attractivity of the set $\mathcal{A}$ in the asymptotic sense is defined as the property that for some $\mu > 0$, the following holds:

$$|x(0)|_{\mathcal{A}} \leq \mu \qquad \Rightarrow \qquad \lim_{t \to \infty} |x(t)|_{\mathcal{A}} = 0$$

This limiting condition characterizes *convergence*: if a maximal solution is complete and its initial distance to $\mathcal{A}$ is less than or equal to μ, then the distance from the solution to $\mathcal{A}$ converges to zero as t tends to ∞. Unlike the case when $\mathcal{A}$ is a singleton, it does not imply that the function $t \mapsto x(t)$ has a limit, but rather, it states that the distance from this function to the set $\mathcal{A}$ converges to zero as a function of time. In fact, $t \mapsto x(t)$ may not have a limit, yet $t \mapsto |x(t)|_{\mathcal{A}}$ might converge to zero as t tends to infinity. Throughout this book, by "the solution x converges to $\mathcal{A}$" it is meant that the distance from the solution to $\mathcal{A}$ converges as (hybrid) time gets large.

---

[1]The distance from a point x to the set $\mathcal{A}$ is denoted $|x|_{\mathcal{A}}$ and is given by $|x|_{\mathcal{A}} = \inf_{y \in \mathcal{A}} |x - y|$. See List of Symbols and Definition A.10.

[2]Maximal and complete solutions are introduced in Definition 2.31.

The attractivity notion is local and the set of initial conditions from where maximal solutions converge to $\mathcal{A}$ is the *basin of attraction* – in the literature, the term *region of attraction* is also used. In the case that the system has maximal solutions that are not complete, then a natural generalization of attractivity called *pre-attractivity* consists of requiring that the limit above holds for complete solutions and that maximal solutions that are not complete have a bounded distance to the set $\mathcal{A}$ – note that the solution itself does not need to be bounded when $\mathcal{A}$ is unbounded. This highlighted by the prefix *pre*.

Due to not requiring that maximal solutions are complete, this notion is called *pre-attractivity*.

In this way, a set $\mathcal{A}$ is said to be *pre-asymptotically stable* for $\dot{x} = f(x)$ when it is both *stable* and *pre-attractive*. This notion allows for systems to have maximal solutions that are not complete – for instance, as explained in §2.1, due to stopping to exist after a finite amount of time or due to having a finite escape time.

To assert (pre-)asymptotic stability, one usually relies on methods to guarantee the ε-δ property, boundedness of the distance to the set, as well as the limiting property outlined above without having to check every solution explicitly. One such method is Lyapunov's second method. It consists of finding a nonnegative scalar-valued function of the state that only vanishes at points in $\mathcal{A}$ and decreases to zero along solutions. Such a function is called a *Lyapunov function*. More precisely, given a set $\mathcal{A}$ and a set $\mathcal{U}$ containing a neighborhood of $\mathcal{A}$, the interest is in finding a continuously differentiable function $V : \mathcal{U} \to \mathbb{R}_{\geq 0}$ that has positive values outside of $\mathcal{A}$ and vanishes at $\mathcal{A}$ – such a function is said to be positive definite with respect to $\mathcal{A}$ (see Definition A.25) – and is such that, for each solution $t \mapsto x(t)$, the function $t \mapsto V(x(t))$ is decreasing when $x(t)$ is not in $\mathcal{A}$. In this way, and when $t \mapsto x(t)$ is complete, since V is positive definite with respect to $\mathcal{A}$, if $V(x(t))$ decreases to zero then the distance between $x(t)$ and $\mathcal{A}$ would converge to zero as t tends to ∞.

By interpreting the time derivative $\frac{d}{dt} V(x(t))$ as the evaluation at $x(t)$ of the infinitesimal quantity[3] $\langle \nabla V(x), f(x) \rangle$, the decrease of $V(x(t))$ as t gets large is guaranteed locally when the infinitesimal condition

$$\langle \nabla V(x), f(x) \rangle < 0$$

holds for each $x \in \mathcal{U} \setminus \mathcal{A}$. In fact, when this condition holds, every complete solution starting from a compact sublevel set $L_V(r)$ of V with $r > 0$ such that $L_V(r) \subset \mathcal{U}$ converges to $\mathcal{A}$. Furthermore, in particular, when $\mathcal{A}$ is such that $f(\mathcal{A}) = \{0\}$, namely, when $\mathcal{A}$ is an equilibrium point or an equilibrium set for $\dot{x} = f(x)$, then solutions starting from $\mathcal{A}$ remain in $\mathcal{A}$, or, equivalently, every solution x from $x_\circ$ such that $V(x_\circ) = 0$ satisfies

$$V(x(t)) = 0 \qquad \forall t \in \operatorname{dom} x$$

---

[3]Given V and $t \mapsto x(t)$ smooth enough, using the chain rule, the time derivative of $V(x(t))$ at $t \in \operatorname{dom} x \setminus \{0\}$ is equal to

$$\frac{\partial V}{\partial x}(x) \cdot f(x) = \langle \nabla V(x), f(x) \rangle$$

evaluated at $x(t)$, where $\cdot$ denotes the scalar product.

The power of this approach is that the infinitesimal condition

$$\langle \nabla V(x), f(x) \rangle < 0$$

which is in terms of the state and needs to hold for each point in a neighborhood outside of $\mathcal{A}$, implies the desired property for all solutions starting from a compact sublevel sets of V that are strictly contained in the said neighborhood, without computing solutions.

A Lyapunov theorem for hybrid systems is given in Theorem 3.19.

Lyapunov's second method for the case of discrete-time systems $x^+ = g(x)$ consists of finding a function V such that the strict infinitesimal condition

$$V(g(x)) - V(x) < 0$$

holds for each $x \in \mathcal{U} \setminus \mathcal{A}$ and, in addition, the weak infinitesimal condition

$$V(g(x)) - V(x) \leq 0$$

for all $x \in \mathcal{A}$. Unlike the continuous-time case, the weak infinitesimal condition at points in $\mathcal{A}$ is to assure that solutions from $\mathcal{A}$ do not jump outside of $\mathcal{A}$. Interestingly, when a function V only satisfies the weak infinitesimal condition but on $\mathcal{U}$ rather than just on $\mathcal{A}$, under continuity of the right-hand side g, it is still possible to characterize the set to which complete and bounded solutions converge to. It requires to study the largest invariant subset of $\mathcal{U}$ given by points x at which the change of V is zero. Such a result, known as *the invariance principle*, is very useful in cases where constructing (strict) Lyapunov functions is difficult. An invariance principle for hybrid systems is given later in this chapter – see Theorem 3.23.

This chapter extends to hybrid dynamical systems the notions and tools outlined above. Since hybrid dynamical systems allow for both flows and jumps, as outlined in § 1.1 and formalized in § 2.3, a solution to such systems is parameterized by ordinary time t to determine the amount of flow time and by a counter $j \in \mathbb{N}$ to keep track of the number of jumps. In this setting, a maximal solution is still a solution that cannot be further extended.

It is important to remind the reader that, for continuous-time systems and discrete-time systems, existence of solutions is fully determined by the properties of the right-hand side. But, as seen in § 2.3, existence of solutions to hybrid closed-loop systems is much more intricate, and, in particular, require the flow map and flow set to satisfy certain geometric conditions that guarantee flow at points where jumps are not possible. Even when F is smooth, existence of solutions to hybrid closed-loop systems is not guaranteed by infinitesimal conditions like the ones stated above for (pre-)asymptotic stability.

3.2 NOTIONS

Most of the dynamical properties of interest for the analysis and design of hybrid systems used in this book are introduced in this section.

3.2.1 Asymptotic Stability

A vast majority of control problems consist of designing a feedback algorithm assuring that the complete solutions to the plant converge to a desired set-point condition *(attractivity)* and, when starting close to it, the solutions remain nearby *(stability)*. Such a property is precisely defined as follows.

Definition 3.1 (Stability, pre-attractivity, and pre-asymptotic stability). *Given a hybrid closed-loop system $\mathcal{H}$ as in (2.19), a nonempty set $\mathcal{A} \subset \mathbb{R}^n$ is said to be*

1. *Stable for $\mathcal{H}$ if for each $\varepsilon > 0$ there exists $\delta > 0$ such that each solution x to $\mathcal{H}$ with*

$$|x(0,0)|_{\mathcal{A}} \leq \delta$$

   satisfies

$$|x(t,j)|_{\mathcal{A}} \leq \varepsilon \qquad \forall (t,j) \in \operatorname{dom} x$$

2. *Pre-Attractive for $\mathcal{H}$ if there exists $\mu > 0$ such that every solution x to $\mathcal{H}$ with*

$$|x(0,0)|_{\mathcal{A}} \leq \mu$$

   is such that $(t,j) \mapsto |x(t,j)|_{\mathcal{A}}$ is bounded and if x is complete then

$$\lim_{(t,j)\in \operatorname{dom} x, \ t+j\to\infty} |x(t,j)|_{\mathcal{A}} = 0$$

3. *Attractive for $\mathcal{H}$ if there exists $\mu > 0$ such that every maximal solution x to $\mathcal{H}$ with*

$$|x(0,0)|_{\mathcal{A}} \leq \mu$$

   is complete and satisfies

$$\lim_{(t,j)\in \operatorname{dom} x, \ t+j\to\infty} |x(t,j)|_{\mathcal{A}} = 0$$

4. *Pre-Asymptotically stable for $\mathcal{H}$ if it is stable and pre-attractive.*

5. *Asymptotically stable for $\mathcal{H}$ if it is stable and attractive.*

Figure 3.2(a) shows a solution that starts from an initial condition with distance to $\mathcal{A}$ less than or equal to δ and stays with distance to $\mathcal{A}$ that is less than or equal to ε, with δ and ε given as in item 1 of Definition 3.1. Figure 3.2(b) depicts a bounded and complete solution that starts within μ of the set $\mathcal{A}$ and converges to $\mathcal{A}$, via flows and jumps, to $\mathcal{A}$; hence, illustrating the notion in items 2 and 3 of Definition 3.1.

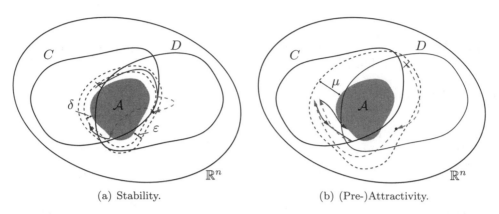

(a) Stability.   (b) (Pre-)Attractivity.

Figure 3.2: Stability and (pre-)attractivity of a nonempty set $\mathcal{A} \subset \mathbb{R}^n$ for a hybrid closed-loop system $\mathcal{H}$.

The reason that a general set $\mathcal{A}$ is considered is due to the fact that in many control problems of interest, the desired set-point condition is not necessarily an isolated point, but rather a set. In fact, the problem of designing a hybrid controller $\mathcal{H}_K$ for a hybrid plant $\mathcal{H}_P$ typically pertains to the stabilization of sets, in particular, due to the state of the hybrid controller including timers that persistently evolve within a bounded time interval and logic variables that take values from discrete sets. The following example illustrates this point in a hybrid closed-loop system resulting from controlling a continuous-time plant with a logic-based hybrid controller.

Example 3.2 (Multi-link pendulum, revisited). *In the control problem about global swing-up of the links of a pendulum system in Example 1.5, the objective is to steer the links to the upright configuration with zero velocity. This goal corresponds to steering the angles ϕ_1 and ϕ_2 to zero and the angular velocities ω_1 and ω_2 to zero; i.e., steer z to the origin. However, if a hybrid controller like the one outlined in Example 1.5 is used, then the state of the closed-loop system would include a logic variable and a timer. The logic variable is denoted by q and the finite set of its possible values by Q. The timer is denoted by τ and the range of values is given by $[0, T^*]$. Then, the set to render asymptotically stable for the hybrid closed-loop system is given by $\mathcal{A} = \{0\} \times Q \times [0, T^*]$. This set does not impose any condition on the logic variable and on the timer, other than they appropriate ranges. When $q = q^* \in Q$ is the value of the logic variable associated with the local controller used around the upright configuration – see item 1 in Example 1.5 – then the set to asymptotically stabilize is $\mathcal{A} = \{0\} \times \{q^*\} \times [0, T^*]$. Furthermore, if the timer is forced to remain at zero when $q = q^*$, then the set to asymptotically stabilize is $\mathcal{A} = \{0\} \times \{q^*\} \times \{0\}$.*

In certain control problems, the state of the plant is to be stabilized to a set on which more than one value is allowed. The following example introduces the problem of controlling the attitude (or orientation) of a rigid body for which the state of the plant is to be stabilized to a two-point set.

Example 3.3 (Rigid body model of a quadrotor). *Topological constraints and underactuation are key challenges in the design of estimation, navigation, and control* *algorithms for aerospace systems. The main obstacles to (robust) asymptotic stabilization of such systems stem from the properties of the manifold that the attitude of a rigid body belongs to. While Euler angles belong to a Euclidean space, which would make analysis simpler, they do not provide a global parameterization of attitude. In fact, no three-parameter parameterization of attitude is globally nonsingular. When attitude is parameterized using rotation matrices, the attitude state is restricted to the manifold*

$$SO(3) = \{R \in \mathbb{R}^{3 \times 3} : R^\top R = RR^\top = I, \ \det R = 1\}$$

This manifold, known as the special-orthogonal manifold, is not diffeomorphic to any Euclidean space. Due to this property, similar to the challenges in stabilizing to a point the system on the circle in § 1.2.3, static continuous state-feedback laws cannot globally stabilize a set-point R_d in $SO(3)$.

Such an obstacle to global stabilization is easier to visualize when attitude is parameterized in terms of unit quaternions. Unit quaternions provide a four-parameter parameterization of attitude given by vectors $\mathfrak{q} = (\mathfrak{n}, \mathfrak{e})$ taking values from

$$\mathbb{S}^3 = \left\{\mathfrak{q} \in \mathbb{R}^4 : |\mathfrak{q}| = 1\right\}$$

Note that for each given set-point rotation matrix $R_d \in SO(3)$ there are two antipodal points, $\mathfrak{q}_d$ and $-\mathfrak{q}_d$, associated with it. Hence, the problem of globally asymptotically stabilizing the attitude of a rigid body to R_d reduces to the stabilization of a two-point set on the four-dimensional unit sphere $\mathbb{S}^3$. This double covering needs to be carefully handled to avoid issues with performance and robustness, in particular, since, as explained in § 1.2.4, nonsmooth or hybrid feedback is required.

A hybrid controller with state η defined as a logic variable taking values from $Q := \{0, 1\}$ can be designed to cope with all of these challenges. For the case when attitude is parameterized by quaternions, and using a hybrid controller with $\eta - q \subset Q$, the set to stabilize for the closed loop is $\mathcal{A} = (\{-\mathfrak{q}_d\} \cup \{-\mathfrak{q}_d\}) \times Q$. The design of a hybrid controller that achieves this property is presented in Example 7.21, in the context of synergistic feedback.

When $\mathcal{A}$ is closed and bounded,[4] the boundedness condition on $(t, j) \mapsto |x(t, j)|_\mathcal{A}$ in the pre-attractivity notion in Definition 2 reduces to the solution $(t, j) \mapsto x(t, j)$ itself being bounded (as a function). Such is the case in the two examples above, in which the set to pre-asymptotically stabilize for the closed loop is compact. However, Definition 2 is much more general as it allows for unbounded sets $\mathcal{A}$. The following example illustrates one such situation.

Example 3.4 (Synchronization over a network, revisited). *Consider the synchronization problem outlined in § 1.2.5 and the hybrid closed-loop system model proposed in Example 2.23. Since the goal is to establish that the states of the systems synchronize, in the original coordinates, this set is given by*

$$\mathcal{A} := \left\{(z, \ell, \tau) \in \mathbb{R}^{n_P N} \times \mathbb{R}^{rN} \times \mathcal{T} : z_1 = z_2 = \ldots = z_N, \ell_1 = \ell_2 = \ldots = \ell_N = 0\right\}$$

---

[4]Recall that a subset of $\mathbb{R}^n$ is compact if it is both closed and bounded. For a summary of set-related notions, see § A.3.

In the error coordinates defined in (2.31) of Example 2.23, the set to stabilize is

$$\mathcal{A} := \left\{ (z, \theta, \tau) \in \mathbb{R}^{n_P N} \times \mathbb{R}^{rN} \times \mathcal{T} : z_1 = z_2 = \ldots = z_N, \theta_1 = \theta_2 = \ldots = \theta_N = 0 \right\}$$

Both of these sets are closed and unbounded; hence, not compact.

Remark 3.5 (*Basin of attraction and the global case*). The basin of pre-attraction of a pre-asymptotically stable set $\mathcal{A}$ for a hybrid closed-loop system $\mathcal{H}$ is the set of initial conditions from where solutions satisfy the pre-attractivity property. This set is denoted as $\mathcal{B}_{\mathcal{A}}^p$. The set $\mathcal{A}$ is said to be *globally pre-attractive* if the basin of pre-attraction is equal to the entire state space. Moreover, the set $\mathcal{A}$ is said to be *globally pre-asymptotically stable* if it is both *stable* and *globally pre-attractive*. Note that points not in $\overline{C} \cup D$ are automatically included in the basin of pre-attraction since solutions do not exist from such points – indeed, there are no solutions from such points that have to satisfy Definition 3.1. Similarly, the basin of attraction of an asymptotically stable set $\mathcal{A}$ for a hybrid closed-loop system $\mathcal{H}$ is denoted as $\mathcal{B}_{\mathcal{A}}$. Similarly, the set $\mathcal{A}$ is said to be *globally attractive* if the basin of attraction contains $\overline{C} \cup D$. Consequently, the set $\mathcal{A}$ is said to be *globally asymptotically stable* if it is both *stable* and *globally attractive*. △

Remark 3.6 (*On the prefix "pre"*). When every maximal solution to $\mathcal{H}$ is complete, the prefix "pre" is omitted as the "classical" asymptotic stability notion is recovered. The classical asymptotic stability notion is typically studied under the assumption that every maximal solution is complete. For instance, for a continuous-time system of the form $\dot{x} = f(x)$ with a pre-asymptotically stable set $\mathcal{A}$, every maximal solution starting from a point nearby $\mathcal{A}$ is complete when f is locally Lipschitz. For the general hybrid closed-loop system $\mathcal{H}$ in (2.19), Proposition 2.34 provides conditions such that every maximal solution is complete – the objective is to rule out cases b and c therein. △

As stated in Remark 3.5, the (pre-)attractivity and (pre-)asymptotic stability notions in Definition 3.1 can be strengthened to be global by requiring that the (pre-)attractivity notions therein hold for all solutions to $\mathcal{H}$. In addition, stability and pre-attractivity can be strengthened to be uniform, in the following sense.

Definition 3.7 (Uniform global stability, pre-attractivity, and pre-asymptotic stability). *Given a hybrid closed-loop system $\mathcal{H}$ as in (2.19), a nonempty set $\mathcal{A} \subset \mathbb{R}^n$ is said to be*

1. *Uniformly globally stable for $\mathcal{H}$ if there exists a class-$\mathcal{K}_\infty$ function[5] α such that every solution x to $\mathcal{H}$ satisfies $|x(t, j)|_{\mathcal{A}} \leq \alpha(|x(0, 0)|_{\mathcal{A}})$ for each $(t, j) \in \operatorname{dom} x$;*

2. *Uniformly globally pre-attractive for $\mathcal{H}$ if for each $\varepsilon > 0$ and $r > 0$ there exists $T > 0$ such that, for every solution x to $\mathcal{H}$ with $|x(0, 0)|_{\mathcal{A}} \leq r$,*

$$(t, j) \in \operatorname{dom} x \quad and \quad t + j \geq T \qquad \Longrightarrow \qquad |x(t, j)|_{\mathcal{A}} \leq \varepsilon$$

3. *Uniformly globally pre-asymptotically stable for $\mathcal{H}$ if it is uniformly globally stable and uniformly globally pre-attractive.*

---

[5] $\mathcal{K}_\infty$-functions are introduced in Definition A.17.

Remark 3.8 (On uniformity). The term *uniform* in *uniform global pre-attractivity* requires solutions to reach a neighborhood of the set $\mathcal{A}$ in hybrid time that is uniform on the set of initial conditions defined by r. For instance, for the synchronization problem in Example 3.4, uniform global pre-attractivity of the set $\mathcal{A}$ (the one in error coordinates given therein) implies the following: with ε, r, and T given in the definition, when z starts close to each other and θ start close to zero, with "close" meaning that $|(z(0,0), \theta(0,0), \tau(0,0))|_{\mathcal{A}} \leq r$ then for each (t,j) in the domain of solution satisfying $t + j \geq T$ the distance $|(z(t,j), \theta(t,j), \tau(t,j))|_{\mathcal{A}}$ is no larger than ε – note that the τ component of the solutions plays no role in the distance to $\mathcal{A}$. Namely, after finite hybrid time, in particular, the synchronization error becomes smaller than ε for all future time. This is a desired property in most applications. Note that this property holds for free for solutions x that are not complete since, in such a case, for T large enough there would be no (t,j)'s in dom x satisfying $t + j \geq T$; hence the implication $(t,j) \in \operatorname{dom} x$ and $t + j \geq T$ imply $|x(t,j)|_{\mathcal{A}} \leq \varepsilon$ holds for free. $\triangle$

Conveniently, through the use of $\mathcal{KL}$-functions,[6] it is possible to capture the properties of solutions required by uniform global stability and by uniform global pre-attractivity in a single inequality. In fact, when each solution x to $\mathcal{H}$ satisfies

$$|x(t,j)|_{\mathcal{A}} \leq \beta(|x(0,0)|_{\mathcal{A}}, t+j) \quad \forall (t,j) \in \operatorname{dom} x \tag{3.1}$$

for some $\mathcal{KL}$ function β, then $\mathcal{A}$ is uniformly globally pre-asymptotically stable. In particular, the solutions to a (continuous-time or discrete-time) linear time-invariant system with an exponentially stable origin satisfy the bound (3.1) with $\beta(r,s) = a\, r \exp(-bs)$ for some positive constants a and b.

A bound like the one in (3.1) also holds when the distance to the set $\mathcal{A}$ is replaced by a proper indicator of $\mathcal{A}$ on its basin of attraction $\mathcal{B}_{\mathcal{A}}^p$, namely, a nonnegative function that is zero only on $\mathcal{A}$ and grows unbounded as the state approaches the boundary of $\mathcal{B}_{\mathcal{A}}^p$.[7] Note that the $\mathcal{KL}$ function involved in the bound would depend on the proper indicator.

The following definition captures the general uniform property in (3.1) for solutions to $\mathcal{H}$.

Definition 3.9 ($\mathcal{KL}$ pre-asymptotic stability). *Let a nonempty closed set $\mathcal{A} \subset \mathbb{R}^n$ and an open set $\mathcal{U} \subset \mathbb{R}^n$ such that $\mathcal{A} \subset \mathcal{U}$ be given. The set $\mathcal{A}$ is said to be $\underline{\mathcal{KL}}$ $\underline{\text{pre-asymptotically stable}}$ on $\mathcal{U}$ for the hybrid closed-loop system $\mathcal{H}$ in (2.19) if for every proper indicator ω of $\mathcal{A}$ on $\mathcal{U}$ there exists a class-$\mathcal{KL}$ function β such that each solution x to $\mathcal{H}$ satisfies*

$$\omega(x(t,j)) \leq \beta(\omega(x(0,0)), t+j) \quad \forall (t,j) \in \operatorname{dom} x \tag{3.2}$$

Remark 3.10 (On relationships between $\mathcal{KL}$ stability and asymptotic stability). It is easy to see that, from its very definition, $\mathcal{KL}$ pre-asymptotic stability implies pre-asymptotic stability. This relationship is general, even when $\mathcal{H}$ does not satisfy

---

[6]$\mathcal{KL}$-functions are introduced in Definition A.18.

[7]Proper indicators are introduced in Definition A.19.

the hybrid basic conditions – see Definition 2.20 – and $\mathcal{A}$ is not compact. In fact, for ω given by the distance to $\mathcal{A}$, due to the properties of β, for each $\varepsilon > 0$ there exists $\delta > 0$ such that $\beta(\delta, 0) \leq \varepsilon$, implying that every solution x to $\mathcal{H}$ with $|x(0,0)|_{\mathcal{A}} \leq \delta$ satisfies

$$|x(t,j)|_{\mathcal{A}} \leq \beta(|x(0,0)|_{\mathcal{A}}, t+j) \leq \varepsilon \quad \forall (t,j) \in \text{dom}\, x$$

Then, the closed set $\mathcal{A}$ is stable. Since $\mathcal{U}$ is open and such that $\mathcal{A} \subset \mathcal{U}$, there exists $\mu > 0$ such that every solution x to $\mathcal{H}$ with $|x(0,0)|_{\mathcal{A}} \leq \mu$ is such that $x(0,0) \in \mathcal{U}$ and, due to (3.2), x is bounded. Furthermore, if the solution x is complete then

$$\lim_{(t,j) \in \text{dom}\, x,\ t+j \to \infty} \beta(|x(0,0)|_{\mathcal{A}}, t+j) = 0$$

which implies that $(t,j) \mapsto |x(t,j)|_{\mathcal{A}}$ converges to zero. Then, $\mathcal{A}$ is pre-attractive. As it is shown in Theorem 3.22 below, when the data of $\mathcal{H}$ satisfies the hybrid basic conditions and $\mathcal{A}$ is compact, then pre-asymptotic stability is equivalent to uniform pre-asymptotic stability. $\triangle$

As pointed out below (3.1), exponential stability of a linear time-invariant system implies $\mathcal{KL}$ asymptotic stability of the origin. The following notion introduces pre-exponential stability of a set for a hybrid closed-loop system.

Definition 3.11 (Global pre-exponential stability). *Given a hybrid closed-loop system $\mathcal{H}$ as in (2.19), a nonempty set $\mathcal{A} \subset \mathbb{R}^n$ is said to be <u>globally pre-exponentially</u> <u>stable</u> for $\mathcal{H}$ if there exist positive constants a and b such that each solution x to $\mathcal{H}$ satisfies*

$$|x(t,j)|_{\mathcal{A}} \leq a \exp(-b(t+j))|x(0,0)|_{\mathcal{A}} \qquad \forall (t,j) \in \text{dom}\, x \qquad (3.3)$$

The basin of pre-exponential attraction of a pre-exponentially stable set $\mathcal{A}$ for a hybrid closed-loop system $\mathcal{H}$ is the set of initial conditions from where solutions satisfy (3.3). This set is denoted as $\mathcal{B}_{\mathcal{A}}^{pe}$. Note that (3.3) needs to hold with the same constants a and b for all solutions to $\mathcal{H}$. From its very definition, pre-exponential stability implies pre-asymptotic stability, uniform pre-asymptotic stability, and also $\mathcal{KL}$ pre-asymptotic stability.

3.2.2   Invariance

Another property of interest in analysis and control design is *set invariance*. In simple words, a set is invariant for a dynamical system if its solutions starting from the set remain in the set. Since solutions may stay in the set in forward or in backward time, invariance notions are classified as *forward invariance* and *backward invariance*, respectively. Furthermore, since at times solutions from an initial condition may not be unique, invariance properties that require only one solution to stay in the set (in forward or in backward time) are of interest. Such notions carry the prefix *weak*. Such notions for hybrid closed-loop systems $\mathcal{H}$ are introduced next.

Definition 3.12 (Weak invariance). *Given a hybrid closed-loop system $\mathcal{H}$ as in (2.19), a nonempty set $K \subset \mathbb{R}^n$ is said to be*

1. *weakly forward pre-invariant for $\mathcal{H}$ if for each $x_\circ \in K$ there exists a maximal solution x to $\mathcal{H}$ from $x_\circ$ such that $x(t, j) \in K$ for all $(t, j) \in \operatorname{dom} x$;*

2. *weakly forward invariant for $\mathcal{H}$ if for each $x_\circ \in K$ there exists a complete solution x to $\mathcal{H}$ from $x_\circ$ such that $x(t, j) \in K$ for all $(t, j) \in \operatorname{dom} x$;*

3. *weakly backward invariant for $\mathcal{H}$ if for each $x'_\circ \in K$ and each $T > 0$ there exist $x_\circ \in K$ and at least one solution x to $\mathcal{H}$ from $x_\circ$ such that for some $(t^*, j^*) \in \operatorname{dom} x$ with $t^* + j^* \geq T$, x satisfies $x(t^*, j^*) = x'_\circ$ and $x(t, j) \in K$ for all $(t, j) \in \operatorname{dom} x$ such that $t + j \leq t^* + j^*$;*

4. *weakly invariant for $\mathcal{H}$ if it is both weakly forward invariant and weakly backward invariant.*

Figure 3.3(a) shows a complete solution that starts from a point $x_\circ$ in the set K and remains in K in forward hybrid time. This solution illustrates weak forward invariance of K in item 1 of Definition 3.12. Note that there exists another solution from $x_\circ$ that leaves the set K. Similarly, Figure 3.3(b) depicts a solution that reaches a point $x'_\circ$ in K at a time (t^*, j^*) for which there exists a point $x_\circ$, also in K. This solution starts from $x_\circ$ and remains in K until it reaches $x'_\circ$, hence, illustrating weak backward invariance of K in item 3 of Definition 3.12. This notion allows for another solution passing through $x'_\circ$ that does not necessarily remains in K.

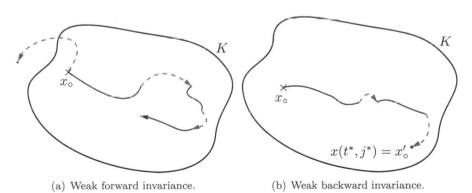

(a) Weak forward invariance.   (b) Weak backward invariance.

Figure 3.3: Weak forward and backward invariance of a set $K \subset \mathbb{R}^n$ for a hybrid closed-loop system $\mathcal{H}$.

When all solutions from K are required to stay in the set K, the set is said to be forward pre-invariant or forward invariant, depending on whether maximal solutions from the set are complete. These notions are formalized in the following definition.

Definition 3.13 (Forward invariance). *Given a hybrid closed-loop system $\mathcal{H}$ as in (2.19), a nonempty set $K \subset \mathbb{R}^n$ is said to be*

1. *forward pre-invariant for $\mathcal{H}$ if each maximal solution x to $\mathcal{H}$ from K satisfies $x(t,j) \in K$ for all $(t,j) \in \text{dom}\, x$;*

2. *forward invariant for $\mathcal{H}$ if each maximal solution x to $\mathcal{H}$ from K is complete and satisfies $x(t,j) \in K$ for all $(t,j) \in \text{dom}\, x$.*

Forward invariance plays a key role in some of the examples in the previous chapters. Two such examples are revisited next.

Example 3.14 (DC/AC inverter, revisited). *In the problem of controlling the DC/AC inverter in Example 1.1, Example 1.4, and Example 2.37, once solutions converge to the tracking band, which is defined as K in (2.40), the control algorithm must guarantee that they stay in it. This requirement translates into the problem of designing a control algorithm that renders the set K defined therein forward invariant for the resulting hybrid closed-loop system. The design of such algorithm is presented in Chapter 11; see Example 11.12.*

Example 3.15 (Rigid body model of a quadrotor, revisited). *The rotational kinematics model of a rigid body to be employed to solve the control problem outlined in Example 3.3 has a control input that is to be assigned by a controller so as to asymptotically stabilize the quaternion state to a reference quaternion value. As explained therein, this reference value is typically obtained from a given set-point rotation matrix $R_d \in SO(3)$, and is either one of the two antipodal points, q_d and $-\mathsf{q}_d$. Such an asymptotic stabilization task needs to be accomplished while keeping the quaternion state within the four-dimensional unit sphere $\mathbb{S}^3$. This latter goal corresponds to rendering the set $\mathbb{S}^3$ forward invariant. The design of this controller is carried out in Example 7.21.*

Weak forward and backward invariance are used in the Hybrid Invariance Principle in Theorem 3.23. Weak and (nonweak) forward invariance are exploited in Chapter 11, in the context of invariance-based control. In that chapter, tools to certify such invariance notions are also provided.

3.2.3 Robustness to Disturbances

The hybrid control algorithms proposed in this book aim at guaranteeing that the specific property of interest holds with some (nonzero) margin of robustness to disturbances. At times it might be possible to guarantee that the actual property for the nominal system still holds – at least practically – when the disturbances are small. Such a property can typically be attained when the specific property for which the controller is designed for is *intrinsically robust*. Conveniently, as established in the forthcoming Theorem 3.26, when a nominal hybrid closed-loop system satisfies the hybrid basic conditions and has a compact set pre-asymptotically stable, then the pre-asymptotic stability property holds – practically and semiglobally – for the hybrid system with disturbances; see § 2.3.5. This notion, which is formalized next, is employed extensively in this book as it is a characterization of robustness of pre-asymptotic stability for a hybrid closed-loop system.

Definition 3.16 (Semiglobal practical robust $\mathcal{KL}$ pre-asymptotic stability). *Given a hybrid closed-loop system $\mathcal{H}$ as in (2.19), a nonempty closed set $\mathcal{A} \subset \mathbb{R}^n$, and an open set $\mathcal{U} \subset \mathbb{R}^n$ such that $\mathcal{A} \subset \mathcal{U}$, the set $\mathcal{A}$ is said to be <u>semiglobally practically robustly $\mathcal{KL}$ pre-asymptotically stable</u> for $\mathcal{H}$ on $\mathcal{U}$ if for every proper indicator ω of $\mathcal{A}$ on $\mathcal{U}$, every function $\beta \in \mathcal{KL}$ such that (3.2) holds for the solutions to $\mathcal{H}$ from $\mathcal{U}$, and every continuous function $\rho^* : \mathbb{R}^n \to \mathbb{R}_{\geq 0}$ that is positive on $\mathcal{U} \setminus \mathcal{A}$ the following holds: for each compact set $K \subset \mathcal{U}$ and each $\varepsilon > 0$, there exists $\delta^* > 0$ such that for each solution x_ρ to $\mathcal{H}^\rho$ in (2.48) with $\rho = \delta^* \rho^*$ and $x_\rho(0,0) \in K$ satisfies*

$$\omega(x_\rho(t,j)) \leq \beta(\omega(x_\rho(0,0)), t+j) + \varepsilon \quad \forall (t,j) \in \operatorname{dom} x_\rho \tag{3.4}$$

The robust pre-asymptotic stability notion in Definition 3.16 relates the size of the disturbances captured by the ρ-perturbation of $\mathcal{H}$ in (2.48) (the hybrid system $\mathcal{H}^\rho$), the set of initial conditions (the set K), and the level of closeness (defined by the parameter ε) to the set $\mathcal{A}$ that complete solutions converge to as $t+j$ gets large. Even though the size of the disturbances might be small – namely, δ^* might be small – the robustness property is very important since very small disturbances can already have a significant effect on the solutions to a hybrid system. In particular, the jump times of a solution x to $\mathcal{H}$ are likely different from those of a solution x_ρ to $\mathcal{H}^\rho$, no matter how small the values of the perturbation function ρ might be. Such a mismatch typically leads to large error between x and x_ρ at a common time t.

When large disturbances affect the system, the design of the controller typically requires information about the disturbances in order to compensate for them. Tools leading to controllers that guarantee such "large" robustness are given in Chapter 10 and Chapter 11.

3.3 ANALYSIS TOOLS

The main tools for certifying pre-asymptotic stability of sets for hybrid dynamical systems used in this book are introduced in this section. These tools are used in most of the forthcoming chapters. Tools for the analysis of other properties of interest for the purpose of specific chapters are introduced as needed, later on. In particular, tools to study passivity are presented in Chapter 9, to study invariance appear in Chapter 11, and to certify finite-time attractivity are given in Chapter 12.

3.3.1 Hybrid Lyapunov Theorem

Lyapunov functions are employed to certify the notions in Definition 3.1. To this end, the basic properties that one such function must satisfy to serve as a Lyapunov function for the hybrid closed-loop system $\mathcal{H}$ are introduced next.

Definition 3.17 (Lyapunov function candidate). *The sets $\mathcal{U}, \mathcal{A} \subset \mathbb{R}^n$ and the function $V : \operatorname{dom} V \to \mathbb{R}$ define a Lyapunov function candidate on $\mathcal{U}$ with respect to $\mathcal{A}$ for the hybrid closed-loop system $\mathcal{H} = (C, F, D, G)$ if the following conditions hold:*

1. $(\overline{C} \cup D \cup G(D)) \cap \mathcal{U} \subset \operatorname{dom} V$;

2. $\mathcal{U}$ contains an open neighborhood of $\mathcal{A} \cap (C \cup D \cup G(D))$;

3. V is continuous on $\mathcal{U}$ and locally Lipschitz on an open set containing $\overline{C} \cap \mathcal{U}$;

4. V is positive definite on $C \cup D \cup G(D)$ with respect to $\mathcal{A}$.

These properties for V assure that if V is decreasing to zero along solutions to $\mathcal{H}$, then the distance between the solutions and the set $\mathcal{A}$ decreases. Then, in particular, in the limit as $t + j$ tends to infinity, such distance converges to zero for complete solutions that remain in $\mathcal{U}$. Positive definiteness of V with respect to $\mathcal{A}$ enables such convergence. The set $\mathcal{U}$ characterizes a region nearby $\mathcal{A}$ including points that belong to the basin of pre-attraction. When $\mathcal{U}$ includes an open neighborhood of $C \cup D \cup G(D)$, then the candidate is for global pre-asymptotic stability of $\mathcal{A}$.

Given a Lyapunov function candidate V on $\mathcal{U}$ with respect to $\mathcal{A}$, along any solution x to $\mathcal{H}$ with values in $\mathcal{U}$ – namely, $\operatorname{rge} x \subset \mathcal{U}$ – the interest is in conditions that guarantee that

$$(t, j) \mapsto V(x(t, j))$$

is nonincreasing and, when t or j get large, converges to zero. To this end, let $(\underline{t}, \underline{j}), (\overline{t}, \overline{j}) \in \operatorname{dom} x$ be such that $\underline{t} + \underline{j} \leq \overline{t} + \overline{j}$. Let $t(j)$ denote the least time t such that $(t, j) \in \operatorname{dom} x$, and $j(t)$ denote the least index j such that $(t, j) \in \operatorname{dom} x$. The change of V from hybrid time $(\underline{t}, \underline{j}) \in \operatorname{dom} x$ to hybrid time $(\overline{t}, \overline{j}) \in \operatorname{dom} x$ is given by

$$V(x(\overline{t}, \overline{j})) - V(x(\underline{t}, \underline{j})) = \int_{\underline{t}}^{\overline{t}} \frac{d}{dt} V(x(t, j(t))) \, dt$$

$$+ \sum_{j=\underline{j}+1}^{\overline{j}} [V(x(t(j), j)) - V(x(t(j), j-1))] \tag{3.5}$$

This expression takes into account the "continuous contribution" to the change in V due to the integration of the time derivative of $t \mapsto V(x(t, j))$ for each j over I_x^j with nonempty interior. It also takes into account the "discrete contribution" due to the difference in the values of V before and after each jump. The integral above expresses the change of V during flow since $t \mapsto V(x(t, j(t)))$ is locally Lipschitz and locally absolutely continuous on every interval on which $t \mapsto j(t)$ is constant.

Suppose that $\mathcal{H} = (C, F, D, G)$ satisfies the hybrid basic conditions. When F is single valued and V is continuously differentiable on a neighborhood containing $C \cap \mathcal{U}$, then the quantity

$$\langle \nabla V(x), F(x) \rangle \tag{3.6}$$

captures the change of V along flows of solutions to $\mathcal{H}$ evolving within $C \cap \mathcal{U}$. When V is only locally Lipschitz on an open neighborhood containing $C \cap \mathcal{U}$, then the change of V along flows in (3.5) can be captured using generalized derivatives. The Clarke generalized directional derivative of V at a point x in the direction of χ is given by

$$V^\circ(x, \chi) = \max_{\nu \in \partial V(x)} \langle \nu, \chi \rangle \tag{3.7}$$

where $\partial V(x)$ is the *Clarke generalized gradient* of V at x. It is a closed, convex, and nonempty set equal to the convex hull of all limits of sequences $\nabla V(x_i)$, where x_i is any sequence converging to x while avoiding an arbitrary set of measure zero containing all the points at which V is not differentiable. When V is locally Lipschitz, ∇V exists almost everywhere on $\text{dom } V$; see the discussion above Definition 2.27. This generalized derivative leads to the infinitesimal quantity

$$\max_{\chi \in F(x)} V^\circ(x, \chi) = \max_{\chi \in F(x)} \max_{\nu \in \partial V(x)} \langle \nu, \chi \rangle \tag{3.8}$$

which, on $C \cap \mathcal{U}$, bounds the change of V along flows of solutions to $\mathcal{H}$.

With the above characterizations of the change of V along flows, given a solution x to $\mathcal{H}$, it follows that

$$\frac{d}{dt} V(x(t, j(t))) \leq \dot{V}(x(t, j(t))) \tag{3.9}$$

for almost all t in I_x^j, with I_x^j having a nonempty interior – namely, the solution flows for some time for the given j. Formally introduced in Definition 3.18, $\dot{V}$ is defined as the quantity in (3.6) or in (3.8), depending on the regularity of V.

When C and F satisfy (A1) and (A2) in the hybrid basic conditions, the bound in (3.9) also holds when F is intersected by the tangent cone of C. In fact, a solution x to $\mathcal{H}$ satisfies $x(t, j) \in T_C(x(t, j))$ for each $(t, j) \in I_x^j \times \{j\}$ with I_x^j having a nonempty interior; see Lemma A.22. Hence, (3.9) holds with $\dot{V}$ given by

$$\max_{\chi \in F(x) \cap T_C(x)} V^\circ(x, \chi) \qquad \forall x \in C \cap \mathcal{U} \tag{3.10}$$

This refinement is included in the definitions of $\dot{V}$ given below.

When $\mathcal{H}$ satisfies the hybrid basic conditions, the change of V at jumps in (3.5) is captured by the infinitesimal quantity

$$\max_{\chi \in G(x)} V(\chi) - V(x) \tag{3.11}$$

for points x in $D \cap \mathcal{U}$. Even without any regularity on V, V evaluated along a solution x satisfies

$$V(x(t_{j+1}, j+1)) - V(x(t_{j+1}, j)) \leq \Delta V(x(t_{j+1}, j)) \tag{3.12}$$

where ΔV is equal to the quantity in (3.11) and the t_j's define the hybrid time domain of x; see Definition 2.26.

Definition 3.18 ($\dot{V}$ and ΔV). *Suppose a hybrid closed-loop system $\mathcal{H} = (C, F, D, G)$ satisfies the hybrid basic conditions, and that sets $\mathcal{U}, \mathcal{A} \subset \mathbb{R}^n$ and a function $V : \operatorname{dom} V \to \mathbb{R}$ defining a Lyapunov function candidate on $\mathcal{U}$ with respect to $\mathcal{A}$ for $\mathcal{H}$ are given.*

- *The change of V along flows is given by*[8]

$$\dot{V}(x) := \max_{\chi \in F(x) \cap T_C(x)} V^\circ(x, \chi) \qquad \forall x \in C \cap \mathcal{U} \qquad (3.13)$$

  When, in addition, V is continuously differentiable on an open set containing $\overline{C} \cap \mathcal{U}$, $\dot{V}$ is given by

$$\dot{V}(x) := \max_{\chi \in F(x) \cap T_C(x)} \langle \nabla V(x), \chi \rangle \qquad \forall x \in C \cap \mathcal{U} \qquad (3.14)$$

  and furthermore, if F is single valued, then $\dot{V}$ is given by

$$\dot{V}(x) := \langle \nabla V(x), F(x) \cap T_C(x) \rangle \qquad \forall x \in C \cap \mathcal{U} \qquad (3.15)$$

- *The change of V at jumps is given by*

$$\Delta V(x) := \max_{\chi \in G(x)} V(\chi) - V(x) \qquad \forall x \in D \cap \mathcal{U} \qquad (3.16)$$

  and furthermore, if G is single valued, then ΔV is given by

$$\Delta V(x) := V(G(x)) - V(x) \qquad \forall x \in D \cap \mathcal{U} \qquad (3.17)$$

With the definition of Lyapunov function candidate in Definition 3.17 and the infinitesimal characterization of its change along solutions in Definition 3.18, a Lyapunov theorem to certify pre-asymptotic stability of a set for a hybrid closed-loop system is given next.

Theorem 3.19 (Hybrid Lyapunov Theorem). *Given sets $\mathcal{U}, \mathcal{A} \subset \mathbb{R}^n$ and a function $V : \operatorname{dom} V \to \mathbb{R}$ defining a Lyapunov function candidate on $\mathcal{U}$ with respect to $\mathcal{A}$ for a hybrid closed-loop system $\mathcal{H} = (C, F, D, G)$ with state $x \in \mathbb{R}^n$, the following hold:*

1. *The set $\mathcal{A}$ is stable for $\mathcal{H}$ if it is compact, $\mathcal{H}$ satisfies the hybrid basic conditions as in Definition 2.20, and $\dot{V}$ and ΔV satisfy*

$$\dot{V}(x) \le 0 \qquad\qquad\qquad \forall x \in C \cap \mathcal{U} \qquad (3.18)$$
$$\Delta V(x) \le 0 \qquad\qquad\qquad \forall x \in D \cap \mathcal{U} \qquad (3.19)$$

---

[8]In all of the definitions of $\dot{V}$, when x is such that $F(x) \cap T_C(x)$ is empty, the value of $\dot{V}(x)$ is taken to be $-\infty$.

2. *The set $\mathcal{A}$ is pre-asymptotically stable for $\mathcal{H}$ if it is compact, $\mathcal{H}$ satisfies the hybrid basic conditions, and one of the following conditions hold:*

   a) *Strict decrease during flows and jumps: $\dot{V}$ and ΔV satisfy (3.18) and (3.19), and*

   $$\dot{V}(x) < 0 \qquad\qquad \forall x \in (C \cap \mathcal{U}) \setminus \mathcal{A} \qquad (3.20)$$
   $$\Delta V(x) < 0 \qquad\qquad \forall x \in (D \cap \mathcal{U}) \setminus \mathcal{A} \qquad (3.21)$$

   b) *Strict decrease during flows and no discrete solutions: $\dot{V}$ and ΔV satisfy (3.18) and (3.19), and*

      i) *$\dot{V}$ satisfies (3.20);*

      ii) *any discrete and complete solution x to $\mathcal{H}$ that remains in $\mathcal{U}$ converges to $\mathcal{A}$; i.e, any complete solution x to $\mathcal{H}$ with $\operatorname{dom} x = \{0\} \times \mathbb{N}$ such that $x(0,j) \in \mathcal{U}$ for all $j \in \mathbb{N}$ satisfies $\lim_{j \to \infty} |x(0,j)|_{\mathcal{A}} = 0$.*

   c) *Strict decrease during jumps and no complete continuous solutions: $\dot{V}$ and ΔV satisfy (3.18) and (3.19), and*

      i) *ΔV satisfies (3.21);*

      ii) *any continuous and complete solution x to $\mathcal{H}$ that remains in $\mathcal{U}$ converges to $\mathcal{A}$; i.e, any complete solution x to $\mathcal{H}$ with $\operatorname{dom} x = \mathbb{R}_{\geq 0} \times \{0\}$ such that $x(t,0) \in \mathcal{U}$ for all $t \in \mathbb{R}_{\geq 0}$ satisfies $\lim_{t \to \infty} |x(t,0)|_{\mathcal{A}} = 0$.*

   d) *Weak decrease during flows and jumps: $\dot{V}$ and ΔV satisfy (3.18) and (3.19), and there exists $r^* > 0$ such that for each $x_\circ \in \mathcal{U}$ with $V(x_\circ) =: r \in (0, r^*)$, there is no complete solution to $\mathcal{H}$ from $x_\circ$ that remains in $V^{-1}(r) \cap \mathcal{U}$; i.e., there is no complete solution x to $\mathcal{H}$ with $x(0,0) = x_\circ$ such that*

   $$x(t,j) \in V^{-1}(r) \cap \mathcal{U} \qquad \forall (t,j) \subset \operatorname{dom} x \qquad (3.22)$$

   e) *Increase balanced by decrease: there exist constants $\lambda_c \in \mathbb{R}$ and $\lambda_d \in \mathbb{R}$ such that*

   $$V(x) < \lambda_c V(x) \qquad\qquad \forall x \in C \cap \mathcal{U} \qquad (3.23)$$
   $$\Delta V(x) \leq (\exp(\lambda_d) - 1) V(x) \qquad\qquad \forall x \in D \cap \mathcal{U} \qquad (3.24)$$

   and there exist positive constants M and γ such that for each solution x to $\mathcal{H}$ that remains in $\mathcal{U}$

   $$(t,j) \in \operatorname{dom} x \quad \implies \quad \lambda_c t + \lambda_d j \leq M - \gamma(t+j) \qquad (3.25)$$

   If $\mathcal{U}$ contains $C \cup D \cup G(D) =: X$ and every sublevel set $L_V(r)$ of V with $r \in V(X)$ is compact, then the pre-asymptotic stability property of $\mathcal{A}$ in items 2a-2e is global.

3. *The set $\mathcal{A}$ is uniformly globally pre-asymptotically stable for $\mathcal{H}$ if it is closed and one of the following conditions hold:*

   a) *Strict decrease during flows and jumps: there exist $\alpha_1, \alpha_2 \in \mathcal{K}_\infty$ and con-*

tinuous functions $\rho_c, \rho_d \in \mathcal{PD}$ *satisfying*

$$\alpha_1(|x|_\mathcal{A}) \leq V(x) \leq \alpha_2(|x|_\mathcal{A}) \qquad \forall x \in C \cup D \cup G(D) \qquad (3.26)$$

$$\dot{V}(x) \leq -\rho_c\left(|x|_\mathcal{A}\right) \qquad \forall x \in C \qquad (3.27)$$

$$\Delta V(x) \leq -\rho_d\left(|x|_\mathcal{A}\right) \qquad \forall x \in D \qquad (3.28)$$

b) *Strict decrease during flows and no increase at jumps: there exist* $\alpha_1, \alpha_2 \in \mathcal{K}_\infty$ *and a continuous function* $\rho_c \in \mathcal{PD}$ *such that* (3.26) *and* (3.27) *hold,* (3.28) *holds with* $\rho_d \equiv 0$, *and, for each* $r > 0$, *there exist* $\gamma_r \in \mathcal{K}_\infty$, $N_r \geq 0$ *such that for every solution* x *to* $\mathcal{H}$

$$|x(0,0)|_\mathcal{A} \in (0, r], \ (t, j) \in \mathrm{dom}\, x, \ t + j \geq T \quad \Longrightarrow \quad t \geq \gamma_r(T) - N_r$$

c) *Strict decrease during jumps and no increase during flows: there exist* $\alpha_1, \alpha_2 \in \mathcal{K}_\infty$ *and a continuous function* $\rho_d \in \mathcal{PD}$ *such that* (3.26) *and* (3.28) *hold,* (3.27) *holds with* $\rho_c \equiv 0$, *and, for each* $r > 0$, *there exist* $\gamma_r \in \mathcal{K}_\infty$, $N_r \geq 0$ *such that for every solution* x *to* $\mathcal{H}$

$$|x(0,0)|_\mathcal{A} \in (0, r], \ (t, j) \in \mathrm{dom}\, x, \ t + j \geq T \quad \Longrightarrow \quad j \geq \gamma_r(T) - N_r$$

d) *Increase balanced by decrease: there exist* $\alpha_1, \alpha_2 \in \mathcal{K}_\infty$ *such that* (3.26) *holds, constants* $\lambda_c \in \mathbb{R}$ *and* $\lambda_d \in \mathbb{R}$ *such that* (3.23) *and* (3.24) *hold with* $\mathcal{U} = \mathbb{R}^n$, *and positive constants* M *and* γ *such that each solution* x *to* $\mathcal{H}$ *satisfies* (3.25).

e) *Bounded time of flow: there exist* $\alpha_1, \alpha_2 \in \mathcal{K}_\infty$, *a continuous function* $\rho_d \in \mathcal{PD}$, *and* $\lambda \in \mathbb{R}$ *such that* (3.26) *and* (3.28) *hold,*

$$\dot{V}(x) \leq \lambda V(x) \qquad \forall x \in C$$

and, for each $r > 0$, *there exist* T_r *such that for every solution* x *to* $\mathcal{H}$

$$|x(0,0)|_\mathcal{A} \in (0, r], \ (t, j) \in \mathrm{dom}\, x \quad \Longrightarrow \quad t \leq T_r$$

f) *Finite number of jumps: there exist* $\alpha_1, \alpha_2 \in \mathcal{K}_\infty$, *a continuous function* $\rho_c \in \mathcal{PD}$, *and* $\lambda \in \mathcal{K}_\infty$ *such that* (3.26) *and* (3.27) *hold,*

$$V(\chi) \leq \lambda(V(x)) \qquad \forall x \in D, \ \forall \chi \in G(x)$$

and there exist $\gamma \in \mathcal{K}$ *and* $J > 0$ *such that for every solution* x *to* $\mathcal{H}$

$$(t, j) \in \mathrm{dom}\, x \quad \Longrightarrow \quad j \leq \gamma(|x(0,0)|_\mathcal{A}) + J$$

4. *The set* $\mathcal{A}$ *is globally pre-exponentially stable for* $\mathcal{H}$ *if* $\mathcal{A}$ *is closed,* V *is such that there exist positive constants* c_1, c_2, p, *and* $\widetilde{\lambda}_c$, *and* $\widetilde{\lambda}_d \in (0, 1]$ *such that one of the items 3a-3d above holds,* α_1 *and* α_2 *given by* $\alpha_1(s) = c_1 s^p$ *and* $\alpha_2(s) = c_2 s^p$ *for each* $s \in \mathbb{R}_{\geq 0}$, *and* ρ_c *and* ρ_d *therein replaced by* $\widetilde{\lambda}_c V$ *and* $\widetilde{\lambda}_d V$, *respectively.*

The conditions guaranteeing pre-asymptotic stability of $\mathcal{A}$ in the Hybrid Lyapunov Theorem guarantee that along each solution, the value of the Lyapunov function V converges to zero as t or j – depending on which conditions in the theorem are satisfied – gets large. Figure 3.4 depicts this situation when V is strictly decreasing, both during flows and at jumps. A function V with such a decrease is said to be a *Lyapunov function*. If the function does not decrease in one of the regimes but still can be used to certify stability or pre-attractivity, then it is said to be a *weak Lyapunov function* or a *Lyapunov-like function*.

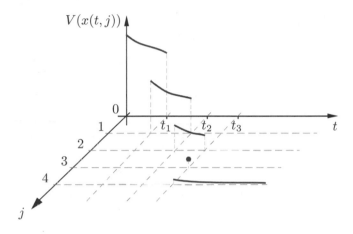

Figure 3.4: The value of a Lyapunov function candidate V along a solution to a hybrid closed-loop system $\mathcal{H}$, in the case that the conditions for strict decrease of V in the Hybrid Lyapunov Theorem in Theorem 3.19 are satisfied.

Remark 3.20 (On Lyapunov conditions). Theorem 3.19 is the main tool that is used in this book to guarantee stability properties of sets. A proof is in Appendix B. The conditions for pre-asymptotic stability of a compact set in item 2a do not involve information about solutions and guarantee a strict decrease of $(t, j) \mapsto V(x(t, j))$ for each solution x to $\mathcal{H}$; see Figure 3.4. The conditions in the other items within item 2 require further information on solutions to assure pre-attractivity of the compact set $\mathcal{A}$. This is the price to pay when using a weak Lyapunov function to certify pre-asymptotic stability of a set. However, some of these conditions in item 2 can be checked using methods that only rely on properties of the data of $\mathcal{H}$. In particular, condition ii) in item 2b can be checked by studying the discrete dynamics of $\mathcal{H}$ alone. In fact, when $\mathcal{H}$ has no discrete solutions, which is assured when $G(D) \cap D$ is empty, condition ii) in item 2b holds for free. Similar observations apply to condition ii) in item 2c. On the other hand, the condition in item 2d is a direct consequence of the invariance principle for hybrid systems, which is presented below, and requires analysis of the solutions to $\mathcal{H}$ from $V^{-1}(r) \cap \mathcal{U}$. It should be noted that the conditions in item 2e allow for the Lyapunov function candidate to increase in one regime and to decrease in the other regime; e.g., it allows V to increase during flows and decrease at jumps when $\lambda_c > 0$ and $\lambda_d \in (0, 1)$. Similar comments apply to the conditions in item 3 for the case when $\mathcal{A}$ is closed. The conditions in items 3 and 4 do not require $\mathcal{H}$ to satisfy the hybrid basic conditions.[9] For simplicity, the

---

[9]In this case, the max's in the definition of $\dot{V}$ and ΔV are replaced by sup's.

conditions in those items are stated for the global case, but replacing C by $C \cap \mathcal{U}$ and D by $D \cap \mathcal{U}$ in the conditions therein lead to local versions.   △

Remark 3.21 (On an estimate of the basin of pre-attraction). When any of the conditions in item 2 of Theorem 3.19 holds, an estimate of the basin of pre-attraction is given by any compact sublevel set of V that is contained in $\mathcal{U}$: if $r \in \mathbb{R}_{\geq 0}$, V and $\mathcal{U}$ are such that $L_V(r)$ is compact and

$$L_V(r) \subset \mathcal{U}$$

then $L_V(r) \subset \mathcal{B}_{\mathcal{A}}^p$. Note that r has to be smaller than r^* in the estimate associated to the conditions in item 2d of Theorem 3.19.   △

The following result establishes that pre-asymptotic stability of a compact set $\mathcal{A}$ for a hybrid closed-loop system satisfying the hybrid basic conditions implies $\mathcal{KL}$ pre-asymptotic stability of $\mathcal{A}$. This result is key in establishing that pre-asymptotic stability of a compact set is robust as defined in Definition 3.16.

Theorem 3.22 (Pre-asymptotic stability implies $\mathcal{KL}$ pre-asymptotic stability). *Suppose that the hybrid closed-loop system $\mathcal{H}$ satisfies the hybrid basic conditions and that a compact set $\mathcal{A}$ is pre-asymptotically stable with basin of pre-attraction $\mathcal{B}_{\mathcal{A}}^p$. Then, $\mathcal{B}_{\mathcal{A}}^p$ is open and $\mathcal{A}$ is $\mathcal{KL}$ pre-asymptotically stable on $\mathcal{B}_{\mathcal{A}}^p$ for $\mathcal{H}$.*

The reverse implication of the claim in Theorem 3.22 is immediate using Definition 3.9. It is left as an exercise for the reader to show that under the assumptions in Theorem 3.22, the compact set $\mathcal{A}$ therein is uniformly (local or global) pre-asymptotically stable. The reader might find Remark 3.10 useful.

3.3.2   Hybrid Invariance Principle

The so-called *invariance principle* is a powerful tool that permits establishing attractivity when only a "weak" Lyapunov-like function is available – meaning that the function does not strictly decrease along both flows and at jumps of the hybrid system. It is also a useful tool to determine where particular solutions of interest converge to. This principle for hybrid closed-loop systems $\mathcal{H} = (C, F, D, G)$ satisfying the hybrid basic conditions is as follows.

Theorem 3.23 (Hybrid Invariance Principle). *Given a hybrid closed-loop system $\mathcal{H} = (C, F, D, G)$ with state $x \in \mathbb{R}^n$ satisfying the hybrid basic conditions, nonempty $\mathcal{U} \subset \mathbb{R}^n$, and a function $V : \operatorname{dom} V \to \mathbb{R}$, suppose that items 1 and 3 in Definition 3.17 are satisfied, and that (3.18) and (3.19) hold. With $X := C \cup D \cup G(D)$, recall the following from List of Symbols:*

$$V^{-1}(r) := \{x \in X \,:\, V(x) = r\} \tag{3.29}$$

$$\dot{V}^{-1}(0) := \left\{x \in C \,:\, \dot{V}(x) = 0\right\} \tag{3.30}$$

$$\Delta V^{-1}(0) := \{x \in D \,:\, \Delta V(x) = 0\} \tag{3.31}$$

Let x be a precompact solution to $\mathcal{H}$ with $\overline{\text{rge}\,x} \subset \mathcal{U}$. Then, for some constant $r \in V(\mathcal{U} \cap X)$, the following hold:

1. The solution x converges to the largest weakly invariant set in

$$V^{-1}(r) \cap \mathcal{U} \cap \left[\dot{V}^{-1}(0) \cup (\Delta V^{-1}(0) \cap G(\Delta V^{-1}(0))) \right] \qquad (3.32)$$

2. The solution x converges to the largest weakly invariant set in

$$V^{-1}(r) \cap \mathcal{U} \cap \Delta V^{-1}(0) \cap G(\Delta V^{-1}(0)) \qquad (3.33)$$

if in addition the solution x is Zeno;

3. The solution x converges to the largest weakly invariant set in

$$V^{-1}(r) \cap \mathcal{U} \cap \dot{V}^{-1}(0) \qquad (3.34)$$

if, in addition, the solution x is such that, for some $\gamma > 0$ and some $J \in \mathbb{N}$, $t_{j+1} - t_j \geq \gamma$ for all $j \geq J$; i.e., the given solution x is such that the elapsed time between consecutive jumps is eventually bounded below by a positive constant γ.

Remark 3.24 (Connections to pre-asymptotic stability). Unlike Theorem 3.19, the invariance principle does not involve a set $\mathcal{A}$, at least explicitly. However, when the sets $\mathcal{U}, \mathcal{A}$ and a function V define a Lyapunov function candidate, Theorem 3.23 implies that every precompact solution that stays in $\mathcal{U}$ converges to some r-level set of V. Figure 3.5 depicts such a situation and the associated sets involved in the Hybrid Invariance Principle. Under the condition in item 4 of Theorem 3.19, Theorem 3.23 implies that the only possible value of r is zero, and since $V^{-1}(0) \cap (C \cup D) \cap \mathcal{U} \subset \mathcal{A}$ when the sets $\mathcal{U}, \mathcal{A}$ and the function V define a Lyapunov function candidate, every such precompact solution converges to $\mathcal{A}$. When $\mathcal{U}$ is a forward invariant set containing an open neighborhood of $\mathcal{A}$, pre-attractivity of $\mathcal{A}$ ensues. Also when the sets $\mathcal{U}, \mathcal{A}$ and a function V define a Lyapunov function candidate, if $\dot{V}(x) < 0$ for each $x \in (C \cap \mathcal{U}) \setminus \mathcal{A}$, then the set $\dot{V}^{-1}(0) \cap \mathcal{U}$ is contained in $\mathcal{A}$. Similarly, when $\Delta V(x) < 0$ for each $x \in (D \cap \mathcal{U}) \setminus \mathcal{A}$ then $\Delta V^{-1}(0) \cap \mathcal{U}$ is contained in $\mathcal{A}$. $\triangle$

3.3.3 Robustness from $\mathcal{KL}$ Pre-Asymptotic Stability

A model of a hybrid plant under the effect of disturbances is in §2.3.5. For a (nominal) hybrid closed-loop system $\mathcal{H} = (C, F, D, G)$, a model with added disturbances of practical interest is given by

$$\mathcal{H}_w : \begin{cases} x + w_m \in C & \dot{x} \in F(x + w_m) + w_F \\ x + w_m \in D & x^+ \in G(x + w_m) + w_G \end{cases} \qquad (3.35)$$

where w_m captures measurement noise, while w_F and w_G capture modeling error and disturbances on the continuous and on the discrete dynamics of $\mathcal{H}$. It turns out that when the nominal hybrid closed-loop system $\mathcal{H}$ satisfies the hybrid ba-

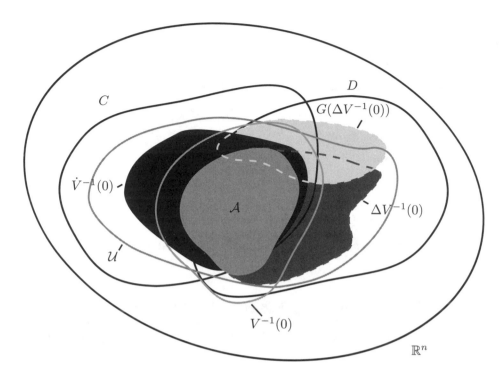

Figure 3.5: The sets involved in the Hybrid Invariance Principle when the sets $\mathcal{U}, \mathcal{A}$ and the function V define a Lyapunov function candidate.

sic conditions and has a pre-asymptotically stable compact set $\mathcal{A}$, if the effect of the disturbances on $\mathcal{H}$ described in $\mathcal{H}_w$ can be captured by the ρ-perturbation of $\mathcal{H}$ given in (2.48), then the original pre-asymptotic stability property of $\mathcal{A}$ for $\mathcal{H}$ is preserved *semiglobally and practically*, as formally stated in Definition 3.16. As argued in § 2.3.5, this ρ-perturbation is quite general and is suitable to capture disturbances arising in most practical settings, such as unmodeled plant dynamics, actuator noise, and measurement noise. The same robustness property holds for the more general class of ρ-perturbation models of $\mathcal{H}$ satisfying the following assumption. It uses the outer limit of a sequence of sets; see Definition A.30.

Assumption 3.25 (Convergence property). *The ρ-perturbation of the hybrid closed-loop system $\mathcal{H} = (C, F, D, G)$ given by $\mathcal{H}^\rho = (C^\rho, F^\rho, D^\rho, G^\rho)$ is such that for any sequence $\{\rho_i\}_{i=1}^\infty$, $1 > \rho_1 > \rho_2 > \ldots > 0$, converging to zero, the following holds:*

C1) *The sequences of sets $\{C^{\rho_i}\}_{i=1}^\infty$ and $\{D^{\rho_i}\}_{i=1}^\infty$ are such that*

$$C^0 \subset C, \qquad D^0 \subset D \tag{3.36}$$

where C^0 and D^0 are the outer limit of $\{D^{\rho_i}\}_{i=1}^\infty$ and $\{C^{\rho_i}\}_{i=1}^\infty$, respectively, *which are given by*

$$C^0 = \limsup_{i\to\infty} C^{\rho_i}, \qquad D^0 = \limsup_{i\to\infty} D^{\rho_i}$$

C2) *The sequences of set-valued mappings* $\{F^{\rho_i}\}_{i=1}^{\infty}$ *and* $\{G^{\rho_i}\}_{i=1}^{\infty}$ *are such that*

$$F^0(x) \subset F(x) \quad \forall x \in C, \qquad G^0(x) \subset G(x) \quad \forall x \in D \qquad (3.37)$$

where F^0 *and* G^0 *are the outer graphical limits of* $\{F^{\rho_i}\}_{i=1}^{\infty}$ *and* $\{G^{\rho_i}\}_{i=1}^{\infty}$, *respectively, which are given by*

$$\mathrm{gph}\, F^0 = \limsup_{i\to\infty} \mathrm{gph}\, F^{\rho_i}, \qquad \mathrm{gph}\, G^0 = \limsup_{i\to\infty} \mathrm{gph}\, G^{\rho_i}$$

C3) *The sequences of set-valued mappings* $\{F^{\rho_i}\}_{i=1}^{\infty}$ *and* $\{G^{\rho_i}\}_{i=1}^{\infty}$ *are locally eventually bounded.*

The following result states that nominal pre-asymptotic stability of a compact set for $\mathcal{H}$ is robust to disturbances that satisfy the properties in Assumption 3.25. In simple words, robustness means that for each compact subset K of the basin of attraction of $\mathcal{A}$ and every desired level of closeness $\varepsilon > 0$, there exists a maximum allowed level of perturbation such that every solution x_ρ to $\mathcal{H}^\rho$ with such perturbation level from $x_\rho(0,0) \in K$ satisfies

$$\omega(x_\rho(t,j)) \le \beta(\omega(x_\rho(0,0)), t+j) + \varepsilon \quad \forall (t,j) \in \mathrm{dom}\, x_\rho$$

In other words, the set $\mathcal{A}$ is semiglobally practically robustly $\mathcal{KL}$ pre-asymptotically stable for $\mathcal{H}$ as in Definition 3.16.

Theorem 3.26 (Semiglobal practical robust $\mathcal{KL}$ pre-asymptotic stability from nominal pre-asymptotic stability). *Given a hybrid closed-loop system $\mathcal{H}$ and a nonempty set $\mathcal{A} \subset \mathbb{R}^n$, suppose*

1. *$\mathcal{H}$ satisfies the hybrid basic conditions;*

2. *$\mathcal{A}$ is a compact pre-asymptotically stable set for $\mathcal{H}$ with basin of pre-attraction $\mathcal{B}_\mathcal{A}$, for which the proper indicator ω of $\mathcal{A}$ on $\mathcal{B}_\mathcal{A}$ and the class-$\mathcal{KL}$ function β are such that (3.2) holds for each solution x to $\mathcal{H}$;*

3. *The ρ-perturbation of $\mathcal{H}$ given by $\mathcal{H}^\rho$ satisfies Assumption 3.25.*

Then, the compact set $\mathcal{A}$ is semiglobally practically robustly $\mathcal{KL}$ pre-asymptotically stable for $\mathcal{H}$.

As suggested by the ρ-perturbation model $\mathcal{H}_\rho$ in (2.48), which, in particular, captures the disturbances in the hybrid plant model in (2.43) and the hybrid closed-loop system model in (3.35), under enough regularity on the data of $\mathcal{H}_\rho$, disturbances of practical interest can be modeled so that the convergence property in Assumption 3.25 holds. In addition, outer and inner perturbations as well as temporal and spatial regularizations satisfy such properties; see Exercise 25 and Exercise 26.

3.4   EXERCISES

Exercise 20 (bouncing ball). Consider the point-mass bouncing ball system in Exercise 2 and the model formulated in Exercise 11.

1. Show that the energy of the system is a Lyapunov function candidate on $\mathbb{R}^2$ for $\mathcal{A} = \{(0,0)\}$.
2. Apply the Hybrid Lyapunov Theorem in Theorem 3.19 to show that $\mathcal{A}$ is stable.
3. Show that the energy of the system does not satisfy the properties in item 2a of Theorem 3.19
4. Show that $\mathcal{A}$ is globally asymptotically stable using an item in Theorem 3.19 other than item 2a.
5. Show that $\mathcal{A}$ is globally asymptotically stable using the Hybrid Invariance Principle in Theorem 3.23.

Exercise 21 (Hybrid control for consensus with events). Consider two agents with point-mass dynamics

$$\dot{z}_i = u_i \qquad i \in \mathcal{V} := \{1,2\} \tag{3.38}$$

that are allowed to exchange information over a network at time instances triggered by a single timer τ with hybrid dynamics

$$\begin{aligned} \dot{\tau} &= -1 & \tau &\in [0, T_2] \\ \tau^+ &\in [T_1, T_2] & \tau &= 0 \end{aligned} \tag{3.39}$$

where $T_2 > 0$ is the maximum time allowed in between the events and $T_1 \in (0, T_2]$ is the minimum time allowed. The jump map is equal to the interval $[T_1, T_2]$ to capture nondeterminism on the amount of time until the next communication event. At such events, both agents are allowed to exchange the value of their state.

1. Given the parameter T_2 of the network, for each $i \in \{1,2\}$, consider the following control algorithm:

$$\begin{aligned} u_i &= \ell_i \\ \dot{\ell}_i &= 0 & \tau &\in [0, T_2] \\ \ell_i^+ &= -\gamma(z_i - z_k) & \tau &= 0 \end{aligned} \tag{3.40}$$

where $\gamma > 0$ is a parameter. Derive a model of the plant, controller, and closed-loop system.

2. Show that if there exists a positive scalar γ and a positive definite symmetric matrix P satisfying

$$A_g^\top \exp(A_f^\top \nu) P \exp(A_f \nu) A_g - P < 0 \quad \forall \nu \in [T_1, T_2], \qquad (3.41)$$

where $A_g = \begin{bmatrix} 0 & 1 \\ 0 & 0 \end{bmatrix}$ and $A_f = \begin{bmatrix} 1 & 0 \\ -\gamma & 0 \end{bmatrix}$ then the set

$$\mathcal{A} := \{(z_1, z_2, \ell_1, \ell_2, \tau) : z_1 = z_2, \ell_1 = \ell_2 = 0, \tau \in [0, T_2]\}$$

is globally exponentially stable.

Exercise 22 (Juggling system). Consider the following simplification of the one-degree-of-freedom juggling system defined in Example 2.32 and revisited in Example 2.35:

$$\mathcal{H}_P : \begin{cases} z \in C_P & \dot{z} = \begin{bmatrix} z_2 \\ -\gamma \end{bmatrix} =: F_P(z) \\ (z, u) \in D_P & z^+ = \begin{bmatrix} z_1 \\ -\lambda z_2 + u \end{bmatrix} =: G_P(z, u) \\ & y = z =: h(z) \end{cases}$$

where

$$C_P = \{z \in \mathbb{R}^2 : z_1 \geq 0\}$$
$$D_P = \{(z, u) \in \mathbb{R}^2 \times \mathbb{R} : z_1 = 0, z_2 \leq 0, u \in [\underline{u}, \overline{u}]\}$$

In this model, $z = (z_1, z_2) \in \mathbb{R}^2$ denotes the vertical position and velocity of the ball, and u the actuation at impacts.

1. Design a controller guaranteeing that the ball converges to rest in finite time.
2. Design another controller guaranteeing that the ball converges to a periodic pattern, in which the time elapsed between impacts is lower bounded by a positive constant.

Exercise 23 (Unmodeled dynamics and measurement noise in state-feedback hybrid control). Suppose that a state-feedback hybrid controller $\mathcal{H}_K$ has been designed for the nominal model of a hybrid plant $\mathcal{H}_P$. Derive a model of the resulting hybrid closed-loop system with disturbances given by unmodeled dynamics on the flow and jump maps of $\mathcal{H}_P$ that are bounded by a constant $\overline{d}$ and on measurement noise of the state of the plant that is bounded by a constant $\overline{m}$. Your model should be such that Assumption 3.25 is satisfied.

Exercise 24 (Uniform pre-asymptotic stability from pre-asymptotic stability). Show that under the assumptions in Theorem 3.22, the set $\mathcal{A}$ is uniformly pre-asymptotically stable.

Exercise 25 (Outer perturbations). Given a hybrid closed-loop system $\mathcal{H} = (C, F, D, G)$, its ρ-perturbation is given by $\mathcal{H}^\rho = (C^\rho, F^\rho, D^\rho, G^\rho)$ in (2.48).

1. Show that $\mathcal{H}^\rho$ satisfies Assumption 3.25.
2. Show that $\mathcal{H}^\rho$ satisfies the hybrid basic conditions when α is continuous.

Exercise 26 (Temporal regularization). Given a hybrid closed-loop system $\mathcal{H} = (C, F, D, G)$, its temporal regularization is given by the hybrid system

$$\mathcal{H}_\tau : \quad \begin{cases} (x, \tau) \in (C \times \mathbb{R}_{\geq 0}) \cup (\mathbb{R}^n \times [0, T^*]) & (\dot{x}, \dot{\tau}) \in F(x) \times \{1\} \\ (x, \tau) \in D \times [0, T^*] & (x^+, \tau^+) \in G(x) \times \{0\} \end{cases} \quad (3.42)$$

where τ is a resettable timer and T^* a constant. System $\mathcal{H}_\tau$ is an augmentation of $\mathcal{H}$, with state taking values from $\mathbb{R}^{n+1}$.

1. Show that every solution to $\mathcal{H}_\tau$ is such that, after the first jump, the time between jump is larger than or equal to T^*.
2. Show that the behavior of the x state component of $\mathcal{H}_\tau$ is the same as that of the state of $\mathcal{H}$.
3. Show that $\mathcal{H}_\tau$ satisfies Assumption 3.25.
4. Show that if $\mathcal{H}$ is well-posed, then $\mathcal{H}_\tau$ is also well-posed.

3.5   NOTES

The stability and attractivity notions in § 3.1 are the well-known Lyapunov stability and attractivity notions extended to general sets; see, e.g., [95, Chapter 3]. The distance in Definition A.10 is well defined due to nonemptiness of $\mathcal{A}$; see [96, Problem 8.12 on page 55]. The stability, pre-attractivity, and pre-asymptotic stability notions in Definition 3.1 follow the classical ones for continuous-time and discrete-time systems. These can also be found in [1, Definition 7.1].

The definitions of uniform global stability, uniform pre-attractivity, and uniform pre-asymptotic stability given in Definition 3.7 are the same as those in [1, Definition 3.6]. The $\mathcal{KL}$ pre-asymptotic stability notion in Definition 3.9 coincides with the notion in [1, Definition 7.10], which follows the classical notion in the literature; see, e.g., [97]. The notion of pre-exponential stability in Definition 3.11 coincides with the one in [98]. It can be interpreted as a special case of $\mathcal{KL}$ pre-asymptotic stability, with the appropriate function β.

Weak invariance for hybrid inclusions is defined and exploited in [9] for the purpose of characterizing the omega-limit set of precompact solutions. Forward invariance is defined in [88] and [99] for the analysis and design of controllers inducing such property. Chapter 11 presents tools to render sets forward (nonweak and weak) invariant using barrier functions.

Definition 3.17 is such that a classical Lyapunov function V (e.g., as defined in [95]) satisfies its properties. Typically, V is chosen to be a continuously differentiable function defined on $\mathbb{R}^n$, so items 1-3 of Definition 3.17 already hold. The function V is also required to be positive definite with respect to the set to stabilize; hence, item 4 holds. Definition 3.17 is essentially [1, Definition 3.16].

Items 1-2 of Theorem 3.19 follow from [9, Theorem 7.6, Corollary 7.7, and Theorem 7.8]. Item 3 follows from [1, Theorem 3.18, Proposition 3.24, Proposition 3.27, and Proposition 3.29]. Theorem 3.23 generalizes the invariance principle for continuous-time and discrete-time to the hybrid setting. It first appeared in [9]. Proofs of the Hybrid Invariance Principle can be found in [9, Theorem 4.7 and Corollary 4.4], [1, Theorem 8.2]. Note that, as in Definition 3.18, the results in these references define functions capturing the change of V during flows and jumps for every point in $\mathbb{R}^n$.

The max over ∂V in V° used in (3.13) exists since, at each x, ∂V is closed and, for functions on finite dimensional spaces, is also upper semicontinuous – see [100, Proposition 2.1.5(d)]. Note that outer semicontinuous mappings have closed values. If the mapping is also locally bounded, the values are compact. For locally bounded set-valued mappings with closed values, outer semicontinuity agrees with what is often referred to as upper semicontinuity.

More details about the ideas in Example 3.3 and Example 3.15 are given in Example 7.21 and in [101].

The link in Theorem 3.22 between pre-asymptotic stability and $\mathcal{KL}$-asymptotic stability of a compact set $\mathcal{A}$ is essentially [1, Theorem 7.12].

The convergence properties in Assumption 3.25 are basically (C1)-(C4) in [6]. The semiglobal practical robustness result in Theorem 3.26 is a combination of [1, Lemma 7.20] and [1, Theorem 7.21]. Essentially, it is [6, Theorem 6.6].

The emphasis on hybrid closed-loop systems $\mathcal{H}$ that satisfy the hybrid basic conditions is due to robustness of pre-asymptotic stability [1]. In the context of feedback control, the hybrid basic conditions enable the following specific robustness properties:

1. *Robustness to fast actuators and sensors*: if the interconnection between a plant and a hybrid feedback controller results in a hybrid closed-loop system having a compact set pre-asymptotically stable when the actuator and sensor dynamics are omitted, or equivalently, are infinitely fast, then the same compact set is pre-asymptotically stable, semiglobally and practically in the finite speed of the actuator and sensor dynamics. In terms of the notion in Definition 3.16, the solutions to the hybrid system with the fast actuators and sensors satisfy (3.4) when the speed of the actuators and sensors is fast enough. This result appears in [102]; see also [103].

2. *Robustness to the discretization of controller*: the sample-and-hold implementation of a hybrid feedback controller preserves (semiglobally and practically) the asymptotic stability properties of a compact set. More precisely, when a hybrid controller is implemented digitally and is interfaced to a continuous-time plant through sample-and-hold devices (with asynchronous events), the same compact set is pre-asymptotically stable, semiglobally and practically, in the rates used for discretization of the algorithm, sampling of the state of the plant, and update of the control input. This result appeared in [104].

3. *Robustness to delays in the events*: pre-asymptotic stability of a compact set is pre-asymptotically stable, semiglobally and practically, in the presence of delayed jumps. More precisely, when the nominal hybrid closed-loop system $\mathcal{H}$ has a pre-asymptotically stable compact set, for small enough delays on the events (or jumps) of $\mathcal{H}$, solutions to a version of $\mathcal{H}$ with delayed jumps converge to a neighborhood of a set of interest related to the aforementioned compact set. This result is established in [105].

Chapter Four

Uniting Control

In certain control applications, control design tools that divide the problem into subproblems for which several control laws can be designed independently and then combined to solve the original problem are prevalent for many reasons. They reduce design and implementation time as well as add modularity and flexibility to the control system. They are also suitable when a single, continuous stabilizing control law does not exist or when its design is not straightforward. Moreover, multiple control laws, when properly designed and applied to the plant, can enhance the robustness properties of the closed-loop system. Such a "divide and conquer" approach to control design is also ubiquitous in control problems where precise control is desired nearby particular operating points while less stringent conditions need to be satisfied in other regions.

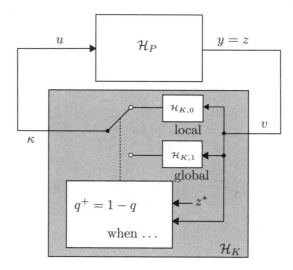

Figure 4.1: Closed-loop system resulting from uniting control.

In this chapter, a hybrid control strategy that uses two feedback controllers and a logic-based algorithm to select the one that should be applied is presented. Figure 4.1 depicts the hybrid control architecture. It corresponds to the strategy for uniting a local and a global controller outlined in § 1.2.6. Such a strategy is particularly useful as, most times, a controller that is capable of globally stabilizing a set-point does not have good performance in the entirety of the state space. On the other hand, local control design techniques can be used to design controllers with satisfactory performance, but only locally. The uniting control strategy is among the simplest hybrid feedback control laws, featuring a logic state $\eta = q$ governed by a logic-based algorithm selecting the appropriate feedback to apply to the plant.

For simplicity of the exposition, this hybrid control strategy is presented for the case when the two individual controllers are given by static state-feedback laws and the plant is a continuous-time system. Extensions of this strategy are motivated in examples and summarized in the notes at the end of the chapter. Furthermore, Chapter 8 provides a hybrid control strategy that is much more general than the one presented here, but for pedagogical reasons is presented later in the book.

4.1 OVERVIEW

The basic uniting control strategy consists of combining two controllers stabilizing the same set-point z^* where one controller performs better near the set-point than far away from it, and another controller performs better far away from it but not nearby. The motivation behind such a strategy stems from the occasional impossibility of robustly and globally stabilizing the set-point z^* with smooth or discontinuous control laws. For instance, when autonomously controlling a vehicle, the presence of obstacles typically leads to a partition of the state space, e.g., the vehicle can only move *left, right, up,* or *down* so as to avoid the obstacle and reach a desired destination; see § 1.2.4. In most cases, the presence of such partitions demands a discontinuous controller to solve the problem, which, due to such discontinuity, might not be robust to small measurement noise.

This chapter considers the problem of stabilizing $\mathcal{H}_P$ to a set-point z^* with the following objectives:

- Render z^* stable and, from nearby z^*, guarantee that solutions converge to z^* using a feedback controller denoted $\mathcal{H}_{K,0}$;

- From points where $\mathcal{H}_{K,0}$ does not guarantee convergence to z^*, use another feedback controller, denoted $\mathcal{H}_{K,1}$, to guarantee that solutions to the plant converge to a small neighborhood of z^* from where controller $\mathcal{H}_{K,0}$ can be used.

Similar to the illustration in § 1.2.6, the controller $\mathcal{H}_{K,0}$ should assure local asymptotic stability of the set-point z^* and the controller $\mathcal{H}_{K,1}$ should guarantee global convergence to nearby z^*. These objectives potentially stem from controller $\mathcal{H}_{K,0}$ producing efficient transient responses but only working nearby z^*, and from controller $\mathcal{H}_{K,1}$ resulting in less efficient transients but actually working globally. To accomplish such objectives, it is required to design a control strategy that globally asymptotically stabilizes z^* while using $\mathcal{H}_{K,0}$ near z^* and $\mathcal{H}_{K,1}$ far from it.

Following this overview of the uniting control strategy, a hybrid controller $\mathcal{H}_K$ that combines $\mathcal{H}_{K,0}$ and $\mathcal{H}_{K,1}$ is given as follows:

- The controller state is $\eta = q \in Q := \{0, 1\}$, where q is a logic variable – see item 1 in § 1.2.1.

- The controller input is $v = z$.

- The control logic for uniting control (UC) is as follows:

(UC-L1) When $q = 0$, apply controller $\mathcal{H}_{K,0}$ as long as the solution to the plant stays close to z^*. If the state of the plant leaves the region where $\mathcal{H}_{K,0}$ guarantees convergence to z^*, update q to 1, namely, according to

$$q^+ = 1$$

(UC-L2) When $q = 1$, apply controller $\mathcal{H}_{K,1}$ until the solution to the plant gets close enough to z^*, to a point where $\mathcal{H}_{K,0}$ can be used. When any such point is reached, update q to 0, namely, according to

$$q^+ = 0$$

For any other condition, the logic variable q remains constant.

Figure 4.2 depicts a diagram implementing the control logic outlined above.

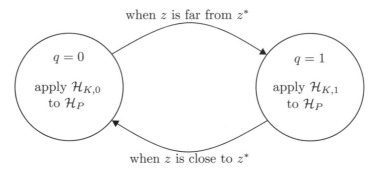

Figure 4.2: Control logic implemented by uniting control.

Before getting technical with the model of the strategy and the associated design methods, the uniting control strategy is illustrated in two examples. The first example pertains to the problem of controlling the position of the magnetic head of a disk drive using two individual controllers.

Example 4.1 (Disk drive control). *The dynamics of the magnetic head of a disk drive can be captured by the second-order system*

$$\dot{p} = \nu, \qquad \dot{\nu} = u \tag{4.1}$$

where $p \in \mathbb{R}^2$ denotes position and $\nu \in \mathbb{R}^2$ velocity of the magnetic head. Design specifications in disk drives translate into precise positioning to a specific location within a track of the disk drive – which is where information resides – and, to access data stored at different locations, to the requirement of rapid transitioning between locations and tracks. Under the name mode-switching control, commercial disk drives use algorithms that combine a track-seeking controller and a track-following controller to accomplish both tasks with desired performance. In mode-switching control algorithms, the track-seeking controller is designed to rapidly steer the magnetic head to a neighborhood of the desired location, denoted p^*, while the track-following controller regulates position and velocity, precisely and robustly, to *enable read/write operations at $p = p^*$ with zero velocity. Then, once the track-seeking controller has steered the state of the magnetic head to a region where the track-following controller is applicable, the algorithm switches to the track-following controller. The control strategy results in a hybrid closed-loop system.*

The mode-switching control strategy outlined above can be modeled and designed using the uniting control strategy. First, the second-order model of the magnetic head is modeled as a plant $\mathcal{H}_P$ with only continuous-time dynamics – namely,

$$\mathcal{H}_P \ : \ \begin{cases} (z, u) \in C_P := \mathbb{R}^4 \times \mathbb{R}^2 & \dot{z} = F_P(z, u) := \begin{bmatrix} z_2 \\ u \end{bmatrix} \\ \qquad\qquad\qquad y = h(z) := z \end{cases} \tag{4.2}$$

where $z = (p, \nu) \in \mathbb{R}^4$ is the state and $u \in \mathbb{R}^2$ is the control input. The control objective is to stabilize the point $z^* := (p^*, 0)$. Since the mode-switching control strategy uses two controllers, the set Q is defined as $\{0, 1\}$. The controller for $q = 0$, namely, $\mathcal{H}_{K,0}$, corresponds to the track-following controller and the controller for $q = 1$, namely, $\mathcal{H}_{K,1}$, corresponds to the track-seeking controller. To achieve the desired control goal using the uniting strategy, the track-following control law is designed to locally asymptotically stabilize the point z^*, while the track-seeking control law is designed to render the point z^* (or a neighborhood around it) globally attractive.

The next example, which is from the optimization literature, pertains to the problem of globally minimizing a function with good performance, both nearby and far away from the set of minimizers. Since the output of the associated plant is not equal to its state, this example illustrates the application of uniting control to the output-feedback case.

Example 4.2 (Heavy ball method). *Consider the problem of finding the critical points of a real-valued continuously differentiable function L using measurements of its gradient, ∇L. These critical points correspond to the roots of $\nabla L(\xi) = 0$, which are considered to be unknown. A method capable of guaranteeing convergence to the set of roots $\{\xi : \nabla L(\xi) = 0\}$ is the heavy ball method. The method consists of running a dynamical system that, under appropriate assumptions, has all of its solutions converging to a critical point of the function L. The dynamical system used in the heavy ball method is given by the second-order system*

$$\ddot{\xi} + \lambda \dot{\xi} + \gamma \nabla L(\xi) = 0 \tag{4.3}$$

where λ and γ are tunable parameters. This system resembles the dynamics of a particle sliding on a surface with profile defined by L, with friction. In such a setting, the parameter λ represents the ratio between the viscous friction coefficient and the mass of the particle, and γ represents the gravity constant.

The performance of the heavy ball method depends highly on the choice of the parameters λ and γ. In particular, for a fixed value of γ, the choice of the "friction parameter" λ significantly affects the asymptotic behavior of the solutions to (4.3). For rather simple choices of the function L, the literature on this method indicates that large values of λ are seen to give rise to slowly converging solutions resembling the steepest descent method, while smaller values give rise to fast solutions with oscillations that get wilder as λ decreases. The top plot in Figure 4.3 shows that for a quadratic function L, when λ is large, the algorithm converges slowly. The middle

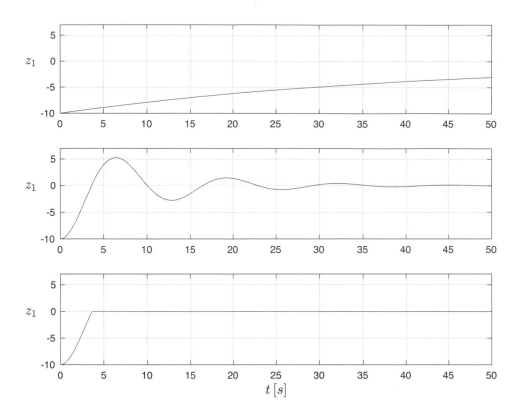

Figure 4.3: Comparison of the performance of the heavy ball method, with large and small values of λ, and with a uniting controller for $L(z_1) = \frac{1}{4}z_1^2$. Top: when λ is large, heavy ball converges very slowly. Middle: when λ is small, heavy ball converges quickly, but with wild oscillations. Bottom: a uniting control strategy yields fast convergence, with no oscillations.

plot shows the case when λ is small. In that case convergence is fast but with large oscillations. Such a compromise between damping the oscillations and converging fast suggests the need of an adaptation algorithm for the value of λ, or rather, two algorithms, one that uses large λ until solutions get close to the set of critical points – so as to converge fast – and then uses a small value of λ to avoid oscillations. For the case of a unique minimizer or of a connected continuum set of minima for L, the latter approach can be implemented using the uniting controller strategy when measurements of L are also available. The bottom plot in Figure 4.3 shows a solution with such a hybrid algorithm. As the plot depicts, the hybrid algorithm assures both fast convergence and no oscillations.

As suggested by the model in (4.3), the heavy ball method can be interpreted as a plant $\mathcal{H}_P$ with continuous-time double integrator dynamics and without jumps. For *the case of a planar double integrator with a single input, such a model is given by*

$$
\mathcal{H}_P \ : \quad
\begin{cases}
(z, u) \in C_P := \mathbb{R}^2 \times \mathbb{R} & \dot{z} = F_P(z, u) := \begin{bmatrix} z_2 \\ u \end{bmatrix} \\[2ex]
& y = h(z) := \begin{bmatrix} z_2 \\ \nabla L(z_1) \end{bmatrix}
\end{cases}
\tag{4.4}
$$

The function L is assumed to be twice continuously differentiable, radially unbounded, and positive definite with respect to its unique global minimizer at the unknown location $z_1^* \in \mathbb{R}$, at which L is equal to zero; see Definition A.27.

To arrive to the model in (4.3) describing the heavy ball method, the individual controllers used for uniting control leading to the response in the bottom plot in Figure 4.3 are designed as static output-feedback laws of the form

$$
\kappa_q(v) := -\lambda_q v_1 - \gamma_q v_2
\tag{4.5}
$$

for each $q \in Q := \{0, 1\}$ and each $v = (v_1, v_2) \in \mathbb{R}^2$, where v_1 is assigned to z_2, which is the first component of y, and v_2 is assigned to $\nabla L(z_1)$, which is the second component of y. The values of the parameters λ_q and γ_q would be different for each $q \in Q$, so as to achieve the fast convergence without oscillations nearby the minimizer shown in Figure 4.3.

4.2 HYBRID CONTROLLER

In this section, a hybrid controller $\mathcal{H}_K$ implementing the logic in §4.1 is constructed. Its construction requires the characterization of the regions of operation for each individual controller and the conditions for toggling q. For simplicity, the plant is assumed to only have continuous-time dynamics and an output equal to its state, resulting in a hybrid plant $\mathcal{H}_P$ given by

$$
\mathcal{H}_P \ : \quad
\begin{cases}
(z, u) \in C_P & \dot{z} \in F_P(z, u) \\
& y = z
\end{cases}
\tag{4.6}
$$

See (2.9). The controllers $\mathcal{H}_{K,0}$ and $\mathcal{H}_{K,1}$ are given by static state-feedback laws κ_0 and κ_1, respectively – see the discussion above Definition 2.8. Extensions are discussed in § 4.6 and in the design of the uniting controller to solve the problem stated in Example 4.2.

To satisfy the stated objectives, the following is assumed. Recall that asymptotic stability and attractivity are introduced in Definition 3.1 – see item 5 and item 3 therein; see also Remark 3.5 and Remark 3.6.

Assumption 4.3 (Conditions for uniting control). *Given $z^* \in \mathbb{R}^{n_P}$ and a plant $\mathcal{H}_P$ as in (4.6), there exist an open set $\mathcal{U}_0 \subset \mathbb{R}^{n_P}$, which contains an open neighborhood of z^*, and a closed set $\mathcal{E}_0 \subset \mathbb{R}^{n_P}$ such that*

(UC-A1) A state-feedback law $\kappa_0 : \mathbb{R}^{n_P} \to \mathbb{R}^{m_P}$ such that $\mathcal{H}_P$ controlled by κ_0 is such that z^* is asymptotically stable with basin of attraction containing $\overline{\mathcal{U}_0}$;

(UC-A2) A state-feedback law $\kappa_1 : \mathbb{R}^{n_P} \to \mathbb{R}^{m_P}$ guaranteeing that every solution z to $\mathcal{H}_P$ controlled by κ_1 starting from $\overline{\Pi_c(C_P)} \setminus \mathcal{E}_0$ reaches $\mathcal{E}_0$ after a finite

amount of flow time or converges to it as t tends to ∞;

(UC-A3) Positive constants δ_0 and δ_0^c, and a closed set $\mathcal{T}_{1,0}$ satisfying

$$\mathcal{E}_0 + \delta_0^c \mathbb{B} \subset \mathcal{T}_{1,0}, \qquad \mathcal{T}_{1,0} + 2\delta_0 \mathbb{B} \subset \mathcal{U}_0 \qquad (4.7)$$

and each solution to the plant with initial condition in $\mathcal{T}_{1,0}$ resulting from applying κ_0 remains in $\mathcal{U}_0$.

The first condition in (4.7) and the property of solutions in (UC-A3) involves properties of both control laws, κ_0 and κ_1. Since $\mathcal{T}_{1,0}$ is closed and $\mathcal{U}_0$ is an open set containing an open neighborhood of z^*, a positive constant δ_0 satisfying (4.7) is always guaranteed to exist when $\mathcal{T}_{1,0} \subset \mathcal{U}_0$. Figure 4.4 depicts these sets and a sample plant solution.

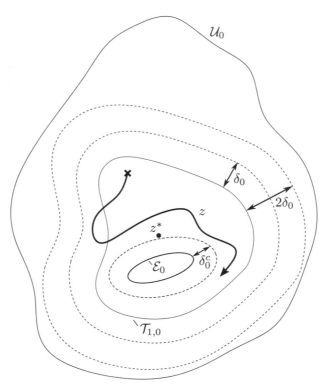

Figure 4.4: Sets associated with the uniting control strategy. The solution z to $\mathcal{H}_P$ controlled by κ_0 from a point in $\mathcal{T}_{1,0}$ indicates that $\mathcal{T}_{1,0}$ might not necessarily be forward (pre-)invariant.

A hybrid controller $\mathcal{H}_K = (C_K, F_K, D_K, G_K, \kappa)$ implementing the control strategy in § 4.1 has state $\eta = q \in Q = \{0, 1\}$, input $v \in \mathbb{R}^{r_P}$, which is assigned to the state of the plant z, output $\zeta \in \mathbb{R}^{m_P}$, and data defined as follows:

The data $(C_K, F_K, D_K, G_K, \kappa)$ is given by

$$C_K := \bigcup_{q \in Q} (C_{K,q} \times \{q\}), \qquad \begin{cases} C_{K,0} := \overline{\mathcal{U}_0} \\ C_{K,1} := \overline{\mathbb{R}^{n_P} \setminus \mathcal{T}_{1,0}} \end{cases} \tag{4.8}$$

$$F_K(z,q) := 0 \qquad \forall (z,q) \in C_K \tag{4.9}$$

$$D_K := \bigcup_{q \in Q} (D_{K,q} \times \{q\}), \qquad \begin{cases} D_{K,0} := \overline{\mathbb{R}^{n_P} \setminus \mathcal{U}_0} \\ D_{K,1} := \mathcal{T}_{1,0} \end{cases} \tag{4.10}$$

$$G_K(z,q) := 1 - q \qquad \forall (z,q) \in D_K \tag{4.11}$$

$$\kappa(z,q) := q\kappa_1(z) + (1-q)\kappa_0(z) \tag{4.12}$$

Conveniently, the output κ can be written as $\kappa(z,q) = \kappa_q(z)$.

This construction is such that the logic variable q is updated from 1 to 0 when z reaches $\mathcal{T}_{1,0}$, which, when item (UC-A2) in Assumption 4.3 holds, is guaranteed to occur in finite time. After such a jump, the control law κ_0 is applied to the plant. Then, due to items (UC-A1) and (UC-A3) in Assumption 4.3, the resulting closed-loop solution is such that z stays in $\mathcal{U}_0$ and converges to z^*.

Remark 4.4 (About assumptions (UC-A1)-(UC-A3)). The hybrid controller $\mathcal{H}_K$ with data in (4.8)-(4.12) allows solutions starting with z at the boundary of $\mathcal{U}_0$ and with $q = 0$ to trigger a jump resetting q to one. However, after such a jump, the controller is designed so that z reaches $\mathcal{T}_{1,0}$, from where, after a jump back to $q = 0$, convergence to z^* is assured by κ_0. The factor two in (4.7) is mainly for the purposes of robustness. It guarantees that after jumps from $q = 1$ to $q = 0$, the solution remains at a distance from the boundary of $\mathcal{U}_0$ that is larger than or equal to δ_0. Since $\mathcal{U}_0$ is open, condition (UC-A3) guarantees that the solutions to the plant from $\mathcal{T}_{1,0}$ and under the effect of κ_0 cannot reach the boundary of $\mathcal{U}_0$. $\triangle$

As shown next, under mild conditions on its data, the proposed hybrid controller is such that it is well-posed as it satisfies the hybrid basic conditions. For more details about the importance of well-posedness in hybrid systems, see § 2.3.2.

Lemma 4.5 (Well-posedness of $\mathcal{H}_K$). *For each $q \in Q$, suppose $z \mapsto \kappa_q(z)$ is continuous. Then, the hybrid controller $\mathcal{H}_K$ in (4.8)-(4.12) satisfies the hybrid basic conditions.[1]*

Proof. By construction and closedness of $\mathcal{T}_{1,0}$ from (UC-A3), the sets $C_{K,0}$, $C_{K,1}$, and $D_{K,0}$ are closed. The set $D_{K,1}$ is closed since $\mathcal{T}_{1,0}$ is closed. Then, C_K and D_K are closed as they are defined as the finite union of closed sets. Also by construction, the maps F_K and G_K are continuous. Finally, the output map κ is continuous since κ_0 and κ_1 are continuous functions of (z,q). $\square$

---

[1]See Definition 2.25 and the discussion below it.

4.3   CLOSED-LOOP SYSTEM

The state $x = (z, q)$ of the hybrid closed-loop system resulting from controlling $\mathcal{H}_P$ in (4.6) with the hybrid controller $\mathcal{H}_K = (C_K, F_K, D_K, G_K, \kappa)$ in (4.8)-(4.12) changes according to

$$\dot{z} \in F_P(z, \kappa_q(z))$$
$$\dot{q} = 0$$

during flows. At jumps, it is updated according to

$$z^+ = z$$
$$q^+ = 1 - q$$

Following the construction in Definition 2.11, the hybrid closed-loop system $\mathcal{H} = (C, F, D, G)$ has state $x = (z, q) \in \mathbb{R}^{n_P} \times Q =: X$ and is given as in (2.19) with

$$C := \{(z, q) \in X : (z, \kappa_q(z)) \in C_P, z \in C_{K,q}\} \tag{4.13}$$

$$F(x) := \begin{bmatrix} F_P(z, \kappa_q(z)) \\ 0 \end{bmatrix} \qquad \forall x \in C \tag{4.14}$$

$$D := \{(z, q) \in X : (z, \kappa_q(z)) \in C_P, z \in D_{K,q}\} \tag{4.15}$$

$$G(x) := \begin{bmatrix} z \\ 1 - q \end{bmatrix} \qquad \forall x \in D \tag{4.16}$$

The next result establishes key properties of this hybrid closed-loop system.

Theorem 4.6 (Uniting control). *Given $z^* \in \mathbb{R}^{n_P}$ and*

$$\mathcal{H}_P : \begin{cases} (z, u) \in C_P & \dot{z} \in F_P(z, u) \\ & y = z \end{cases}$$

suppose Assumption 4.3 holds. Let the hybrid closed-loop system $\mathcal{H}$ have data (C, F, D, G) as in (4.13)-(4.16).

1. *If $\mathcal{H}_P$ satisfies the hybrid basic conditions and, for each $q \in Q$, $z \mapsto \kappa_q(z)$ is continuous then $\mathcal{H}$ satisfies the hybrid basic conditions.*

2. *Every maximal solution to $\mathcal{H}$ from $C \cup D$ is complete and exhibits no more than two jumps.*

3. *The set*

$$\mathcal{A} := \{z^*\} \times \{0\} \tag{4.17}$$

    is globally asymptotically stable for $\mathcal{H}$. Furthermore, when the conditions in item 1 hold, the asymptotic stability property of $\mathcal{A}$ is robust in the sense of Definition 3.16.

Proof. The flow map F is outer semicontinuous, convex valued, and locally bounded relative to C since, due to $\mathcal{H}_P$ satisfying the hybrid basic conditions, F_P satisfies (A2$_P$) in Definition 2.25 and $z \mapsto \kappa_q(z)$ is continuous for each $q \in Q$; see Exercise 97. Then, being the stack of F_P and the zero function, F satisfies (A2) in Definition 2.20.

The jump map G satisfies (A3) therein by construction. In fact, the map G given in (4.16) is a linear affine function of the state x. Due to continuity of $z \mapsto \kappa_q(z)$ for each q and closedness of C_P from $\mathcal{H}_P$ satisfying the hybrid basic conditions, the flow set C is closed and, with the closedness of $D_{K,0}$, the jump set D is closed. Then, $\mathcal{H}$ satisfies the hybrid basic conditions and item 1 holds.

Next, it is shown that every maximal solution to $\mathcal{H}$ from $C \cup D$ is complete. Proceeding by contradiction, suppose that there exists a maximal solution x with $x(0,0) \in C \cup D$ that is not complete. Let $(T, J) = \sup \operatorname{dom} x$ and note that since x is not complete, $T + J < \infty$. From Proposition 2.34, either item b or item c therein has to hold. Since by Assumption 4.3, the feedback κ_0 assures asymptotic stability of the point z^* with basin of attraction that contains $\mathcal{U}_0$, maximal solutions to the closed loop under the effect of κ_0 are bounded (and complete). The same assumption assures that maximal solutions to the closed loop under the effect of κ_1 are bounded (though not necessarily complete). Hence, item b in Proposition 2.34 does not hold. Item c is ruled out using item 3 in Proposition 2.34. In fact, let $x = (z, q) \in D$. If $q = 0$, then by definition of D it follows that $z \in D_{K,0} = \overline{\mathbb{R}^{n_P} \setminus \mathcal{U}_0}$ and $(z, \kappa_0(z)) \in C_P$. Since $G(x) = (z, 1)$ and, since by the properties of $\mathcal{U}_0$ and $\mathcal{T}_{1,0}$ in (4.7) the set $\mathcal{T}_{1,0}$ is (strictly) contained in $\mathcal{U}_0$, $D_{K,0} \subset C_{K,1} = \overline{\mathbb{R}^{n_P} \setminus \mathcal{T}_{1,0}}$, which leads to $z \in \overline{\mathbb{R}^{n_P} \setminus \mathcal{T}_{1,0}}$. Similarly, if $q = 1$, by definition of D it follows that $z \in D_{K,1} = \mathcal{T}_{1,0}$ and $(z, \kappa_1(z)) \in C_P$. Since $G(x) = (z, 0)$, by the definition of C it follows that $z \in C_{K,0} = \overline{\mathcal{U}_0}$. Then, using (4.7), the set $\mathcal{T}_{1,0}$ is (strictly) contained in $\mathcal{U}_0$, $D_{K,0} \subset C_{K,1} = \overline{\mathbb{R}^{n_P} \setminus \mathcal{T}_{1,0}}$ and, consequently, $z \in \overline{\mathbb{R}^{n_P} \setminus \mathcal{T}_{1,0}}$. So in either case, z is in $\overline{\mathbb{R}^{n_P} \setminus \mathcal{T}_{1,0}}$ or in $\overline{\mathbb{R}^{n_P} \setminus \mathcal{T}_{1,0}}$, and $(z, \kappa_q(z)) \in C_P$. Hence, $G(D) \subset C \cup D$.

The proof of item 2 is completed by showing that each solution x to the closed loop experiences at most two jumps. Pick a solution x to $\mathcal{H}$ and note that $x(0,0) \in C \cup D$. Only the following three cases are possible:

- If $x(0,0) \in C_{K,1} \times \{1\}$, the solution x reaches $D_{K,1}$ in finite hybrid time since, by item (UC-A2) in Assumption 4.3, the plant component z of the solution reaches $\mathcal{E}_0$ after finite flow time or as t tends to ∞, from points not in $\mathcal{E}_0$. By items (UC-A1) and (UC-A3) in Assumption 4.3, after that jump, the solution x remains flowing in $(C_{K,0} \setminus D_{K,0}) \times \{0\}$ for all future hybrid time in its domain.

- If $x(0,0) \in D_{K,1} \times \{1\}$, the solution x exhibits the same behavior as in the first bullet of this list.

- If $x(0,0) \in C_{K,0} \times \{0\}$, then the following two cases are possible for the solution x. If $x(0,0) \in \mathcal{T}_{1,0} \times \{0\}$ then, by items (UC-A1) and (UC-A3) in Assumption 4.3, the solution remains flowing in $(C_{K,0} \cap \mathcal{U}_0) \times \{0\}$ for all future hybrid time in its domain. If $x(0,0) \in (C_{K,0} \setminus \mathcal{T}_{1,0}) \times \{0\}$, then the solution x may either flow forever or reach the boundary of $\mathcal{U}_0$ triggering a jump resetting its q component to 1. From such new value, the solution has the property in the first bullet of this list.

As a consequence, every maximal solution has at most two jumps.

Attractivity in item 3 of the claim follows from the attractivity property of κ_1 in item (UC-A2) in Assumption 4.3 and the solution-based arguments above. Stability in item 3 follows from the property induced by κ_0.

Finally, robustness of the asymptotic stability of $\mathcal{A}$ follows from Theorem 3.26, since the hybrid closed-loop system satisfies the hybrid basic conditions and $\mathcal{A}$ is a (globally) asymptotically stable compact set for the nominal closed-loop $\mathcal{H}$. $\qquad\square$

Remark 4.7 (*When the set-point z^* is a set*). Theorem 4.6 is stated for the case of a set-point z^* given by a singleton. The same result holds for the case when z^* is a compact set, after appropriate adjustments of the assumptions. More precisely, denoting $\mathcal{A}^*$ as the set-point, the state-feedback law κ_0 has to render the set $\mathcal{A}^*$ asymptotically stable. The set $\mathcal{U}_0$ can be chosen to be an open neighborhood of $\mathcal{A}^*$ by, for example, defining it using the distance to the set $\mathcal{A}^*$ as $\mathcal{U}_0 := \{z \in \mathbb{R}^{n_P} : |z|_{\mathcal{A}^*} < \epsilon\}$ for some $\epsilon > 0$. $\hspace{8cm}\triangle$

The asymptotic stability property guaranteed by Theorem 4.6 hinges upon the design of the feedbacks κ_0 and κ_1, and their associated sets, satisfying the conditions in Assumption 4.3. Such a design can be performed using Lyapunov functions as introduced in Definition 3.17. This is the purpose of the next section.

4.4   DESIGN

To define the hybrid controller $\mathcal{H}_K$ in (4.8)-(4.12), the feedback laws κ_0 and κ_1 used in the flow map of the controller and the sets $\mathcal{U}_0$ and $\mathcal{T}_{1,0}$ used in the flow set and jump set of the controller have to be designed. The conditions to satisfy are those in Assumption 4.3. These conditions depend on the given point z^* and have to hold for some set $\mathcal{E}_0$ and some positive constants δ_0 and δ_0^c. In particular, as indicated in Remark 4.4, the value of δ_0 plays a key role on the robustness of the asymptotic stability property of the hybrid closed-loop system. This point is illustrated in Example 4.11 given below.

The design of κ_0 inducing asymptotic stability of z^* can be performed using Lyapunov functions. Such a design amounts to finding smooth enough functions V_0 and κ_0, with V_0 positive definite with respect to z^* such that $\dot{V}_0(z) < 0$ for all points $z \neq z^*$ in a neighborhood of z^*. When such a function is available, then $\mathcal{U}_0$ satisfying (UC-A1) could be defined in terms of V_0; e.g., a compact sublevel set of V_0.

The following design procedure is suitable when V_0 certifies local asymptotic stability with a state-feedback law κ_0, and a feedback law κ_1 induces global attractivity of a set contained in a small enough compact sublevel set of V_0.

Proposition 4.8 (Controller synthesis of κ_0 via Lyapunov-based design). *Suppose $\mathcal{H}_P$ as in (4.6) satisfies the hybrid basic conditions and that there exist a locally Lipschitz function $V_0 : \operatorname{dom} V_0 \to \mathbb{R}$ that is positive definite with respect to z^*, with $\operatorname{dom} V_0$ containing the intersection between an open neighborhood $\mathcal{U}$ of z^* and $\Pi_c(C_P)$, and a continuous function $\kappa_0 : \mathbb{R}^{n_P} \to \mathbb{R}^{m_P}$ such that* [2]

$$\dot{V}_0(z) \leq 0 \qquad \forall z \in \mathcal{U} \; : \; (z, \kappa_0(z)) \in C_P \tag{4.18}$$

$$\dot{V}_0(z) < 0 \qquad \forall z \in \mathcal{U} \setminus \{z^*\} \; : \; (z, \kappa_0(z)) \in C_P \tag{4.19}$$

---

[2]Recall that from (3.13), $\dot{V}_0$ at z is given by

$$\dot{V}_0(z) = \max_{\chi \in F_P(z, \kappa_0(z)) \cap T_{\Pi_c(C_P)}(z)} V_0^\circ(z, \chi)$$

Let $c_0 > 0$ be such that

$$\mathcal{U}_0 := \{z \in \mathbb{R}^{n_P} : V_0(z) < c_0\} \tag{4.20}$$

is bounded and $\overline{\mathcal{U}_0} \subset \mathcal{U}$. Furthermore, suppose that every maximal solution to

$$\dot{z} \in F_P(z, \kappa_0(z)) \qquad (z, \kappa_0(z)) \in C_P \; : \; z \in \overline{\mathcal{U}_0} \tag{4.21}$$

is complete. Then, for any continuous feedback law $\kappa_1 : \mathbb{R}^{n_P} \to \mathbb{R}^{m_P}$ such that for a nonempty closed set

$$\mathcal{E}_0 \subset \{z \in \mathbb{R}^{n_P} : V_0(z) < c_1\} \tag{4.22}$$

solutions to $\mathcal{H}_P$ controlled by κ_1 starting from $\overline{\Pi_c(C_P)} \setminus \mathcal{E}_0$ reach $\mathcal{E}_0$ after a finite amount of flow time or converge to it as t tends to ∞, where

$$0 < c_1 < c_0 \tag{4.23}$$

defining

$$\mathcal{T}_{1,0} := \{z \in \mathbb{R}^{n_P} : V_0(z) \leq c_{10}\}, \qquad c_{10} \in (c_1, c_0) \tag{4.24}$$

it follows that conditions (UC-A1)-(UC-A3) in Assumption 4.3 hold.

Proof. The functions κ_0 and κ_1, and the sets $\mathcal{U}_0$ and $\mathcal{T}_{1,0}$ given in the statement of the result satisfy Assumption 4.3. The function V_0 being locally Lipschitz on $\mathcal{U}$ and positive definite with respect to z^* imply that, according to Definition 3.17, V_0 is a Lyapunov function candidate on $\mathcal{U}$ with respect to the (singleton) set $\{z^*\}$ for $(C_P, F_P(\cdot, \kappa_0(\cdot)), \emptyset, \star)$. Since κ_0 is such that (4.18)-(4.19) holds, by item 1 of Theorem 3.19, z^* is pre-asymptotically stable for $(C_P, F_P(\cdot, \kappa_0(\cdot)), \emptyset, \star)$. Hence, (UC-A1) is satisfied since maximal solutions to (4.21) are complete. Item (UC-A2) holds directly by the assumption on κ_1. Finally, item (UC-A3) holds since $\mathcal{E}_0 \subset \{z \in \mathbb{R}^{n_P} : V_0(z) < c_1\}$ belongs to the interior of the bounded set $\mathcal{T}_{1,0} = \{z \in \mathbb{R}^{n_P} : V_0(z) \leq c_{10}\}$, which implies that there exists $\delta_0^c > 0$ such that $\mathcal{E}_0 + \delta_0^c \mathbb{B} \subset \mathcal{T}_{1,0}$. This property also implies that $\mathcal{T}_{1,0}$ is a subset of the bounded set $\mathcal{U}_0 = \{z \in \mathbb{R}^{n_P} : V_0(z) < c_0\}$ due to $c_1 < c_{10} < c_0$, which, in turn, implies that there exists $\delta_0 > 0$ such that $\mathcal{T}_{1,0} + 2\delta_0 \mathbb{B} \subset \mathcal{U}_0$. $\qquad \square$

The requirement in Proposition 4.8 for the design of the feedback law κ_1 does not include stability of any particular set, but rather, only attractivity. At times, it might be convenient to design that feedback using weak Lyapunov functions that, by exploiting the Hybrid Invariance Principle in Theorem 3.23, guarantees the desired attractivity property of the set $\mathcal{E}_0$.

Proposition 4.9 (Controller synthesis of κ_1 via the Hybrid Invariance Principle). *Suppose $\mathcal{H}_P$ as in (4.6) satisfies the hybrid basic conditions and that there exist V_0 and $\mathcal{U}$ such that (4.18)-(4.19) holds, and $\mathcal{U}_0$ in (4.20) is bounded and $\overline{\mathcal{U}_0} \subset \mathcal{U}$. Furthermore, suppose there exist a locally Lipschitz function $V_1 : \text{dom } V_1 \to \mathbb{R}$ that, with $\text{dom } V_1$ closed and containing a neighborhood of $\Pi_c(C_P) \setminus \mathcal{U}_0$, and a continuous function $\kappa_1 : \mathbb{R}^{n_P} \to \mathbb{R}^{m_P}$ such that*

$$\dot{V}_1(z) \leq 0 \qquad \forall z \in \text{dom } V_1 \; : \; (z, \kappa_1(z)) \in C_P \tag{4.25}$$

and the largest weakly invariant set in

$$\{z \,:\, V_1(z) = r\} \cap \left\{z \,:\, \dot{V}_1(z) = 0\right\} \cap \Pi_c(C_P) \cap \operatorname{dom} V_1, \qquad r \in \mathbb{R} \qquad (4.26)$$

denoted $\mathcal{M}_1$, is a compact subset of $\mathcal{U}_0$. With $\mathcal{E}_0$ a compact set such that $\mathcal{M}_1 \subset \mathcal{E}_0 \subset \mathcal{U}_0$, if every maximal solution to

$$\dot{z} \in F_P(z, \kappa_1(z)) \qquad (z, \kappa_1(z)) \in C_P \,:\, z \in \operatorname{dom} V_1 \qquad (4.27)$$

is complete or ends at a point in $\mathcal{E}_0$, then (UC-A2) holds and there exist $\mathcal{T}_{1,0}$ and c_1 satisfying (4.22), (4.23), and such that (UC-A3) holds.

Proof. The plant $\mathcal{H}_P$ controlled by the state-feedback law κ_1 restricted to $\operatorname{dom} V_1$ is given by (4.27). The resulting closed-loop system $\mathcal{H}_1$ has data $(C_1, F_1, \emptyset, \star)$, where

$$C_1 := \{z \in \operatorname{dom} V_1 \,:\, (z, \kappa_1(z)) \in C_P\}$$

and F_1 is defined as $F_1(z) := F_P(z, \kappa_1(z))$ for each $z \in C_1$. Hence, since $\mathcal{H}_P$ satisfies the hybrid basic conditions as in Definition 2.25, κ_1 is continuous, and $\operatorname{dom} V_1$ is closed, $\mathcal{H}_1$ satisfies the hybrid basic conditions as in Definition 2.20; see Exercise 95 and Exercise 100. The function V_1 is such that items 1 and 3 in Definition 3.17 hold with $\mathcal{U} = C_1$ – note that an extension of V_1 outside of C_1 is always possible; see §A.4. The property in (4.25) implies that (3.18) holds. Then, every precompact solution to $\mathcal{H}_1$ – or equivalently, to (4.27) – converges to the largest weakly invariant set in (4.26). This set is denoted $\mathcal{M}_1$ and is a compact subset of $\mathcal{U}_0$. Under the additional assumption on maximal solutions to (4.27), maximal solutions that are not complete also converge to a point in $\mathcal{E}_0$, after finite flow time. Hence, (UC-A2) holds. With $\mathcal{E}_0$ as in the assumption, since $\mathcal{E}_0 \subset \mathcal{U}_0$ and $\mathcal{U}_0$ is open, there exists c_1 satisfying (4.22) and (4.23), and also there exists a compact set $\mathcal{T}_{1,0}$ satisfying condition (UC-A3) for some positive constants δ_0 and δ_0^c. $\qquad\blacksquare$

A similar design method using the Hybrid Invariance Principle for κ_0 is possible. Such a design is given in the forthcoming Example 4.12.

The joint design of κ_0 and κ_1 using a strict Lyapunov function is certainly possible, via Theorem 3.19. The following result provides one such design procedure.

Proposition 4.10 (Controller synthesis of κ_0 and κ_1 via Lyapunov-based design).
Suppose $\mathcal{H}_P$ as in (4.6) satisfies the hybrid basic conditions and that there exist a locally Lipschitz function $V_0 : \operatorname{dom} V_0 \to \mathbb{R}$ that is positive definite with respect to z^*, with $\operatorname{dom} V_0$ containing the intersection between an open neighborhood $\mathcal{U}$ of z^* and $\Pi_c(C_P)$, a positive definite function $\rho_0 : \operatorname{dom} V_0 \to \mathbb{R}_{\geq 0}$, and a continuous function $\kappa_0 : \mathbb{R}^{n_P} \to \mathbb{R}^{m_P}$ such that

$$\dot{V}_0(z) \leq -\rho_0(|z - z^*|) \qquad \forall z \in \mathcal{U} \,:\, (z, \kappa_0(z)) \in C_P \qquad (4.28)$$

Furthermore, suppose there exists a locally Lipschitz function $V_1 : \operatorname{dom} V_1 \to \mathbb{R}$ that is positive definite with respect to z^* and has compact sublevel sets, with $\operatorname{dom} V_1$ containing a neighborhood of $\Pi_c(C_P)$, a positive definite function $\rho_1 : \operatorname{dom} V_1 \to \mathbb{R}_{\geq 0}$, and a continuous function $\kappa_1 : \mathbb{R}^{n_P} \to \mathbb{R}^{m_P}$ such that

$$\dot{V}_1(z) \leq -\rho_1(|z - z^*|) \qquad \forall z \,:\, (z, \kappa_1(z)) \in C_P \qquad (4.29)$$

and every maximal solution to $\mathcal{H}_P$ controlled by κ_1 is complete. Then, for any $c_0 > c_{10} > 0$ defining $\mathcal{U}_0$ and $\mathcal{T}_{1,0}$ in (4.20) and in (4.24), respectively, such that

$$\overline{\mathcal{U}_0} \subset \mathcal{U}$$

and every maximal solution to $\mathcal{H}_P$ controlled by κ_0 starting from $\overline{\mathcal{U}_0}$ is complete, it follows that there exists a compact set $\mathcal{E}_0$ such that the conditions (UC-A1)-(UC-A3) in Assumption 4.3 hold. Alternatively, $\mathcal{T}_{1,0}$ can be chosen as a sublevel set of V_1 that is a subset of $\mathcal{U}_0$.

Next, the design tools presented above and some of their immediate extensions are applied to the uniting control problems in the examples presented in § 4.1.

Example 4.11 (Disk drive control, revisited). *The design tools presented in this section are employed to solve the uniting control problem formulated in Example 4.1. Suppose these controllers are given by the static state-feedback laws κ_0 and κ_1, which are continuous and represent the local and global controller, respectively, for which functions V_0 and V_1 certifying the properties stated in Proposition 4.10 hold. Since (4.2) is a double integrator and the state is measured, these functions are guaranteed to exist. The design is as follows:*

* *The set $\mathcal{U}_0$ is defined by a strict c_0-sublevel set of V_0 with $c_0 > 0$ such that it is contained in the basin of attraction $\mathcal{B}^0_{z^*}$ induced by κ_0, i.e., using (4.20),*

$$\mathcal{U}_0 := \left\{ z \in \mathbb{R}^4 : V_0(z) < c_0 \right\}, \qquad \overline{\mathcal{U}_0} \subset \mathcal{B}^0_{z^*}$$

  This is the construction given in Proposition 4.8.
* *The set $\mathcal{T}_{1,0}$ is defined by a $c_{1,0}$-sublevel set of V_1 with $c_{1,0} > 0$ chosen so that $\mathcal{T}_{1,0}$ is contained in the interior of $\mathcal{U}_0$, i.e., following Proposition 4.10,*

$$\mathcal{T}_{1,0} := \left\{ z \in \mathbb{R}^4 : V_1(z) \le c_{1,0} \right\} \subset \left\{ z \in \mathbb{R}^4 : V_0(z) < c_0 \right\} = \mathcal{U}_0 \qquad (4.30)$$

  This choice of $\mathcal{T}_{1,0}$ is possible since $\mathcal{U}_0$ is bounded and the sublevel sets of V_1 are compact. To satisfy the conditions in Proposition 4.10, the set $\mathcal{E}_0$ is chosen as a c_1-sublevel set of V_1 with $c_1 \in (0, c_{1,0})$, i.e.,

$$\mathcal{E}_0 := \left\{ z \subset \mathbb{R}^4 : V_1(z) \le c_1 \right\} \qquad (4.31)$$

The conditions in Assumption 4.3 hold. Condition (UC-A1) holds since κ_0 is a local stabilizer and $\overline{\mathcal{U}_0}$ is a subset of its basin of attraction. Condition (UC-A2) is satisfied since κ_1 renders globally attractive the point z^* with V_1 decreasing along solutions – in fact, it renders z^* globally asymptotically stable. So solutions to the plant under the effect of κ_1 converge (in finite time) to $\mathcal{E}_0$ since it is a sublevel set of V_1 and z^* is in the interior of $\mathcal{E}_0$. Finally, condition (UC-A3) holds since: i) $\mathcal{E}_0$ and $\mathcal{T}_{1,0}$ are both compact sublevel sets of V_1 with levels c_1 and $c_{1,0}$ satisfying $0 < c_1 < c_{1,0}$; and ii) $\mathcal{T}_{1,0}$ is contained in the interior of $\mathcal{U}_0$. In fact, since $0 < c_1 < c_{1,0}$, the sets $\mathcal{U}_0$, $\mathcal{E}_0$, and $\mathcal{T}_{1,0}$ are compact, and

$$\mathcal{E}_0 + \delta_0^c \mathbb{B} = \left\{ z \in \mathbb{R}^4 : V_1(z) \le c_1 \right\} + \delta_0^c \mathbb{B} \subset \mathcal{T}_{1,0} = \left\{ z \in \mathbb{R}^4 : V_1(z) \le c_{1,0} \right\}$$

holds when δ_0^c *is chosen as*

$$\delta_0^c \in \left(0, \min_{\chi_1 \in V_1^{-1}(c_1), \chi_2 \in V_1^{-1}(c_{1,0})} |\chi_1 - \chi_2|\right]$$

Similarly,

$$\mathcal{T}_{1,0} + 2\delta_0 \mathbb{B} = \left\{z \in \mathbb{R}^4 : V_1(z) \leq c_{1,0}\right\} + 2\delta_0 \mathbb{B} \subset \mathcal{U}_0 = \left\{z \in \mathbb{R}^4 : V_0(z) < c_0\right\}$$

holds when δ_0 *is chosen as*

$$\delta_0 \in \left(0, \frac{1}{2} \min_{\chi_1 \in V_1^{-1}(c_{1,0}), \chi_2 \in V_0^{-1}(c_0)} |\chi_1 - \chi_2|\right]$$

The definitions above synthesize the uniting controller in (4.8)-(4.12). *The resulting closed-loop system when applied to* (4.2) *has state* $x = (z, q) \in \mathbb{R}^4 \times Q$ *and is given by*

$$\mathcal{H} : \begin{cases} (z, q) \in (\{z \in \mathbb{R}^4 : V_0(z) \leq c_0\} \times \{0\}) \cup & \begin{bmatrix} \dot{z} \\ \dot{q} \end{bmatrix} = \begin{bmatrix} \begin{bmatrix} z_2 \\ \kappa_q(z) \end{bmatrix} \\ 0 \end{bmatrix} \\ \quad (\{z \in \mathbb{R}^4 : V_1(z) \geq c_{1,0}\} \times \{1\}) & \\ (z, q) \in (\{z \in \mathbb{R}^4 : V_0(z) \geq c_0\} \times \{0\}) \cup & \begin{bmatrix} z^+ \\ q^+ \end{bmatrix} = \begin{bmatrix} z \\ 1 - q \end{bmatrix} \\ \quad (\{z \in \mathbb{R}^4 : V_1(z) \leq c_{1,0}\} \times \{1\}) & \end{cases}$$

$$(4.32)$$

With this construction, solutions using the track-following control law κ_0 *that start in* $\mathcal{T}_{1,0} = \{z \in \mathbb{R}^4 : V_1(z) \leq c_{1,0}\}$ *do not reach the boundary of* $\mathcal{U}_0$. *Figure 4.5 illustrates these sets. To prevent chattering between the two controllers in the intersection of* $C_{K,0} = \{z \in \mathbb{R}^4 : V_0(z) \leq c_0\}$ *and* $C_{K,1} = \{z \in \mathbb{R}^4 : V_1(z) \geq c_{1,0}\}$, *the uniting algorithm allows mode switching when* z *is in the closed complement of these sets. In other words, a switch from the track-seeking mode* $(q = 1)$ *to the track-following mode* $(q = 0)$ *may occur when* $z \in \mathcal{T}_{1,0}$, *while a switch from the track-following mode to the track-seeking mode may occur when* $z \in \{z \in \mathbb{R}^4 : V_0(z) \geq c_0\}$. *Finally, in nominal conditions, the maximum number of jumps a solution to* (4.32) *experiences is two. In fact, if there are more than one jump for a closed-loop solution* x *then one of the first two jumps must be from* $q = 1$ *to* $q = 0$. *When such a jump occurs, it must be that the* z *component of the solution is in* $\mathcal{T}_{1,0}$ *and, since* z *does not change at jumps,* z *is in* $\mathcal{T}_{1,0}$ *after the jump. Since a solution to* (4.2) *under the effect of* κ_0 *that starts in* $\mathcal{T}_{1,0}$ *cannot reach* $\{z \in \mathbb{R}^4 : V_0(z) \geq c_0\}$, *the solution* x *does not jump again after a jump from* $q = 1$ *to* $q = 0$.

Example 4.12 (Heavy ball method, revisited). *The design of the parameters of the optimization algorithm in Example 4.2 is performed using the function*

$$V_q(z) := \gamma_q L(z_1) + \frac{1}{2} z_2^2 \tag{4.33}$$

defined for each $q \in Q$ *and each* $z \in \mathbb{R}^2$. *Suppose that* L *is positive definite with respect to* $\mathcal{A}^* = \{z \in \mathbb{R}^2 : \nabla L(z_1) = z_2 = 0\} = \{(z_1^*, 0)\}$ *and has compact sublevel sets. Furthermore, by construction, for each* $q \in Q$, *the composition of* F_P *with* κ_q *in* (4.5) *is continuously differentiable leading to a closed-loop system given by a*

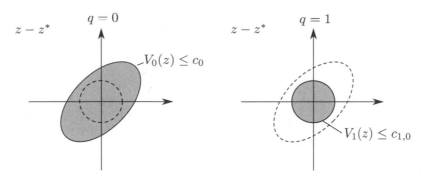

Figure 4.5: Depiction of sets associated to the uniting hybrid controller for the disk drive control problem in Example 4.11 when $z - z^*$ is treated as two dimensional.

continuous-time system $\mathcal{H}_q$ with data $(\mathbb{R}^2, F_P(\cdot, \kappa_q(\cdot)), \emptyset, \star)$. It follows that for each $q \in Q$, V_q satisfies

$$\dot{V}_q(z) = \langle \nabla V_q(z_1), F_P(z, \kappa_q(y)) \rangle = -\lambda_q z_2^2 \qquad \forall z \in \mathbb{R}^2 \qquad (4.34)$$

By an application of Theorem 3.19, for each $q \in Q$ and using $\mathcal{U} = \mathbb{R}^2$, $\mathcal{A} = \mathcal{A}^* = \left\{ z \in \mathbb{R}^2 : \nabla L(z_1) = z_2 = 0 \right\} = \{(z_1^*, 0)\}$, and V_q in (4.33), the set $\mathcal{A}^*$ is stable for the closed-loop system $\dot{z} = F_P(z, \kappa_q(y))$ if γ_q is positive and λ_q is nonnegative. When λ_q is positive, the largest weakly invariant set that is contained in

$$\left\{ z \in \mathbb{R}^2 : V_q(z) = r_q \right\} \cap \left\{ z \in \mathbb{R}^2 : \dot{V}_q(z) = 0 \right\}, \qquad r_q \geq 0 \qquad (4.35)$$

for the closed-loop system $\dot{z} = F_P(z, \kappa_q(h(z)))$ is equal to $\mathcal{A}^*$. This property of the largest weakly invariant set is also true for each $q \in Q$, and, very importantly, holds for $r_q = 0$. To see this, notice that $\left\{ z \in \mathbb{R}^2 : \dot{V}_q(z) = 0 \right\} = \{z \in \mathbb{R}^2 : z_2 = 0\}$, and that after setting z_2 to zero in the closed-loop system, the z_1 component of its solutions satisfy $0 = \gamma_q \nabla L(z_1)$, which since $\gamma_q > 0$ and L has only one critical point at z_1^*, leads to $z_1 = z_1^*$. Then, the only maximal solution that starts and stays in (4.35) is the solution from $(z_1^*, 0)$, which remains at $(z_1^*, 0)$ for all time. As argued above, maximal solutions to $\mathcal{H}_q$ are already bounded. Completeness of such solutions follows from the fact that $z \mapsto F_P(z, \kappa_q(h(z)))$ is continuously differentiable; e.g., via Proposition 2.34. Then, an application of item 2d of Theorem 3.19 implies that, for each $q \in Q$, $\mathcal{A}^*$ is globally asymptotically stable for $\mathcal{H}_q$.

Having these properties of the individual controllers established, the design of the uniting control strategy follows using the constructions provided in the design tools, as in Example 4.1:

- *The set $\mathcal{U}_0$ is defined by the strict c_0-sublevel set of V_0 given in (4.20), with $c_0 > 0$. The set $\mathcal{U}_0$ defines the region of the space where parameters λ_0 and γ_0 of the heavy ball method are to be used. In this design, λ_0 might be small to avoid oscillations when converging to $\mathcal{A}^*$. Note that $\overline{\mathcal{U}_0}$ is contained in the basin of attraction induced by κ_0 due to the global asymptotic stability property it guarantees. Due to the use of the Hybrid Invariance Principle, this design for κ_0 and its associated sets uses the ideas in Proposition 4.9.*

- *The set $\mathcal{T}_{1,0}$ is defined by a $c_{1,0}$-sublevel of V_1 with $c_{1,0} > 0$ chosen so that $\mathcal{T}_{1,0}$ is contained in the interior of $\mathcal{U}_0$. This construction is given in (4.30). This choice of $\mathcal{T}_{1,0}$ is possible since $\mathcal{U}_0$ and the sublevel sets of V_1 are compact. The set $\mathcal{E}_0$ is chosen as a c_1-sublevel of V_1 with $c_1 \in (0, c_{1,0})$. This construction is given in (4.31) and follows from Proposition 4.9.*

The conditions in Assumption 4.3 hold via Proposition 4.9 (and its immediate extension for the design of κ_0). In fact, conditions (UC-A1) and (UC-A2) hold since both κ_0 and κ_1 globally asymptotically stabilize $\mathcal{A}^*$. So maximal solutions to the plant under the effect of κ_1, namely, maximal solutions to $\mathcal{H}_1$, converge to $\mathcal{E}_0$ since it is a sublevel set of V_1 . Condition (UC-A3) holds since the sets are constructed as in Proposition 4.9. The resulting closed loop with this uniting controller is essentially the same as the one in Example 4.1 – see (4.32) – with only difference being that z during flows is governed by $F_P(z, \kappa_q(z))$ and the plant output is h in (4.4).

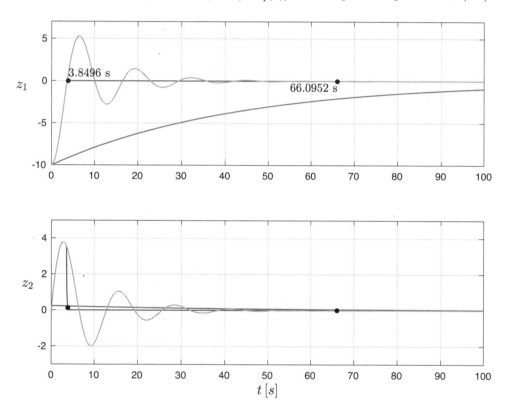

Figure 4.6: A comparison of the $z = (z_1, z_2)$ components of the solutions to the closed-loop systems $\mathcal{H}_1$, which uses (λ_1, γ_1) and is oscillatory, $\mathcal{H}_0$, which uses (λ_0, γ_0) and is slow, and $\mathcal{H}$, which is due to the hybrid algorithm. For comparison, the times at which each solution is almost at its steady state are shown.

To show the effectiveness of the hybrid algorithm, the individual optimization algorithms are compared. The function L employed is $L(z_1) = \frac{1}{4}z_1^2$, leading to $\mathcal{A}^* = \{(z_1^*, 0)\} = (0, 0)$. None of the algorithms is designed using the actual expression of the function L, or of the location of its minima, but they have access to the values of L and ∇L at the current value of z_1. The local algorithm, κ_0, leading to the closed *loop $\mathcal{H}_0$, uses a large λ_0 value to produce slow convergence with no oscillations. The simulation shown in Figure 4.6 for this algorithm employs $\lambda_0 = 10.5$ and $\gamma_0 = \frac{1}{2}$. The resulting state z is the oscillatory trajectory, which settles to within a 1% margin of $\mathcal{A}^*$ in about 193.1 sec (not depicted). The global algorithm, κ_1, leading to the closed loop $\mathcal{H}_1$, uses a small λ value to produce fast convergence with oscillations. The simulation shown in Figure 4.6 for this algorithm uses $\lambda_1 = \frac{1}{5}$ and $\gamma_1 = \frac{1}{2}$, and corresponds to the slow trajectory that settles to within a 1% margin of $\mathcal{A}_1$ in about 45.9 sec. The uniting controller implemented by the hybrid closed-loop system $\mathcal{H}$ switches between the values λ_0 and λ_1 to ensure fast convergence to the minima without oscillations. The parameters of the hybrid algorithm are $c_0 = 12.5$ and $c_{1,0} = 6.3$. The state z resulting from simulating $\mathcal{H}$ is shown in Figure 4.6, which is the one that settles to within a 1% margin of $\mathcal{A}^*$ in about 3.6sec.*

The following example pertains to obstacle avoidance. A control problem where multiple obstacles need to be avoided to guarantee safety is depicted in the cover photo of this book. As also shown in Figure 4.7, the goal of each kayak is to *flow* down the river, via a waterfall, without hitting the large boulders sticking off the water or the bedrock below the water in shallow parts of the river. In addition the kayaks need to eventually *jump* into a waterfall, safely. The following example also illustrates how the proposed uniting control strategy can be extended in several directions. It considers the case when the individual controllers render attractive and asymptotically stable different set-points. It also motivates the use of hybrid controllers as individual controllers, which is a topic that is addressed in Chapter 8.

Figure 4.7: Kayakers avoiding multiple obstacles while flowing in a river and jumping into a waterfall. *Image credit: Kayakers on the Soča River, Slovenia / Janossy Gergely / Shutterstock.*

Example 4.13 (Obstacle avoidance with target). *Consider the problem of globally stabilizing the position of an autonomous vehicle to a target location while avoiding an obstacle. Figure 4.8 depicts such a setting. The target location is × and the* *obstacle is denoted as $\mathcal{O}$. The goal is to drive the autonomous vehicle, which for simplicity is modeled as a fully actuated point-mass, from any initial position to the location of the target while avoiding the obstacle. This goal is similar to that of the kayakers shown in the cover photo of this book, also shown in Figure 4.7. The kayaker that is already down the river has cleared the boulders seen at the top of the figure. Those boulders could be represented by the set $\mathcal{O}$. The kayaker is entering the waterfall from a safe location, which can be represented by $\times$.*

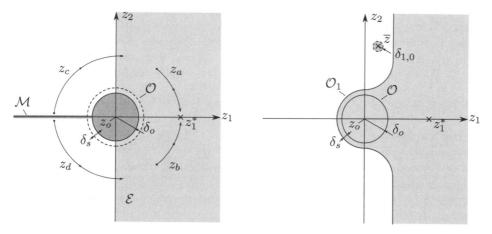

Figure 4.8: Sets and parameters associated with the obstacle avoidance control problem in Example 4.13.

Suppose that there exists a static state-feedback control law, denoted $\widetilde{\kappa}_1$, that achieves stability of and "global" convergence to the target – in the sense that for every point not in $\mathcal{O}$, the position of the vehicle converges to the target. Furthermore, and for simplicity of exposition, suppose that the solutions to the resulting closed-loop system are unique and once they reach some set $\mathcal{E}$, like the one depicted in Figure 4.8, they converge to the target when the static state-feedback law $\widetilde{\kappa}_0$ is used. Due to the topological obstructions associated to this problem (see also § 1.2.4), as illustrated in Figure 4.8, there exists a set of points $\mathcal{M}$ such that for initial conditions above or below it, solutions converge to the right-half plane of the position space, either from above the obstacle or from below it. As a consequence, the static state-feedback control law is a discontinuous function of the state. It follows that for initial conditions arbitrarily close to the line $\mathcal{M}$, there exists a noise signal w_y for the measurements of the vehicle position such that there exist closed-loop solutions that stay in a neighborhood of the line $\mathcal{M}$ for all time. Eventually, if the forward velocity of the vehicle is positive, such solutions would correspond to the vehicle crashing onto the obstacle. This issues is explained in § 1.2.4.

A uniting controller rendering the target location globally asymptotically stable for the position of the autonomous vehicle is designed. The point-mass vehicle model is given by a plant $\mathcal{H}_P$ with continuous dynamics defined by the data

$$F_P(z, u) := u \qquad \forall (z, u) \in C_P := \mathbb{R}^2 \times \mathbb{R}^2, \qquad h := Id \qquad (4.36)$$

and without discrete dynamics, where $z \in \mathbb{R}^2$ is the state denoting the position of the vehicle on the plane and $u \in \mathbb{R}^2$ is the control input. The target is at the location z^*. The center of the obstacle $\mathcal{O}$ is at the location z_o and is modeled as a (closed) ball of radius $\delta_o > 0$. Without loss of generality, these locations are taken *as* $z^* = (z_1^*, 0)$ *with* $z_1^* > \delta_o$ *and* $z_o = (0,0)$. *Then,* $\mathcal{O} := \delta_o \mathbb{B}$ *and, according to Figure 4.8,* $\mathcal{E}$ *is taken as* $\overline{(\mathbb{R}_{\geq 0} \times \mathbb{R}) \setminus \mathcal{O}}$. *Next, the individual controllers,* $\mathcal{H}_{K,0}$ *and* $\mathcal{H}_{K,1}$, *are designed.*

The "local" controller $\mathcal{H}_{K,0}$ *consists of a static state-feedback law* κ_0 *asymptotically stabilizing* z^* *with basin of attraction containing* $\mathcal{E}$ *and rendering* $\mathcal{E}$ *forward invariant. One such control law may is the static state-feedback control law* $\widetilde{\kappa}_0$ *mentioned above. Another suitable design of the control law* κ_0 *is given next. Define*

$$V_0(z) := \frac{1}{2}(z - z^*)^\top (z - z^*) + B(d_o(z)) \qquad \forall z \in \overline{\mathbb{R}^2 \setminus (\mathcal{O} + \delta_s \mathbb{B})} \qquad (4.37)$$

where $\delta_s > 0$, *given* $\delta_v > \delta_s$, $B : \mathbb{R}_{>0} \to \mathbb{R}$ *is a function defined as*

$$B(\xi) := \begin{cases} (\xi - \delta_v)^2 \ln \frac{1}{\xi} & \text{if } \xi \in (0, \delta_v] \\ 0 & \text{if } \xi > \delta_v \end{cases}$$

and $d_o : \mathbb{R}^2 \to \mathbb{R}_{\geq 0}$ *is defined as*

$$d_o(z) := \begin{cases} |z| - \delta_o & \text{if } |z| > \delta_o \\ 0 & \text{otherwise} \end{cases}$$

The construction of these functions is such that $z \mapsto B(d_o(z))$ *is continuously differentiable on an open neighborhood of* $\overline{\mathbb{R}^2 \setminus (\mathcal{O} + \delta_s \mathbb{B})}$. *The function* d_o *measures the distance from any point in* $\mathbb{R}^2$ *to the obstacle* $\mathcal{O}$. *The parameter* δ_s *is a safety margin defining the minimum safety distance to the obstacle. In this way,* V_0 *in (4.37) is well defined. The parameter* δ_v *determines when the term* $B(d_o(z))$ *in (4.37) vanishes: when* $d_o(z) > \delta_v$, *meaning that the vehicle is at least a distance* δ_v *from the obstacle, then this term is zero. With* V_0 *as defined above being continuously differentiable,* $\mathcal{H}_{K,0} = (C_{K,0}, F_{K,0}, D_{K,0}, G_{K,0}, \kappa_0)$ *is given by* κ_0 *defined as the gradient descent law*

$$\kappa_0(z) := -\nabla V_0(z) \qquad \forall z \in \overline{\mathcal{E} \setminus (\mathcal{O} + \delta_s \mathbb{B})} =: C_{K,0} \qquad (4.38)$$

and no discrete dynamics – namely, $D_{K,0}$ *is empty and* $G_{K,0}$ *is arbitrary. The closed-loop system resulting with this control law is a gradient system on* $C_{K,0}$ *with* z^* *(globally) asymptotically stable.*

The "global" controller, $\mathcal{H}_{K,1}$, *is designed as a static state-feedback* κ_1 *that steers* z *to a particular point in* $C_{K,0}$ *while avoiding the obstacle. One such control law is given as follows: as depicted in Figure 4.8,*

- *Let the parameter* $\overline{z} = (\overline{z}_1, \overline{z}_2)$ *belong to the interior of* $\mathcal{E}$ *with* $\overline{z}_2 > \delta_o + \delta_s$ *and* $\overline{z}_1 \in (0, \delta_o)$.
- *Pick* $\delta_1 > 0$ *such that*

$$\overline{z} + \delta_1 \mathbb{B} \subset \mathcal{E} \qquad (4.39)$$

- *Define*
$$\kappa_1(z) := -\nabla V_1(z) \qquad \forall z \in C_{K,1} := \overline{\mathbb{R}^2 \setminus \mathcal{O}_1}$$

where $V_1(z) := \frac{1}{2}(z - \overline{z})^\top (z - \overline{z}) + B(d_1(z))$ *for each* $z \in C_{K,1}$, *with* $d_1(z)$ *being the distance from* z *to* $\mathcal{O}_1$, *which is the set of points defined as*

$$\mathcal{O}_1 = \mathcal{O} \cup ((\mathbb{R}_{\geq 0} \times \mathbb{R}) \setminus (\mathcal{N}(-1) \cup \mathcal{N}(1))) \qquad (4.40)$$
$$\mathcal{N}(r) = (r\{(0, 2(\delta_o + \delta_s))\} + (\delta_o + \delta_s)\mathbb{B}) \cup ([0, \delta_o + \delta_s] \times r[2(\delta_o + \delta_s), \infty))$$

This controller is such that the resulting closed loop is also a gradient system on $C_{K,1}$ with $\bar{z}$ (globally) asymptotically stable.

With the individual controllers $\mathcal{H}_{K,0}$ and $\mathcal{H}_{K,1}$ designed, the sets $\mathcal{U}_0$ and $\mathcal{T}_{1,0}$ of the uniting control strategy are chosen as

$$\mathcal{U}_0 := \left\{ z \in \mathbb{R}^2 : z_1 > -1 \right\} \supset C_{K,0}, \qquad \mathcal{T}_{1,0} := \bar{z} + \delta_{1,0} \mathbb{B} \qquad (4.41)$$

with $\delta_{1,0} > 0$ such that $\mathcal{T}_{1,0}$ is in the interior of $\mathcal{E}$. Item (UC-A1) of Assumption 4.3 holds since every maximal solution from $C_{K,0}$ is complete and converges (exponentially fast) to z^*. Item (UC-A2) holds with $\mathcal{E}_0 = \bar{z} + \delta'_{1,0} \mathbb{B}$, $\delta'_{1,0} \in (0, \delta_{1,0})$. Finally, the conditions in (4.7) of item (UC-A3) hold by construction.

Note that while the control law κ_1 proposed above is a static state-feedback control law, one such control law would not solve the problem globally (and robustly) when one insists on having the vehicle reaching set $\mathcal{E}$ from above when initially $z_1 > 0$ and from below when initially $z_1 < 0$. This difficulty is due to the topological obstruction mentioned above and motivates the use of a hybrid controller $\mathcal{H}_{K,1}$ in the uniting strategy in (4.8)-(4.12) in place of the static one therein; see Exercise 33.

4.5 EXERCISES

Exercise 27 (Global stabilization on the line). Consider the scalar system

$$\dot{z} = u, \qquad y = h(z) := \begin{cases} 2 - \cos\left(\frac{\pi}{2}(z-1)\right) & \text{if } z \geq 1 \\ z & \text{if } |z| < 1 \\ -2 - \cos\left(\frac{\pi}{2}(z-1)\right) & \text{if } z \leq -1 \end{cases} \qquad (4.42)$$

Design a controller that globally asymptotically stabilizes its origin. *Hint: use linear controller when $|z|$ is small.*

Exercise 28 (Global stabilization with dual specifications). Consider the single-input single-output system governed by

$$\ddot{y} + \dot{y} + y = u \qquad (4.43)$$

where $u \in \mathbb{R}$ is the input and $y \in \mathbb{R}$ is the output. Design a controller that satisfies the following specifications:

- The overshoot of the step response to the closed-loop system is less than or equal to 20 % of the final value; and

- The rise time to the closed-loop system is less than or equal to 300msec;

for initial conditions starting at zero. Validate your design numerically.

Exercise 29 (Global stabilization on the unit circle). Consider the system on the unit circle given in § 1.2.3.

1. For $z^* = \mathbf{1}$, where $\mathbf{1}$ denotes the vector $(1,0) \in \mathbb{S}^1$ – see List of Symbols – and the hybrid controller given in Exercise 15:

   a) Show that it can be rewritten as a uniting controller with logic state $q \in$

$\{0, 1\}$ such that the hybrid closed-loop system satisfies the hybrid basic conditions.

b) Design the controller so that the set $\mathcal{A} := \{z^*\} \times \{0\}$ is globally asymptotically stable.

c) Simulate the hybrid closed-loop system from initial conditions for $(z, q) \in \mathbb{S}^1 \times \{0, 1\}$ given by $((-1, 0), 1)$, $((0, 1), 0)$, $((0, -1), 0)$, and $((1, 0), 1)$.

2. Extend the hybrid controller to globally asymptotically stabilize a given point $z^* \in \mathbb{S}^1$.

Exercise 30 (Uniting control with local output feedback). Consider the system

$$\dot{z} = \begin{bmatrix} -z_1 - z_2 z_1^2 + u_1 \\ -z_2 + z_1^2 + u_2 \end{bmatrix}, \tag{4.44}$$

where $z \in \mathbb{R}^2$ is the state and $u = (u_1, u_2) \in \mathbb{R}^2$ is the control input.

1. Design a static feedback controller that locally asymptotically stabilizes the origin using measurements of z_1 only. *Hint: choose V_0 as a function that is quadratic in the state of the plant.*

2. Design a static state-feedback controller that globally asymptotically stabilizes the origin – in this case, the controller can use measurements of z.

3. Design a hybrid controller that unites the two controllers so that the locally stabilizing one is used in a neighborhood of the origin and the globally stabilizing one everywhere else. Define the output of the plant providing the minimum amount of information about the state of the plant.

4. Validate your design numerically.

Exercise 31 (Uniting control for the double integrator). Consider the double integrator system on the plane given by

$$\dot{z}_1 = z_2, \quad \dot{z}_2 = u, \tag{4.45}$$

where $z = (z_1, z_2)$ is the state and u is the control input with the constraint $|u| \leq 1$.

1. Design a static state-feedback law that steers the state to the origin in minimum time. *Hint: consider a discontinuous control law $\kappa_0 : \mathbb{R}^2 \to \{-1, 1\}$.*

2. Design a static state-feedback law that locally asymptotically stabilizes the origin.

3. Design a hybrid controller that unites the two controllers so that the locally stabilizing one is used in a neighborhood of the origin and the globally stabilizing one everywhere else.

4. Comment on the hybrid closed-loop system satisfying the hybrid basic conditions.

5. Validate your design numerically.

Exercise 32 (Uniting control for the pendulum system). Consider the pendulum system given by the nonlinear model

$$\dot{\theta} = \omega, \qquad \dot{\omega} = -\frac{\gamma}{\ell_p} \sin(\theta) + u$$

where $\theta \in [-\pi/2, \pi/2]$ denotes the angle, $\omega \in \mathbb{R}$ the angular velocity, and u the torque input. The positive constants γ and ℓ_p denote the gravity constant and the length of the pendulum.

1. Design a static state-feedback law that globally asymptotically stabilizes a given $\theta^* \in (\pi/2, \pi/2)$ with zero angular velocity.

2. Design a static state-feedback law that locally asymptotically stabilizes the origin with minimum overshoot.

3. Design a hybrid controller that unites the two controllers so that the locally stabilizing one is used in a neighborhood of the origin and the globally stabilizing one everywhere else.

4. Validate your design numerically.

Exercise 33 (Obstacle avoidance with target). For the uniting problem in Example 4.13:

1. Show that the control law κ_0 in (4.38) globally asymptotically stabilizes z^*.

2. Show that the control law κ_1 in (4.38) globally asymptotically stabilizes $\bar{z}$.

3. Validate the design of the uniting controller numerically.

4. Redesign the control algorithms so that the vehicle reaches set $\mathcal{E}$ from above the obstacle when $z_1 > 0$ and from below it when $z_1 < 0$. Make sure the resulting closed-loop system satisfies the hybrid basic conditions. *Hint: Add a logic variable to $\mathcal{H}_{K,0}$ so that the set of points $\{-\bar{z}\} \cup \{\bar{z}\}$ is attractive from points not in $\mathcal{O}_1$ as in (4.40).*

Exercise 34 (Global stabilization of a differential inclusion). Consider the differential inclusion

$$\dot{z} \in F_P(z, u) := [-\rho_1(z), \rho_1(z)] + \rho_2(u)$$

where $z, u \in \mathbb{R}$. The function ρ_1 is continuously differentiable and, for some $\gamma \in (0, 1)$, such that

$$\rho_1(z) = 1 \quad \text{if} \quad |z| > \frac{1}{\gamma}, \qquad \rho_1(z) = 0 \quad \text{if} \quad |z| < \gamma$$

The function ρ_2 is the saturation function given by

$$\rho_2(u) = u \quad \text{if} \quad |z| < 2\gamma, \qquad \rho_2(u) = 2\gamma \text{sign}(z) \quad \text{if} \quad |z| \geq 2\gamma$$

Design a controller that globally asymptotically stabilizes the origin of the differential inclusion. *Hint: Use a linear controller when $|z|$ is small and design a controller to dominate ρ_1 when $|z|$ is large.*

Exercise 35 (Global stabilization using a linear quadratic regulator). Consider the system in Exercise 32. For its linearization around the origin, namely, zero angle and angular velocity, unitary parameters, and with the input constrained to $[-1, 1]$, design a globally stabilizing uniting controller satisfying the following properties:

- Nearby the origin, the uniting controller uses a linear quadratic regulator that is designed to guarantee that the input does not reach the boundary of $[-1, 1]$;

- Far away from the origin, uses a feedback controller that guarantees convergence as fast as possible to the neighborhood where the linear quadratic regulator operates.

4.6 NOTES

Early precursors of the idea of uniting two controllers include the work in [34], where a dynamic time-invariant controller is proposed to combine a predesigned local controller with a global set-point stabilizing controller. Appearing around a similar time, [106] proposed a continuous static time-invariant controller given by a state-feedback law that, when a "continuous path" between the local and global controllers exists, combines the two controllers by smoothly "blending" them. In [107], several solutions to the uniting control problem are proposed, specifically, using static, dynamic (with hysteresis), and periodic controllers. In [36], this problem is solved by patching together a local optimal controller and a global controller designed using backstepping. It should be noted that the strategies in [34], [108], and [109] piece the individual controllers together using logic variables, leading to a control scheme with mixed discrete/continuous dynamics. Related control algorithms include the trajectory-based approach for the design of robust multi-objective controllers that regulate a particular output to zero while keeping another output within a prescribed limit that was introduced in [110]. In the context of performance, a trajectory-based approach was also employed in [111] to generate dwell-time and hysteresis-based control strategies that guarantee an input-output stability property characterizing closed-loop system performance.

The uniting control strategy introduced in this chapter is based on the references mentioned above. The construction of the hybrid controller proposed in this chapter is an alternative to the one in [1, Example 3.23] using nested Lyapunov functions. An extension to the case of asymptotically stabilizing the plant to a general set $\mathcal{A}^*$ rather than a point z^* is immediate. The case of $\mathcal{H}_P$ having hybrid dynamics can be treated similarly and follows using natural extensions of the assumptions. The design conditions in Proposition 4.8, Proposition 4.9, and Proposition 4.10 provide specific design conditions guaranteeing that each of the feedback controllers can be united, but other designs are possible.

A version of the model in Example 4.1 appeared in [11]. The model for the heavy ball algorithm in Example 4.2 is standard and follows from the optimization literature; see [112] and references therein. The model and control problem in Example 4.13 is inspired by [32].

The uniting control strategy in this chapter has also been extended in [38] to the case when, rather than state-feedback, only output-feedback continuous-time controllers are available. More recently, in [113], the extension for the case when the individual controllers are hybrid and use only output measurements was proposed. Further extensions of the uniting strategy are presented in the next two chapters.

Chapter Five

Event-Triggered Control

Numerous control algorithms are hybrid due to the presence of events. Such events might be triggered when the state variables, the input, and the output of the plant or of the controller satisfy certain conditions. At the events, variables of the control algorithm might be instantaneously updated to new values. For instance, the events in a sample-and-hold controller correspond to the sampling of the output of the plant and to the update of its control input to a new value; see Example 1.3. These events are typically triggered periodically by a timer. In network control systems, the events correspond to transmission and reception of information. These events are typically triggered by network protocols and occur at independent time instances that are not necessarily periodic; see Example 2.24. In event-triggered control strategies, events occur when internal variables, inputs, or outputs indicate that a particular quantity has reached a threshold; see Example 2.10. The quantity, the threshold, and the feedback law are usually jointly designed to maximize the time elapsed in between events – called the *inter-event times* – so as to optimize the computational resources in the system.

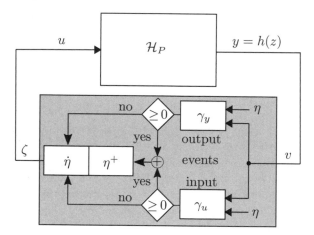

Figure 5.1: Closed-loop system resulting from event-triggered control.

In this chapter, a general hybrid control strategy for event-triggered feedback control is presented. Figure 5.1 depicts the associated hybrid closed-loop system. The strategy allows for different types of events in the algorithm: events triggered by the state of the plant, by the state of the controller, by time, inputs, or outputs. The jump set of the hybrid controller models the conditions triggering the jumps using *event-triggering functions*. The flow set of the controller is designed as the complement of the jump set, so as to allow continuous change of the system variables when no events occur. The jump map of the hybrid controller instantaneously updates the state at the events via *reset functions*. The flow map of the controller governs the change of its variables during flows, some of which may remain constant (e.g., memory states), grow linearly with time (e.g., timers), or change nonlinearly according to a differential equation or inclusion.

5.1 OVERVIEW

The most elementary algorithm with events is perhaps the sample-and-hold controller that, instead of being treated as a discrete-time system, includes a model of the mechanism triggering the sample and hold events. One such controller is presented in Example 1.3. As explained therein, one such controller has a state η with components being a memory variable ℓ_v storing the value of the input applied to the plant, which is to be held constant during flows, and a resettable timer τ, which triggers the events. Given a sampling period $T^* > 0$, the events in the system in Example 1.3 are triggered when

$$\tau \geq T^*$$

Following the construction in Example 2.10, this triggering condition can be captured by the *event-triggering function* defined as

$$\gamma(y,\eta) := \tau - T^*$$

since

$$\gamma(y,\eta) \geq 0 \tag{5.1}$$

if and only if

$$\tau \geq T^*$$

When such an event occurs, the timer is to be reset to zero and the memory state is to be reset to the value of the control law evaluated at the most recent measurement of the output of the plant. For the case when the control law is a static state-feedback law, this reset function defines the jump map of the hybrid controller as

$$G_K(z,\eta) = \begin{bmatrix} \kappa_c(y) \\ 0 \end{bmatrix}$$

where κ_c denotes the static output-feedback law.

When the control law is dynamic and measures y continuously, namely, when it is given by the output-feedback continuous-time controller

$$\dot{\eta}_c = F_c(y,\eta_c) \qquad \zeta_c = \kappa_c(y,\eta_c) \tag{5.2}$$

where η_c represents the state of the dynamic control law, a similar event-triggered implementation is possible by including the state η_c as a component of the state η of the controller. The event-triggered implementation of such a dynamic controller requires the use of additional timers and memory states.

The general hybrid control algorithm proposed in this chapter consists of an event-triggered implementation of a generic feedback controller, covering static feedback, as κ_c, and dynamic feedback, as in (5.2). It features generic event triggering conditions, which can be written as in (5.1) using event-triggering functions. The general model allows for two main types of events:

- *input events* triggered by an event-triggering function denoted as γ_u, and

- *output events* triggered by an event-triggering function denoted as γ_y.

In addition to the output and input of the plant, the proposed conditions allow for the components of η to be involved in the triggering of the events. The model proposed below also includes an auxiliary state that permits capturing dynamical control laws, like the one in (5.2), and their discretization, as well as any other mechanism triggering the events using timers, memory states, and logic variables.

Using these elements, the general event-triggered hybrid control strategy in this chapter is defined as follows:

- The controller state is $\eta = (\ell_y, \ell_u, \chi) \in X_K$, where

  - ℓ_y and ℓ_u are states that store the output of the plant and the value of control law, respectively. They may remain constant during flows and are only typically updated at the events, much like memory states, but could also be updated during flows so as to provide a more accurate representation of such quantities.

  - χ is an auxiliary state variable that is included to aid the implementation of the event-triggered strategy. In the sample-and-hold algorithm above, the state χ consists of a timer τ used to trigger the events; see Example 1.3. Furthermore, if the control law is dynamic, like the one in (5.2), the state χ would include state components that capture the evolution of η_c therein.

- The controller input is v, which is assigned to the output y of the plant.

- The control logic for event triggered (ET) control is as follows: given event-triggering functions γ_y and γ_u, reset functions G_y and G_u, and maps F_χ and G_χ,

  (ET-L1) When the value of γ_y is larger than or equal to zero, an output event is triggered. At such event, the memory state ℓ_y is updated according to the reset function G_y evaluated at the current value of the output of the plant and of the state of the controller. Since, in general, the reset function might be set valued, at such events the jumps are modeled as

  $$\ell_y^+ \in G_y(y, \eta) \qquad \text{when } \gamma_y(y, \eta) \geq 0$$

  (ET-L2) When the value of γ_u is larger than or equal to zero, then the memory state ℓ_u is updated according to the reset function G_u evaluated at the current value of the output of the plant and of the state of the controller. In general, this update law assumes the form

$$\ell_u^+ \in G_u(y, \eta) \qquad \text{when } \gamma_u(y, \eta) \geq 0$$

(ET-L3) At each of the events described in (ET-L1) and (ET-L2), the auxiliary variable χ is reset according to G_χ. In this way, the update of χ at jumps is governed by

$$\chi^+ \in G_\chi(y, \eta) \qquad \text{when } \gamma_y(y, \eta) \geq 0 \text{ or when } \gamma_u(y, \eta) \geq 0$$

(ET-L4) In between events, flows are governed as follows:

– The states ℓ_y and ℓ_u evolve according to

$$\dot{\ell}_y \in \hat{F}_y(\eta), \quad \dot{\ell}_u \in \hat{F}_u(\eta)$$

where $\hat{F}_y$ and $\hat{F}_u$ are referred to as *holding functions*. When ℓ_y and ℓ_u are held constant between events, the maps $\hat{F}_y$ and $\hat{F}_u$ are identically zero, in which case ℓ_y and ℓ_u act as true memory states.

– The auxiliary state χ evolves according to

$$\dot{\chi} \in F_\chi(y, \eta)$$

The dependency of $\hat{F}_y$, $\hat{F}_u$, and F_χ on η is for generality, so as to allow for the continuous evolution of the states ℓ_y, ℓ_u, and χ to depend on any of the components of η. In addition, since χ plays a role in triggering the events, its continuous dynamics may depend on y as well.

The update laws defined above are captured by the jump maps $G_{K,y}$ and $G_{K,u}$ resetting the controller state η.

When the event conditions in (ET-L1) and (ET-L2) occur simultaneously, then both resets may occur, either simultaneously or sequentially; see Example 5.2. As explained in Chapter 1, the set-valued maps in the update laws for ℓ_y, ℓ_u, and χ in items (ET-L1)-(ET-L3) and in the flows of ℓ_y, ℓ_u, and χ in item (ET-L4) permit modeling uncertainty and control mechanisms implemented by nondeterministic models. Figure 5.2 depicts the logic in the proposed event-triggered control strategy.

As stated above, a (rather elementary) instance of event-triggered control is given in Example 1.3 when modeling a sample-and-hold implementation of a static output-feedback law, while an event-triggered controller model that is closely related to the one presented in this chapter is given in Example 2.10. The following examples introduce other event-triggered control algorithms that can be captured by the general model presented in this chapter. The first example considers an event-triggered implementation of a static state-feedback controller.

Example 5.1 (Event-triggered implementation of a static state-feedback control law). *Consider an event-triggered implementation of the static state-feedback controller defined by $\zeta = \kappa_c(z)$ for asymptotic stabilization of the origin of the*

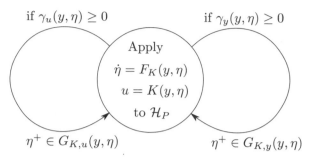

Figure 5.2: Logic implemented by event-triggered control.

continuous-time plant $\dot{z} = F_P(z, u)$, $y = z$. *The feedback law* κ_c *is such that there exist a continuously differentiable function* V_P *and* $\mathcal{K}_\infty$ *functions* α_1, α_2, ρ_1, *and* ρ_2 *satisfying*

$$\alpha_1(|z|) \leq V_P(z) \leq \alpha_2(|z|) \qquad\qquad\qquad \forall z \in \mathbb{R}^{n_P}$$
$$\langle \nabla V_P(z), F_P(z, \kappa_c(z + e)) \rangle \leq -\rho_1(|z|) + \rho_2(|e|) \qquad \forall z \in \mathbb{R}^{n_P}, \forall e \in \mathbb{R}^{n_P} \qquad (5.3)$$

When these inequalities hold, the closed-loop system resulting from using κ_c *is input-to-state stable with respect to* e *(relative to the origin). An event-triggering implementation of* κ_c *is as follows. Pick* $\sigma \in (0, 1)$:

- *The memory state* ℓ_y *stores the value of the plant state* z *at each event and remains constant in between events.*

- *The plant input is assigned to* $\kappa_c(\ell_y)$.

- *Whenever* $\rho_2(|\ell_y - z|) \geq \sigma \rho_1(|z|)$, ℓ_y *is updated to* z.

The intuition behind this strategy is as follows. Since the events are triggered when $\rho_2(|\ell_y - z|) \geq \sigma \rho_1(|z|)$, *flows are only allowed when*

$$\rho_2(|\ell_y - z|) \leq \sigma \rho_1(|z|) \quad \Longleftrightarrow \quad -\sigma \rho_1(|z|) + \rho_2(|\ell_y - z|) \leq 0$$

Since $\sigma \in (0, 1)$, V_P *is nonincreasing along solutions, during flow. Treating the term* $\ell_y - z$ *as* e *in* (5.3), *the second inequality in* (5.3) *holds.*

   This strategy can be captured by an event-triggered hybrid controller with state η *given by the memory state* ℓ_y *and a single event type, at which the state of the plant is sampled and its input updated. Setting* $v = z$, *and since* $y = z$, *the associated event-triggering function is given by*

$$\gamma_y(z, \eta) = \rho_2(|\ell_y - z|) - \sigma \rho_1(|z|) \qquad\qquad (5.4)$$

and the reset function by

$$G_y(z, \eta) = z \qquad\qquad (5.5)$$

Since ℓ_y *remains constant during flows, the map* $\hat{F}_y$ *is taken to be zero. A complete model of this strategy is presented in Example 5.3.*

   The next example extends the sample-and-hold model in Example 1.3 to the case when two events are considered, one corresponding to sampling the state of the plant and the other to updating the input to the plant.

Example 5.2 (Aperiodic sample-and-hold control with two events). *Consider a sample-and-hold implementation of the dynamic controller in (5.2) allowing for independent sample and hold events, which do not necessarily occur periodically. One such implementation is as follows:*

- *At sampling events, the state of the plant is sampled and stored in the controller state ℓ_y. These are output events. This state remains constant until the next sampling event. Following the model in Example 2.24, the sampling events are triggered when an internal timer τ_y has elapsed for at least $T^*_{y,1}$ seconds and at most $T^*_{y,2}$ seconds, where $0 < T^*_{y,1} \leq T^*_{y,2}$.*

- *At holding events, the controller state ℓ_u is updated to a value given by a discretization of the output-feedback continuous-time controller (5.2). These are input events. This state remains constant until the next holding event. At all times, the input to the plant u is assigned to ℓ_u. Similar to the sampling events, these events are triggered when an internal timer τ_u has elapsed for at least $T^*_{u,1}$ seconds and at most $T^*_{u,2}$ seconds, with $0 < T^*_{u,1} \leq T^*_{u,2}$.*

This nondeterministic sample-and-hold implementation of a continuous-time controller can be modeled using the event-triggered hybrid controller presented in this chapter. The main ingredients are presented next and the full model in Example 5.6.

The controller state η is given by (ℓ_y, ℓ_u, χ) with the auxiliary state defined as $\chi = (\tau_y, \tau_u, \eta_c) \in [0, T^*_{y,2}] \times [0, T^*_{u,2}] \times \mathbb{R}^{n_c}$, where τ_y and τ_u are to be used in triggering the sample and hold events, while η_c is used to implement a discretization of (5.2). Then, the set X_K is defined as $\mathbb{R}^{r_P} \times \mathbb{R}^{m_P} \times [0, T^*_{y,2}] \times [0, T^*_{u,2}] \times \mathbb{R}^{n_c}$. The triggering of the individual, nondeterministic events is captured using two independent event-triggering functions and by appropriately defining the dynamics of the timer states. To trigger the events, and following the constructions in Example 2.23, the dynamics of the timer is defined as

$$\begin{aligned} \dot{\tau}_y &= -1 & \text{when } \gamma_y(y, \eta) \leq 0 \\ \tau_y^+ &\in [T^*_{y,1}, T^*_{y,2}] & \text{when } \gamma_y(y, \eta) \geq 0 \end{aligned} \tag{5.6}$$

where

$$\gamma_y(y, \eta) := -\tau_y \qquad \forall (y, \eta) \ : \ \tau_y \in [0, T^*_{y,2}] \tag{5.7}$$

As depicted in Figure 5.3, the jump map G_χ is to be defined to reset τ_y to a point in $[T^*_{y,1}, T^*_{y,2}]$ when $\gamma_y(y, \eta)$ is larger than or equal to zero. In this way, the times at which the sample events occur satisfy (2.28) with parameters $T^*_{y,1}$ and $T^*_{y,2}$. The event-triggering function γ_u and the map G_χ is defined similarly, so as to trigger events when $\tau_u = 0$ and to reset τ_u to a point in $[T^*_{u,1}, T^*_{u,2}]$.

The proposed implementation is such that the value of the state ℓ_y is updated to y at the sampling events and ℓ_u is reset to a discrete-time equivalent of the dynamic controller in (5.2) at the holding events. Denoting the discretization of F_c with step size $s > 0$ as F_c^s, the reset function for ℓ_u updates ℓ_u to $\kappa_c(\ell_y, F_c^s(\eta))$, for which an additional timer state playing the role of the dynamic step size is needed. More details are provided in Example 5.6.

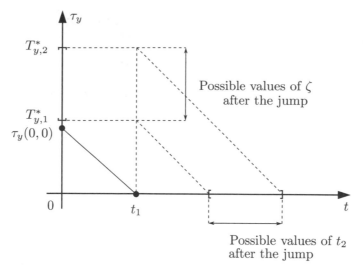

Figure 5.3: The τ_y component of a solution to the nondeterministic timer model in (5.6) triggering a sampling event at $(t, j) = (t_1, 0)$. After the first jump, the next event occurs no sooner than $T_{y,1}^*$ and no later than $T_{y,2}^*$ seconds.

5.2 HYBRID CONTROLLER

The logic in § 5.1 is implemented with an event-triggered hybrid controller $\mathcal{H}_K$. The case of a plant $\mathcal{H}_P$ with continuous dynamics only and no constraints, namely, $\dot{z} \in F_P(z, u)$, $y = h(z)$, is considered. The hybrid controller for event-triggered control is now defined following the outline in § 5.1.

The controller has state $\eta = (\ell_y, \ell_u, \chi) \in X_K := Y \times U \times X_A$, where the sets Y, U, and X_A are defined based on the particular event-triggered control strategy to implement. Its input is $v \in Y$, which is assigned to the output y of the plant, its output is $\zeta \in U$, and its data $(C_K, F_K, D_K, G_K, \kappa)$ is defined as follows. The flow set C_K is defined as

$$C_K := C_{K,y} \cap C_{K,u} \tag{5.8}$$

where $C_{K,y}$ and $C_{K,u}$ define the conditions allowing flows:

$$\begin{aligned} C_{K,y} &:= \{(y, \eta) \in Y \times X_K : \gamma_y(y, \eta) \leq 0\} \\ C_{K,u} &:= \{(y, \eta) \in Y \times X_K : \gamma_u(y, \eta) \leq 0\} \end{aligned} \tag{5.9}$$

During flows, the state η of the controller is updated according to the right-hand side F_K defined as

$$F_K(y, \eta) := \begin{bmatrix} \hat{F}_y(\eta) \\ \hat{F}_u(\eta) \\ F_\chi(y, \eta) \end{bmatrix} \qquad \forall (y, \eta) \in C_K \tag{5.10}$$

where $(\hat{F}_y, \hat{F}_u)$ define the continuous update of the states (ℓ_y, ℓ_u) and F_χ that of the auxiliary state χ. The events are triggered by solutions reaching the jump set

$$D_K := D_{K,y} \cup D_{K,u} \tag{5.11}$$

with $D_{K,y}$ and $D_{K,u}$ capturing the individual jump conditions as

$$\begin{aligned}
D_{K,y} &:= \{(y,\eta) \in Y \times X_K : \gamma_y(y,\eta) \geq 0\} \\
D_{K,u} &:= \{(y,\eta) \in Y \times X_K : \gamma_u(y,\eta) \geq 0\}
\end{aligned} \tag{5.12}$$

The jump map G_K is defined as

$$G_K(y,\eta) := G_{K,y}(y,\eta) \cup G_{K,u}(y,\eta) \qquad \forall (y,\eta) \in D_K \tag{5.13}$$

where the first piece in its definition executes the reset functions associated with events occurring due to $D_{K,y}$ and the second piece performs the resets due to $D_{K,u}$. These maps are defined as

$$G_{K,y}(y,\eta) := \begin{bmatrix} G_y(y,\eta) \\ \ell_u \\ G_\chi(y,\eta) \end{bmatrix} \qquad \forall (y,\eta) \in D_{K,y} \tag{5.14}$$

and empty elsewhere, and

$$G_{K,u}(y,\eta) := \begin{bmatrix} \ell_y \\ G_u(y,\eta) \\ G_\chi(y,\eta) \end{bmatrix} \qquad \forall (y,\eta) \in D_{K,u} \tag{5.15}$$

and empty elsewhere. The output of the controller is given by $(y,\eta) \mapsto \kappa(y,\eta)$, which is also to be defined based on the particular event-triggered control strategy to implement.

This construction is illustrated with the strategy in Example 5.1.

Example 5.3 (Event-triggered implementation of a static state-feedback control law, revisited). *The constructions in Example 5.1 already introduce the main elements defining the data of the event-triggered controller $\mathcal{H}_K$ given in (5.9)-(5.15). The state η is defined as the memory state $\ell_y \in X_K := \mathbb{R}^{n_P}$. The output ζ is ℓ_u. The event-triggering function γ_y is given in (5.4) and the reset function is given in (5.5). Since the memory state remains constant in between events, $\hat{F}_y(\eta) \equiv 0$. Note that since the model does not use an auxiliary state χ, then maps associated to χ do not need to be defined. Summing up, the resulting hybrid controller has state $\eta = \ell_y$ and data given by*

$$\begin{aligned}
C_K &:= \{(z,\ell_y) \in \mathbb{R}^{n_P} \times X_K : -\sigma\rho_1(|z|) + \rho_2(|\ell_y - z|) \leq 0\} \\
F_K(\ell_y) &:= 0 \qquad \forall \ell_y : (z,\ell_y) \in C_K \\
D_K &:= \{(z,\ell_y) \in \mathbb{R}^{n_P} \times X_K : -\sigma\rho_1(|z|) + \rho_2(|\ell_y - z|) \geq 0\} \\
G_K(z) &= z \qquad \forall z : (z,\ell_y) \in D_K \\
\kappa(\ell_y) &:= \kappa_c(\ell_y) \qquad \forall \ell_y \in \mathbb{R}^{n_P}
\end{aligned}$$

To assure that the pre-asymptotic stability obtained via the proposed event-triggered controller $\mathcal{H}_K$ is robust in the sense of Definition 3.16, the following is assumed for its data.

Assumption 5.4 (Regularity properties for event-triggered control). *The maps γ_y, γ_u, $\hat{F}_y$, $\hat{F}_u$, F_χ, G_y, G_u, G_χ, and κ, and the sets Y and X_K defining the data $(C_K, F_K, D_K, G_K, \kappa)$ of the event-triggered hybrid controller in (5.9)-(5.15) satisfy the following properties:*

(ET-A1) *The event-triggering functions $\gamma_y : Y \times X_K \to \mathbb{R}$ and $\gamma_u : Y \times X_K \to \mathbb{R}$ are continuous.*

(ET-A2) *$\hat{F}_y : X_K \rightrightarrows Y$ is outer semicontinuous, nonempty, and locally bounded relative to C_K and, for each $(y, \eta) \in C_K$, $\hat{F}_y(\eta)$ is convex.*

(ET-A3) *$\hat{F}_u : X_K \rightrightarrows U$ is outer semicontinuous, nonempty, and locally bounded relative to C_K and, for each $(y, \eta) \in C_K$, $\hat{F}_u(\eta)$ is convex.*

(ET-A4) *$F_\chi : Y \times X_K \rightrightarrows X_A$ is outer semicontinuous, nonempty, and locally bounded relative to C_K and, for each $(y, \eta) \in C_K$, $F_\chi(\eta)$ is convex.*

(ET-A5) *The maps $G_y : X_K \rightrightarrows Y$, $G_u : X_K \rightrightarrows U$, and $G_\chi : X_K \rightrightarrows X_A$ are outer semicontinuous, nonempty, and locally bounded relative to $D_{K,y}$, $D_{K,u}$, and D_K, respectively.*

(ET-A6) *The map $\kappa : Y \times X_K \to U$ is continuous.*

(ET-A7) *The sets Y and X_K are closed.*

Item (ET-A1) is a mild assumption on the event-triggering functions that is needed, along with closedness of the sets in item (ET-A7), to assure that the set of points where flows and jumps can occur is closed. When the maps in items (ET-A2)-(ET-A6) are single valued, the properties therein hold when the maps are continuous. These conditions are needed so that the resulting hybrid closed-loop system satisfies the hybrid basic conditions, and, hence, is well-posed. Section 2.3.2 provides more details about the importance of well-posedness in hybrid systems.

Lemma 5.5 (Well-posedness of $\mathcal{H}_K$). *Suppose that Assumption 5.4 holds. Then, the hybrid controller $\mathcal{H}_K$ in (5.9)-(5.15) satisfies the hybrid basic conditions.*[1]

Proof. Closedness of C_K and D_K follows from closedness of the sets Y and X_K, and continuity of γ_y and γ_u; see Exercise 100. Outer semicontinuity, local boundedness, and the property of convex values of F_K follow directly from the properties of $\hat{F}_y$, $\hat{F}_u$, and F_χ. Finally, outer semicontinuity and local boundedness of G_K is a direct consequence of the assumptions and Lemma A.33. □

Lemma 5.5 is illustrated in the event-triggered implementation of the event-triggered strategy in Example 5.2.

Example 5.6 (Aperiodic sample-and-hold control with two events, revisited). *The sample-and-hold implementation of the dynamic output-feedback controller outlined in Example 5.2 is captured by $\mathcal{H}_K$ in (5.9)-(5.15) with state $\eta = (\ell_y, \ell_u, \chi) = (\ell_y, \ell_u, (\tau_y, \tau_u, \tau_s, \eta_c)) \in X_K := Y \times U \times [0, T_{y,2}^*] \times [0, T_{u,2}^*] \times [0, T_{u,2}^*] \times \mathbb{R}^{n_c}$, $U := \mathbb{R}^{m_P}$, input $v = y \in Y := \mathbb{R}^{r_P}$, output $\zeta = \kappa(\eta) = \ell_u$, and parameters $0 < T_{y,1}^* \le T_{y,2}^*$,*

---

[1]See Definition 2.25 and the discussion below it.

$0 < T_{u,1}^* \leq T_{u,2}^*$. Since X_K and Y are closed, (ET-A7) holds. The timer state τ_s is included to keep track of time in between input events. This information is needed in the implementation of the discretization of the dynamic controller in (5.2). The event-triggering functions γ_y and γ_u depend only on the components τ_y and τ_u of η, respectively, and are given by

$$\gamma_y(\tau_y) := -\tau_y \qquad \forall \tau_y \in [0, T_{y,2}^*]$$
$$\gamma_u(\tau_u) := -\tau_u \qquad \forall \tau_u \in [0, T_{u,2}^*] \tag{5.16}$$

from where the sets $D_{K,y}$ and $D_{K,u}$ are defined as in (5.12). For simplicity, these sets are written only as a function of the timer states, which due to the definition of X_K reduce to τ_y or τ_u being zero. Since γ_y and γ_u are continuous, (ET-A1) holds.

The strategy in Example 5.2 implements a discretization of the continuous output-feedback controller in (5.2), so the state η_c remains constant during flows – this is captured by the last entry of F_χ in (5.20) below. The update of the controller variables at the two possible events is as follows:

- At jumps due to a sampling event, the state ℓ_y is updated via G_y to the current value of the plant output y. Then, since the controller input v is assigned to the plant output y, G_y is defined as

$$G_y(y, \eta) := y \qquad \forall (y, \eta) : \tau_y = 0 \tag{5.17}$$

and empty at points such that $\tau_y > 0$.

- At jumps due to the holding events, the value of ℓ_u is updated via G_u to the (likely new) value of the control signal provided by the discretized controller, which is defined as

$$G_u(\eta) := \kappa_c(\ell_y, F_c^s(\eta)) \qquad \forall \eta : \tau_u = 0 \tag{5.18}$$

and empty for points such that $\tau_u > 0$. At such events, the value of η_c is updated via the discretization function F_c^s. A particular construction of F_c^s using a forward Euler discretization scheme is $F_c^s(\eta) = \eta_c + \tau_s F_c(\ell_y, \eta_c)$, where τ_s plays the role of the step size s.

- At each holding event, the timer state τ_u is reset to a point in $[T_{u,1}^*, T_{u,2}^*]$ and τ_s is reset to zero so as to count time in between such events. Similarly, at each sampling event, the timer state τ_y is reset to a point in $[T_{y,1}^*, T_{y,2}^*]$.

Then, the jump map G_χ governing $\chi = (\tau_y, \tau_u, \eta_c)$ is defined as[2]

$$G_\chi(\eta) := \begin{cases} ([T_{u,1}^*, T_{u,2}^*], \tau_y, 0, \eta_c) & \text{if } \tau_u = 0, \tau_y > 0 \\[2mm] (\tau_u, [T_{y,1}^*, T_{y,2}^*], 0, F_c^s(\eta)) & \text{if } \tau_u > 0, \tau_y = 0 \\[2mm] \left\{ \begin{bmatrix} [T_{u,1}^*, T_{u,2}^*] \\ \tau_y \\ 0 \\ \eta_c \end{bmatrix}, \begin{bmatrix} \tau_u \\ [T_{y,1}^*, T_{y,2}^*] \\ 0 \\ F_c^s(\eta) \end{bmatrix} \right\} & \text{if } \tau_u = 0, \tau_y = 0 \end{cases} \tag{5.19}$$

---

[2]This definition of G_χ follows the construction in Lemma A.33.

for each $\eta \in D_K$. When F_c^s is continuous, G_χ satisfies (ET-A5). Outer semicontinuity of G_χ follows from Lemma A.33.

Due to the fact that the components (τ_y, τ_u, τ_s) of χ evolve continuously as timers, the map F_χ is given by the constant function

$$F_\chi(\eta) := \begin{bmatrix} -1 \\ -1 \\ 1 \\ 0 \end{bmatrix} \tag{5.20}$$

Note that τ_y and τ_u count down and τ_s counts up. It follows that (ET-A4) holds. Since the states ℓ_y and ℓ_u remain constant during flows, the functions $\hat{F}_y$ and $\hat{F}_u$ are given as

$$\hat{F}_y(\eta) := 0, \qquad \hat{F}_u(\eta) := 0 \qquad \forall \eta \in C_K \tag{5.21}$$

Hence, (ET-A2) and (ET-A3) hold.

Combining the above definitions, the resulting hybrid controller has state $\eta = (\ell_y, \ell_u, \tau_y, \tau_u, \tau_s, \eta_c)$ and data given by

$$C_K = X_K$$
$$F_K(\eta) = (0, 0, -1, -1, 1, 0) \qquad \forall \eta \in C_K$$
$$D_K = \{(y, \eta) \in Y \times X_K : \tau_y = 0\} \cup \{(y, \eta) \in Y \times X_K : \tau_u = 0\}$$

$$G_K(y, \eta) := \begin{cases} \begin{bmatrix} y \\ \ell_u \\ [T_{u,1}^*, T_{u,2}^*] \\ \tau_y \\ 0 \\ \eta_c \end{bmatrix} & \text{if } \tau_u = 0, \tau_y > 0 \\[2em] \begin{bmatrix} \ell_y \\ \kappa_c(\ell_y, F_c^s(\eta)) \\ \tau_u \\ [T_{y,1}^*, T_{y,2}^*] \\ 0 \\ F_c^s(\eta) \end{bmatrix} & \text{if } \tau_u > 0, \tau_y = 0 \\[2em] \left\{ \begin{bmatrix} y \\ \ell_u \\ [T_{u,1}^*, T_{u,2}^*] \\ \tau_y \\ 0 \\ \eta_c \end{bmatrix}, \begin{bmatrix} \ell_y \\ \kappa_c(\ell_y, F_c^s(\eta)) \\ \tau_u \\ [T_{y,1}^*, T_{y,2}^*] \\ 0 \\ F_c^s(\eta) \end{bmatrix} \right\} & \text{if } \tau_u = 0, \tau_y = 0 \end{cases}$$

$$\kappa(\eta) = \ell_u$$

where G_K is defined for each $(y, \eta) \in D_K$. Since κ is continuous, (ET-A6) holds.

It is left as an exercise to the reader to simplify this model by removing the state τ_s. A way to avoid having to add this extra state is to turn τ_u into a timer that counts up, instead of counting time down as it currently does. With that change, the timer τ_u can be used in the discretization of F_c.

5.3   CLOSED-LOOP SYSTEM

Using the hybrid controller $\mathcal{H}_K$ in (5.9)-(5.15) to control the plant $\dot{z} \in F_P(z, u)$, $y = h(z)$, the hybrid closed-loop system has a state given by $x = (z, \eta) \in X := X_P \times X_K$, with $\eta = (\ell_y, \ell_u, \chi)$ the controller state. It evolves according to

$$\dot{z} \in F_P(z, \kappa(y, \eta))$$
$$\dot{\ell}_y \in \hat{F}_y(\eta), \quad \dot{\ell}_u \in \hat{F}_u(\eta), \quad \dot{\chi} \in F_\chi(y, \eta)$$

during flows. At jumps, z remains constant, so it is updated by

$$z^+ = z$$

while ℓ_y, ℓ_u, and χ are updated according to

$$\ell_y^+ \in G_y(y, \eta), \quad \ell_u^+ = \ell_u, \quad \chi^+ \in G_\chi(y, \eta) \qquad \text{when } \gamma_y(y, \eta) \geq 0$$
$$\ell_y^+ = \ell_y, \quad \ell_u^+ \in G_u(y, \eta), \quad \chi^+ \in G_\chi(y, \eta) \qquad \text{when } \gamma_u(y, \eta) \geq 0$$

When both conditions hold simultaneously, either reset is possible.

Following the construction in Definition 2.11, the hybrid closed-loop system $\mathcal{H}$ has state $x = (z, \eta) = (z, \ell_y, \ell_u, \chi) \in X$ with data (C, F, D, G) given by

$$C := \{(z, \eta) \in X : (y, \eta) \in C_{K,y}\} \cap \{(z, \eta) \in X : (y, \eta) \in C_{K,u}\} \quad (5.22)$$

$$F(x) := \begin{bmatrix} F_P(z, \kappa(y, \eta)) \\ \hat{F}_y(\eta) \\ \hat{F}_u(\eta) \\ F_\chi(y, \eta) \end{bmatrix} \qquad \forall x \in C \qquad (5.23)$$

$$D := \{(z, \eta) \in X : (y, \eta) \in D_{K,y}\} \cup \{(z, \eta) \in X : (y, \eta) \in D_{K,u}\} \quad (5.24)$$

$$G(x) := \begin{bmatrix} z \\ G_K(y, \eta) \end{bmatrix} \qquad \forall x \in D \qquad (5.25)$$

where $y = h(z)$, and the definitions of $C_{K,y}$, $C_{K,u}$, $D_{K,y}$, $D_{K,u}$, and G_K are given in (5.9)-(5.15).

As shown in Examples 1.3, 2.10, 5.1, and 5.2, the resulting hybrid closed-loop system simplifies significantly when modeling particular event-triggered hybrid controllers. Before addressing stability, completeness of solutions, lower bounds on inter-event times, and robustness for particular event-triggered hybrid control strategies, the next result determines conditions guaranteeing that the closed loop satisfies the hybrid basic conditions. A proof follows immediately from an application of Lemma 5.5 and Lemma 2.21, and properties in Exercise 100. For pedagogical reasons, a proof that presents the steps in those lemmas is provided next.

Lemma 5.7 (Well-posedness of $\mathcal{H}$). *Suppose the plant $\mathcal{H}_P$ given by $\dot{z} \in F_P(z, u)$, $y = h(z)$ satisfies the hybrid basic conditions and that Assumption 5.4 holds. Then, the hybrid closed-loop system $\mathcal{H}$ with data as in (5.22)-(5.25) satisfies the hybrid basic conditions.*[3]

---

[3]See Definition 2.20.

Proof. To show that C and D are closed and, hence, satisfy (A1), an application of Lemma 5.5 establishes that the sets $C_{K,y}$, $C_{K,u}$, $D_{K,y}$, and $D_{K,u}$ are closed. The first set in the definition of C is closed due to closedness of $C_{K,y}$ and continuity of h guaranteed by (A4$_P$). The same argument applies to the second set in the definition of C; hence, C is closed. The proof that D is closed follows similarly.

The flow map F of $\mathcal{H}$ satisfies (A2) since F_P satisfies (A2$_P$), κ satisfies (ET-A6), and the maps $\hat{F}_y$, $\hat{F}_u$, F_χ satisfy (ET-A2), (ET-A3), and (ET-A4), respectively. Outer semicontinuity of F is due to the composition of F_P with κ being outer semicontinuous and due to outer semicontinuity of the maps defining its other components; see Exercise 97. The proof that the map F has convex values is a direct consequence from the properties of its components. Similarly, locally boundedness of F relative to C is due to locally boundedness of its individual components. With κ continuous due to (ET-A6), F_P being locally bounded implies that the first component of F is locally bounded relative to C. Locally boundedness of its other components follows directly from (ET-A2), (ET-A3), and (ET-A4). Nonemptiness of F on C holds since its components are nonempty on C.

The proof concludes by showing that G satisfies (A3). Due to the definition of G and of G_K, to show that G is outer semicontinuous relative to D it suffices to show that $G_{K,y}$ in (5.14) and $G_{K,u}$ in (5.15) are outer semicontinuous relative to D; see Exercise 97. Since G_y and G_χ are outer semicontinuous relative to $D_{K,y}$ due to (ET-A5), $G_{K,y}$ is outer semicontinuous. Similarly, since G_u and G_χ are outer semicontinuous relative to $D_{K,u}$ also due to (ET-A5), $G_{K,y}$ is outer semicontinuous. Locally boundedness and nonemptiness of G on D follows similarly to the proof for F; see Lemma 2.21 and its proof. □

5.4   DESIGN

Design conditions for the event-triggered controller are formulated in this section. One desired property corresponds to maximal solutions to the closed-loop system existing for arbitrary large t or j, namely, that maximal solutions are complete. General conditions assuring such completeness property is presented first, in §5.4.1. Another property that is critical in real-world applications of event-triggered hybrid control is the existence of a lower bound on the inter-event times. In particular, this property rules out the existence of Zeno solutions. Conditions guaranteeing such a property are presented in §5.4.2. Sufficient conditions guaranteeing pre-asymptotic stability of a set of interest are provided in §5.4.3.

5.4.1   Completeness of Maximal Solutions

The data of the event-triggered hybrid controller in (5.9)-(5.15) needs to be properly designed to guarantee that for every point in $C \cup D$, every maximal solution to the closed-loop system is complete. In particular, the event-triggering function and the reset functions need to be chosen to guarantee that either flows or jumps within $C \cup D$ are always possible. For instance, if in Example 5.6 the parameter $T_{u,1}^*$ were to be negative, then, due to the definition of the jump map in (5.19), it would be possible for a maximal solution to have its τ_u component take a negative value after a jump, from which neither flow nor jump would be possible. The following result provides a set of conditions guaranteeing that such pathology does not occur by assuring that every maximal solution to the hybrid closed-loop system is complete. Recall that Lemma 5.7 states when $\mathcal{H}$ satisfies the hybrid basic conditions.

Proposition 5.8 (Completeness of maximal solutions). *Suppose the hybrid closed-loop system $\mathcal{H}$ with data in (5.22)-(5.25) satisfies the hybrid basic conditions. Then, from each point in $C \cup D$, there exists a nontrivial solution to $\mathcal{H}$ if*

$$F(\xi) \cap T_C(\xi) \neq \emptyset \qquad \forall \xi \in \{(z,\eta) \in X : \gamma_y(h(z),\eta) < 0, \gamma_u(h(z),\eta) < 0\} \quad (5.26)$$

Furthermore, every maximal solution to $\mathcal{H}$ from $C \cup D$ is complete if

1. *Case b in Proposition 2.34 does not hold for every maximal solution to $\mathcal{H}$; and*

2. $G(D) \subset C \cup D$.

Proof. Since $\mathcal{H}$ satisfies the hybrid basic conditions, Proposition 2.34 is applied to show the stated property of solutions. By definition of D, if $\xi = (\xi_z, \xi_\eta) \in D$ then either $\gamma_y(h(\xi_z), \xi_\eta) \geq 0$ or $\gamma_u(h(\xi_z), \xi_\eta) \geq 0$. Note that (5.26) implies that (VC) in Proposition 2.34 holds for each $\xi \in C \setminus D$. Then, there exists a nontrivial solution to $\mathcal{H}$ from every initial point in $C \cup D$. To show that every maximal solution is complete, note that case b in Proposition 2.34 does not hold by assumption. Case c therein is ruled out via item 3 in Proposition 2.34 since $G(x) \subset C \cup D$ for each $x \in D$. Then, by Proposition 2.34, every maximal solution to $\mathcal{H}$ is complete. $\qquad \square$

When $C \cup D = \mathbb{R}^n$ and $\mathcal{H}$ satisfies the hybrid basic conditions, the set $C \setminus D$ is open. Then, since for every point in the interior of a set its tangent cone is the entire state space, condition (5.26) holds for free. The following example exploits this fact when illustrating Proposition 5.8.

Example 5.9 (Event-triggered implementation of a static state-feedback control law, revisited). *The hybrid closed-loop system resulting from controlling the plant $\dot{z} = F_P(z, u)$ with the controller in Example 5.1 has state $x = (z, \eta)$ with $\eta = \ell_y$ and dynamics*

$$\mathcal{H} : \begin{cases} \rho_2(|\ell_y - z|) \leq \sigma \rho_1(|z|) & \begin{bmatrix} \dot{z} \\ \dot{\ell}_y \end{bmatrix} - \begin{bmatrix} F_P(z, \kappa_c(\ell_y)) \\ 0 \end{bmatrix} \\ \rho_2(|\ell_y - z|) \geq \sigma \rho_1(|z|) & \begin{bmatrix} z^+ \\ \ell_y^+ \end{bmatrix} = \begin{bmatrix} z \\ z \end{bmatrix} \end{cases} \quad (5.27)$$

Note that $C \cup D = \mathbb{R}^{n_P} \times \mathbb{R}^{n_P}$ and that from Example 5.1, the functions ρ_1 and ρ_2 are continuous. When F_P and κ_c are such that the map $(z, \ell_y) \mapsto F_P(z, \kappa_c(\ell_y))$ is locally Lipschitz, then from every point in $C \cup D$ there exists a nontrivial solution and every maximal solution to $\mathcal{H}$ is complete. It should be noted that solutions from the origin are Zeno. In fact, when $z = \ell_y = 0$, then the function condition in (5.27) becomes $\rho_2(0) \geq \sigma \rho_1(0)$, which is satisfied since ρ_1 and ρ_2 are class-$\mathcal{K}_\infty$ functions. Since the jump map resets (z, ℓ_y) to zero, there exists a discrete complete solution – hence, Zeno – from the origin.

5.4.2   Minimum Time in Between Events

Due to limited computational power in physical platforms, most implementations of event-triggered control require a minimum time in between consecutive events. In fact, closed-loop systems with an event-triggered control algorithm that induces Zeno solutions would require arbitrarily fast computations. As a consequence, event-triggered algorithms that guarantee a positive lower bound on the time between consecutive events are desired. The result in this section provides sufficient conditions on the system data to guarantee a lower bound on the inter-event time for all solutions. This result is motivated by the following example.

Example 5.10 (Nonrobustness of inter-event times with vanishing noise). *Consider the scalar point-mass system*

$$\dot{z} = u \qquad (z, u) \in \mathbb{R} \times \mathbb{R} \qquad (5.28)$$

with output $y = z$, controlled by the static state-feedback law

$$v = \kappa_c(z) := -z \qquad \forall z \in \mathbb{R} \qquad (5.29)$$

implemented using the event-triggered strategy in Example 5.1 with $\rho_1(s) = \rho_2(s) = s^2$ for each $s \geq 0$. Condition (5.3) holds with $V_P(z) = z^\top z$. With such choices, the closed loop (5.27) in Example 5.9 is given by

$$\mathcal{H} \; : \; \begin{cases} |\ell_y - z| \leq \sqrt{\sigma}|z| & \begin{bmatrix} \dot{z} \\ \dot{\ell}_y \end{bmatrix} = \begin{bmatrix} -\ell_y \\ 0 \end{bmatrix} \\[2ex] |\ell_y - z| \geq \sqrt{\sigma}|z| & \begin{bmatrix} z^+ \\ \ell_y^+ \end{bmatrix} = \begin{bmatrix} z \\ z \end{bmatrix} \end{cases} \qquad (5.30)$$

Now suppose that u is affected by actuator noise w_u and, hence, the first entry of the flow map of (5.30) is given by $-\ell_y + w_u$. Figure 5.4 illustrates (an approximation of) the maximal solution to the closed loop when the noise is $w_u = -\ell_y/\sqrt{|\ell_y|}$, $\sqrt{\sigma} = 1/2$, and the initial condition is $z(0,0) = \ell_y(0,0) = 1/2$. Even though the norm of the noise vanishes to zero due to ℓ_y converging to zero, the resulting solution is Zeno, with Zeno time at $t_f \approx 0.9497$.

This property is generic. In fact, let z_j denote the value of the plant state z at the j-th event, $j \in \mathbb{N}$, with $z_0 > 0$ being the chosen initial condition for z and ℓ_y. From the choice of the disturbance w_u and the fact that ℓ_y is updated to z at every jump, the change of z after the j-th event is given by $\dot{z} = -z_j - z_j/\sqrt{|z_j|}$. Since $\ell_y(0,0) = z(0,0) = z_0 > 0$, there is no jump at $(t,j) = (0,0)$ and the plant state z initially continuously decreases until for some $t_1 > 0$, $|\ell_y(t_1,0) - z(t_1,0)| = \sqrt{\sigma}|z(t_1,0)|$ — or, equivalently, $|z_0 - z_1| = \sqrt{\sigma}|z_1|$. Since $z_0 - z_1 > 0$, it follows that $z_1 = z_0/(\sqrt{\sigma}+1)$. By induction, it can be shown that $z_j = z_0/(\sqrt{\sigma}+1)^j$ and $z_{j+1} = z_j/(\sqrt{\sigma}+1)$ for each $j \in \mathbb{N}$. Then, using the dynamics of z, the time $t_{j+1} - t_j$ for z to flow from z_j to z_{j+1} satisfies

$$z_j + \left(-z_j - \frac{z_j}{\sqrt{|z_j|}} \right)(t_{j+1} - t_j) = z_{j+1}$$

from where

$$t_{j+1} - t_j = \frac{\sqrt{\sigma}}{\sqrt{\sigma}+1} \frac{z_j}{z_j + z_j |z_j|^{-1/2}}$$

Using $z_j = z_0/(\sqrt{\sigma}+1)^j$, *it follows that*

$$\sum_{j=0}^{\infty}(t_{j+1} - t_j) = \frac{\sqrt{\sigma}}{\sqrt{\sigma}+1} \sum_{j=0}^{\infty} \frac{z_0}{z_0 + (\sqrt{\sigma}+1)^{j/2}} \qquad (5.31)$$

which converges. Hence, the measurement noise w_u *induces a solution with inter-event times converging to zero. Such a solution is Zeno. Figure 5.4 denotes such a solution. Associated simulation files are at* @BookSite/Simulation/ETZeno.

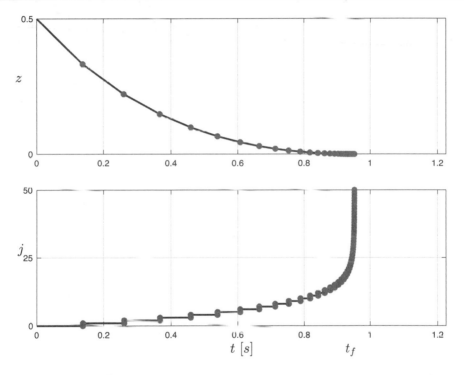

Figure 5.4: Simulated solution of the hybrid closed-loop system in Example 5.1 with vanishing disturbance w_u on u. The top plot shows the state z of the plant associated to the solution, as a function of t. The bottom plot shows that the number of jumps in the solution, also as a function of t, grows unbounded.

According to § 5.1, the general event-triggered hybrid controller $\mathcal{H}_K$ in (5.9)-(5.15) can exhibit two types of events: *input events* triggered by the event-triggering function γ_u and *output events* triggered by the event-triggering function γ_y. Constructive conditions assuring a uniform lower bound on the time between consecutive events of the same event type are provided next. Note that, since the conditions triggering these events are independent in general, the time elapsed between each type of event could be very small, even zero. In other words, an input event and an output event may still occur at the same ordinary time instant.

Before stating the result, the following notation is introduced for convenience. Let x be a solution to the hybrid closed-loop system with the event-triggered hybrid controller. Given a solution x to the hybrid closed-loop system, denote by E_u the collection of points in dom x at which input events occurs, and by E_y the collection of points in dom x at which output events occurs. Then, for the given solution x, the minimum inter-event times for such events are defined as

$$\Delta t_u(x) = \inf \{t' - t \,:\, (t, j), (t', j') \in E_u, j < j'\} \tag{5.32}$$

and

$$\Delta t_y(x) = \inf \{t' - t \,:\, (t, j), (t', j') \in E_y, j < j'\} \tag{5.33}$$

The following result provides conditions so that the inter-event times are uniformly lower bounded by a positive constant. Figure 5.5 provides a pictorial description of these conditions.

Proposition 5.11 (Positive lower bound on inter-event times). *Suppose $\mathcal{H}$ with data in (5.22)-(5.25) satisfies the hybrid basic conditions and that every maximal solution is bounded. Then, the following hold:*

1. *For every maximal solution x to $\mathcal{H}$ there exists $\lambda_u > 0$ such that $\Delta t_u(x) \geq \lambda_u$ if and only if*

$$D_{K,u} \cap G_{K,u}(D_{K,u}) = \emptyset \tag{5.34}$$

2. *For every maximal solution x to $\mathcal{H}$ there exists $\lambda_y > 0$ such that $\Delta t_y(x) \geq \lambda_y$ if and only if*

$$D_{K,y} \cap G_{K,y}(D_{K,y}) = \emptyset \tag{5.35}$$

Proof. Necessity of the second claim is shown first – the proofs for the first claim follow similarly. Let x be a maximal solution to $\mathcal{H}$. Since by assumption x is bounded, using assumption (A2), $F(\mathrm{rge}\, x)$ is bounded. It follows that there exists $\delta > 0$ such that $|\dot{x}(t, j)| \leq \delta$ for all $(t, j) \in \mathrm{dom}\, x$. As defined right above (5.32), let E_y be the collection of points (t, j) in dom x at which jumps are triggered by output events. Then, the set $\overline{x(E_y)}$ is compact and a subset of $C \cup D$. In addition, by construction of $D_{K,y}$, the set $D_{K,y}$ is closed and $\overline{x(E_y)} \subset D_{K,y}$ by definition of E_y. Moreover, $G(\overline{x(E_y)}) \subset G(D_{K,y})$ is closed since the jump map G is outer semicontinuous by assumption (A3). Then, due to (5.35), it follows that $\overline{x(E_y)} \cap G(\overline{x(E_y)}) = \emptyset$, and the distance between $\overline{x(E_y)}$ and $G(\overline{x(E_y)})$ is positive. Denote this distance by $\varepsilon > 0$. Then, the time interval between two consecutive output events is lower bounded by ε/δ, that is, $\Delta t_y(x) \geq \lambda_y := \frac{\varepsilon}{\delta}$.

Sufficiency is shown using contradiction: assume that $D_{K,y} \cap G(D_{K,y}) \neq \emptyset$ holds and construct a maximal solution to $\mathcal{H}$ such that Δt_y is zero. To this end, define $D'_{K,y} := D_{K,y} \cap G(D_{K,y})$. By definition of solutions to $\mathcal{H}$, there exists a solution x from $D'_{K,y}$ with $(0, 0), (0, 1) \in E_y$. Thus, $\Delta t_y(x)$ given as in (5.33) is equal to zero. Hence, there does not exist $\lambda_y > 0$ such that $\Delta t_y(x) \geq \lambda_y$, which is a contradiction. $\square$

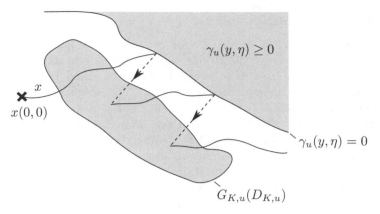

Figure 5.5: Jump set and image of the jump map guaranteeing a finite amount of flow time in between events.

Proposition 5.11 leads to semiglobal and uniform lower bounds on the inter-event times. In fact, on compact sets of initial conditions to $\mathcal{H}$, the constants λ_u and λ_y in items 1 and 2 therein can be chosen as the minimum for each bounded solution from a given compact set of interest.

It is immediate to check that the closed-loop system obtained with the event-triggered control algorithm in Example 5.10 does not satisfy (5.35). In fact, that closed-loop system has a Zeno solution at the origin; i.e., from z and ℓ_y equal to zero. On the other hand, the sample-and-hold control example with two events in Example 5.2 and Example 5.6 does satisfy (5.34) and (5.35).

Example 5.12 (Aperiodic sample-and-hold control with two events). *Consider the event-triggered controller for the sample-and-hold implementation in Example 5.6. In particular, the jump set $D_{K,y}$ is given by*

$$\{(y,\eta) \in Y \times X_K : \eta = (\ell_y, \ell_u, (\tau_y, \tau_u, \tau_s, \eta_c)), \tau_y \leq 0\}$$

which, with the definition of X_K, triggers jumps when $\tau_y = 0$. Since $G_{K,y}$ resets the timer to a point in $[T_{y,1}^*, T_{y,2}^*]$ where $0 < T_{y,1}^* \leq T_{y,2}^*$, condition (5.35) holds: $(y,\eta) \in D_{K,y}$ implies $\tau_y = 0$, and $\chi' \in G_{K,y}(y,\eta)$, $\chi' = (\tau_y', \tau_u', \tau_s', \eta_c')$, implies $\tau_y' > 0$. The proof that condition (5.34) holds follows similarly. When the resulting hybrid closed-loop system satisfies the hybrid basic conditions and every maximal solution to it is bounded, Proposition 5.11 guarantees a positive lower bound on the inter-event times for each of its solutions.

Remark 5.13 (*Lower bound via temporal regularization*). The conditions in Proposition 5.11 guarantee a lower bound on the inter-event times through the design of the event-triggering and reset functions. When the conditions therein are not enforced at the design stage, the closed-loop system may have Zeno solutions from initial conditions in $\mathcal{A}$. This is the case for the strategy in Example 5.1; see Example 5.9. A way to guarantee such a lower bound when the conditions in Proposition 5.11 do not hold is to *temporally regularize* the closed-loop system. A temporal regularization consists of the addition of a timer with dynamics that allow events to occur only after a particular positive amount of time has elapsed since the last event. Precise details are given at the end of § 5.4.3. △

5.4.3 Pre-Asymptotic Stability

Before discussing sufficient conditions for pre-asymptotic stability under event-triggered hybrid control, two particular cases of conditions triggering the events that have practical interest are presented. Event-triggered hybrid control with events triggered by a timer is first introduced. This strategy is presented for a linear time-invariant plant. For this class of systems, the case of proportional linear control is formulated, followed by a hybrid implementation of the widely used proportional-integral-derivative (PID) controller. For both such cases, controller design conditions are proposed. These conditions follow the so-called direct design approach; see § 1.2.5. Finally, event-triggered hybrid control with events triggered by an event-triggering function involving a Lyapunov function is presented. The Lyapunov function used certifies a nominal stability property induced by a static state-feedback law. Emulation-based design conditions are proposed for this kind of event.

5.4.3.1 Timer-triggered Events

Consider the continuous-time plant $\dot{z} = F_P(z, u)$, $y = h(z)$ with data

$$F_P(z, u) := Az + Bu, \qquad h(z) := Mz \qquad \forall (z, u) \in \mathbb{R}^{n_P} \times \mathbb{R}^{m_P} \tag{5.36}$$

where A, B, and M are matrices of appropriate dimensions. Following Example 5.2, an event-triggered hybrid controller implementing a proportional output-feedback controller $\kappa_c(y) = K_P y$ with (single) aperiodic events is designed. In this way, the state of the event-triggered hybrid controller is given by $\eta := (\ell_u, \chi)$ with the auxiliary state defined as $\chi = \tau_u \in [0, T_{u,2}^*]$, where τ_u is a timer that triggers a single event, at which the output of the plant is sampled and the input to it is updated. The set X_K is given by $Y \times [0, T_{u,2}^*]$ with $Y := \mathbb{R}^{r_P}$. As in Example 5.2, the dynamics of the timer are governed as in (5.6) with an event-triggering function γ_u given as in (5.7) (with y replaced by u). In this way, the events are triggered when

$$\tau_u = 0 \tag{5.37}$$

The memory state ℓ_u is reset to the value of the control law at each event. Similar to the model in Example 1.3, the controller $\mathcal{H}_K$ implementing this event-triggered control strategy has data given by

$$F_K(\eta) = \begin{bmatrix} 0 \\ -1 \end{bmatrix} \qquad\qquad \forall \eta \in C_K := \left\{ \eta \in X_K \ : \ \tau_u \in [0, T_{u,2}^*] \right\}$$

$$G_K(y, \eta) = \begin{bmatrix} K_P y \\ [T_{u,1}^*, T_{u,2}^*] \end{bmatrix} \qquad \forall (y, \eta) \in D_K := \{ (y, \eta) \in Y \times X_K \ : \ \tau_u = 0 \}$$

$$\kappa(y, \eta) = \ell_u$$

where the controller input v has been assigned to $y = Mz$.

With this hybrid proportional output-feedback controller for the plant with data in (5.36), the hybrid closed-loop system has state $x = (z, \eta)$, with $\eta = (\ell_u, \tau_u)$ and dynamics

$$
\mathcal{H} : \begin{cases}
\tau_u \in [0, T_{u,2}^*] & \begin{bmatrix} \dot{z} \\ \dot{\ell}_u \\ \dot{\tau}_u \end{bmatrix} = \begin{bmatrix} Az + B\ell_u \\ 0 \\ -1 \end{bmatrix} =: F(x) \\[2em]
\tau_u = 0 & \begin{bmatrix} z^+ \\ \ell_u^+ \\ \tau_u^+ \end{bmatrix} \in \begin{bmatrix} z \\ K_P M z \\ [T_{u,1}^*, T_{u,2}^*] \end{bmatrix} =: G(x)
\end{cases}
\tag{5.38}
$$

Its flow set is $C = \mathbb{R}^{n_P} \times C_K$ and its jump set is $D = \{(z, \eta) \in \mathbb{R}^{n_P} \times X_K : \tau_u = 0\}$. The output y of the plant is measured at the events only, and, when those occur, its value is used to update memory state ℓ_u via the proportional output-feedback feedback law. That is, at each event triggered by the event-triggered controller, ℓ_u is updated to $K_P y = K_P M z$, where K_P is the proportional gain matrix.

To design this gain, the linear structure of (5.38) is exploited. More precisely, the fact that, during flows, the dynamics of the state components z and ℓ_u satisfy

$$
\begin{bmatrix} \dot{z} \\ \dot{\ell}_u \end{bmatrix} = \begin{bmatrix} A & B \\ 0 & 0 \end{bmatrix} \begin{bmatrix} z \\ \ell_u \end{bmatrix} =: A_f \begin{bmatrix} z \\ \ell_u \end{bmatrix}
\tag{5.39}
$$

permits characterizing the change of these state components by

$$
\exp(A_f \tau_u) \begin{bmatrix} z \\ \ell_u \end{bmatrix}
\tag{5.40}
$$

from a point $x = (z, \ell_u, \tau_u)$ from where flows are possible. Conveniently, the continuous change of (5.40) in the directions of the flow map of (5.38) is zero since

$$
\left\langle \nabla \left\{ \exp(A_f \tau_u) \begin{bmatrix} z \\ \ell_u \end{bmatrix} \right\}, F(x) \right\rangle = \exp(A_f \tau_u) A_f \begin{bmatrix} z \\ \ell_u \end{bmatrix} + \exp(A_f \tau_u) A_f \begin{bmatrix} z \\ \ell_u \end{bmatrix} (-1)
\tag{5.41}
$$

Moreover, the fact that

$$
\begin{bmatrix} z^+ \\ \ell_u^+ \end{bmatrix} = \begin{bmatrix} I & 0 \\ K_P M & 0 \end{bmatrix} \begin{bmatrix} z \\ \ell_u \end{bmatrix} =: A_y \begin{bmatrix} z \\ \ell_u \end{bmatrix}
\tag{5.42}
$$

at jumps implies that the change of the state components (z, ℓ_u) after a jump followed by an interval of flow is given by

$$
\exp(A_f \nu) A_g
\tag{5.43}
$$

where ν belongs to $[T_{u,1}^*, T_{u,2}^*]$. These observations lead to the result below, which is stated under the following assumption.

Assumption 5.14 (Conditions for timer-triggered events). *Given positive scalars $T_{u,1}^*$ and $T_{u,2}^*$ such that $T_{u,1}^* \leq T_{u,2}^*$, assume the following:*

(ET1) There exist matrices K_P and $P = P^\top > 0$ satisfying

$$
\Gamma(\nu)^\top P \Gamma(\nu) - P < 0 \qquad \forall \nu \in [T_{u,1}^*, T_{u,2}^*]
\tag{5.44}
$$

where $\Gamma(\nu) := \exp(A_f\nu)A_g$, and the matrices A_f and A_g are given in (5.39) and (5.42), respectively.

Theorem 5.15 (Pre-asymptotic stability of event-triggered control with timer-triggered events)**.** *Let $T^*_{u,1}$ and $T^*_{u,2}$ be positive scalars such that $T^*_{u,1} \leq T^*_{u,2}$. Suppose Assumption 5.14 holds. Then, the compact set*

$$\mathcal{A} := \{0\} \times \{0\} \times [0, T^*_{u,2}] \tag{5.45}$$

is globally asymptotically stable for the hybrid closed-loop system $\mathcal{H}$ in (5.38). Furthermore, when F_P is continuous, the pre-asymptotic stability property of $\mathcal{A}$ is robust in the sense of Definition 3.16.

Proof. Consider the partition of the state x as (x_1, x_2) with $x_1 = (z, \ell_u)$ and $x_2 = \tau_u$. Due to the definition of the set $\mathcal{A}$ in (5.45), it follows that the distance from x to the set $\mathcal{A}$ satisfies $|x|_{\mathcal{A}} = |x_1|$. Now, consider the Lyapunov function candidate

$$V(x) = W(\exp(A_f\tau_u)x_1) \qquad \forall x \in \mathbb{R}^{n_P} \times \mathbb{R}^{m_P} \times [0, T^*_{u,2}] \tag{5.46}$$

where $W(x_1) = x_1^\top P x_1$ for each $x_1 \in \mathbb{R}^{n_P} \times \mathbb{R}^{m_P}$. The function V satisfies

$$\underline{c}|x|^2_{\mathcal{A}} \leq V(x) \leq \bar{c}|x|^2_{\mathcal{A}} \quad \forall x \in C \cup D \cup G(D) \tag{5.47}$$

where $0 < \underline{c} \leq \bar{c}$ with

$$\begin{aligned}
\bar{c} &= \max_{\nu \in [0, T^*_{u,2}]} \lambda_{\max}(\exp(A_f^\top \nu)P\exp(A_f\nu)) \\
\underline{c} &= \min_{\nu \in [0, T^*_{u,2}]} \lambda_{\min}(\exp(A_f^\top \nu)P\exp(A_f\nu))
\end{aligned} \tag{5.48}$$

The sets C and D are defined below (5.38). Note that (5.46) is indeed a Lyapunov function candidate according to Definition 3.17; namely, V is continuously differentiable and the set $C \cup D \cup G(D)$ is contained in the domain of V.

The change of V during flows is characterized by the inner product between the gradient of V in the directions of F at each point in C. Using (5.41) and the definition of F in (5.38), it follows that

$$\dot{V}(x) = \langle \nabla V(x), F(x) \rangle = 0 \qquad \forall x \in C$$

To determine the change of V at jumps, note that jumps occur when $\tau_u = 0$. It follows that for each $x \in D$ and each $\chi = (\chi_z, \chi_{\ell_u}, \nu) \in G(x)$, the change in V satisfies

$$\begin{aligned}
\Delta V(x) = V(\chi) - V(x) &= W(\exp(A_f\nu)A_g x_1) - W(x_1) \\
&= x_1^\top (\Gamma(\nu)^\top P\Gamma(\nu) - P)x_1 \\
&\leq -\epsilon|x_1|^2 = -\epsilon|x|^2_{\mathcal{A}}
\end{aligned}$$

where $\epsilon > 0$ is guaranteed to exist in light of the continuity of (5.44). Then, from item 1 in Theorem 3.19 with $\mathcal{U}$ containing an open neighborhood of $\mathbb{R}^{n_P} \times \mathbb{R}^{m_P} \times [0, T^*_{u,2}]$, the set $\mathcal{A}$ is stable for $\mathcal{H}$.

Global pre-attractivity follows from item 2c in Theorem 3.19. In fact, $\Delta V(x) < 0$

for all $x \in D \setminus \mathcal{A}$ using any set $\mathcal{U}$ as above. Due to the fact that $\dot{\tau}_u = -1$ and jumps occur when $\tau_u = 0$, the closed loop does not have any continuous and complete solution. Hence, $\mathcal{A}$ is globally pre-attractive for $\mathcal{H}$.

Finally, maximal solutions to (5.38) are complete by an application of Proposition 2.34. In fact, in such a case, for every point x in D the jump map resets the state x to a point in C from where flows are possible. Flows are possible for each x with the τ_u component in $(0, T^*_{u,2}]$ since $\dot{\tau}_u = -1$. Then, (VC) holds for each $C \setminus D$. Due to F being linear and $G(C \cup D) \subset C \setminus D$, items 2 and 3 in Proposition 2.34 establish completeness of maximal solutions. $\qquad\square$

Theorem 5.15 provides a design condition for the proportional feedback gain K_P in terms of P, the system matrices A, B, and M, and the parameters $T^*_{u,1}$ and $T^*_{u,2}$ of the event-triggered hybrid controller. Certainly, when the events occur periodically, this condition needs to hold only for $\nu = T^*_{u,1} = T^*_{u,2}$. But even in such a case, the design parameter K_P is multiplied by P, which is also an unknown. The following result linearizes (5.44) by using matrix inequalities techniques.

Proposition 5.16 (Linearization of (5.44)). *Given matrices A, B, and M defining (5.36), matrices A_f and A_g as in (5.39) and (5.42), respectively, positive scalar parameters $T^*_{u,1} \leq T^*_{u,2}$, and a matrix $P = P^\top > 0$, the condition in (5.44) holds if there exists a matrix F such that*

$$\begin{bmatrix} -(F + F^\top) & FA_g & \exp\left(A_f^\top \nu\right)P \\ A_g^\top F^\top & -P & 0 \\ P\exp\left(A_f\nu\right) & 0 & -P \end{bmatrix} < 0 \qquad \forall \nu \in [T^*_{u,1}, T^*_{u,2}] \qquad (5.49)$$

Proof. Define

$$Z(\nu) = \begin{bmatrix} \exp\left(A_f^\top \nu\right)P\exp\left(A_f\nu\right) & 0 \\ 0 & -P \end{bmatrix}, \quad S = \begin{bmatrix} A_g \\ I \end{bmatrix}, \quad R = \begin{bmatrix} 0 \\ I \end{bmatrix} \qquad (5.50)$$

Then, condition (5.44) can be rewritten as

$$S^\top Z(\nu)S < 0 \qquad \forall \nu \in [T^*_{u,1}, T^*_{u,2}] \qquad (5.51)$$

The positive definiteness of P can be equivalently expressed as

$$R^\top Z(\nu)R < 0 \qquad \forall \nu \in [T^*_{u,1}, T^*_{u,2}] \qquad (5.52)$$

Now, using the projection lemma, inequalities (5.51) and (5.52) hold if and only if there exists a matrix F such that

$$\begin{bmatrix} \exp\left(A_f^\top \nu\right)P\exp\left(A_f\nu\right) - (F + F^\top) & FA_g \\ A_g^\top F^\top & -P \end{bmatrix} < 0 \qquad \forall \nu \in [T^*_{u,1}, T^*_{u,2}] \qquad (5.53)$$

By Schur's complement, one can obtain

$$\begin{bmatrix} -(F + F^\top) & FA_g & \exp\left(A_f^\top \nu\right) \\ A_g^\top F^\top & -P & 0 \\ \exp\left(A_f\nu\right) & 0 & -P^{-1} \end{bmatrix} < 0 \qquad (5.54)$$

The inequality in condition (5.49) follows by multiplying (5.54) from the left and from the right by

$$\begin{bmatrix} I & 0 & 0 \\ 0 & I & 0 \\ 0 & 0 & P \end{bmatrix}.$$

□

The alternative form of (5.44) given by Proposition 5.16 is linear in P, F, and A_g. However, it still needs to be checked for infinitely many values of ν, that is, for each $\nu \in [T_{u,1}^*, T_{u,2}^*]$. A way to reduce this condition to finite many inequalities consists of embedding $\exp\left(A_f[T_{u,1}^*, T_{u,2}^*]\right)$ into the convex hull of finitely many matrices. The following result provides one such embedding.

Proposition 5.17 (Embedding of (5.49)). *Let $T_{u,1}^*$ and $T_{u,2}^*$ be positive scalars such that $T_{u,1}^* \leq T_{u,2}^*$. For some $\bar{w} \in \mathbb{N}_{\geq 1}$, let the collection of matrices $\{E_1, E_2, \ldots, E_{\bar{w}}\}$ satisfy*

$$\exp(A_f[T_{u,1}^*, T_{u,2}^*]) \subset \mathrm{con}\{E_1, E_2, \ldots, E_{\bar{w}}\} \tag{5.55}$$

If there exist matrices $J = \begin{bmatrix} J_{11} & 0 \\ J_{21} & 0 \end{bmatrix}$ and $F = \begin{bmatrix} F_{11} & F_{12} \\ F_{21} & 0 \end{bmatrix}$ with F_{12} invertible and $J_{21} = F_{21}$, and a positive definite symmetric matrix P such that

$$\begin{bmatrix} -(F + F^\top) & J & E_i^\top P \\ J^\top & -P & 0 \\ PE_i & 0 & -P \end{bmatrix} < 0 \qquad \forall i \in \{1, 2, \ldots, \bar{w}\} \tag{5.56}$$

then the matrices P, A_f, and A_g with K_P such that $K_P M = F_{12}^{-1}(J_{11} - F_{11})$ satisfy condition (5.44).

It is straightforward to show that, under condition (5.55), (5.56) implies (5.49). The structure of J and F are obtained using the form of A_g in the general solution to the linearization obtained from using $J = FA_g$ in (5.49).

The design results above can be extended to the general case when the nominal feedback controller includes integral and derivative actions. For the case of output feedback, such a controller typically assigns the input u to the plant via the standard proportional-integral-derivative (PID) control action

$$u = K_P y + K_I \int_0^t y(s)ds + K_D \dot{y}$$

where K_P, K_I, and K_D are parameters to be designed. In state-space control design, an integral controller requires the introduction of an auxiliary state to store the integral of y. In an event-triggered control implementation of such a controller, the integral can be approximated using the memory state ℓ_y storing the most recent measurement of the output and an auxiliary state ℓ_Ix storing an approximation of the running integral of y. Between sensor measurements, the integral state ℓ_I evolves according to $\dot{\ell}_I = \ell_y$. The integral action is then implemented as $K_I \ell_I$. The input to the plant is assigned to ℓ_u, which in this implementation is updated at each output event.

For the state-feedback case,[4] which is when the output map of the plant with data (5.36) is the identity, the derivative action can be implemented as a function of the measurements. To see this, note that if $u = K_D \dot{z}$ and $I - K_D B$ is invertible, then $K_D \dot{y} = K_D \dot{z} = K_D(Az + Bu)$, from where the derivative action becomes $(I - K_D B)^{-1} K_D A z$. Extending this expression to the general PID control action, the derivative action becomes $(I - K_D B)^{-1} K_D (Az + B K_P z + B K_I \ell_I)$.

Combining the expressions above, an event-triggered hybrid controller $\mathcal{H}_K$ implementing a PID control action for the plant in (5.36) but with $h(z) = z$ has state $\eta = (\ell_y, \ell_u, \ell_I, \tau_u)$ and data defined as

$$
F_K(\eta) = \begin{bmatrix} 0 \\ 0 \\ \ell_y \\ -1 \end{bmatrix} \qquad \forall \eta \in C_K := \left\{ \eta \in X_K : \tau_u \in [0, T^*_{u,2}] \right\}
$$

$$
G_K(z, \eta) = \begin{bmatrix} z \\ (\widetilde{K}_P + \widetilde{K}_D)z + \widetilde{K}_I \ell_I \\ \ell_I \\ [T^*_{u,1}, T^*_{u,2}] \end{bmatrix} \qquad \forall \eta \in D_K = \left\{ \eta \in X_K : \tau_u = 0 \right\}
$$

$$
\kappa(z, \eta) = \ell_u
$$

(5.57)

where

$$
\begin{aligned}
\widetilde{K}_P &= K_P + (I - K_D B)^{-1} K_D B K_P \\
\widetilde{K}_I &= K_I + (I - K_D B)^{-1} K_D B K_I \\
\widetilde{K}_D &= (I - K_D B)^{-1} K_D A
\end{aligned}
$$

(5.58)

and the set X_K is given by $Y \times [0, T^*_{u,2}]$ with $Y := \mathbb{R}^{n_P}$. Note that the controller input v has been assigned to z.

The resulting closed-loop system has state $x = (z, \ell_y, \ell_u, \ell_I, \tau_u) \in X := \mathbb{R}^{n_P} \times \mathbb{R}^{n_P} \times \mathbb{R}^{m_P} \times \mathbb{R}^{n_P} \times [0, T^*_{u,2}]$ and data

$$
F(x) := \begin{bmatrix} A_f x_1 \\ -1 \end{bmatrix} \qquad \forall x \in C := X
$$

$$
G(x) := \begin{bmatrix} A_g x_1 \\ [T^*_{u,1}, T^*_{u,2}] \end{bmatrix} \qquad \forall x \in D := \left\{ x \in X : \tau_u = 0 \right\}
$$

(5.59)

where, in this case, $x_1 = (z, \ell_y, \ell_u, \ell_I)$, and the matrices A_f and A_g are given by

$$
A_f = \begin{bmatrix} A & 0 & B & 0 \\ 0 & 0 & 0 & 0 \\ 0 & 0 & 0 & 0 \\ 0 & I & 0 & 0 \end{bmatrix}, \qquad A_g = \begin{bmatrix} I & 0 & 0 & 0 \\ I & 0 & 0 & 0 \\ \widetilde{K}_P + \widetilde{K}_D & 0 & 0 & \widetilde{K}_I \\ 0 & 0 & 0 & I \end{bmatrix}
$$

(5.60)

The definitions of $\widetilde{K}_P$, $\widetilde{K}_I$, and $\widetilde{K}_D$ depend on (K_P, K_D), (K_I, K_D), and K_D, respectively. For the scalar case, the expressions are so that K_P and K_I can be chosen to yield desired values of $\widetilde{K}_P$ and $\widetilde{K}_I$, even though K_D plays a role in their

---

[4]When $y = Mz$, then the derivative of y is needed to be able to write the controller as a function of the available measurements for the purpose of control.

definition. The design of these gains should be so that

$$\mathcal{A} = \{x \in X \; : \; z = \ell_y = \ell_u = 0\}$$

is globally pre-asymptotically stable for the closed loop. Such design can be performed using similar design conditions as in results above. This is left as an exercise; see Exercise 41.

5.4.3.2 Lyapunov-based Triggered Events

Consider the continuous-time plant[5] $\dot{z} = F_P(z, u)$ and a static state-feedback law $\zeta = \kappa_c(z)$ rendering $\mathcal{A}^* \subset \mathbb{R}^{n_P}$ globally pre-asymptotically stable. This is the setting in Example 5.1. The event-triggered implementation of this static state-feedback law given in Example 5.3 is employed to design, via an emulation-based approach, an event-triggered hybrid controller. The proposed controller has the same structure as the one in Example 5.3, but with an event-triggering function constructed using the Lyapunov function associated to the static state-feedback law.

Assumption 5.18 (Conditions for Lyapunov-based events). *Given a compact set $\mathcal{A}^* \subset \mathbb{R}^{n_P}$ and a plant $\mathcal{H}_P$ given by $\dot{z} = F_P(z, u)$, suppose there exist*

(ET2) *A continuous function $\kappa_c : \mathbb{R}^{n_P} \to \mathbb{R}^{m_P}$, a function $V : \mathbb{R}^{n_P} \to \mathbb{R}_{\geq 0}$ that is continuously differentiable and positive definite with respect to $\mathcal{A}^*$, and a continuous function $\rho : \mathbb{R}^{n_P} \to \mathbb{R}_{\geq 0}$ that is positive definite with respect to $\mathcal{A}^*$ such that*

$$\langle \nabla V(z), F_P(z, \kappa_c(z)) \rangle \leq -\rho(z) \qquad \forall z \in \mathbb{R}^{n_P} \tag{5.61}$$

The proposed event-triggered hybrid controller uses only a memory state ℓ_y that stores the value of the plant state z at each event. In between events, it applies $\kappa_c(\ell_y)$ to the plant. Following Example 5.3, the hybrid controller is given by

$$\mathcal{H}_K \; : \; \begin{cases} \gamma_y(z, \ell_y) \leq 0 & \dot{\ell}_y = 0 \\ \gamma_y(z, \ell_y) \geq 0 & \ell_y^+ = z \end{cases} \tag{5.62}$$

with output $\zeta = \kappa(\ell_y) := \kappa_c(\ell_y)$, where the event-triggering function γ_y is defined as

$$\gamma_y(z, \ell_y) := \langle \nabla V(z), F_P(z, \kappa_c(\ell_y)) \rangle + \mu \rho(z) \tag{5.63}$$

for each (z, ℓ_y), where $\mu \in (0, 1)$ is a design parameter. Note that v has been assigned to z. Following Example 5.9, the hybrid closed-loop system is obtained by further assigning u to ζ. It results in

$$\mathcal{H} \; : \; \begin{cases} x \in C & \begin{bmatrix} \dot{z} \\ \dot{\ell}_y \end{bmatrix} = \begin{bmatrix} F_P(z, \kappa_c(\ell_y)) \\ 0 \end{bmatrix} =: F(x) \\ x \in D & \begin{bmatrix} z^+ \\ \ell_y^+ \end{bmatrix} = \begin{bmatrix} z \\ z \end{bmatrix} =: G(x) \end{cases} \tag{5.64}$$

---

[5]The single-valued case is treated here for simplicity, but the set-valued case follows similarly.

where $x = (z, \ell_y) \in \mathbb{R}^{n_P} \times \mathbb{R}^{n_P}$ and

$$C := \{(z, \ell_y) : \gamma_y(z, \ell_y) \le 0\}, \qquad D := \{(z, \ell_y) : \gamma_y(z, \ell_y) \ge 0\} \qquad (5.65)$$

The following result presents sufficient conditions for pre-asymptotic stability of a closed set for this hybrid closed-loop system.

Theorem 5.19 (Pre-asymptotic stability of event-triggered control with Lyapunov-based triggered events). *Given a compact set $\mathcal{A}^* \subset \mathbb{R}^{n_P}$ and a plant $\mathcal{H}_P$ given by $\dot{z} = F_P(z, u)$, suppose Assumption 5.18 holds. If $\mathcal{H}_P$ satisfies the hybrid basic conditions, then for each controller parameter $\mu \in (0, 1)$ the following hold:*

1. *The hybrid closed-loop system $\mathcal{H}$ in (5.64) satisfies the hybrid basic conditions.*

2. *The closed set $\mathcal{A} := \mathcal{A}^* \times \mathbb{R}^{n_P}$ is pre-asymptotically stable for $\mathcal{H}$.*

Proof. Item 1 is shown using the properties of $\mathcal{H}_P$ from the assumption; the properties of κ_c, V, and ρ from Assumption 5.18; and the properties of the particular construction of γ_y. Since, by Assumption 5.18, V is continuously differentiable and ρ is continuous, and F_P is continuous from $\mathcal{H}_P$ satisfying the hybrid basic conditions, $(v, \ell_y) \mapsto \gamma_y(v, \ell_y) = \langle \nabla V(v), F_P(v, \kappa_c(\ell_y)) \rangle + \mu\rho(v)$ is continuous on its domain of definition; see Exercise 95. Then, since κ_c is continuous by Assumption 5.18, the sets C and D are closed; see Exercise 100. Since F_P is continuous and, from Assumption 5.18, κ_c is continuous, then F is continuous. Continuity of G follows directly from its definition. Finally, by construction, F_P and G_P are defined on C and on D, respectively. Hence, $\mathcal{H}$ satisfies the hybrid basic conditions. Alternative, one can apply Lemma 5.7.

To show item 2, consider the Lyapunov function candidate on $\mathbb{R}^{n_P}$ with respect to $\mathcal{A}$ given by $\widetilde{V}(x) = V(z)$ for each $x \in \mathbb{R}^{n_P} \times \mathbb{R}^{n_P}$. By construction of C, for each $(x, \ell_y) \in C$ it follows that $\gamma_y(z, \ell_y) \le 0$. Then, from the definition of γ_y in (5.63),

$$\left\langle \nabla \widetilde{V}(z), F_P(z, \kappa_c(\ell_y)) \right\rangle \le -\mu\rho(z) \qquad \forall (z, \ell_y) \in C \qquad (5.66)$$

At jumps, since G does not change the state component z, it follows that

$$\widetilde{V}(G(x)) - \widetilde{V}(x) \qquad \forall x = (z, \ell_y) \in D \qquad (5.67)$$

Then, by item 1 in Theorem 3.19 with $\mathcal{U} = \mathbb{R}^{n_P} \times \mathbb{R}^{n_P}$ the set $\mathcal{A}$ is stable for $\mathcal{H}$.

To show pre-attractivity, note that (5.66) and (5.67) yield

$$\dot{\widetilde{V}}^{-1}(0) \subset \mathcal{A}, \qquad \Delta\widetilde{V}^{-1}(0) = D \qquad (5.68)$$

where $\dot{\widetilde{V}}^{-1}$ and $\Delta\widetilde{V}^{-1}(0)$ are defined in (3.30) and (3.31), respectively. Then, since $\mathcal{H}$ satisfies the hybrid basic conditions, item 1 of Theorem 3.23 implies that every precompact solution to $\mathcal{H}$ converges to the largest weakly invariant subset of

$$\widetilde{V}^{-1}(r) \cap (\mathcal{A} \cup (D \cap G(D))) \qquad (5.69)$$

for some $r \ge 0$. To determine $D \cap G(D)$, note that each $x = (z, \ell_y) \in D$ satisfies

$$\gamma_y(z, \ell_y) \ge 0 \quad \Longleftrightarrow \quad \left\langle \nabla \widetilde{V}(z), F_P(z, \kappa_c(\ell_y)) \right\rangle \ge -\mu\rho(z) \qquad (5.70)$$

From the definition of G, it follows that after a jump, $x = (z, \ell_y) \in D$ is mapped to (z, z). Every such z satisfies property (5.61) in (ET1). Then, with (5.70), for x to belong to both D and $G(D)$ it has to satisfy

$$-\mu\rho(z) \le \left\langle \nabla \widetilde{V}(z), F_P(z, \kappa_c(\ell_y)) \right\rangle \le -\rho(z)$$

which, due to $\mu \in (0, 1)$, is only possible when $z \in \mathcal{A}^*$. Then, (5.69) can only hold with $r = 0$, meaning that it is equal to $\mathcal{A}$ due to $\widetilde{V}$ vanishing on that set only. To conclude pre-attractivity of $\mathcal{A}$, what is left to show is that there exists $\mu > 0$ such that every maximal solution x to $\mathcal{H}$ with $|x(0,0)|_{\mathcal{A}} \le \mu$ has a bounded distance to $\mathcal{A}$, namely, the z component of each such solution is bounded. By positive definiteness of V with respect to $\mathcal{A}^*$, since the second component of $\mathcal{A}$ is $\mathbb{R}^{n_P}$, there exists $\mu' > 0$ such that $L_V(\mu')$ is compact. Then, since $\mathcal{A}^*$ is compact, stability of $\mathcal{A}$ implies boundedness of the z component of each solution from that set. $\qquad\square$

The properties established by Theorem 5.19 rely on the strict decrease of V during flows, as (5.66) indicates, along with the fact that due to (5.67), V remains constant at jumps. The Hybrid Lyapunov Theorem in Theorem 3.19 can be employed to design an event-triggered controller for which V may grow in one of the regimes. One such controller is given as follows:

$$\mathcal{H}_K : \begin{cases} \gamma_u(z, \ell_u) \le 0, \ \tau_u \in [0, T^*_{u,2}] & \dot{\ell}_u = 0, \ \dot{\tau}_u = 1 \\ \gamma_u(z, \ell_u) \ge 0, \ \tau_u \in [T^*_{u,1}, T^*_{u,2}] & \ell^+_u = \kappa_c(z), \ \tau^+_u = 0 \end{cases} \quad (5.71)$$

with output $\zeta = \kappa(\ell_u) := \ell_u$, where the event-triggering function γ_u is defined as

$$\gamma_u(z, \ell_u) := \langle \nabla V_P(z), F_P(z, \ell_u) \rangle - \mu V_P(z) \quad (5.72)$$

for each (z, ℓ_u). The parameters $T^*_{u,1}$ and $T^*_{u,2}$ are positive and satisfy $T^*_{u,1} \le T^*_{u,2}$, while μ is a constant parameter of the controller. The function V_P is continuously differentiable, positive definite, and radially unbounded. As a difference to the controller in (5.62), this controller triggers input events and uses ℓ_u. The hybrid closed-loop system is obtained by assigning u to ζ. It results in

$$\mathcal{H} : \begin{cases} x \in C & \begin{bmatrix} \dot{z} \\ \dot{\ell}_u \\ \dot{\tau}_u \end{bmatrix} = \begin{bmatrix} F_P(z, \ell_u) \\ 0 \\ 1 \end{bmatrix} =: F(x) \\[20pt] x \in D & \begin{bmatrix} z^+ \\ \ell^+_u \\ \dot{\tau}^+_u \end{bmatrix} = \begin{bmatrix} z \\ \kappa_c(z) \\ 0 \end{bmatrix} =: G(x) \end{cases} \quad (5.73)$$

where $x = (z, \ell_u, \tau_u) \in \mathbb{R}^{n_P} \times \mathbb{R}^{n_P} \times [0, T^*_{u,2}] =: X$ and

$$\begin{aligned} C &:= \left\{ (z, \ell_u, \tau_u) : \gamma_y(z, \ell_u) \le 0, \tau_u \in [0, T^*_{u,2}] \right\} \\ D &:= \left\{ (z, \ell_u, \tau_u) : \gamma_y(z, \ell_u) \ge 0, \tau_u \in [T^*_{u,1}, T^*_{u,2}] \right\} \end{aligned} \quad (5.74)$$

Using the Lyapunov function candidate

$$V(z, \ell_u, \tau_u) = \exp(\sigma\tau_u)V_P(z) \qquad \forall (z, \ell_u, \tau_u) \in X$$

with $\sigma > 0$ such that

$$\mu + \sigma < \sigma \frac{T_{u,1}^*}{T_{u,2}^*}$$

item 3d in Theorem 3.19 certifies global pre-asymptotic stability of the closed set $\mathcal{A} := \{x \in X : z = 0\}$ for the hybrid closed-loop system in (5.73). Interestingly, the function V satisfies

$$\dot{V} \leq (\mu + \sigma)V$$

during flows, which may lead to a growth of V when $\mu + \sigma$ is positive. At jumps, the function V is guaranteed to decrease. Checking these properties are left as an exercise for the reader.

5.4.3.3 Pre-asymptotic Stability for General Strategies

The event-triggered hybrid controller in (5.9)-(5.15) is general enough to capture the majority of algorithms with events available in the literature. The Hybrid Lyapunov Theorem in Theorem 3.19 and the Hybrid Invariance Principle in Theorem 3.23 are applicable in most settings. The use of these results has been detailed in the proofs of Theorem 5.15 and Theorem 5.19, which serve as a guideline on how to apply them to closed-loop systems resulting from other event-triggering control strategies. A summary of specific strategies in the literature that have already been analyzed with such tools is given in §5.6.

When an event-triggered controller is designed to pre-asymptotically stabilize a desired compact set without enforcing a positive lower bound on the inter-event times, a uniform and positive such lower bound can be guaranteed through temporal regularization. Following Remark 5.13, for the closed-loop system $\mathcal{H} = (C, F, D, G)$ with state x resulting from using an event-triggered controller, temporal regularization requires appending a timer state τ with positive threshold T^*. The augmented version of the closed-loop system is denoted $\widetilde{\mathcal{H}}$, has state $\widetilde{x} = (x, \tau) \in \mathbb{R}^n \times \mathbb{R}_{\geq 0}$, and dynamics

$$\begin{aligned}
\widetilde{x} &\in (C \times \mathbb{R}_{\geq 0}) \cup (\mathbb{R}^n \times [0, T^*]) & \dot{\widetilde{x}} &\in F(x) \times \rho(\tau) \\
\widetilde{x} &\in D \times [T^*, \infty) & \widetilde{x}^+ &\in G(x) \times \{0\}
\end{aligned} \qquad (5.75)$$

where ρ is designed to have τ converge to $[0, T^*]$. A particular choice is $\rho(\tau) = 1$ for each $\tau \in [0, T^*)$, $\rho(\tau) = [0, 1]$ for $\tau = T^*$, and $\rho(\tau) = -\tau + T^*$ for each $\tau > T^*$. Note that when $T^* = 0$, the x component of the solutions to $\widetilde{\mathcal{H}}$ matches those of $\mathcal{H}$. The following result states key properties for $\widetilde{\mathcal{H}}$.

Theorem 5.20 (Temporal regularization for event-triggered control). *Suppose that the hybrid closed-loop system $\mathcal{H}$ with data in (5.22)-(5.25) satisfies the hybrid basic conditions, and that the set $\mathcal{A} \subset \mathbb{R}^n$ is compact and pre-asymptotically stable for $\mathcal{H}$ with basin of pre-attraction $\mathcal{B}_\mathcal{A}$. Then, the set $\mathcal{A} \times [0, T^*]$ is semiglobally practically (in the parameter T^*) pre-asymptotically stable on $\mathcal{B}_\mathcal{A} \times \mathbb{R}_{\geq 0}$; i.e., there exists a* *class-$\mathcal{KL}$ function $\widetilde{\beta}$ such that for each compact set $K_x \times K_\tau \subset \mathcal{B}_{\mathcal{A}} \times \mathbb{R}_{\geq 0}$ and each $\varepsilon > 0$ there exists $T^{*\prime} > 0$ such that for each $T^* \in [0, T^{*\prime})$, every solution $\widetilde{x}$ to $\widetilde{\mathcal{H}}$ in (5.75) with $\widetilde{x}(0,0) \in K_x \times K_\tau$ satisfies*

$$|\widetilde{x}(t,j)|_{\mathcal{A} \times [0,T^*]} \leq \widetilde{\beta}(|\widetilde{x}(0,0)|_{\mathcal{A} \times [0,T^*]}, t+j) + \varepsilon \qquad \forall (t,j) \in \operatorname{dom} \widetilde{x} \qquad (5.76)$$

As the result states, when the hybrid closed-loop system satisfies the hybrid basic conditions and the event-triggered controller pre-asymptotically stabilizes a compact set but induces Zeno, the stability property can be preserved (semiglobally and practically) by setting the temporal regularization parameter T^* small enough so that solutions converge to a neighborhood of $\mathcal{A}$ of desired size ε.

5.5 EXERCISES

Exercise 36 (Global stabilization using sample-and-hold control). Consider the continuous-time plant $\dot{z} = F_P(z, u)$, $y = h(z)$ with

$$F_P(z, u) := \begin{bmatrix} 0 & 1 \\ 0 & 0 \end{bmatrix} z + \begin{bmatrix} 0 \\ 1 \end{bmatrix} u, \qquad h(z) := z \qquad \forall (z, u) \in \mathbb{R}^2 \times \mathbb{R} \qquad (5.77)$$

Suppose the measurements of y are available at isolated time instances separated by $T^* > 0$.

1. Design a controller such that the following holds for the resulting closed-loop system:

   a) The hybrid basic conditions are satisfied.
   b) Every maximal solution is complete.
   c) A compact set with plant component equal to the origin is globally asymptotically stable.

2. Redesign your controller so that it globally asymptotically tracks the desired set-point output given by the constant $(z_{r,1}, 0) \in \mathbb{R}^2$ with a rise time for the plant output that is less than or equal to t_r. Determine conditions relating T^*, $z_{r,1}$, and t_r.

Validate your designs numerically for the case where $z_{r,1} = 1$ and $t_r = 0.01$ sec.

Exercise 37 (Global stabilization using event-triggered control). Consider the continuous-time plant $\dot{z} = F_P(z, u)$ in (5.77).

1. Show that there exist a matrix $K_P \in \mathbb{R}^{2 \times 2}$, a quadratic function V_P, and linear functions α_1, α_2, ρ_1, and ρ_2 such that (5.3) is satisfied with $\kappa_c(z) = K_P z$.

2. Pick suitable functions ρ_1 and ρ_2 to define the data of the hybrid controller in Example 5.3 so that the resulting hybrid closed-loop system has the following properties:

    a) The hybrid basic conditions are satisfied.

    b) Every maximal solution is complete.

    c) A closed set with plant component equal to the origin is globally asymptotically stable.

    d) Are there matrices K_P and functions V_P, α_1, α_2, ρ_1, and ρ_2 so each solution to the closed-loop system has nonzero (not necessarily uniform) lower bound on the inter-event times?

3. Validate your design numerically.

4. Add a desired set-point output desired set-point output given by the constant $(z_{r,1}, 0) \in \mathbb{R}^2$ and tune the functions ρ_1 and ρ_2 to achieve a rise time for the plant output that is as close as possible to t_r. Compare your numerical results with those obtained in item 2 of Exercise 36.

Exercise 38 (Event-triggered control under temporal regularization). Consider the plant and event-triggered controller in Example 5.10.

1. Show that the maximal solution to the hybrid closed-loop system in (5.30) with initial condition is $z(0,0) = \ell_y(0,0) = 1$ under the effect of measurement noise $w_u = -\ell_y / \sqrt{|\ell_y|}$ and with $\sqrt{\sigma} = 1/2$ is such that (5.31) is finite. What is the largest lower bound on inter-event times? Does this property also hold for other maximal solutions to the closed loop therein?

2. Show that the hybrid closed-loop system $\mathcal{H}$ in (5.30) has the compact set $\mathcal{A} := \{0\} \times \{0\}$ globally asymptotically stable.

3. Augment $\mathcal{H}$ in (5.30) using the temporal regularization presented at the end of § 5.4.3. Pick $K_x = 10\mathbb{B}$ and $K_\tau = [0, 1]$, and numerically establish a relationship between $\varepsilon > 0$ and $T^* \in [0, T^{*\prime})$ in Theorem 5.20 as follows:

    a) Uniformly discretize the sets K_x, K_τ, and $[0, T^{*\prime})$ with a desired precision.

    b) For each element in the discretized version of $[0, T^{*\prime})$ defining the value of T^*:

       • Simulate the augmented hybrid closed-loop system for each initial condition in $K_x \times K_\tau$ for long enough hybrid time to identify an upper bound on ε such that (5.76) holds.

       • Document the value of T^* and associated upper bound on ε in a table.

    c) Provide both linear and polynomial best fits for the relationship between ε and τ; that is, determine $\varepsilon = \varphi(\tau)$ with $\varphi(s)$ equal to $a + bs$ and $a + bs + cs^2 + \ldots$.

Exercise 39 (Event-triggered control, revisited). Consider the discretized version of an event-triggered implementation of a PI controller in Exercise 17.

1. Define the data of the hybrid controller $\mathcal{H}_K$ in (5.9)-(5.15) to model the discretized version of the event-triggered implementation of a PI controller.

2. Show that the resulting hybrid closed-loop system satisfies the hybrid basic conditions.

3. Show that there exists a uniform positive lower bound on the inter-event times of the solutions to the closed-loop system.

4. For the plant given by a scalar system $\dot{z} = Az + Bu$, $y = Mz$ with $A = B = M = 1$, use the ideas in Theorem 5.15 to design the parameters of the event-triggered controller so that the origin for the plant is globally asymptotically stable. Validate your design numerically.

Exercise 40 (Dynamic output-feedback event-triggering control). Consider the continuous-time plant

$$
\begin{aligned}
\dot{x}_p &= A_p x_p + B_p u, \qquad x_p \in \mathbb{R}^{n_p} \\
y &= C_p x_p
\end{aligned}
$$

the dynamic controller

$$
\begin{aligned}
\dot{x}_c &= A_c x_c + B_c y, \qquad x_c \in \mathbb{R}^{n_c} \\
u &= C_c x_c + D_c y
\end{aligned}
$$

and the following event-triggered implementation of the dynamic controller:

- Memory variables $\hat{u}$ and $\hat{y}$ store the latest values of the output of the controller u and of the output of the plant y, respectively.

- Defining $e_y = \hat{y} - y$ and $e_u = \hat{u} - u$, the events updating $\hat{u}$ and $\hat{y}$ are triggered when

$$
|e_y|^2 \geq \sigma_y |y|^2 + \varepsilon_y \qquad \text{or} \qquad |e_u|^2 \geq \sigma_u |u|^2 + \varepsilon_u
$$

for some positive parameters σ_y, σ_u, ε_y, and ε_u.

The matrices $A_p, B_p, C_p, A_c, B_c, C_c$, and D_c have appropriate dimension.

1. Define the data of the hybrid controller $\mathcal{H}_K$ in (5.9)-(5.15) to model the event-triggered implementation of the dynamic controller outline above.

2. Show that the resulting hybrid closed-loop system satisfies the hybrid basic conditions.

3. Show that there exists a uniform positive lower bound on the inter-event times of the solutions to the closed-loop system.

4. Design the parameters σ_y, σ_u, ε_y, and ε_u and the matrices A_c, B_c, C_c, and D_c of the hybrid controller so that the hybrid closed-loop system has a globally asymptotically stable closed set with plant component equal to the origin.

5. Validate your results numerically for the plant defined by $A_p = \begin{bmatrix} 1 & 2 \\ -2 & 1 \end{bmatrix}$, $B_p = \begin{bmatrix} 0 \\ 2 \end{bmatrix}$, $C_p = [2\ 0]$, $A_c = \begin{bmatrix} -2 & 2 \\ -3 & -2 \end{bmatrix}$, $D_p = 0$, and the dynamic controller by $B_c = \begin{bmatrix} \frac{3}{2} \\ \frac{1}{4} \end{bmatrix}$, $C_c = [-\frac{1}{4}\ -\frac{3}{2}]$, and $D_c = 0$.

Exercise 41 (Event-triggered PID). Consider the model of the PID with events in (5.57).

1. Show that every maximal solution to the hybrid closed-loop system is complete.

2. Formulate the problem of designing the gains $\widetilde{K}_P$, $\widetilde{K}_I$, and $\widetilde{K}_D$ as the problem of asymptotically stabilizing a compact set.

3. Provide sufficient conditions for global pre-asymptotic stability of the desired compact set.

4. Apply the design conditions obtained in item 3 to the plant $\mathcal{H}_P$ with data

$$F_P(z, u) = Az + Bu, \qquad h(z) = Mz, \qquad C_P = \mathbb{R}^{n_P} \times \mathbb{R}^{m_P}, \qquad D_P = \emptyset$$

with G_P arbitrary and

$$A = \begin{bmatrix} 0 & 1 \\ -1 & 0 \end{bmatrix}, \qquad B = \begin{bmatrix} 0 \\ 1 \end{bmatrix}, \qquad M = \begin{bmatrix} 1 & 0 \end{bmatrix}$$

Pick the gains of the controller to meet the following specifications:

- Rise time less than or equal to 2 seconds;

- Overshoot less than or equal to 15%.

Exercise 42 (Global stabilization using sample-and-hold control). Consider the continuous-time plant

$$\dot{x}_p = f_p(x_p, u), \qquad x_p \in \mathbb{R}^{n_p}$$

the dynamic controller

$$\dot{x}_c = f_c(x_c, x_p), \qquad x_c \in \mathbb{R}^{n_c}$$
$$u = g_c(x_c, x_p),$$

and the following event-triggered implementation of the dynamic controller:

- Memory variables $\hat{u}$ and $\hat{x}_p$ store the latest values of the output of the controller u and of the state of the plant x_p, respectively.

- Defining $e = (\hat{x}_p - x_p, \hat{u} - u)$, the events updating $\hat{u}$ and $\hat{y}$ are triggered when an auxiliary state χ satisfies $\chi = \underline{c} > 0$, which resets χ to $\overline{c}$, where $\underline{c} < \overline{c}$. During flows, the auxiliary state χ evolves according to

$$\dot{\chi} = -2L(x_p, x_c, e)\chi - \chi^2 - \gamma(x_p, x_c, e)$$

where L and γ are continuous functions that satisfy the following: for some continuously differentiable functions $V : \mathbb{R}^{n_p} \times \mathbb{R}^{n_c} \to \mathbb{R}$ and $W : \mathbb{R}^{n_e} \to \mathbb{R}$; class-$\mathcal{K}_\infty$ functions $\underline{\alpha}_V, \overline{\alpha}_V, \underline{\alpha}_W$, and $\overline{\alpha}_W$; and a class-$\mathcal{K}$ function ρ satisfy for all $e \in \mathbb{R}^{n_e}$ and all $(x_c, x_p) \in \mathbb{R}^{n_p} \times \mathbb{R}^{n_c}$

$$\underline{\alpha}_W(|e|) \leq W(e) \leq \overline{\alpha}_W(|e|), \quad \dot{W}(e) \leq L(x_p, x_c, e)W(e) + H(x_p, x_c)$$

and

$$\underline{\alpha}_V(|(x_p, x_c)|) \leq V(x_p, x_c) \leq \overline{\alpha}_V(|(x_p, x_c)|)$$
$$\dot{V}(x_p, x_c) \leq -\rho(|(x_c, x_p)|) - \rho(|e|) - H(x_c, x_p)^2 + \gamma(x_c, x_p, e)W(e)^2$$

The functions f_p, f_c, and g_c are given.

1. Define the data of the hybrid controller $\mathcal{H}_K$ in (5.9)-(5.15) to model the event-triggered implementation of the dynamic controller.

2. Show that the resulting hybrid closed-loop system satisfies the hybrid basic conditions.

3. Show that every maximal solution to the closed loop is complete.

4. Show that the compact set

$$\mathcal{A} := \{(x_p, x_c, e, \chi) : x_p = 0, x_c = 0, e = 0, \chi \in [\underline{c}, \overline{c}]\}$$

is globally pre-asymptotically stable.
Hint: use $V(x_p, x_c) + \chi W(e)^2$ as a Lyapunov function candidate.

5. Show that each maximal solution to the closed loop has a positive lower bound on the inter-event times.

5.6   NOTES

The hybrid model of the sample-and-hold controller in §5.1 is essentially the one in [1, Example 1.4]. The general model of event-triggered controllers introduced in §5.1 and presented in detail in §5.2, as well as similar versions to the results presented in this chapter appeared in [80] and [114]. The conditions and control strategy in Example 5.1 follow those in [115], where an event-triggered algorithm for scheduling tasks in embedded processors is proposed. The nondeterministic timer model used in (5.6) to model aperiodic events was employed in [116] to model the intermittent availability of measurements in the context of an observer problem – similar models were used in the context of network control; see, e.g., [117].

Proposition 5.8 providing sufficient conditions for maximal solutions being complete exploits the general result on existence of solutions in Proposition 2.34. Conditions for that purpose in [118, Theorem 1] require $G(D) \subset C \cup D$ and $F(x) \subset T_C(x)$ for each $x \in C \setminus D$.

Proposition 5.11 provides a lower bound on the inter-event times and is a slight variation of [9, Lemma 2.7]; see also item 1(ii) in Proposition 2.34. The presence of the Zeno solution at $\mathcal{A} = \{0\}$ in the hybrid closed-loop system in Example 5.10 is due to the hybrid closed-loop system therein not satisfying the conditions in Proposition 5.11. Such issue has been pointed out in the literature (see, e.g., [114]) and has motivated the introduction of the following dwell-time notion that holds only when the solutions are outside the set $\mathcal{A}$: solutions to $\mathcal{H}$ with data in (5.22)-(5.25) have a *uniform semiglobal dwell-time outside* $\mathcal{A}$, with $\mathcal{A}$ forward invariant, if for any $\delta > 0$ there exists $T > 0$ such that for any solution x to $\mathcal{H}$

$$\left. \begin{array}{l} |x(0,0)|_{\mathcal{A}} \leq \delta, \quad (s,i),(t,j) \in \mathrm{dom}\, x \\ \qquad s + i \leq t + j, \quad x(t,j) \notin \mathcal{A} \end{array} \right\} \quad \Longrightarrow \quad j - i \leq \frac{t-s}{T} + 1$$

It can be shown that this property holds for the solutions to the hybrid closed loop in (5.30) that start outside the origin and are not affected by noise.

The "dwell-time outside $\mathcal{A}$" notion outlined above was introduced in [118]. Sufficient conditions to guarantee it can be found therein; see [118, Proposition 1]. More interestingly, [118] proposes two event-triggered control strategies that exploit the flexibility provided by the auxiliary state χ, which is part of the state η of the hybrid controller proposed in this chapter – note that the states ℓ_y and ℓ_u are denoted as $\hat{x}_p$ and $\hat{u}$ in [118], respectively. For instance, the strategy in [118, Section V.B] uses the auxiliary state χ as a timer that flows according to a nonlinear differential equation and that, at jumps, is reset to a constant. The event-triggered control strategy in Exercise 42 follows the construction in [118, Section V.B]. Remarkably, under appropriate conditions, the resulting hybrid closed-loop system has solutions that satisfy the "dwell-time outside $\mathcal{A}$" property, uniformly and semiglobally.

The lack of robustness of the inter-event times in the system in Example 5.10, which was reported in [114], was previously observed in [119] – see Theorem IV.1 therein, in addition to other insightful discussions about inter-event times.

Sufficient conditions for pre-asymptotic stability of event-triggered control algorithms with events triggered by timers that are similar to those in Theorem 5.15, Proposition 5.16, and Proposition 5.17 appeared in [116] and [120], the latter pertaining to an observer-based controller with two types of events. Sufficient conditions for pre-asymptotic stability of event-triggered algorithms with events triggered by Lyapunov-based conditions have been widely studied in the literature. The particular one provided in Theorem 5.19 appeared in [80], and several other related results are available in the vast literature of event-triggered control, some of which are cited next.

For linear systems with dynamic output feedback, the stability and input/output performance of event-triggered control strategies are studied in [121]. The survey paper [122] collects many more event-triggered control strategies, classifies them into different categories, such as event-triggered and self-triggered, and highlights key properties they guarantee. Moreover, the recent application of event-triggered control to a plethora of different problems, such as the stabilization of control affine systems [123, 124], attitude control [125], and quadrotor stabilization [126] further highlight the importance of the development of analysis and synthesis tools for event-triggered control systems. The general formulation proposed in this chapter captures closed-loop systems resulting from using both static and dynamic output (or state) asynchronous event-triggered feedback laws. It also allows for local events triggered by part of the state components [127, 128], which may involve memory states storing the most recent controller and output values. Approaches that model the closed-loop systems in a similar manner as done in this chapter include [129], [122], and [118].

Chapter Six

Throw-Catch Control

The uniting control strategy in Chapter 4 provides a solution to the asymptotic stabilization problem when a static-state feedback controller, $\mathcal{H}_{K,1}$, inducing a global attractivity property is available. One such controller bringing all solutions to a desired region of operation may not always exist or it might be difficult to design. For instance, as outlined in § 1.2.4, topological obstructions might prohibit the design of a static state-feedback law satisfying the requirements in item (UC-A2) in Assumption 4.3. Such types of obstructions motivate the extension of uniting control presented in this chapter, which is called *throw-catch control*.

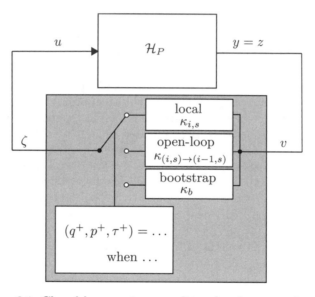

Figure 6.1: Closed-loop system resulting for throw-catch control.

The hybrid control strategy in this chapter extends the uniting control strategy in Chapter 4 by allowing for the use of more than two controllers, some of which might be feedback controllers and some others might be open-loop controllers. The solution outlined in Example 1.5 to globally swinging up the pendubot uses such a combination of controllers. Figure 6.1 depicts the associated hybrid control architecture. The local state-feedback laws are capable of *catching* the state around isolated points (or sets). The open-loop control laws are able to steer the state between those points. The bootstrap controller is capable of bringing the state to points where throw and catch is possible.

The throw-catch strategy is particularly suitable for settings where it is possible to design state-feedback control laws that locally asymptotically stabilize isolated points in the state space, and to "connect" those points via open-loop control laws, following an order that steers the solutions to the desired equilibrium point or set. An additional feedback controller capable of driving the plant solution to nearby points where the local controllers work would be needed to induce an attractivity property that is similar to that of the "global" controller in uniting control. Such a controller is referred to as "bootstrap" due to bringing the system to points where convergence to the desired configuration happens. As in the uniting control strategy, using measurements of the state of the plant, a logic-based algorithm selects the control law to be applied in real (hybrid) time among the several ones available, namely, static state-feedback controllers, open-loop control laws, and the bootstrap controller.

6.1 OVERVIEW

Given a plant $\mathcal{H}_P$, suppose the goal is to render a point or a set, denoted $\mathcal{A}_0$, globally asymptotically stable by exploiting the "divide and conquer" advantage provided by hybrid feedback control, with the added value of robustness. While designing a single, continuous static state-feedback controller might be a possibility, as stated in Chapter 4, the design of such a controller for certain applications is not always easy, and in some situations, it is impossible. One of the main reasons to such impossibility is that for certain plants, discontinuous state-feedback controllers are needed to steer the solutions of the nominal system to $\mathcal{A}_0$. Unfortunately, when measurement noise is introduced, closed-loop solutions may remain far away from $\mathcal{A}_0$, even when the size of the measurement noise is arbitrarily small; see § 1.2.4.

The main idea in the control strategy presented in this chapter is as follows. Suppose that, for the given plant, it is known how to steer its solutions from a neighborhood $\mathcal{S}_1$ of some set $\mathcal{A}_1$ to a neighborhood $\mathcal{E}_0$ of the set $\mathcal{A}_0$, via a n open-loop control law given by the time function $\kappa_{1\to0}$. Moreover, suppose that local state-feedback controllers κ_0 and κ_1 can be designed to locally asymptotically stabilize the sets $\mathcal{A}_0$ and $\mathcal{A}_1$, respectively. Finally, suppose that there exists a state-feedback law κ_b that is capable of steering the plant solutions to points that are either in $\mathcal{S}_1$ or in $\mathcal{E}_0$, when they start from points that are far from $\mathcal{A}_1$ and $\mathcal{A}_0$.

With these elements, the throw-catch strategy proposed in this chapter performs the following tasks:

- *Catch nearby* $\mathcal{A}_1$: If the state z is near $\mathcal{A}_1$, then apply κ_1.

- *Throw from nearby* $\mathcal{A}_1$ *to nearby* $\mathcal{A}_0$: While applying κ_1, when the state z reaches $\mathcal{S}_1$, then apply $\kappa_{1\to0}$.

- *Catch nearby* $\mathcal{A}_0$: While applying $\kappa_{1\to0}$, when the state z reaches $\mathcal{E}_0$, then apply κ_0.

- *Recovery:* If the state z is neither near $\mathcal{A}_1$ nor $\mathcal{A}_0$, then apply κ_b.

Figure 6.2 depicts a sample solution when applying this strategy to a plant $\mathcal{H}_P$ given by a continuous-time system. In "Catch nearby $\mathcal{A}_1$" and "Catch nearby $\mathcal{A}_0$" outlined above, the state of the plant is steered to nearby $\mathcal{A}_1$ and $\mathcal{A}_0$, respectively. This logic can be seen as "catching" the state, and consequently, it is referred to as *catch mode*. In "Throw from nearby $\mathcal{A}_1$ to nearby $\mathcal{A}_0$" the state of the plant is transferred from points that are nearby $\mathcal{A}_1$ to points that are nearby $\mathcal{A}_0$. This task can be interpreted as "throwing" the state from an initial location nearby $\mathcal{A}_1$ to another location nearby $\mathcal{A}_0$. This stage is called *throw mode*, giving rise to the name *throw-catch* for this strategy. In "Recovery" the state of the plant is steered to a point nearby $\mathcal{A}_1 \cup \mathcal{A}_0$ so that catch mode or throw mode can be employed. This mode is called *recovery mode*.

Additional logic is added to the strategy to improve robustness when using the open-loop control law during throw mode. Since in that mode the controller does not use any information about the state of the system, if due to exogenous disturbances the control law $\kappa_{1\to0}$ fails to steer the state of the plant to the neighborhood $\mathcal{E}_0$ of $\mathcal{A}_0$, then the control logic should steer the state back to a location from which another control law can be applied. This capability is included as part of recovery mode in the throw-catch strategy.

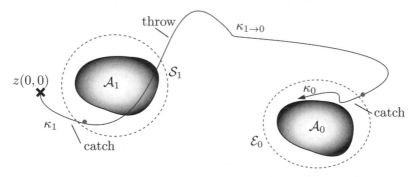

Figure 6.2: A solution to the plant starting at $z(0,0)$ under the effect of the local state-feedback law κ_1 (catch) reaches the set $\mathcal{S}_1$ from where the controller switches to the open-loop controller $\kappa_{1\to0}$ (throw). Once the state is in $\mathcal{E}_0$, the local control law κ_0 (catch) is applied.

The throw-catch hybrid control strategy can be used to stabilize plants with nonlinear dynamics, with multiple equilibrium points, and when the information available for feedback control is limited. One such plant is the pendubot shown in Figure 1.6. For this system, local stabilization around the resting (both links "down") and upright (both links "up") equilibrium points of the system is possible via static state feedback. These two control laws can be employed during catch mode. Open-loop control laws that are capable of driving the state away from points where one link is "down" and another is "up" – as Figure 1.6 shows, this system has two such equilibrium points – to points nearby the resting equilibrium point can be designed. These open-loop control laws can be used in throw mode and can be designed, for instance, by solving two-point boundary value problems, or by trial and error. The following example revisits the swing up problem for the pendubot in the context of throw-catch control.

Example 6.1 (Global swing up of the pendubot). *Consider the problem of globally asymptotically stabilizing the upright configuration of the pendubot system shown in Figure 1.6. A mathematical model of this nonlinear continuous-time plant is outlined in Example 1.5. The target set to globally asymptotically stabilize is given the state values describing the static upright condition for the pendubot, i.e., both links "up." According to the definition of the plant state and the coordinates used therein, such a configuration corresponds to points $z = (\phi_1, \omega_1, \phi_2, \omega_2) \in \mathbb{R}^4$ with zero angular velocities ω_1 and ω_2, and angles ϕ_1 and ϕ_2 equal to zero (or any multiple of 2π). To avoid dealing with infinitely many equilibrium points or restricting the range of the angles to $(-\pi, \pi)$, ϕ_1 and ϕ_2 are given by the angle of the vectors ξ_1 and ξ_2 on the unit circle $\mathbb{S}^1$, respectively. Using this embedding technique, the problem of globally stabilizing the pendubot to the swing-up configuration is equivalent to globally stabilizing the plant to the compact set defined by points $(\xi_1, \omega_1, \xi_2, \omega_2)$ such that $\xi_1 = \xi_2 = (1, 0) = \mathbf{1} \in \mathbb{R}^2$ and $\omega_1 = \omega_2 = 0$. The design below is in the original coordinates $z = (\phi_1, \omega_1, \phi_2, \omega_2)$ as that is more intuitive, but the resulting hybrid controller can be rewritten in the coordinates resulting from the embedding on $\mathbb{S}^1$; the same embedding is used in the model of a particle evolving in the unit circle given in § 1.2.3.*

According to the model in Example 1.5 and the definitions in Figure 1.7, the pendubot has the following main equilibrium points:

- *Resting (both links "down") equilibrium point: denoted z_r^* and given by the point $\phi_1 = \phi_2 = \pi$, $\omega_1 = \omega_2 = 0$. In the coordinates corresponding to the embedding, this point is given by $\xi_1 = \xi_2 = -\mathbf{1}$, $\omega_1 = \omega_2 = 0$.*

- *Upright (both links "up") equilibrium point: denoted z_u^* and given by the point $\phi_1 = \omega_1 = \phi_2 = \omega_2 = 0$. In the coordinates with the embedding, this point corresponds to $\xi_1 = \xi_2 = \mathbf{1}$, $\omega_1 = \omega_2 = 0$.*

- *Upright/resting (one link "up" and the other "down") equilibrium point: denoted z_{ur}^* and given by $\phi_1 = \omega_1 = \omega_2 = 0$, $\phi_2 = \pi$. In the coordinates with the embedding, this point corresponds to $\xi_1 = \mathbf{1}$, $\omega_1 = \omega_2 = 0$, $\xi_2 = -\mathbf{1}$.*

- *Resting/upright (one link "down" and the other "up") equilibrium point: denoted z_{ru}^* and given by $\phi_1 = \pi$, $\omega_1 = \phi_2 = \omega_2 = 0$. In the coordinates with the embedding, this point corresponds to $\xi_1 = -\mathbf{1}$, $\omega_1 = \omega_2 = 0$, $\xi_2 = \mathbf{1}$.*

A hybrid controller that implements the throw-catch strategy for global swing-up of the pendubot is designed in Example 6.5, Example 6.6, and Example 6.11. This controller executes the tasks outlined in Example 1.5. Figure 1.8 depicts a hybrid automaton implementing the throw-catch control logic for global swing-up of the pendubot.

The problem of steering autonomous vehicles with limited information is another example where the throw-catch strategy can be applied, as the following example illustrates. In such a scenario, sensors on the vehicles usually have a limited area of coverage. Consequently, relative measurements to a target location are only available in a neighborhood of it. As a consequence, position feedback control is not a global solution. On the other hand, open-loop control laws capable of steering the vehicle from nearby points of one location to nearby points of another location, where position measurements are available, can be combined with feedback laws.

Figure 6.3 describes this scenario involving a single vehicle.

Example 6.2 (Vehicle control with limited information). *Consider the problem of steering the position of a vehicle to a target location under the following sensing constraints:*

- *Relative position information between the vehicle and the target location is only available nearby the target location.*

- *Relative position information to an intermediate location, which is different from the target location, is always available.*

- *Information on how to reach a neighborhood of the target location is available nearby the intermediate location.*

Due to these limitations, position feedback control does not provide a global solution to this steering problem. Intuitively, an algorithm solving this problem globally requires using the information on how to reach the target location that is available at the intermediate location. The throw-catch strategy proposed in this chapter is employed to solve this problem. Figure 6.3 depicts the setting and parts of the control strategy. A throw-catch control strategy for the vehicle is presented in Example 6.4 and Example 6.10.

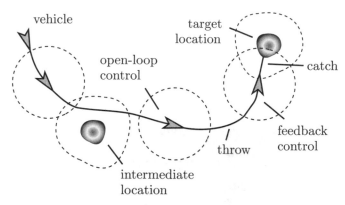

Figure 6.3: Control of an autonomous vehicle with limited position measurements. Relative position between the vehicle and the target location is only available nearby the target location. The information at the intermediate location is sufficient to steer the vehicle towards the target, using an open-loop control law.

With the overview and examples given above, the control objectives for the throw-catch strategy extend those in § 4.1 to the stabilization of a set – rather than of an isolated point – with multiple control laws, including both state-feedback and open-loop controllers – rather than only two state-feedback controllers. The control problem for the stabilization of a plant $\mathcal{H}_P$ to the compact set $\mathcal{A}^*$ has the following objectives:

- Render $\mathcal{A}^*$ stable and, from nearby $\mathcal{A}^*$, guarantee that solutions converge to $\mathcal{A}^*$ using the feedback controller $\mathcal{H}_{K,0}$;
- From points where $\mathcal{H}_{K,0}$ does not guarantee convergence to $\mathcal{A}^*$, coordinate

a family of controllers to guarantee that solutions to the plant converge to a small neighborhood of $\mathcal{A}^*$ from where controller $\mathcal{H}_{K,0}$ can be used.

As in the uniting control strategy in § 4.1, the local controller $\mathcal{H}_{K,0}$ assures local asymptotic stability of the set $\mathcal{A}^*$ – see the "local" controller used in uniting control. The "global" property guaranteed by the "global" controller in uniting control is achieved by multiple controllers, via an appropriate logic coordinating them.

6.2 HYBRID CONTROLLER

In this section, a hybrid controller $\mathcal{H}_K$ implementing the throw-catch strategy is constructed. The hybrid controller uses controller $\mathcal{H}_{K,0}$ nearby $\mathcal{A}^*$, from a subset of its basin of attraction $\mathcal{B}_{\mathcal{A}^*}$. From other points, since convergence to $\mathcal{A}^*$ using $\mathcal{H}_{K,0}$ may not be possible, the hybrid controller coordinates a family of controllers to steer the state of the plant to nearby $\mathcal{A}^*$. This setting is depicted in Figure 6.1. For simplicity of the exposition, the throw-catch control strategy is presented for the case when the the plant is an unconstrained continuous-time system with output equal to its state as in (2.9), namely,

$$\mathcal{H}_P \ : \ \begin{cases} (z,u) \in \mathbb{R}^{n_P} \times \mathbb{R}^{m_P} & \dot{z} \in F_P(z,u) \\ & y = z \end{cases} \tag{6.1}$$

The following assumption formalizes the properties induced by each of the control laws involved in throw-catch control.

Assumption 6.3 (Family of controllers for throw-catch control). *Given a compact set $\mathcal{A}^* \subset \mathbb{R}^{n_P}$ and a plant $\mathcal{H}_P$ as in (6.1), there exist*

(TC-A1) A state-feedback law $\kappa_0 : \mathbb{R}^{n_P} \to \mathbb{R}^{m_P}$ such that $\mathcal{H}_P$ controlled by κ_0 is such that $\mathcal{A}_0 := \mathcal{A}^*$ is asymptotically stable with basin of attraction $\mathcal{B}_{\mathcal{A}_0}$.

Given integers $p_{\max} \geq 1$ and $q_{\max,s} \geq 1$ for each $s \in P := \{1, 2, \ldots, p_{\max}\}$, defining

- *for each $s \in P$, $Q_s := \{1, 2, \ldots, q_{\max,s}\}$, $\widetilde{Q}_s := \{1, 2, \ldots, q_{\max,s} - 1\}$, $\mathcal{A}_{0,s} := \mathcal{A}^*$, $\mathcal{B}_{\mathcal{A}_{0,s}} := \mathcal{B}_{\mathcal{A}^*}$, and $\kappa_{0,s} := \kappa_0$;*

- *$R := \bigcup_{s \in P} (Q_s \times \{s\})$ and $\widetilde{R} := \{(i, s) \in R : i < q_{\max,s}\}$;*

there exist

(TC-A2) For each $(i, s) \in \widetilde{R}$, compact sets $\mathcal{A}_{i,s} \subset \mathbb{R}^{n_P}$ such that $\mathcal{A}^* \cap \mathcal{A}_{i,s} = \emptyset$ and $\mathcal{A}_{i',s} \cap \mathcal{A}_{i'',s} = \emptyset$ for each $(i', s), (i'', s) \in \widetilde{R}$, $i' \neq i''$, and state-feedback control laws $\kappa_{i,s} : \mathbb{R}^{n_P} \to \mathbb{R}^{m_P}$ for $\mathcal{H}_P$ such that the set $\mathcal{A}_{i,s}$ is asymptotically stable with basin of attraction $\mathcal{B}_{\mathcal{A}_{i,s}}$;

(TC-A3) For each $(i, s) \in R$, open-loop control laws $\kappa_{(i,s) \to (i-1,s)} : \mathbb{R}_{\geq 0} \to \mathbb{R}^{m_P}$ that are capable of steering trajectories of $\mathcal{H}_P$ from each point in the set $\mathcal{S}_{i,s}$ to a point in the set $\mathcal{E}_{i-1,s}$ in finite time with maximum time

$T_{i,s}^* \geq 0$, where $\mathcal{S}_{i,s}$ is such that it contains an open neighborhood of $\mathcal{A}_{i,s}$. This construction guarantees, for each $(i, s) \in R$, the existence of a closed set $\widetilde{\mathcal{S}}_{i,s}$ containing an open neighborhood of $\mathcal{A}_{i,s}$ and a positive constant $\delta_{i,s}^t$ satisfying

$$\widetilde{\mathcal{S}}_{i,s} + \delta_{i,s}^t \mathbb{B} \subset \mathcal{S}_{i,s} \tag{6.2}$$

(TC-A4) For each $(i, s) \in \widetilde{R} \cup (\{0\} \times P)$, there exist

- an open set $\mathcal{U}_{i,s}$ containing an open neighborhood of $\mathcal{A}_{i,s}$ such that $\overline{\mathcal{U}_{i,s}} \subset \mathcal{B}_{\mathcal{A}_{i,s}}$ and

- a positive constant $\delta_{i,s}$ and a compact set $\mathcal{T}_{i,s}$ such that

$$\mathcal{T}_{i,s} + 2\delta_{i,s} \mathbb{B} \subset \mathcal{U}_{i,s} \tag{6.3}$$

and each solution to the plant $\mathcal{H}_P$ controlled by $\kappa_{i,s}$ with initial condition in $\mathcal{T}_{i,s}$ remains in $\mathcal{U}_{i,s}$;

and with $\mathcal{E}_{i,s}$ associated to the open-loop control law steering the state of the plant from $\mathcal{S}_{i+1,s}$ to $\mathcal{E}_{i,s}$, there exists a positive constant $\delta_{i,s}^c$ such that $\mathcal{E}_{i,s}$ in (TC-A3) and $\mathcal{T}_{i,s}$ satisfy

$$\mathcal{E}_{i,s} + \delta_{i,s}^c \mathbb{B} \subset \mathcal{T}_{i,s} \tag{6.4}$$

(TC-A5) A state-feedback law $\kappa_b : \mathbb{R}^{n_P} \to \mathbb{R}^{m_P}$ such that each solution z to

$$\dot{z} \in F_P(z, \kappa_b(z))$$

converges to the set $\mathcal{E}_b$ from initial conditions in a closed set $\mathcal{B}_b$ satisfying

$$\bigcup_{(q,p)\in\widetilde{R}\cup(\{0\}\times P)} \partial \mathcal{U}_{q,p} + \bigcup_{(q,p)\in\widetilde{R}} \mathcal{R}_{\leq\widetilde{T}_{q,p}^*}^{\kappa(q,p)\to(q-1,p)}(\widetilde{\mathcal{S}}_{q,p}) \subset \mathcal{B}_b \tag{6.5}$$

where $\widetilde{T}_{q,p}^* > T_{q,p}^*$ and, for some positive constant δ_b, the set $\mathcal{E}_b$ satisfies

$$\mathcal{E}_b + \delta_b \mathbb{B} \subset \left(\bigcup_{(q,p)\in R} \widetilde{\mathcal{S}}_{q,p} \right) \cup \left(\bigcup_{(q,p)\in R\cup(\{0\}\times P)} \mathcal{T}_{q,p} \right) \tag{6.6}$$

and $\mathcal{R}_{\leq\widetilde{T}_{q,p}^*}^{\kappa(q,p)\to(q-1,p)}(\widetilde{\mathcal{S}}_{q,p})$ is the reachable set from $\widetilde{\mathcal{S}}_{q,p}$ over the time horizon $[0, \widetilde{T}_{q,p}^*]$ under the effect of the open-loop control law $\kappa_{(q,p)\to(q-1,p)}$.

The constructions and conditions in Assumption 6.3 are used to build the throw-catch as follows. The state-feedback controller κ_0 in (TC-A1) renders the set $\mathcal{A}^* (= \mathcal{A}_0)$ locally asymptotically stable. The throw-catch strategy uses this controller nearby the desired set of points to locally asymptotically stabilize $\mathcal{A}^*$. The sets $\mathcal{A}_{i,s}$ in (TC-A2) can be thought of as the nodes of a directed graph with $p_{\max}$ branches. In this way, the set P labels the branches. Figure 6.4 depicts such a graph. In most applications, the sets $\mathcal{A}_{i,s}$ are given by isolated points, in particular, equilibrium points. For each $s \in P$, the s-th branch starts at $\mathcal{A}_{q_{\max,s},s}$ and has $q_{\max,s} + 1$ nodes, with the last node being $\mathcal{A}_{0,s} = \mathcal{A}^*$. Each such branch has an edge from $\mathcal{A}_{q_{\max,s},s}$ to

$\mathcal{A}_{q_{\max,s}-1,s}$, from $\mathcal{A}_{q_{\max,s}-1,s}$ to $\mathcal{A}_{q_{\max,s}-2,s}$, and so on, until it reaches the last node, which is common to all of the branches, and equal to $\mathcal{A}^*$. A local state feedback is already available for the last node in each branch, which is the one with zero index. The set R collects all of the nodes in the graph except the last one in each branch. The only difference between the sets R and $\widetilde{R}$ is that the latter one does not include the nodes with largest index in each branch. Condition (TC-A1) assures the existence of static state-feedback laws $\kappa_{i,s}$ that locally asymptotically stabilize the nodes collected by $\widetilde{R}$.

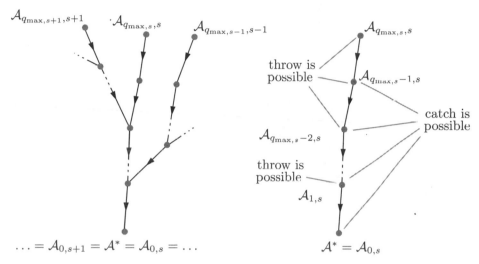

Figure 6.4: General case of directed graph (left) and s-th branch (right).

With (TC-A3) in Assumption 6.3, for each node (i,s) in the graph with $i > 0$, there exists an open-loop control law $\kappa_{(i,s)\to(i-1,s)}$ that transfers the state from nearby points of $\mathcal{A}_{i,s}$ to nearby points of $\mathcal{A}_{i-1,s}$. These open-loop control laws are functions of time that can be recorded in memory. Once nearby $\mathcal{A}_{i-1,s}$, the local state-feedback controllers in (TC-A2) would assure that the state can be steered arbitrarily close to the set $\mathcal{A}_{i-1,s}$, to a point in $\mathcal{S}_{i-1,s}$ from where the open-loop control law can be applied. This property is assured by the $\kappa_{i,s}$'s locally asymptotically stabilizing the $\mathcal{A}_{i,s}$'s, which also assures that maximal solutions exist for arbitrarily long ordinary time. Figure 6.5 depicts these sets.

The state-feedback law κ_b in (TC-A5) in Assumption 6.3 guarantees that the trajectories are steered back to nearby the graph from points at which neither the local state-feedback stabilizers nor the open-loop control laws work. The set $\mathcal{B}_b$ in (TC-A5) collects the set of points from where the bootstrap controller can be initialized and is capable of steering the state of the plant to the set $\mathcal{E}_b$ (in finite time or asymptotically). The requirement for the set $\mathcal{E}_b$ in (6.6) is so that the bootstrap controller takes the plant state to a point from which a state-feedback or open-loop control law can take over. Condition (6.5) assures that the set $\mathcal{B}_b$ contains all of the points that, due to the effect of disturbances, the local static state-feedback controllers may take the plant state to, from where recovery is needed – this is the first set to the left of $\subset$ in (6.5) – as well as the set of points that the open-loop control laws may take the plant state during a failed throw – this is the set given in terms of the finite-time reachable set.

With Assumption 6.3, the control logic between branches is as follows. For the branch $s \in P$, from points nearby $\mathcal{A}_{q_{max,s},s}$ apply the open-loop control law $\kappa_{(q_{max,s},s) \to (q_{max,s}-1,s)}$ until the plant state is nearby $\mathcal{A}_{q_{max,s}-1,s}$. Then, apply the local state-feedback law $\kappa_{q_{max,s}-1,s}$ until the plant state is in the set from where throws to $\mathcal{A}_{q_{max,s}-2,s}$ are possible. Then, apply the open-loop control law $\kappa_{(q_{max,s}-1,s) \to (q_{max,s}-2,s)}$, and repeat.

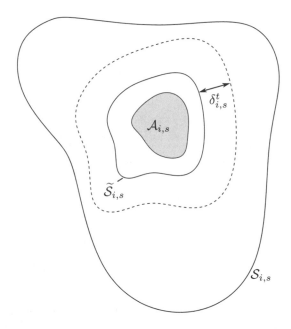

Figure 6.5: Sets associated with the set $\mathcal{A}_{i,s}$ in throw-catch control for throw mode.

The above constructions and conditions are illustrated in the examples.

Example 6.4 (Vehicle control with limited information, revisited). *In this example, the control problem introduced in Example 6.2 is solved using the throw-catch strategy. For simplicity, the vehicle is modeled by the point-mass system*

$$\dot{z} = u \qquad z, u \in \mathbb{R}^2 \tag{6.7}$$

where z denotes the position of the vehicle and u its control input. Denoting the target location as z^* and the intermediate location as $z^\#$, a control strategy consists of employing a feedback law to steer the vehicle to nearby $z^\#$ when it starts far away from z^*. Then, when nearby $z^\#$, employ an open-loop control law to bring it close to z^*. Once close enough to that location, relative position measurements are available and a feedback law to stabilize the vehicle to the target location z^* can be applied. Let the parameter $\rho^* > 0$ define the region $z^* + \rho^*\mathbb{B}$ on which measurements of $z - z^*$ are continuously available. The vector $\nu \in \mathbb{R}^2$ denotes the direction of motion to reach $z^* + \rho^*\mathbb{B}$ from points nearby $z^\#$. The design of the throw-catch strategy requires the definition of the following parameters:

- *The set to asymptotically stabilize $\mathcal{A}^*$: since the goal is to steer z to z^*, then $\mathcal{A}^* := \{z^*\}$.*

- *Parameter $p_{\max}$ defining the number of branches needed in the graph associated with the strategy: since there is only one intermediate position for the vehicle to go, a single branch would suffice to solve the problem. Then, $p_{\max} = 1$ and $\mathcal{A}_{0,1} = \mathcal{A}^*$.*

- *Parameter $q_{\max,1}$ defining the number of nodes in the single branch corresponding to $p = p_{\max} = 1$: the proposed setup requires $q_{\max,1} = 1$ due to having two nodes in the branch. In this way, $Q_1 = \{1\}$ and the graph has a single branch and two nodes.*

- *The choices of $p_{\max}$ and $q_{\max,1}$ result in $R = \{1\} \times \{1\}$ and in $\widetilde{R} = \emptyset$. Hence, one open-loop control law, $\kappa_{(1,1)\to(0,1)}$, one local stabilizer, κ_0, and the bootstrap controller, κ_b, are to be designed.*

- *The set $\mathcal{A}_{1,1}$ defining the second node of the graph: this set is defined as the intermediate location since from nearby that set, the vehicle can be steered to nearby z^* by assigning the input u to the vector ν.*

With these constructions, the design of the control laws and associated sets are to be designed. Their design is given in Example 6.10.

Example 6.5 (Global swing-up of the pendubot, revisited). *The graph needed to apply the throw-catch strategy to the pendubot in Example 6.1 is constructed next. Based on the outline therein, the following definitions are employed:*

$$
\begin{aligned}
p_{\max} &:= 2, \qquad q_{\max,s} := 2 \qquad \forall s \in P := \{1, 2\} \\
\mathcal{A}_{0,s} &:= \{z_u^*\} \ (=: \mathcal{A}^*) \qquad\qquad \forall s \in P \\
\mathcal{A}_{1,1} &= \mathcal{A}_{1,2} := \{z_r^*\}, \quad \mathcal{A}_{2,1} := \{z_{ur}^*\}, \quad \mathcal{A}_{2,2} := \{z_{ru}^*\}
\end{aligned}
\tag{6.8}
$$

The resulting graph has two branches and three nodes in each branch; see Figure 6.9. These constructions lead to

$$
R = (\{1\} \times \{1\}) \cup (\{1\} \times \{2\}) \cup (\{2\} \times \{1\}) \cup (\{2\} \times \{2\})
\tag{6.9}
$$

The control laws and associated sets are designed in Example 6.6.

An outline of the control logic implemented by $\mathcal{H}_K$ to coordinate the state-feedback laws κ_0, $\kappa_{i,s}$, and κ_b, and the open-loop controllers $\kappa_{(i,s)\to(i-1,s)}$ is as follows:

- The controller state is $\eta = (q, p, \tau)$, where q and p are logic variables and τ is a timer. The logic variable p indicates the branch that the hybrid controller is currently at. The logic variable q determines the current node in the current branch, and whether a state-feedback law or an open-loop law is being applied to the plant.

- The controller input is v, which is assigned to z.

- The control logic for throw-catch (TC) control is as follows:

(TC-L1) When the state of the plant is close to $\mathcal{A}^*$, apply control law κ_0 as long as the solution to the plant stays close to $\mathcal{A}^*$;

(TC-L2) Similarly, when a state-feedback control law is being applied nearby $\mathcal{A}_{q,p}$ for some $(q,p) \in \widetilde{R}$, apply that control law until the solution to the plant gets close enough to $\mathcal{A}_{q,p}$, to a point where $\kappa_{(q,p)\to(q-1,p)}$ can be used. When any such point is reached, update q so as to activate $\kappa_{(q,p)\to(q-1,p)}$;

(TC-L3) When an open-loop control law is being applied from nearby $\mathcal{A}_{q,p}$, apply that control law until the solution to the plant gets close enough to $\mathcal{A}_{q-1,p}$, to a point where $\kappa_{q-1,p}$ can be employed. When any such point is reached, update q so as to activate the state feedback $\kappa_{q-1,p}$;

(TC-L4) If the state of the plant is not in the region of operation of the current state-feedback controller or the open-loop control law has not steered the solution to nearby a node in the graph within the expected amount of time, update (q,p) to a value that activates the bootstrap state-feedback controller κ_b;

(TC-L5) When the bootstrap state-feedback law κ_b is being applied, apply that feedback until the state of the plant gets close enough to the graph, to a point where a state-feedback control law $\kappa_{i,s}$ or an open-loop control law $\kappa_{(i,s)\to(i-1,s)}$ for some node in the graph can be applied;

(TC-L6) For any other condition, the logic variables q and p remain constant, and the timer state τ keeps track of ordinary time for its use in the application of the open-loop control laws.

Figure 6.6 depicts a diagram implementing the control logic outlined above.

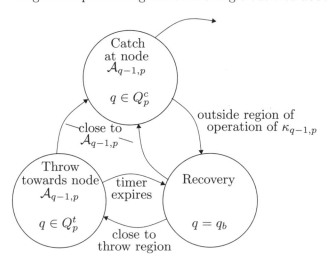

Figure 6.6: Logic implemented by the throw-catch control strategy.

Example 6.6 (Global swing up of the pendubot, revisited). *The trow-catch logic introduced above implements the tasks defined in Example 1.5 to achieve global swing up of the pendubot system. With the constructions in Example 6.5,*

1. The <u>Catch nearby z_u^*</u> task is implemented by item *(TC-L1)*, which applies the static state-feedback controller κ_0 to locally asymptotically stabilize the point z_u^*, which plays the role of the (singleton) set $\mathcal{A}^*$;

2. The tasks <u>Throw from z_{ur}^* and z_{ru}^*</u> and <u>Throw from z_r^*</u> are implemented by item *(TC-L2)* and item *(TC-L3)*. The points z_{ur}^*, z_{ru}^*, and z_r^* belong to the sets $\widetilde{\mathcal{S}}_{i,s}$ from where throws are allowed to start. In particular, from points nearby z_r^*, which define one such set $\widetilde{\mathcal{S}}_{i,s}$, the throw-catch strategy applies an open-loop control law that steers the pendubot to a neighborhood of z_u^*, which defines a set $\mathcal{E}_{i,s}$ in the graph.

3. The task <u>Recover from throw</u> is implemented by item *(TC-L4)* and item *(TC-L5)* via the application of the feedback κ_b.

In addition, the throw-catch logic implements item (TC-L2), which is a catch from nearby z_r^* that helps make sure the state is close to resting, so the throw to nearby z_u^* is successful. The design of a hybrid controller implementing this logic is performed in Example 6.11.

A hybrid controller $\mathcal{H}_K$ implementing the logic in §6.2 is proposed. This controller coordinates the individual control laws in Assumption 6.3. For that purpose, for each $s \in P$, define

$$Q := \{0\} \cup \left(\bigcup_{s \in P} Q_s^t \cup Q_s^c \right) \cup \{q_b\}, \qquad Q_s^t := Q_s, \qquad Q_s^c := -\widetilde{Q}_s$$

as the allowed values for q. The zero value for q in the first element of Q is to denote when κ_0 is being used. The values in the set Q_s^t are to identify the open-loop control law that is being applied: when $q \in Q_p^t$, then $\kappa_{(q,p) \to (q-1,p)}$ is being applied to the plant. The set Q_s^c is to indicate that a state-feedback control law $\kappa_{i,s}$ obtained from (TC-A2) is being applied. This set is defined as the elements in $\widetilde{Q}_s \ (\subset Q_s^t)$ with a minus sign in front – recall that $\widetilde{Q}_s$ does not include zero. In this way, q being negative denotes that the system is "catching" the state at some node with index larger than zero. When $q = 0$, the controller is "catching" the state around the target set $\mathcal{A}^*$. On the other hand, q being positive corresponds to the controller "throwing" the state to nearby the next node. This choice of values for q permits using a single logic variable to coordinate the use of the open-loop control laws and of the state-feedback laws. In this way, when $q \in Q_p^c$, then the state-feedback law $\kappa_{|q|,p}$ from (TC-A2) is applied. Finally, the value q_b in Q is to denote the use of the bootstrap state-feedback control law. This value is defined as $q_b := 1 + \max_{s \in P} q_{\max,s}$, which, by construction, is the largest value that q can take.

With the above definitions, the values that the variables (q, p) can take are collected in the following subset of $Q \times P$:

$$(\{0\} \times P) \cup \left(\bigcup_{p \in P} (Q_p^t \cup Q_p^c) \times \{p\} \right) \cup (\{q_b\} \times P) \tag{6.10}$$

The constructions above conveniently capture the *logic state* or *mode* of the hybrid controller $\mathcal{H}_K$:

- The controller is in *catch mode* at the $|q|$-th node of the p-th branch when $q \in Q_p^c \cup \{0\}$. The state q is nonpositive when the controller is in this mode.

- The controller is in *throw mode* at the q-th node of the p-th branch when $q \in Q_p^t$. The state q is positive when the controller is in this mode.

- The controller is in *recovery mode* when $q = q_b$. The state q is assumes the largest possible value when in this mode.

A hybrid controller $\mathcal{H}_K = (C_K, F_K, D_K, G_K, \kappa)$ implementing the throw-catch control strategy has state $\eta = (q, p, \tau)$, where (q, p) takes values in the set defined in (6.10) and τ in $[0, T^*]$, where $T^* = \max_{(q,p) \in R} \widetilde{T}_{q,p}^*$ and the controller parameters $\widetilde{T}_{q,p}^*$ satisfy $\widetilde{T}_{q,p}^* > T_{q,p}^*$. These conditions on η define the set X_K. Its input is $v \in \mathbb{R}^{n_P}$, which is assigned to z, and its output is $\zeta \in \mathbb{R}^{m_P}$. Its data is defined as follows:

$$C_K := C_{K,0} \cup C_{K,1} \cup C_{K,2} \cup C_{K,3} \cup C_{K,4} \tag{6.11}$$

$$
\begin{cases}
C_{K,0} := \left\{ (z, \eta) \in \mathbb{R}^{n_P} \times X_K : q = 0, \ z \in \overline{\mathcal{U}_{0,p}} \right\} \\
C_{K,1} := \left\{ (z, \eta) \in \mathbb{R}^{n_P} \times X_K : q \in Q_p^t, \ z \in \overline{\mathbb{R}^{n_P} \setminus \mathcal{T}_{q-1,p}} \cap \mathcal{B}_b, \ \tau \in [0, \widetilde{T}_{q,p}^*] \right\} \\
C_{K,2} := \left\{ (z, \eta) \in \mathbb{R}^{n_P} \times X_K : q \in Q_p^c, \ z \in \mathcal{U}_{|q|,p} \setminus \widetilde{\mathcal{S}}_{|q|,p} \right\} \\
C_{K,3} := \left\{ (z, \eta) \in \mathbb{R}^{n_P} \times X_K : q = q_b, \ z \in \overline{\mathcal{B}_b \setminus \mathcal{E}_b} \right\}
\end{cases}
$$

$$F_K(z, \eta) := \begin{bmatrix} 0 \\ 0 \\ \rho_\tau(q, p) \end{bmatrix} \qquad \forall (z, \eta) \in C_K \tag{6.12}$$

$$D_K := D_{K,0} \cup D_{K,1} \cup D_{K,2} \cup D_{K,3} \tag{6.13}$$

$$
\begin{cases}
D_{K,0} := \left\{ (z, \eta) \in \mathbb{R}^{n_P} \times X_K : q = 0, \ z \in \overline{\mathbb{R}^{n_P} \setminus \mathcal{U}_{0,p}} \cap \mathcal{B}_b \right\} \\
D_{K,1} := \left\{ (z, \eta) \in \mathbb{R}^{n_P} \times X_K : q \in Q_p^t, \ z \in \mathcal{T}_{q-1,p}, \ \tau \in [0, T_{q,p}^*] \right\} \cup \\
\qquad \left\{ (z, \eta) \in \mathbb{R}^{n_P} \times X_K : q = q_b, \ \exists (q', p') \text{ s.t. } q' \in Q_{p'}^t \cup \{0\}, \ z \in \mathcal{T}_{q',p'} \right\} \\
D_{K,2} := \left\{ (z, \eta) \in \mathbb{R}^{n_P} \times X_K : q \in Q_p^c, \ z \in \widetilde{\mathcal{S}}_{|q|,p} \right\} \cup \\
\qquad \left\{ (z, \eta) \in \mathbb{R}^{n_P} \times X_K : q = q_b, \ \exists (q', p') \text{ s.t. } q' \in Q_{p'}^t, \ z \in \widetilde{\mathcal{S}}_{q',p'} \right\} \\
D_{K,3} := \left\{ (z, \eta) \in \mathbb{R}^{n_P} \times X_K : q \in Q_p^t, \ z \in \mathcal{B}_b, \ \tau = \widetilde{T}_{q,p}^* \right\} \cup \\
\qquad \left\{ (z, \eta) \in \mathbb{R}^{n_P} \times X_K : q \in Q_p^c, \ z \in \overline{\mathbb{R}^{n_P} \setminus \mathcal{U}_{|q|,p}} \cap \mathcal{B}_b \right\}
\end{cases}
$$

$$G_K(z, \eta) := \begin{bmatrix} G'_K(z, \eta) \\ 0 \end{bmatrix} \qquad \forall (z, \eta) \in D_K \tag{6.14}$$

$$\kappa(z, \eta) := \begin{cases} \kappa_{|q|,p}(z) & \text{if } q \in Q_p^c \cup \{0\} \\ \kappa_{(q,p) \to (q-1,p)}(\tau) & \text{if } q \in Q_p^t \\ \kappa_b(z) & \text{if } q = q_b \end{cases} \tag{6.15}$$

$$\rho_\tau(q, p) := \begin{cases} 1 & \text{if } q \in Q_p^t \\ 0 & \text{otherwise} \end{cases} \tag{6.16}$$

$$G'_K(z, \eta) := G'_{K,1}(z, \eta) \cup G'_{K,2}(z, \eta) \cup G'_{K,3}(z, \eta) \tag{6.17}$$

where $G'_{K,1}$, $G'_{K,2}$, and $G'_{K,3}$ are defined as

$$
G'_{K,1}(z,\eta) := \begin{cases} (-q+1, p) & \text{if } q \in Q_p^t \\[2mm] \{(-|q'|, p') : (q', p') \in R,\ z \in \mathcal{T}_{q',p'}\} & \text{if } q = q_b \\ & \hspace{1cm} \forall (z, \eta) \in D_{K,1}, \end{cases}
$$

$$
G'_{K,2}(z,\eta) := \begin{cases} (|q|, p) & \text{if } q \in Q_p^c \\[2mm] \left\{(q', p') : (q', p') \in R,\ z \in \widetilde{\mathcal{S}}_{q',p'}\right\} & \text{if } q = q_b \\ & \hspace{1cm} \forall (z, \eta) \in D_{K,2}, \end{cases}
$$

$$
G'_{K,3}(z,\eta) := (q_b, p) \hspace{3cm} \forall (z, \eta) \in D_{K,3} \cup D_{K,0},
$$

and empty elsewhere. The sets $\mathcal{A}_{q,p}$, $\mathcal{U}_{q,p}$, $\mathcal{S}_{q,p}$, $\mathcal{T}_{q,p}$, $\mathcal{B}_h$, the control laws $\kappa_{i,s}$, $\kappa_{(i,s)\to(i-1,s)}$, κ_b, and the constants $T^*_{q,p}$ come from Assumption 6.3.

The proposed construction of $\mathcal{H}_K$ implements the logic in (TC-L1)-(TC-L6).

- The set $C_{K,0}$ implements (TC-L1) as it allows flows with the control law κ_0 in the loop, as long as z remains in $\mathcal{U}_{0,p}$, which, by item (TC-A4) in Assumption 6.3, is a subset of the basin of attraction induced by κ_0. If z is not in that set or reaches its boundary, $D_{K,0}$ triggers a jump that activates the bootstrap control law.

- The logic in (TC-L2) is implemented by $C_{K,2}$ by allowing the use of $\kappa_{|q|,p}$ as long as z remains in $\mathcal{U}_{|q|,p}$ and has not reached a point from where $\kappa_{(q,p)\to(q-1,p)}$ can be applied – these are the points collected by $\widetilde{\mathcal{S}}_{|q|,p}$. Since the sets $\widetilde{\mathcal{S}}_{i,s}$ satisfy (6.2) with positive constants $\delta_{i,s}^t$, these sets are contained in the interior of $\mathcal{S}_{i,s}$ and are points from where the associated open-loop control law can be activated. The first set involved in the definition of $D_{K,2}$ triggers an event that, via the definition of $G'_{K,2}$, updates q so as to activate the open-loop control law used during throw mode – see the second entry of κ in (6.15). Note that since at jumps the timer τ is reset to zero, the timer plays the role of time in the application of the open-loop (time dependent) control law.

- The implementation of the throw logic in (TC-L3) is completed by the first set involved in the definition of $D_{K,1}$. This piece of the jump set triggers a jump from throw to catch mode when z reaches $\mathcal{T}_{q-1,p}$, which is a set that $\kappa_{(q,p)\to(q-1,p)}$ steers z to. Note that the size of this set is related to the size of the positive constant $\delta_{q-1,p}^c$ and, due to (6.4), is a subset of the basin of attraction induced by $\kappa_{|q|,p}$ – see item (TC-A4) in Assumption 6.3.

- The logic in (TC-L4) is implemented by $D_{K,0}$ and $D_{K,3}$. In fact, these sets trigger jumps when z leaves or reaches the boundary of the region of operation of the state-feedback controllers – $D_{K,0}$ and the second set in the definition of $D_{K,3}$ implement this – or when an open-loop control law does not reach the expected set of points in the planned amount of time during throw mode.

- The second set involved in the definition of $D_{K,1}$ and the second set in $D_{K,2}$, along with $C_{K,1}$, implement the logic in (TC-L5). The definition of $C_{K,1}$ allows flows in throw mode from points not in the set where catch mode for the next controller is activated, which are points in the set $\mathbb{R}^{n_P} \setminus \overline{\mathcal{T}_{q-1,p}}$. The intersection by $\mathcal{B}_b$ therein is to guarantee that the system can recover from "failure" when the open-loop control law $\kappa_{(i,s)\rightarrow(i-1,s)}$ is applied – the second set to the right of $\subset$ in condition (6.5) guarantees that the state of the plant is within $\mathcal{B}_b$ during throws.

- The flow map F_K implements the logic in (TC-L6).

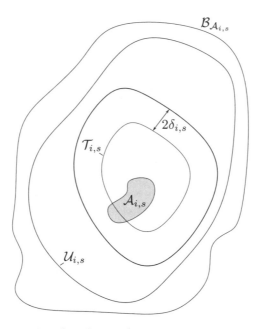

Figure 6.7: Sets associated with set $\mathcal{A}_{i,s}$ in throw-catch control for catch mode.

6.3   CLOSED-LOOP SYSTEM

Along the lines of the model in § 4.3, the hybrid closed-loop system resulting from controlling $\mathcal{H}_P$ as in (6.1) with the hybrid controller $\mathcal{H}_K = (C_K, F_K, D_K, G_K, \kappa)$ in (6.11)-(6.15) has state $x = (z, \eta) = (z, q, p, \tau) \in X := \mathbb{R}^{n_P} \times X_K$, where X_K is defined in the shaded text box below (6.10). The state changes according to

$$
\begin{aligned}
\dot{z} &\in F_P(z, \kappa(z, \eta)) \\
(\dot{q}, \dot{p}) &= 0 \\
\dot{\tau} &= \rho_\tau(q, p)
\end{aligned}
$$

during flows. At jumps, the state is updated according to

$$
\begin{aligned}
z^+ &= z \\
(q^+, p^+) &\in G'_K(z, \eta) \\
\tau^+ &= 0
\end{aligned}
$$

The data (C, F, D, G) of the hybrid closed-loop system $\mathcal{H}$ constructed according to Definition 2.11 is given by

$$
F(x) := \begin{bmatrix} F_P(z, \kappa(\eta, z)) \\ 0 \\ 0 \\ \rho_\tau(q, p) \end{bmatrix} \qquad \forall x \in C := C_K \tag{6.18}
$$

$$
G(x) := \begin{bmatrix} z \\ G'_K(z, \eta) \\ 0 \end{bmatrix} \qquad \forall x \in D := D_K \tag{6.19}
$$

where C_K, D_K, ρ_τ, κ, and G'_K are defined in (6.11)-(6.17).

The next result establishes key properties of this hybrid closed-loop system.

Theorem 6.7 (Throw-catch control). *Given a compact set $\mathcal{A}^* \subset \mathbb{R}^{n_P}$ and*

$$
\mathcal{H}_P : \quad \begin{cases} (z, u) \in \mathbb{R}^{n_P} \times \mathbb{R}^{m_P} & \dot{z} \in F_P(z, u) \\ & y = z \end{cases}
$$

suppose Assumption 6.3 holds. Let the hybrid closed-loop system $\mathcal{H}$ have data as in (6.18)-(6.19).

1. *If $\mathcal{H}_P$ satisfies the hybrid basic conditions in Definition 2.20, and the maps κ_0, $\kappa_{i,s}$, $\kappa_{(i,s) \to (i-1,s)}$, and κ_b are continuous, then $\mathcal{H}$ satisfies the hybrid basic conditions.*

2. *Every maximal solution to $\mathcal{H}$ from $C \cup D$ is complete.*

3. *The compact set*

$$
\mathcal{A} := \mathcal{A}^* \times \{0\} \times P \times [0, T^*] \tag{6.20}
$$

is globally asymptotically stable for $\mathcal{H}$. Furthermore, when the conditions in item 1 hold, the asymptotic stability property of $\mathcal{A}$ is robust in the sense of Definition 3.16.

Proof. When $\mathcal{H}_P$ satisfies the hybrid basic conditions, F_P satisfies (A2$_P$) in Definition 2.25. When κ_0, $\kappa_{i,s}$, $\kappa_{(i,s) \to (i-1,s)}$, and κ_b are continuous, the map $(\eta, z) \mapsto \kappa(\eta, z)$ is continuous on $\mathbb{R}^{n_P} \times Q \times P \times [0, T^*]$. Furthermore, $(q, p) \mapsto \rho_\tau(q, p)$ is continuous on $Q \times P$. Then, the flow map F is single valued and continuous; hence, it satisfies (A2) in Definition 2.20. The set C_K is closed since it is the finite union of closed sets – closedness of $C_{K,3}$ is a direct consequence of closedness of X_K and of $\mathcal{B}_b$, which is guaranteed by item (TC-A5) in Assumption 6.3. The sets $D_{K,0}$, $D_{K,1}$, $D_{K,2}$, and $D_{K,3}$ are closed by construction. Indeed, the set $D_{K,1}$ is closed since $\mathcal{T}_{q-1,p}$ is closed from item (TC-A4) in Assumption 6.3 and the set $D_{K,2}$ is closed since $\widetilde{S}_{|q|,p}$ is closed from item (TC-A3) in Assumption 6.3. To conclude the proof that $\mathcal{H}$ satisfies the hybrid basic conditions, note that the jump map G satisfies (A3) due to being the union of finitely many outer semicontinuous and locally bounded maps; see Lemma A.33.

To show that every maximal solution is complete, proceeding by contradiction, suppose there exists a maximal solution x to $\mathcal{H}$ with $x(0,0) \in C \cup D$ for which $(T, J) = \sup \mathrm{dom}\, x$ is such that $T + J < \infty$. Since such a solution is maximal but not complete, there are two possibilities: i) it exhibits a finite escape time, or ii) it ends at a point in $C \cup D \cup G(D)$ from where it cannot continue.[1]

Case i) cannot happen due to boundedness of x. In fact, when in catch mode, boundedness of solutions is guaranteed by the state-feedback control laws $\kappa_{i,s}$ locally asymptotically stabilizing compact sets $\mathcal{A}_{i,s}$, since $\mathcal{H}_K$ applies them from subsets $\mathcal{T}_{i,s}$ of $\mathcal{B}_{\mathcal{A}_{i,s}}$ satisfying (TC-A4). When in throw mode, boundedness is guaranteed by the very fact that $\kappa_{(i,s) \to (i-1,s)}$ transfers the state from $\widetilde{S}_{i,s}$ to a set $\mathcal{E}_{i,s}$ contained in the set $\mathcal{T}_{i,s}$, which is bounded. So solutions are also bounded until they reach $\mathcal{T}_{i,s}$. When in recovery mode, boundedness is assured by the fact that solutions to the plant from $\mathcal{B}_b$ under the effect of κ_b converge to $\mathcal{E}_b$, which is contained in a set of points from where throw and catch mode can be activated – the latter set being bounded implies that $\mathcal{E}_b$ and the solution is bounded.

Item c is ruled out using item 3 in Proposition 2.34. Now, consider case ii). If the solution ends at a point in D, then it is not maximal since a jump is possible. So, it has to be the case that it ends at a point in $G(D) \setminus (C \cup D)$ after a jump, or at a point in $C \setminus D$ after flow. To rule out the possibility of ending after flow, it is shown that $G(D) \setminus (C \cup D)$ is empty. Pick $x = (z, q, p, \tau) \in D$. The following cases are possible:

1. If $x \in D_{K,0}$, then $z \in \overline{\mathbb{R}^{n_P} \setminus \mathcal{U}_{0,p}} \cap \mathcal{B}_b$ and $G'_K(\eta, z) = G'_{K,3}(\eta, z) = q_b$. Then $G(x) \in C_{K,3}$.

2. Now, suppose $x \in D_{K,1}$. If x belongs to the first set involved in the definition of $D_{K,1}$ then $q \in Q_p^t$, $z \in \mathcal{T}_{q-1,p}$, and $\tau \in [0, T_{q,p}^*]$. In this case, $G'_K(\eta, z) = G'_{K,1}(\eta, z) = (-q + 1, p)$ – note that $G'_{K,3}(\eta, z)$ is empty due to $T_{q,p}^* < \widetilde{T}_{q,p}^*$. If x belongs to the second set in the definition of $D_{K,1}$ then $q = q_b$ and there exists (q', p') such that $q' \in Q_{p'}^t \cup \{0\}$ for which $z \in \mathcal{T}_{q',p'}$. Then, $G'_K(\eta, z) = G'_{K,1}(\eta, z) \cup G'_{K,2}(\eta, z)$. When there does not exist (q'', p'') such that $z \in \widetilde{S}_{q'',p''}$, $G'_K(\eta, z) = (-|q'|, p')$, which puts the system in catch mode, and implies that $G(x) \in C_{K,2}$ if $z \in \overline{\mathcal{U}_{|q'|,p'} \setminus \widetilde{S}_{|q'|,p'}}$ or $G(x)$ belongs to the first set in the definition of $D_{K,2}$ if $z \in \widetilde{S}_{|q'|,p'}$. When there exists (q'', p'') such that $z \in \widetilde{S}_{q'',p''}$, $G'_K(z, \eta) = \{(-|q'|, p')\} \cup \{(q'', p'')\}$ – this is due to x also being in the second set in $D_{K,2}$. In this latter case, $G(x) \subset C_{K,2} \cup D_{K,2}$, as in the previous case, or $G(x) \subset C_{K,1}$.

3. Next, suppose $x \in D_{K,2}$. If x belongs to the first set involved in the definition of $D_{K,2}$ then $q \in Q_p^c$ and $z \in \widetilde{S}_{|q|,p}$. In this case, $G'_K(\eta, z) = G'_{K,2}(\eta, z) = (|q|, p)$ and $G(x) \in C_{K,1}$. If x belongs to the second set in the definition of $D_{K,2}$ then $q = q_b$ and there exists (q', p') such that $q' \in Q_{p'}^t$ and $z \in \widetilde{S}_{q',p'}$. A similar analysis as for the case when x is in the second set in the definition of $D_{K,1}$ leads to $G(x) \subset C_{K,2} \cup D_{K,2} \cup C_{K,1}$.

---

[1] Note that Proposition 2.34 does not apply without the hybrid basic conditions.

4. Finally, suppose $x \in D_{K,3}$. If x belongs to either the first set or the second set in the definition of $D_{K,3}$ then $G'_K(\eta, z) = G'_{K,3}(\eta, z) = (q_b, p)$. Since $z \in \mathcal{B}_b$ and $G(x) = (z, q_b, p, 0)$ in either case, then $G(x) \in C_{K,3}$.

This rules out the possibility of the solution ending after a jump. The solution cannot end after flow since from every point in $C \setminus D$ flow is possible in light of the construction of C_K and the properties of the control laws.

The set $\mathcal{A}$ is stable for $\mathcal{H}$ since the feedback κ_0 is used over a small enough neighborhood of $\mathcal{A}$ with z nearby $\mathcal{A}^*$ and with q equal to zero. In fact, the plant controlled by κ_0 renders $\mathcal{A}^*$ stable for the resulting closed loop, by assumption. Then, since $\mathcal{A}$ in (6.20) only restricts z to be in $\mathcal{A}^*$ and q to be zero, the set $\mathcal{A}$ is stable for $\mathcal{H}$.

Attractivity of $\mathcal{A}$ is established using the properties induced by the individual control laws and the logic implemented by the hybrid controller $\mathcal{H}_K$, as follows. Let x be a maximal solution to the closed loop system $\mathcal{H}$. First note that if $x(0, 0)$ is such that $z(0, 0) \in \mathcal{B}_b$, then, by the properties of κ_b in item (TC-A5) in Assumption 6.3, the solution reaches in finite flow time a point in the set that is to the right of $\subset$ in (6.6). Now, suppose that $x(0, 0) \in C_{K,0}$. Then, the solution either converges to $\mathcal{A}_0$ due to starting from a point in $\mathcal{B}_{\mathcal{A}_0}$ or exhibits a jump, either at $(0, 0)$ if $x(0, 0)$ is at the boundary of $C_{K,0}$ and flows are not possible, or at $(t, 0)$ for some $t > 0$ due to $x(t, 0)$ reaching the boundary of $C_{K,0}$ from where flows are not possible. If the solution x exhibits a jump, then, due to (6.5), the solution is such that z is in $\mathcal{B}_b$, from where, after a finite amount of flow time, it reaches a point in the set that is to the right of $\subset$ in (6.6). If it reaches a set $\mathcal{T}_{q',p'}$ for some (q', p'), then a jump to catch mode for that particular node is triggered. Once in catch mode, due to the properties of $\mathcal{T}_{i,s}$ in (6.3), the solution remains in $\mathcal{U}_{q',p'}$ and converges in finite time to $\widetilde{\mathcal{S}}_{q',p'}$ or, if $q' = 0$, converges to $\mathcal{A}_0 = \mathcal{A}^*$ asymptotically. Convergence to $\mathcal{A}^*$ for the case when the solution reaches $\widetilde{\mathcal{S}}_{q,p}$ (or $\widetilde{\mathcal{S}}_{q',p'}$) in finite time follows from the set-to-set property induced by $\kappa_{(q,p) \to (q-1,p)}$ stated in item (TC-A3) in Assumption 6.3 and by repeating the argument above from points in $C_{K,2}$. $\square$

6.4   DESIGN

The synthesis of the hybrid controller $\mathcal{H}_K$ in (6.11)-(6.15) requires the design of the individual control laws, namely, the state-feedback laws κ_0, $\kappa_{i,s}$, and κ_b, as well as the open-loop laws $\kappa_{(i,s) \to (i-1,s)}$ with the associated parameters $T^*_{i,s}$ and $\widetilde{T}^*_{i,s}$. It also requires the design of the sets $\mathcal{A}_{i,s}$, $\mathcal{U}_{i,s}$, $\mathcal{T}_{i,s}$, $\mathcal{S}_{i,s}$, $\mathcal{E}_{i,s}$, and $\mathcal{B}_b$. This section presents particular designs for these elements defining the data of $\mathcal{H}_K$.

6.4.1   Design of Local Stabilizer κ_0

As in § 4.4, the design of κ_0 assuring asymptotic stability of $\mathcal{A}_0$ can be performed using Lyapunov-based techniques. It amounts to finding smooth enough functions V_0 and κ_0, with V_0 positive definite with respect to $\mathcal{A}_0$ such that $\dot{V}_0(z) < 0$ for all points $z \notin \mathcal{A}_0$ in a neighborhood of $\mathcal{A}_0$. In such a case, $\mathcal{U}_0$ satisfying (TC-A1) can be defined as a sublevel set of V_0. In a similar spirit as Proposition 4.8, the following result presents conditions to design κ_0 using a Lyapunov function.

Proposition 6.8 (Controller synthesis of κ_0, $\mathcal{U}_0$, and $\mathcal{T}_{0,s}$ from Lyapunov-based design conditions). *Suppose $\mathcal{H}_P$ as in (6.1) satisfies the hybrid basic conditions. Given a compact set $\mathcal{A}_0 \subset \mathbb{R}^{n_P}$, suppose there exist a locally Lipschitz function $V_0 : \mathrm{dom}\, V_0 \to \mathbb{R}$ that is positive definite with respect to $\mathcal{A}_0$, with $\mathrm{dom}\, V_0$ containing a neighborhood of $\mathcal{A}_0$, and a continuous function $\kappa_0 : \mathbb{R}^{n_P} \to \mathbb{R}^{m_P}$ such that[2] for some open neighborhood $\mathcal{U} \subset \mathrm{dom}\, V_0$ of $\mathcal{A}_0$*

$$\dot{V}_0(z) \leq 0 \qquad \forall z \in \mathcal{U} \tag{6.21}$$

$$\dot{V}_0(z) < 0 \qquad \forall z \in \mathcal{U} \setminus \mathcal{A}_0 \tag{6.22}$$

Let $c_0 > 0$ be such that

$$\mathcal{U}_0 := \{z \in \mathbb{R}^{n_P} : V_0(z) < c_0\} \tag{6.23}$$

is bounded and $\overline{\mathcal{U}_0} \subset \mathcal{U}$. Furthermore, suppose that every maximal solution to

$$\dot{z} \in F_P(z, \kappa_0(z)) \qquad z \in \overline{\mathcal{U}_0} \tag{6.24}$$

is complete. The following hold:

1. *The conditions in (TC-A1) are satisfied with κ_0 given above;*

2. *For each $s \in P$, the choices $\mathcal{U}_{0,s} := \mathcal{U}_0$ and $\mathcal{T}_{0,s} := \{z \in \mathbb{R}^{n_P} : V_0(z) \leq c_0'\}$ with $c_0' \in (0, c_0)$ are such that there exists a positive constant $\delta_{0,s}$ such that (6.3) in item (TC-A4) holds for $i = 0$.*

Proof. The proof follows similar steps to those of the proof of Proposition 4.8. The function V_0 is a Lyapunov function candidate on $\mathcal{U}$ with respect to $\mathcal{A}_0$ for $(\mathbb{R}^{n_P} \times \mathbb{R}^{m_P}, F_P(\cdot, \kappa_0(\cdot)), \emptyset, \star)$ since it is locally Lipschitz on $\mathcal{U}$ and positive definite with respect to $\mathcal{A}_0$ – see Definition 3.17. Since κ_0 is such that (6.21) and (6.22) hold, and the closed-loop system resulting from controlling $\mathcal{H}_P$ via κ_0 satisfies the hybrid basic conditions, by item 2 of Theorem 3.19, $\mathcal{A}_0$ is pre-asymptotically stable for $(\mathbb{R}^{n_P} \times \mathbb{R}^{m_P}, F_P(\cdot, \kappa_0(\cdot)), \emptyset, \star)$. Since $\mathcal{U}$ is an open set containing the compact set $\mathcal{A}_0$, there exists a positive constant c_0 such that $\mathcal{U}_0$ in (6.23) is bounded and $\overline{\mathcal{U}_0} \subset \mathcal{U}$. Note that this particular construction is such that $\mathcal{U}_0$ is also forward pre-invariant. When every maximal solution to (6.24) is complete then κ_0 and $\mathcal{B}_{\mathcal{A}_0}$ satisfy (TC-A1). With c_0 chosen, pick $c_0' \in (0, c_0)$ and define, for each $s \in P$, $\mathcal{T}_{0,s} := \{z \in \mathbb{R}^{n_P} : V_0(z) \leq c_0'\}$. By construction, since V_0 is continuous, $\mathcal{T}_{0,s}$ is a compact subset of $\mathcal{U}_0$. Since $\mathcal{U}_0$ is open, there exists $\delta_{0,s} > 0$ such that $\mathcal{T}_{0,s} + 2\delta_{0,s}\mathbb{B} \subset \mathcal{U}_{0,s} := \mathcal{U}_0$. Then $\mathcal{T}_{0,s}$ and $\delta_{0,s}$ are such that item (TC-A4) is satisfied for $i = 0$. $\square$

6.4.2 Design of Local Stabilizers $\kappa_{i,s}$ and Sets $\mathcal{A}_{i,s}$

The sets $\mathcal{A}_{i,s}$ are designed to be compact, to have empty intersection with the set $\mathcal{A}_0$, and such that, for each $s \in P$, $\mathcal{A}_{i,s} \cap \mathcal{A}_{k,s}$ is empty for each i and k. The choice of these sets depends on the dynamics of the plant and its stabilizability

---

[2]Recall that, from the definition of the derivative of a Lyapunov function in (3.13),

$$\dot{V}_0(z) = \max_{\chi \in F_P(z, \kappa_0(z))} V_0^{\circ}(z, \chi)$$

properties. In particular, their choice depends on the regions of the state space for which local state-feedback stabilizers can be designed. The definition of such sets also depends on the existence of open-loop control laws $\kappa_{(i,s)\to(i-1,s)}$ being able to steer the trajectories from the set $\mathcal{S}_{i,s}$ to the set $\mathcal{E}_{i-1,s}$. A Lyapunov-based approach for the design of $\kappa_{i,s}$ and the associated sets $\mathcal{U}_{i,s}$, $\mathcal{T}_{i,s}$ is essentially already given in Proposition 6.8 for the design of κ_0. The following result is a direct adaptation of Proposition 6.8. See also § 4.4.

Proposition 6.9 (Controller synthesis of $\kappa_{i,s}$, $\mathcal{U}_{i,s}$, and $\mathcal{T}_{i,s}$ from Lyapunov-based design conditions). *Suppose $\mathcal{H}_P$ as in (6.1) satisfies the hybrid basic conditions. Given a compact set $\mathcal{A}_0 \subset \mathbb{R}^{n_P}$ and, for each $(i,s) \in R$, compact sets $\mathcal{A}_{i,s} \subset \mathbb{R}^{n_P}$, suppose that for each $s \in P$ the sets $\mathcal{A}_{i,s}$ are disjoint, and that for each $(i,s) \in R$ the sets satisfy $\mathcal{A}_0 \cap \mathcal{A}_{i,s} = \emptyset$. For each $(i,s) \in R$, suppose there exist a locally Lipschitz function $V_{i,s} : \mathrm{dom}\, V_{i,s} \to \mathbb{R}$ that is positive definite with respect to $\mathcal{A}_{i,s}$, with $\mathrm{dom}\, V_{i,s}$ containing a neighborhood of $\mathcal{A}_{i,s}$, and a continuous function $\kappa_{i,s} : \mathbb{R}^{n_P} \to \mathbb{R}^{m_P}$ such that for some open neighborhood $\mathcal{U}_{i,s} \subset \mathrm{dom}\, V_{i,s}$ of $\mathcal{A}_{i,s}$,*

$$\dot{V}_{i,s}(z) \leq 0 \qquad \forall z \in \mathcal{U}_{i,s} \tag{6.25}$$

$$\dot{V}_{i,s}(z) < 0 \qquad \vee z \in \mathcal{U}_{i,s} \setminus \mathcal{A}_{i,s} \tag{6.26}$$

For each $(i,s) \in R$, let $c_{i,s} > 0$ be such that

$$\mathcal{U}_{i,s} := \{z \in \mathbb{R}^{n_P} : V_{i,s}(z) < c_{i,s}\} \tag{6.27}$$

$\mathcal{U}_{i,s}$ is bounded, $\overline{\mathcal{U}_{i,s}} \subset \mathcal{U}$, and such that every maximal solution to

$$\dot{z} \in F_P(z, \kappa_{i,s}(z)) \qquad z \in \overline{\mathcal{U}_{i,s}} \tag{6.28}$$

is complete. The following hold:

1. *The conditions in (TC-A2) are satisfied with $\kappa_{i,s}$ given above;*

2. *The choice $\mathcal{T}_{i,s} := \{z \in \mathbb{R}^{n_P} : V_{i,s}(z) \leq c'_{i,s}\}$ with $c'_{i,s} \in (0, c_{i,s})$ is such that there exists a positive constant $\delta_{i,s}$ such that (6.3) in item (TC-A4) holds.*

6.4.3   Design of Open-Loop Control Laws

A design procedure for the sets $\mathcal{S}_{i,s}$, $\mathcal{E}_{i,s}$ and the open-loop law $\kappa_{(i,s)\to(i-1,s)}$ is based on solving the following problem:

(TC-OL) For each $(i,s) \in R$, given $\mathcal{A}_{i,s}$ and $\mathcal{A}_{i-1,s}$ find a continuous function $t \mapsto \alpha_{i,s}(t)$ such that there exist positive constants $\delta'_{i,s}$ and $T^*_{i,s}$ for which the following holds: for every $z' \in \mathcal{A}_{i,s}$, every $t \mapsto z(t)$ defined over $[0, T^*_{i,s}]$ with $z(0) = z'$ satisfies

$$\dot{z}(t) \in F_P(z(t), \alpha_{i,s}(t)) \quad \text{for almost all } t \in [0, T^*_{i,s}] \tag{6.29}$$

and, for some $t' \in [0, T^*_{i,s}]$,

$$z(t') \in \mathcal{A}_{i-1,s} + \delta'_{i,s}\mathbb{B} \tag{6.30}$$

When functions $\alpha_{i,s}$, and constants $\delta'_{i,s}$ and $T^*_{i,s}$ in (TC-OL) exist, then, for each $(i,s) \in R$, there exists a positive constant $\delta''_{i,s}$ such that

1. The functions $\kappa_{(i,s)\to(i-1,s)} := \alpha_{i,s}$ and the sets $\mathcal{S}_{i,s} := \mathcal{A}_{i,s} + \delta''_{i,s}\mathbb{B}$, $\mathcal{E}_{i,s} := \mathcal{A}_{i-1,s} + \delta'_{i,s}\mathbb{B}$ satisfy item (TC-A3);

2. Any closed set $\widetilde{\mathcal{S}}_{i,s}$ belonging to the interior of $\mathcal{S}_{i,s}$ and containing a neighborhood of $\mathcal{A}_{i,s}$ satisfies (6.2) for some positive constant $\delta^t_{i,s}$.

When the plant satisfies the hybrid basic conditions, the existence of the positive constant $\delta''_{i,s}$ is guaranteed by continuity of F_P and of $\kappa_{(i,s)\to(i-1,s)}$, and the fact that the problem has a finite horizon. When $\mathcal{A}_{i,s}$ and $\mathcal{A}_{i-1,s}$ are singletons, then (TC-OL) reduces to a two-point boundary value problem.

6.4.4   Design of Bootstrap Controller and Sets

The state-feedback law κ_b is to be designed to render a set $\mathcal{E}_b$ satisfying (6.6) attractive from $\mathcal{B}_b$. This control law can be designed using a strict Lyapunov function as in Proposition 4.8 or using a weak Lyapunov function as in Proposition 4.9. The conditions in Proposition 4.8 can be directly applied by replacing z^* by $\mathcal{E}_b$, with $\mathcal{E}_b$ compact.

The constants $q_{\max}$ and $p_{\max}$ used in the definition of Q and P determine the number of nodes and branches of the graph and are also considered as design parameters. Their values depend on the desired size of the basin of attraction and on the capability of designing state-feedback and open-loop control laws for the particular application at hand.

The design methods presented in this section are illustrated in the examples.

Example 6.10 (Vehicle control with limited information, revisited). *The throw-catch strategy is designed to solve the control problem introduced in Example 6.2 and further developed in Example 6.4. A design of the static state-feedback control law κ_0 required in (TC-A1) to locally asymptotically stabilize $\mathcal{A}_{0,1} = \{z^*\}$ is given by minus the gradient of the quadratic function*

$$V_0(z) = \frac{1}{2}(z - z^*)^\top (z - z^*)$$

that is,

$$\kappa_0(z) := -\nabla V_0(z) = -(z - z^*)$$

The basin of attraction induced by this control law includes $z^* + \rho^*\mathbb{B}$, where ρ^*, is introduced in Example 6.4. Since $\widetilde{R}$ is empty due to the graph having a single branch with two nodes, (TC-A2) holds for free.

To satisfy (TC-A3), with ν being the direction of motion to reach $z^* + \rho^*\mathbb{B}$ from points nearby $z^\#$, the constant open-loop control law given by $\kappa_{(1,1)\to(0,1)} = \nu$ steers the state z of the solutions to the plant to $z^* + \rho^*\mathbb{B}$ when, in particular, the start set $\mathcal{S}_{1,1}$ is a ball centered at $z^\#$ with radius smaller than or equal to ρ^*. An appropriate choice of this set is $\mathcal{S}_{1,1} := z^\# + \frac{\rho^*}{4}\mathbb{B}$ and for the end set is $\mathcal{E}_{0,1} := z^* + \frac{\rho^*}{2}\mathbb{B}$. The value of the parameter $T^*_{1,1}$ is given by the worst case travel time from $\mathcal{S}_{1,1}$ to $\mathcal{E}_{0,1}$, which corresponds to the time required to travel from the point in $\mathcal{S}_{1,1}$ that is farthest to $\mathcal{E}_{0,1}$ to the point in $\mathcal{E}_{0,1}$ that is farthest to $\mathcal{S}_{1,1}$. This time corresponds to

$T_{1,1}^* = (|z^* - z^\#| + 3\rho^*/4)/|\nu|$. *Finally, the set* $\widetilde{S}_{1,1}$ *is chosen as* $\widetilde{S}_{1,1} = z^\# + \frac{\rho^*}{8}\mathbb{B}$, *for which (6.2) holds with* $\delta_{1,1}^t = \frac{\rho^*}{8}$.

Since $\widetilde{R}$ *is empty, (TC-A4) is satisfied by* $\mathcal{U}_{0,1} := z^* + \rho^*\mathbb{B}^\circ$, $\mathcal{T}_{0,1} := z^* + \frac{3\rho^*}{4}\mathbb{B}$, $\delta_{0,1} = \frac{\rho^*}{8}$, *and* $\delta_{0,1}^c = \frac{\rho^*}{4}$. *In fact, these choices are such that* $\mathcal{U}_{0,1}$ *contains a neighborhood of* $\mathcal{A}_0$,

$$\overline{\mathcal{U}_{0,1}} \subset \mathcal{B}_{\mathcal{A}_0}, \qquad \mathcal{T}_{0,1} + 2\delta_{0,1}\mathbb{B} \subset \mathcal{U}_{0,1}, \qquad \mathcal{E}_{0,1} + \delta_{0,1}^c\mathbb{B} \subset \mathcal{T}_{0,1}$$

and each solution to the plant under the effect of κ_0 *with initial condition in* $\mathcal{T}_{0,1}$ *remains in* $\mathcal{T}_{0,1}$ – *hence, it remains in* $\mathcal{U}_{0,1}$.

Finally, a design of the bootstrap controller required in (TC-A5) is given by minus the gradient of the quadratic function

$$V_b(z) = \frac{1}{2}(z - z^\#)^\top (z - z^\#)$$

that is,

$$\kappa_b(z) := -\nabla V_b(z) = -(z - z^\#)$$

The set $\mathcal{E}_b = z^\# + \frac{\rho^*}{16}\mathbb{B}$ *and* $\mathcal{B}_b = \mathbb{R}^2$ *satisfy the conditions in (TC-A5).*

Since the graph associated to the above throw-catch strategy has a single branch, the logic variable p *is not needed. The resulting hybrid closed-loop system with this control strategy has state* $x = (z, \eta) = (z, q, \tau) \in \mathbb{R}^2 \times \{0, 1, 2\} \times [0, T^*]$, $T^* = \widetilde{T}_{1,1}^* > T_{1,1}^*$, *and dynamics given as follows:*

$$\begin{bmatrix} \dot{z} \\ \dot{q} \\ \dot{\tau} \end{bmatrix} = \begin{bmatrix} \kappa(z, \eta) \\ 0 \\ \rho_\tau(q) \end{bmatrix} \quad \begin{aligned} x \in C = & \left\{ (z, \eta) : q = 0, \ z \in \overline{\mathcal{U}_{0,1}} \right\} \cup \\ & \left\{ (z, \eta) : q = 1, \ z \in \overline{\mathbb{R}^2 \setminus \mathcal{T}_{0,1}}, \ \tau \in [0, \widetilde{T}_{1,1}^*] \right\} \cup \\ & \left\{ (z, \eta) : q = 2, \ z \in \overline{\mathbb{R}^2 \setminus \mathcal{E}_h} \right\} \end{aligned}$$

$$\begin{bmatrix} z^+ \\ q^+ \\ \tau^+ \end{bmatrix} = \begin{bmatrix} z \\ G'_K(z, \eta) \\ 0 \end{bmatrix} \quad \begin{aligned} x \in D = & \left\{ (z, \eta) : q = 0, \ z \in \overline{\mathbb{R}^2 \setminus \mathcal{U}_{0,1}} \right\} \cup \\ & \left\{ (z, \eta) : q = 1, \ z \in \mathcal{T}_{0,1}, \ \tau \in [0, T_{1,1}^*] \right\} \cup \\ & \left\{ (z, \eta) : q = 2, \ z \in \mathcal{T}_{0,1} \right\} \cup \\ & \left\{ (z, \eta) : q = 2, \ z \subset \widetilde{S}_{1,1} \right\} \cup \\ & \left\{ (z, \eta) : q = 1, \ \tau \geq \widetilde{T}_{1,1}^* \right\} \end{aligned}$$

where $G'_K(z, \eta) := G'_{K,1}(z, \eta) \cup G'_{K,2}(z, \eta) \cup G'_{K,3}(z, \eta)$ *with*

$$G'_{K,1}(z, \eta) := 0 \quad \forall (z, \eta) \in \left\{ (z, \eta) : q = 1, \ z \in \mathcal{T}_{0,1}, \ \tau \in [0, T_{1,1}^*] \right\} \cup \\ \left\{ (z, \eta) : q = 2, \ z \in \mathcal{T}_{0,1} \right\}$$

$$G'_{K,2}(z, \eta) := 1 \quad \forall (z, \eta) \in \left\{ (z, \eta) : q = 2, \ z \in \widetilde{S}_{1,1} \right\}$$

$$G'_{K,3}(z, \eta) := 2 \quad \forall (z, \eta) \in \left\{ (z, \eta) : q = 1, \ \tau \geq \widetilde{T}_{1,1}^* \right\} \cup \\ \left\{ (z, \eta) : q = 0, \ z \in \overline{\mathbb{R}^2 \setminus \mathcal{U}_{0,1}} \right\}$$

and empty elsewhere. The output of the hybrid controller is given by

$$\kappa(z,\eta) := \begin{cases} -(z - z^*) & \text{if } q = 0 \\ \nu & \text{if } q = 1 \\ -(z - z^\#) & \text{if } q = 2 \end{cases}$$

and the right-hand side for the timer state by $\rho_\tau(q) = 1$ if $q = 1$ and $\rho_\tau(q) = 0$ if $q \in \{0, 2\}$.

Figure 6.8 shows a simulation of the closed-loop system resulting from controlling the vehicle with the proposed hybrid controller. The initial state of the vehicle is such that it is far away from $z^* := \{(0,0)\}$ and at a point where the bootstrap controller κ_b is applicable so as to steer the vehicle to nearby $z^\# := \{(-2, 7)\}$. With this controller, the vehicle is steered towards the region where the open loop controller is applicable. As the figure shows, after the open loop maneuver is executed, the position of the vehicle converges towards z^* by virtue of the effect of the local controller. Associated simulation files are at @BookSite/Simulation/Obstacle.

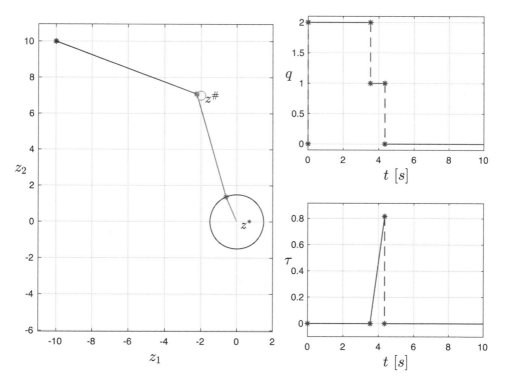

Figure 6.8: A solution to the closed-loop system resulting from controlling a vehicle with limited information. The value of the parameter ρ^* used is 2. The sets $\widetilde{\mathcal{S}}_{1,1}$ and $\mathcal{T}_{1,0}$ are depicted.

Example 6.11 (Global swing-up of the pendubot, revisited). *Using the constructions in Example 6.6 and Example 6.11, the following objects are to be designed to globally asymptotically stabilize the swing-up configuration of the pendubot via throw-catch control:*

- *A static state-feedback law κ_0 locally asymptotically stabilizing $\mathcal{A}^* = \{z_u^*\}$ so as to satisfy (TC-A1) is to be designed. A design of such a control law is given by linearization and pole placement. To that end, let*

$$A_u := \left.\frac{\partial F_P(z, u)}{\partial z}\right|_{z=z_u^*, u=0}, \quad B_u := \left.\frac{\partial F_P(z, u)}{\partial u}\right|_{z=z_u^*, u=0}$$

and choose matrices $K_u \in \mathbb{R}^4$ and $P_u \in \mathbb{R}^{4\times 4}$, $P_u = P_u^\top > 0$, such that

$$(A_u - B_u K_u^\top)^\top P_u + P_u(A_u - B_u K_u^\top) < 0$$

Then, since z_u^* is the origin, let $\kappa_0(z) := K_u^\top z$. Note that matrices K_u and P_u satisfying the Lyapunov inequality above exist since (A_u, B_u) is a controllable pair. With this construction, estimates of the basin of attraction $\mathcal{B}_{\mathcal{A}_0}$ induced by κ_0 can be determined using sublevel sets of the Lyapunov function $V_u(z) := z^\top P_u z$.

With $p_{\max}$, $q_{\max,s}$, $\mathcal{A}_{0,s} := \mathcal{A}^*$, and R as defined in (6.8)-(6.9), design the following:

- *A static state-feedback control law locally asymptotically stabilizing z_r^* so as to satisfy (TC-A2). In fact, according to (TC-A2), for each $(i, s) \in \tilde{R}$, namely, for each $(i, s) \in (\{1\}\times\{1\})\cup(\{1\}\times\{2\})$, the sets $\mathcal{A}_{i,s}$ defined as $\mathcal{A}_{1,1} = \mathcal{A}_{1,2} = \{z_r^*\}$ do not intersect with $\mathcal{A}^*$. Then, a single static state-feedback control law $\kappa_{1,1}(= \kappa_{1,2})$ stabilizing z_r^* can be designed by linearization and pole placement, as in the design of κ_0. With A_r and B_r denoting the system matrices of the linear time-invariant system obtained when linearizing the pendulum system around z_r^*, choose matrices $K_r \in \mathbb{R}^4$ and $P_r \in \mathbb{R}^{4\times 4}$, $P_r = P_r^\top > 0$, such that*

$$(A_r - B_r K_r^\top)^\top P_r + P_r(A_r \quad B_r K_r^\top) < 0$$

and let $\kappa_{1,1}(z) = \kappa_{1,2}(z) := K_r^\top(z - z_r^*)$. Estimates of the basin of attraction $\mathcal{B}_{\mathcal{A}_{1,1}}(= \mathcal{B}_{\mathcal{A}_{1,2}})$ can be determined using sublevel sets of the Lyapunov function $V_r(z) := (z - z_r^*)^\top P_r(z - z_r^*)$.

- *Three open-loop control laws, denoted $\kappa_{(1,1)\to(0,1)} = \kappa_{(1,2)\to(0,2)}$, $\kappa_{(2,1)\to(1,1)}$, and $\kappa_{(2,2)\to(1,2)}$, with associated start and end sets given by $(\mathcal{S}_{1,1}, \mathcal{E}_{0,1}) = (\mathcal{S}_{1,2}, \mathcal{E}_{0,2})$, $(\mathcal{S}_{2,1}, \mathcal{E}_{1,1})$, and $(\mathcal{S}_{2,2}, \mathcal{E}_{1,2})$ as well as maximum time $T_{1,1}^* = T_{1,2}^*$, $T_{2,1}^*$, and $T_{2,2}^*$, respectively, steering points from the start sets to the end sets. To satisfy (TC-A3), the sets $\mathcal{S}_{1,1}$, $\mathcal{S}_{2,1}$, and $\mathcal{S}_{2,2}$ have to contain an open neighborhood of $\mathcal{A}_{1,1}$, $\mathcal{A}_{2,1}$, and $\mathcal{A}_{2,2}$, respectively. A design of $\kappa_{(1,1)\to(0,1)}$ is given by a piecewise-continuous function of time $t \mapsto \alpha_{r\to u}(t)$ such that for the initial condition $z_0 = z_r^*$, and initial time $t = 0$, the unique solution to $\dot{z} = F_P(z, \alpha_{r\to u}(t))$ reaches a small neighborhood of $\mathcal{A}_{0,s}$ after some finite time. Then, by continuity with respect to initial conditions to the pendubot system and since the input is piecewise continuous, there exists an open neighborhood $\mathcal{S}_{1,1}$ of $\mathcal{A}_{1,1}$ and an open neighborhood $\mathcal{E}_{0,1}$ of $\mathcal{A}_{0,s}$ such that solutions to $\dot{z} = F_P(z, \alpha_{r\to u}(t))$ starting from $\mathcal{S}_{1,1}$ reach $\mathcal{E}_{0,1}$ in finite time, being such minimum time $T_{1,1}^*$. Control laws $\kappa_{(2,1)\to(1,1)}$ and $\kappa_{(2,2)\to(1,2)}$ are designed similarly. A convenient technique to synthesize these functions of time consists of defining a basis function parameterizing the control law and then determining its parameters by trial and error. A more systematic approach is* *to solve a two-point boundary value problem with boundary conditions corre-sponding to $\mathcal{A}_{1,1}, \mathcal{A}^*, \mathcal{A}_{2,1}$, and $\mathcal{A}_{2,2}$, as appropriate for each open-loop control law.*

- *To satisfy (TC-A4), design sets $\mathcal{U}_0$ and $\mathcal{U}_1$ containing a neighborhood of $\mathcal{A}_0$ and of $\mathcal{A}_{1,1}$, respectively, closed sets $\mathcal{T}_0$ and $\mathcal{T}_1$, and positive constants δ_0, δ_1, δ_0^c, and δ_1^c such that the following hold:*

  - *$\mathcal{T}_0 + 2\delta_0 \mathbb{B} \subset \mathcal{U}_0$, each solution to the plant under the effect of κ_0 that starts in $\mathcal{T}_0$ remains in $\mathcal{U}_0 + \delta_0 \mathbb{B}$, and $\mathcal{E}_{0,1} + \delta_0^c \mathbb{B} \subset \mathcal{T}_0$. A particular design consists of picking $\mathcal{T}_0$ as a sublevel set of V_u that is large enough so that it contains an open neighborhood of $\mathcal{E}_{0,1}$. Note that this may require redesigning $\kappa_{(1,1)\to(0,1)}$ so that $\mathcal{E}_{0,1}$ is small enough. With such a design, the conditions hold for any positive δ_0 and for small enough and positive δ_0^c.*

  - *$\mathcal{T}_1 + 2\delta_1 \mathbb{B} \subset \mathcal{U}_1$, each solution to the plant under the effect of $\kappa_{1,1}$ that starts in $\mathcal{T}_1$ remains in $\mathcal{U}_1 + \delta_1 \mathbb{B}$, and $\mathcal{E}_{1,1} + \delta_1^c \mathbb{B} \subset \mathcal{T}_1$. The design of $\mathcal{T}_1$, δ_1, and δ_1^v follows the design of $\mathcal{T}_0$, δ_0, and δ_0^c.*

  Then, (TC-A4) holds with $\mathcal{U}_{0,1} = \mathcal{U}_{0,2} := \mathcal{U}_0$, $\mathcal{U}_{1,1} = \mathcal{U}_{1,2} := \mathcal{U}_1$, $\mathcal{T}_{0,1} = \mathcal{T}_{0,2} := \mathcal{T}_0$, $\mathcal{T}_{1,1} = \mathcal{T}_{1,2} := \mathcal{T}_1$, $\delta_{0,1} = \delta_{0,2} := \delta_0$, $\delta_{1,1} = \delta_{1,2} := \delta_1$, $\delta_{0,1}^c = \delta_{0,2}^c := \delta_0^c$, and $\delta_{1,1}^c = \delta_{1,2}^c := \delta_1^c$.

- *A static state-feedback control law κ_b to steer solutions to the plant starting from points not in $\mathcal{A}_{1,1} \cup \mathcal{A}_{0,s} \cup \mathcal{A}_{2,1} \cup \mathcal{A}_{2,2}$ to a small enough neighborhood of it. One such controller is $\kappa_b \equiv 0$, as the natural damping present in the system steer the solutions to $\mathcal{A}_{1,1} \cup \mathcal{A}_{0,s} \cup \mathcal{A}_{2,1} \cup \mathcal{A}_{2,2}$ with zero control input. Such a choice satisfies (TC-A5). A more sophisticated control law that removes energy from the system much faster can also be designed.*

The basic tasks that the throw-catch control strategy performs are as follows:

- *For points nearby $\mathcal{A}_{1,1}$, apply the state feedback $\kappa_{1,1}(= \kappa_{1,2})$ to steer the state to the set $\mathcal{S}_{1,1}$ associated with $\kappa_{(1,1)\to(0,1)}$, and then apply $\kappa_{(1,1)\to(0,1)}$ to steer the solutions to a neighborhood of $\mathcal{A}_{0,s}$;*

- *For points nearby $\mathcal{A}_{0,s}$, apply the state feedback κ_0 to stabilize the solutions to $\mathcal{A}_{0,s}$;*

- *For points nearby $\mathcal{A}_{2,1}$ or $\mathcal{A}_{2,2}$, apply the open-loop control laws $\kappa_{(2,1)\to(1,1)}$ or $\kappa_{(2,2)\to(1,2)}$, respectively, to steer the solutions to a neighborhood of $\mathcal{A}_{1,1}$;*

- *For any other point in $\mathbb{R}^4$, apply the law κ_b to steer the solutions to a neighborhood of $\mathcal{A}_{1,1} \cup \mathcal{A}_{0,s} \cup \mathcal{A}_{2,1} \cup \mathcal{A}_{2,2}$.*

Figure 6.9 shows the combination of these tasks to accomplish global asymptotic stabilization of the point $\mathcal{A}_{0,s}$ of the pendubot. As the figure suggests, the control strategy, interpreted as a graph with nodes given by the equilibrium points, has two paths given by

$$\mathcal{A}_{2,1} \to \mathcal{A}_{1,1} \to \mathcal{A}_{0,1} \quad and \quad \mathcal{A}_{2,2} \to \mathcal{A}_{1,2} \to \mathcal{A}_{0,2}$$

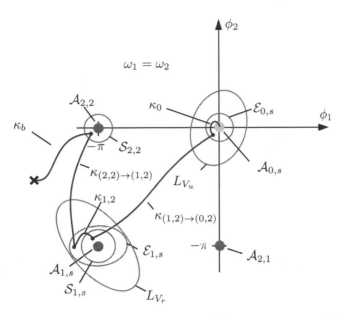

Figure 6.9: Control strategy for robust global stabilization of the pendubot to the point $\mathcal{A}_{0,s}$.

Figure 6.10 shows a simulation of the closed-loop system resulting from controlling the pendubot with the proposed hybrid controller. The initial state of the pendubot is such that it is far away from the regions where the open-loop laws and the local stabilizer κ_0 are applicable. Therefore, the hybrid controller initially switches to recovery mode and applies κ_b. With this controller, the angles of the pendulums reach a neighborhood of $-\pi$ and the angular velocities a neighborhood of 0. Then, the hybrid controller switches to throw mode in the first path and from node $(1,1)$ to node $(0,1)$. The open-loop control law applied is $\kappa_{(1,1)\to(0,1)}$ which steers the state z to a neighborhood of the origin. In that event, a switch to the local stabilizer κ_0 follows, and the state z converges to the origin asymptotically. Associated simulation files are at @BookSite/Simulation/Pendubot.

6.5   EXERCISES

Exercise 43 (Global swing up of a pendulum). Consider the pendulum system given by the nonlinear model

$$\dot{\theta} = \omega, \qquad \dot{\omega} = -\frac{\gamma}{\ell_p}\sin(\theta) + u$$

where $\theta \in (-\pi, \pi]$ denotes the angle, $\omega \in \mathbb{R}$ the angular velocity, and u the torque input. The positive constants γ and ℓ_p denote the gravity constant and the length of the pendulum.

   1. Embed the angle θ in the unit circle by replacing θ by $\xi \in \mathbb{S}^1$. Then, rewrite

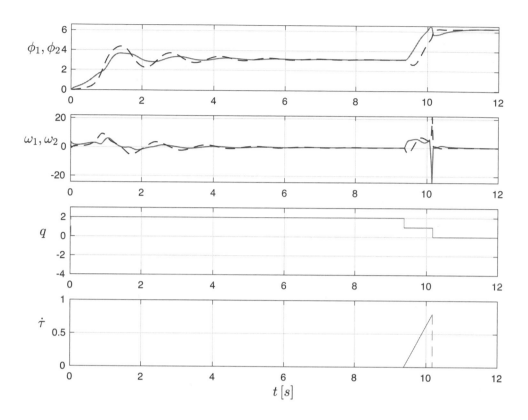

Figure 6.10: Angular position (ϕ_1 solid, ϕ_2 dashed) and angular velocity (ω_1 solid, ω_2 dashed) of the links of the pendubot with the proposed hybrid controller. Initially, the pendulum starts nearby upright, but the initial velocity is so large that the local controller κ_0 is not capable of steering the angular position and angular velocity back to $\mathcal{A}_{1,1}$. As it can be seen from the change in q, the hybrid controller applies the dynamic feedback controller that removes energy from the system and the state converges to a neighborhood of $\mathcal{A}_{1,1}$. From there, the hybrid controller swings up the links to a point nearby $\mathcal{A}_{0,s}$ using $\kappa_{(1,1)\to(0,1)}$. Note that the timer keeps track of the time that $\kappa_{(1,1)\to(0,1)}$ is applied. From there, the state of the system converges to $\mathcal{A}_{0,s}$ – note that ϕ_1 and ϕ_2 converge to 2π, which is physically equivalent to zero.

the model of the pendulum system in the coordinates (ξ, ω).

2. Design a static state-feedback law that locally asymptotically stabilizes ξ to $\mathbf{1}$, with zero angular velocity.

3. Design a static state-feedback law that asymptotically stabilizes ξ to $-\mathbf{1}$, with zero angular velocity.

4. Design an open-loop control law capable of steering ξ from $-\mathbf{1}$ to $\xi = \mathbf{1}$, both with zero angular velocity.

5. Design a hybrid controller that combines the control laws defined above to globally stabilize the plant state (ξ, ω) to $(\mathbf{1}, 0)$.

6. Rewrite the hybrid controller in the original coordinates (θ, ω).

7. Validate your design numerically.

Exercise 44 (Second-order nonlinear system). Consider the second-order nonlinear system

$$\ddot{\xi} + |\xi| = u \qquad \xi, u \in \mathbb{R}$$

1. Show that $u = 0$ is such that from initial values of $(\xi, \dot{\xi})$ given by $(0, 1)$ and by $(0, -1)$, each maximal solution to the system converges to a point $(\xi', \dot{\xi}')$ with $\xi' = -2$ after four seconds.

2. Design a static state-feedback control law that locally asymptotically stabilizes $(\xi, \dot{\xi})$ to $(-2, 0)$ with a basin of attraction that contains $(\xi', \dot{\xi}')$ but not $\{(0, 1)\} \cup \{(0, -1)\}$.

3. Using these controllers, design a hybrid controller that asymptotically stabilizes $(-2, 0)$ and characterize the basin of attraction.

4. Validate your design numerically.

Exercise 45 (Point on a circle). Consider the system on the unit circle given in §1.2.3.

1. Design a static state-feedback controller that locally asymptotically stabilizes $z^* = \mathbf{1}$ with basin of attraction containing $\mathbb{S}^1 \cap \{z \in \mathbb{R}^2 : z_1 > 0\}$.

2. Design an open-loop control law that steers the state from $-\mathbf{1}$ to $\mathbf{1}$.

3. Design a controller locally asymptotically stabilizing the point $-\mathbf{1}$ with basin of attraction containing $\mathbb{S}^1 \cap \{z \in \mathbb{R}^2 : z_1 < 0\}$.

4. Using these controllers, design a hybrid controller that globally asymptotically stabilizes the vehicle state to ξ^*.

5. Compare your design with the hybrid controller proposed in Exercise 15.

6. Validate your design numerically.

Exercise 46 (Global swing up of a pendulum on a cart). Consider the model of a pendulum on a cart given by

$$\ddot{p} = \frac{1}{m_c/m_p + \sin^2 \theta} \left(\frac{\tilde{u}}{m_p} + \dot{\theta}^2 \ell_p \sin \theta - \gamma \sin \theta \cos \theta \right)$$

$$\ddot{\theta} = \frac{1}{\ell_p} \frac{1}{m_c/m_p + \sin^2 \theta} \left(-\frac{\tilde{u}}{m_p} \cos \theta - \dot{\theta}^2 \ell_p \cos \theta \sin \theta + \frac{m_c + m_p}{m_p} \gamma \sin \theta \right)$$

where $\theta \in (-\pi, \pi]$ denotes the angle, $p \in \mathbb{R}$ the position of the cart, and $\tilde{u}$ the force applied to the cart, which is considered to be the input. The positive constants γ, ℓ_p, m_p, and m_c denote the gravity constant, the length of the pendulum, the mass of the pendulum, and the mass of the cart, respectively.

1. Show that the model obtained by inverting the relationship between the input

and the acceleration is given by

$$\ddot{p} = u, \qquad \ddot{\theta} = -\frac{\gamma}{\ell_p}\sin\theta - \frac{u}{\ell_p}\cos\theta$$

where u is the new velocity input.

2. Embed the angle θ in the unit circle by replacing θ by $\xi \in \mathbb{S}^1$. Then, rewrite the model of the pendulum system in the coordinates (p, ν, ξ, ω), where ν is the velocity of the cart and ω is the angular velocity of the pendulum.

3. Design a static state-feedback law that locally asymptotically stabilizes the state (p, ν, ξ, ω) to $(0, 0, \mathbf{1}, 0)$, where $\xi = \mathbf{1}$ corresponds to $\theta = 0$.

4. Design a static state-feedback law that asymptotically stabilizes (p, ν, ξ, ω) to $(0, 0, -\mathbf{1}, 0)$ from every point but $(0, 0, \mathbf{1}, 0)$.

5. Design an open-loop control law capable of steering (p, ν, ξ, ω) from the point $(0, 0, -\mathbf{1}, 0)$ to $(0, 0, \mathbf{1}, 0)$.

6. Using these controllers, design a hybrid controller that combines the control laws defined above to globally stabilize the plant state (p, ν, ξ, ω) to $(0, 0, \mathbf{1}, 0)$.

7. Rewrite the hybrid controller in the original coordinates $(p, \dot{p}, \theta, \dot{\theta})$.

8. Validate your design numerically.

Exercise 47 (Steering a vehicle to a target). For the point-mass vehicle model in Example 6.7, design an algorithm that combines the following strategies to globally asymptotically stabilize the target point $\xi^* \in \mathbb{R}^2$. Given a collection of isolated and disjoint points $\xi^1, \xi^2, \ldots, \xi^N$ with $N > 1$ and $\xi^N = \xi^*$,

1. Design local asymptotically static state-feedback stabilizers for each of the points.

2. Design open-loop control laws capable of steering the vehicle from each point $\xi^1, \xi^2, \ldots, \xi^{N-1}$ to the target point ξ^*.

3. Design a bootstrap controller capable of steering the vehicle to one location among $\xi^1, \xi^2, \ldots, \xi^{N-1}$.

4. Using these controllers, design a hybrid controller that globally asymptotically stabilizes the vehicle state to ξ^*.

5. Validate your design numerically.

6. Redesign your hybrid controller so that the sequence of visited points is as follows: if the vehicle gets close to ξ^i, $i \in \{1, 2, \ldots, N-1\}$, then steer it to ξ^{i+1} first, then to ξ^{i+1}, and so on, until reaching $\xi^N = \xi^*$.

Exercise 48 (Steering a vehicle to a target while avoiding an obstacle). Extend the hybrid controller design in Exercise 47 to the case where there is an obstacle on the plane at location $\bar{\xi}$ such that a ball of radius $\delta > 0$ around it is disjoint from the collection of points $\xi^1, \xi^2, \ldots, \xi^N$. Hint: design the bootstrap controller to avoid a ball of radius smaller than δ around the obstacle location.

Exercise 49 (Rendezvous and docking control for spacecrafts). Consider the problem of controlling the motion of one spacecraft, called the *chaser*, relative to another spacecraft, called the *target*, with the goal being that the chaser docks onto the target and after that, both are steered to a predefined location. The relative translational motion between these two spacecrafts is defined by the so-called Clohessy-Wiltshire equations, which are

$$\ddot{\xi}_x - 2n_\xi \dot{\xi}_y - 3n_\xi^2 \xi_x = \frac{F_x}{m_c}, \qquad \ddot{\xi}_y + 2n_\xi \dot{\xi}_x = \frac{F_y}{m_c} \qquad (6.31)$$

where $(\xi_x, \xi_y) \in \mathbb{R}^2$ and $(\dot{\xi}_x, \dot{\xi}_y)$ are the planar position and velocity, respectively. The inputs F_x and F_y are the control forces in the ξ_x and ξ_y directions, respectively. The parameters in the model are the mass of the chaser spacecraft m_c and n_ξ is defined as $n_\xi := \sqrt{\frac{\tilde{\mu}}{r_o^3}}$, where $\tilde{\mu}$ is the gravitational parameter of the Earth and r_o is the orbit radius of the target spacecraft.

1. Design a static state-feedback control law that globally asymptotically stabilizes the origin of (6.31).

2. Design an open-loop control law capable of steering the docked spacecrafts to a final location $(0, \bar{\xi}_y) \in \mathbb{R}^2$. For the model of the docked spacecrafts, use (6.31) with m_c replaced by $m_c + m_t$, where m_t is the mass of the target spacecraft.

3. Design a static state-feedback control law to locally asymptotically stabilize the docked spacecrafts to $(0, \bar{\xi}_y)$, where $\bar{\xi}_y > 0$.

4. Using these controllers, design a hybrid controller that globally asymptotically stabilizes the vehicle state to $(0, \bar{\xi}_y)$ by first steering the chaser to dock on the target, and once docked, steers both of them to the final location $(0, \bar{\xi}_y) \in \mathbb{R}^2$. Given a small enough threshold $\delta > 0$, $|(\xi_x, \xi_y)| \leq \delta$ is representative of the two spacecrafts being docked.

5. Validate your design numerically using the following values for the constants: $\tilde{\mu} = 3.986 \times 10^{14}$ m^3/s^2, $r_o = 7,100,000$ m, $m_c = 500$ Kg, and $m_t = 2000$ Kg.

Exercise 50 (Three-link pendulum). Using the modeling and control design procedure in Examples 6.1, 6.5, and 6.6, design a hybrid control algorithm that globally asymptotically stabilizes the swing up configuration of the three-link pendulum. A model of a three-link pendulum on a cart is given in [130].

6.6 NOTES

The hybrid control strategy presented in this chapter is inspired by early work on heuristically combining multiple feedbacks using "funnels" in the context of robotic applications [131]. The work therein is motivated by tasks requiring *dynamic dexterity*, which is defined as the ability for robots to perform work on the environment through the proper change of kinetic and potential energy. The main idea in [131] is to sequentially compose robot behaviors based on the energy associated with each behavior. Though the work is empirical, the idea is shown to work well in the presence of obstacles, since the behavior of the robot is expected to change as it approaches obstacles. By partitioning the state space into cells and defining a feedback controller to each cell, a logic-based algorithm that selects the appropriate controller is proposed. As formalized in this chapter for the throw-catch strategy, the basin of attraction of the overall control algorithm in [131] includes the basin of attraction of the individual controllers.

A simplified version of the throw-catch strategy introduced in this chapter first appeared in [22]; see also [132, Chapter 6.2]. As indicated in this chapter, the results providing design conditions in Proposition 6.8 and Proposition 6.9 are just particular instances leading to individual controllers with the needed properties for throw-catch control. The design of the open-loop laws is only touched upon briefly, mainly in examples, but techniques available in the literature can be employed to design such laws. In particular, the design of such open-loop signals can be performed by solving two-point boundary value problems, steering the state from the starting set $\mathcal{S}_{i,s}$ to the ending set $\mathcal{E}_{i-1,s}$, for each node (i, s) of the graph.

The throw-catch strategy in this chapter was extended in [133] to accommodate for the use of controllers that guarantee tracking of reference trajectories, rather than only set-point stabilization. Motivation to such an extension emerged from the motion planning problem when the specified task to accomplish is complex and online computation of reference trajectories that can be tracked by the plant is not always feasible. The extension in [133] combines the motion planning framework in [134], which consists of concatenating a finite number of *motion primitives* selected from a predefined library and the ideas in the original strategy in [22]. Extensions of the throw-catch strategy to the output feedback case are not available in the literature, but possible through the use of observers. The throw-catch strategy was extended in [135] to allow for a richer class of dynamic behavior, in particular, allowing for solutions that periodically visit neighborhoods of the sets $\mathcal{A}_{i,s}$. The extension of the results in this chapter to the case when $\mathcal{H}_P$ has jumps does not impose a technical challenge. Indeed, the ideas extend immediately, but the notation would get much more involved.

In [22], the throw-catch strategy was motivated by the problem of globally swinging up the pendubot – namely, as in Example 6.1, global asymptotic stabilization of the upright configuration. While several control strategies to swing-up the pendubot appeared in the literature, including energy pumping [136], trajectory tracking [137], and jerk control [138], [22] appears to be the first global solution to the problem. A follow-up article presenting experimental results obtained using the throw-catch strategy on a real-world computer controlled pendubot system appeared in [23]. In [23], the parameters of the pendubot system are determined experimentally and the individual controllers designed using linearization of the dynamics and feedback linearization. As a difference to the design in Example 6.6, a proportional-derivative controller is employed for tracking the trajectories that the open-loop laws are to generate. Since only ϕ_1 and ϕ_2 are measured through angle encoders, an observer is included in the control algorithm to estimate the velocities.

Chapter Seven

Synergistic Control

This chapter presents a control strategy that, according to the value of several Lyapunov functions, selects the state-feedback law to apply to the plant. The basic idea consists of designing a *family* of Lyapunov (or Lyapunov-like) functions and "gradient-like" state-feedback law pairs that make the Lyapunov functions decrease along solutions, except at certain points where their gradient vanish, or more generally, where solutions reach a local minima. The pairs in the family are designed so that at such problematic points, there exists another pair whose Lyapunov function has a smaller value. In this way, a switch to the state-feedback law associated to that Lyapunov function with smaller value guarantees a decrease of the overall Lyapunov function for the closed-loop system, in turn, leading to asymptotic convergence to the desired set.

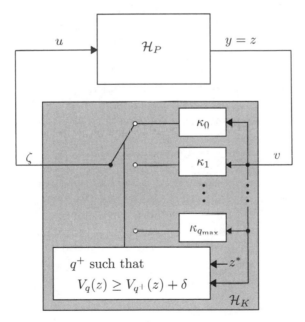

Figure 7.1: Closed-loop system resulting from synergistic control.

A family of pairs of Lyapunov functions and state-feedback laws satisfying the properties above is said to be *synergistic*. As Figure 7.1 shows, a synergistic family of such functions gives rise to a hybrid control algorithm that chooses – using hysteresis – the Lyapunov function and its corresponding feedback control law so as to globally asymptotically stabilize a desired set. Sufficient conditions for analysis and design of the synergistic pairs guaranteeing global asymptotic stability of a given compact set are provided in this chapter.

7.1   OVERVIEW

The main idea in synergistic hybrid feedback control is to use the state-feedback law that has the smallest value of its associated Lyapunov function among those available in a family of predesigned feedbacks and Lyapunov function pairs. With the goal being to asymptotically stabilize a compact set, each such function vanishes on that compact set, is positive elsewhere, and is nonincreasing along solutions to the plant. Using a logic variable $q \in Q := \{0, 1, \ldots, q_{\max}\}$ to index a finite number of state-feedback law and associated Lyapunov function pairs, this hybrid control strategy selects a state-feedback law κ_q as follows.

When the static state-feedback law κ_q is being applied and, for some positive parameter δ, the value of its associated Lyapunov function V_q is such that there exists $p \in Q$ for which

$$V_q(z) \geq V_p(z) + \delta \qquad (7.1)$$

then the strategy updates q to p and, after that jump, applies to the plant the feedback law associated to the new value of q. Condition (7.1) determines the situation when switching to the state-feedback law associated to p leads to a decrease in the function

$$(z, q) \mapsto V_q(z)$$

at least by δ.

A key ingredient exploited in the design of such strategy is that the condition in (7.1) is only required to hold when the individual Lyapunov-like functions stop decreasing, e.g., at critical points. Roughly speaking, when not all of the functions stop to decrease at the same points, then the condition in (7.1) "pushes" the state of the plant away from those critical points by using a different state-feedback law. Functions V_q and their associated state-feedback laws κ_q acting during flows are referred to as *synergistic Lyapunov function and state-feedback pairs*. Figure 7.2 depicts the values of V_q along a solution using the synergistic control strategy. As depicted therein, at time t_1 the value of the Lyapunov function for $q = 2$ is smaller, by an amount δ, than the value of the Lyapunov function for $q = 0$. To benefit from this mismatch, the hybrid controller selects the state feedback associate to $q = 2$. The parameter δ can be a constant or a function of the state.

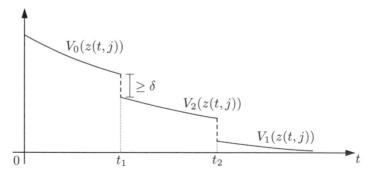

Figure 7.2: Value of V_q along a solution using the synergistic control strategy.

With this overview of the synergistic control strategy, a hybrid controller $\mathcal{H}_K$ using synergistic Lyapunov functions V_q and static state-feedback pairs κ_q is given as follows:

- The controller state is $\eta = q \in Q := \{0, 1, \dots, q_{max}\}$, where q is a logic variable.

- The controller input is $v = z$.

- The control logic for synergistic control (SC) is as follows: given $\delta : \mathbb{R}^{n_P} \times Q \to \mathbb{R}_{\geq 0}$

  (SC-L1) When controller κ_q is being applied to the plant, apply it as long as

  $$V_q(z) < V_p(z) + \delta(z, q) \qquad \forall p \in Q$$

  The logic variable q remains constant when this condition holds.

  (SC-L2) When

  $$V_q(z) \geq V_p(z) + \delta(z, q)$$

  for some (one, or more than one) $p \in Q$, then update q to one such value of p, namely, update q according to

  $$q^+ \in \{p \in Q : V_q(z) \geq V_p(z) + \delta(z, q)\}$$

Figure 7.3 depicts a diagram implementing the control logic outlined above.

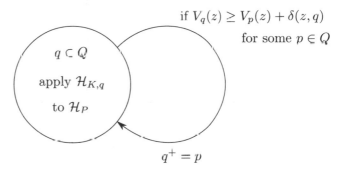

Figure 7.3: Logic implemented by synergistic feedback control.

The following examples motivate the use of synergistic feedback for robust and global asymptotic stabilization of compact sets.

Example 7.1 (Global stabilization of a two-point set on the line). *For the point-mass scalar system*

$$\dot{z} = u \qquad (z, u) \in \mathbb{R} \times \mathbb{R}$$

consider the problem of globally asymptotically stabilizing the two-point set $\{-z^*\} \cup \{z^*\}$ with robustness to small noise on the measurements of z, where $z^* > 0$; see § 1.2.4. To render this set globally attractive, the control algorithm has to steer solutions starting from the interval $(-z^*, z^*)$ to either $-z^*$ or z^*. In particular, due to the stability requirement, solutions starting nearby $-z^*$ and to the right of it *must converge to* $-z^*$, *while solutions starting nearby* z^* *and to the left of it must converge to* z^*. *As a consequence, if the controller is a static state-feedback law, then it is unavoidably discontinuous at a point in the interval* $(-z^*, z^*)$. *One such controller is*

$$u = \kappa(z) := \begin{cases} -\nabla V_0(z) & \text{if } V_0(z) \leq V_1(z) \\ -\nabla V_1(z) & \text{if } V_0(z) > V_1(z) \end{cases} \qquad \forall z \in \mathbb{R} \qquad (7.2)$$

with the functions V_0 *and* V_1 *defined on* $\mathbb{R}$ *as*

$$V_0(z) = \frac{1}{2}(z + z^*)^2, \qquad V_1(z) = \frac{1}{2}(z - z^*)^2$$

Figure 7.4 depicts functions V_0 *and* V_1, *and the possible motions guaranteed by the associated state-feedback laws given by minus their gradients. Since the set of points* z *such that* $V_0(z) \leq V_1(z)$ *is equal to points* $z \leq 0$, *and the set of* z's *such that* $V_0(z) > V_1(z)$ *satisfy* $z > 0$, *arbitrarily small noise from initial conditions that are arbitrarily close to zero would induce solutions that stay nearby zero for all time. In fact, following* § 1.2.4, *to induce such solutions that do not converge, it suffices to design the measurement noise* w_y *such that, for some* $\varepsilon > 0$, $w_y \in \{-\varepsilon, \varepsilon\}$ *and is so that when* $z > 0$ *then* $z + w_y < 0$, *while when* $z < 0$ *then* $z + w_y > 0$. *The synergistic hybrid control strategy introduced in this chapter can be employed to solve the stated stabilization problem, globally and with a quantifiable margin of robustness.*

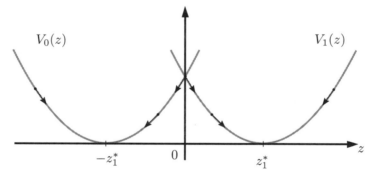

Figure 7.4: Global stabilization of a two-point set using synergistic control.

Example 7.2 (Global stabilization of rigid body kinematics). *Continuing from Example 3.3, the kinematic equations describing rotational motion of a rigid body with attached orthonormal (body) frame and evolving within an orthonormal (inertial) reference frame are given by*

$$\dot{R} = R\widehat{S}(\omega) \qquad (7.3)$$

where R *is the rotation matrix that maps vectors in body frame to inertial frame and* $\omega \in \mathbb{R}^3$ *denotes the angular velocity of the body frame with respect to the inertial frame, expressed in body frame coordinates. As outlined in Example 3.3, the Special Orthogonal Group of order 3 is given by*

$$SO(3) := \{R \in \mathbb{R}^{3 \times 3} : R^\top R = RR^\top = I, \det(R) = 1\} \qquad (7.4)$$

and the Lie Algebra of the SO(3) group by

$$\mathfrak{so}(3) := \left\{ M \in \mathbb{R}^{3\times3} : M = -M^\top \right\} \qquad (7.5)$$

The operator $\widehat{S} : \mathbb{R}^3 \to \mathfrak{so}(3)$ denotes the bijection between $\mathbb{R}^3$ and $\mathfrak{so}(3)$ such that $\widehat{S}(x)y = x \times y$ for any $x, y \in \mathbb{R}^3$, namely,

$$\widehat{S}(\omega) := \begin{bmatrix} 0 & -\omega_3 & \omega_2 \\ \omega_3 & 0 & -\omega_1 \\ -\omega_2 & \omega_1 & 0 \end{bmatrix}, \qquad \widehat{S}^{-1}\left(\begin{bmatrix} 0 & -\omega_3 & \omega_2 \\ \omega_3 & 0 & -\omega_1 \\ -\omega_2 & \omega_1 & 0 \end{bmatrix} \right) := \omega \qquad (7.6)$$

where $S^{-1} : \mathfrak{so}(3) \to \mathbb{R}^3$ denotes its inverse. With the kinematic equation modeling the attitude of a rigid body, the control goal is to globally asymptotically stabilize the orientation of the plant

$$\begin{aligned} \dot{R} &= F_P(R, u) := R\widehat{S}(u) \qquad (R, u) \in SO(3) \times \mathbb{R}^3 \\ y &= R \end{aligned} \qquad (7.7)$$

to the identity, that is, to globally asymptotically stabilize (7.3) to I with measurements of R. Note that this problem corresponds to having $R_d = I$ in Example 3.3. Similar to the challenges in robustly and globally stabilizing a point on a manifold pointed out in § 1.2.3 in the context of attitude control, a continuous static state-feedback law cannot solve this stabilization problem. In Example 7.8 a class of synergistic hybrid controllers solving the problem is provided. In Example 7.16 this problem is revisited and a synergy pair is constructively designed.

7.2   HYBRID CONTROLLER

A hybrid controller $\mathcal{H}_K$ implementing the logic in § 7.1 is presented in this section. To satisfy the objective of stabilizing a compact set, the Lyapunov function and state-feedback laws pairs are assumed to exist so that an asymptotically stabilizing hybrid controller coordinating them can be designed. The individual state-feedback laws κ_q and the associated functions V_q mentioned in § 7.1 are used to define

$$\kappa(q, z) := \kappa_q(z) \qquad \text{and} \qquad V(z, q) := V_q(z)$$

In this way, the logic variable q plays the role of an argument in these functions. Using this notation, the family of Lyapunov function and state-feedback laws pairs is denoted

$$(V, \kappa)$$

For simplicity, the plant is assumed to only have continuous-time dynamics and an output equal to its state, resulting in a hybrid plant $\mathcal{H}_P$ given by

$$\mathcal{H}_P \; : \; \begin{cases} (z, u) \in \mathbb{R}^{n_P} \times \mathbb{R}^{m_P} & \dot{z} \in F_P(z, u) \\ & y = z \end{cases} \qquad (7.8)$$

The case when the hybrid plant exhibits jumps can be dealt similarly, though it carries a larger notational burden.

Definition 7.3 (Synergistic Lyapunov function and state-feedback pair). *Given a plant $\mathcal{H}_P$ as in (7.8), closed sets $X \subset \mathbb{R}^{n_P} \times Q$ and $\mathcal{M} \subset X$, and a compact set $\mathcal{A} \subset \mathcal{M}$, a continuously differentiable function $V : X \to \mathbb{R}_{\geq 0}$ and a continuous function $\kappa : X \to \mathbb{R}^{m_P}$ define a <u>synergistic Lyapunov function and state-feedback pair candidate (V, κ) relative to $(\mathcal{A}, \mathcal{M})$ for $\mathcal{H}_P$</u> if*

1. *$\mathcal{A} = \{(z, q) \in \mathcal{M} : V(z, q) = 0\}$;*

2. *For each $c \geq 0$, the sublevel set $\{(z, q) \in \mathcal{M} : V(z, q) \leq c\}$ is compact; and*

3. *For each $(z, q) \in \mathcal{M}$,*

$$\langle \nabla_z V(z, q), \chi \rangle \leq 0 \qquad \forall \chi \in F_P(z, \kappa(z, q)) \tag{7.9}$$

   where ∇_z denotes the gradient with respect to z.

With the definitions

4. *The set $\Psi \subset \mathcal{E}$ is the largest weakly invariant set for the system*

$$\dot{z} \in F_P(z, \kappa(z, q)), \quad \dot{q} = 0 \qquad (z, q) \in \mathcal{E} \tag{7.10}$$

   where

$$\mathcal{E} := \{(z, q) \in \mathcal{M} : \exists \chi \in F_P(z, \kappa(z, q)) \text{ s.t. } \langle \nabla_z V(z, q), \chi \rangle = 0\}$$

5. *For each $(z, q) \in X$,*

$$\mu_V(z, q) := V(z, q) - \min_{p \in Q} V(z, p) \tag{7.11}$$

the candidate pair (V, κ) is called a synergistic Lyapunov function and state-feedback pair relative to $(\mathcal{A}, \mathcal{M})$ for $\mathcal{H}_P$ if

$$\mu_V(z, q) > 0 \qquad \forall (z, q) \in \left(\Psi \cup \overline{X \setminus \mathcal{M}}\right) \setminus \mathcal{A} \tag{7.12}$$

in which case μ_V is called the synergy gap at (z, q) and any function $\delta : X \to \mathbb{R}_{\geq 0}$ such that

$$\mu_V(z, q) > \delta(z, q) > 0 \qquad \forall (z, q) \in \left(\Psi \cup \overline{X \setminus \mathcal{M}}\right) \setminus \mathcal{A} \tag{7.13}$$

is its positive lower bound.

The set X in Definition 7.3 corresponds to the region of operation for the closed loop. It can also be used to accommodate for state constraints. The set $\mathcal{A}$ has a particular structure: it is given by the zero level set of

$$(z, q) \mapsto V(z, q) = V_q(z)$$

The set $\mathcal{M}$ is the set of points where V is nonincreasing during flows and, due to this, its definition depends on the choice of V and κ. With the construction of the set Ψ in item 4, the condition in (7.13) forces jumps to occur at points where solutions may stop making progress towards $\mathcal{A}$ via flows. In simple terms, a pair

(V, κ) is synergistic relative to $(\mathcal{A}, \mathcal{M})$ if V is nonincreasing during flows and at points where its decrease stops, resetting the value of q to a new value in $Q \setminus \{q\}$ leads to a decrease in the value of V. The decrease after switching is guaranteed by condition (7.12). Indeed, at points (z, q) at which the synergy gap μ_V is positive, q can be updated so as to reduce the value of V. The design (function) parameter δ in (7.13), which has to be positive and lower bound the synergy gap, provides a precise characterization of such decrease. The following assumption enforces this property.

Assumption 7.4 (Conditions for synergistic control). *Given closed sets $X \subset \mathbb{R}^{n_P} \times Q$ and $\mathcal{M} \subset X$, and a compact set $\mathcal{A} \subset \mathcal{M}$, there exists a synergistic Lyapunov function and state-feedback pair (V, κ) relative to $(\mathcal{A}, \mathcal{M})$ with synergy gap having a lower bound given by a continuous function $\delta : X \to \mathbb{R}_{\geq 0}$ that is positive on $X \setminus \mathcal{A}$.*

A hybrid controller $\mathcal{H}_K = (C_K, F_K, D_K, G_K, \kappa)$ implementing the control strategy in § 7.1 is defined as follows. Its state is $\eta = q \in Q = \{0, 1, \ldots, q_{\max}\}$, its input is $v \in \mathbb{R}^{r_P}$, which is assigned to the state of the plant z, its output is $\zeta \in \mathbb{R}^{m_P}$ and is assigned to the input of the plant u. With such assignments, the data of $\mathcal{H}_K$ is defined as

$$C_K := \bigcup_{q \in Q} (\{q\} \times C_{K,q}), \quad C_{K,q} := \{z : (z, q) \in X, \mu_V(z, q) \leq \delta(z, q)\}$$

$$F_K(z, q) := 0 \qquad \forall (z, q) \in C_K$$

$$D_K := \bigcup_{q \in Q} (\{q\} \times D_{K,q}), \quad D_{K,q} := \{z : (z, q) \in X, \mu_V(z, q) \geq \delta(z, q)\}$$

$$G_K(z, q) := \{p \in Q : \mu_V(z, p) = 0\} \qquad \forall (z, q) \in D_K$$

$$(7.14)$$

The output map of the controller is given by the static state-feedback law associated to the current value of q, which is κ itself in the family (V, κ):

$$u = \kappa(z, q) = \kappa_q(z)$$

The jump map G_K does not depend on q.

The jump set triggers events when the synergy gap is larger than or equal to the lower bound δ. Using the definition of the synergy gap in (7.11), this condition corresponds to

$$V(z, q) \geq \min_{p \in Q} V(z, p) + \delta(z, q)$$

In this way, jumps are triggered at points in $(\Psi \cup \overline{X \setminus \mathcal{M}}) \setminus \mathcal{A}$ when there exists $p \in Q$ such that $V(z, q) > V(z, p)$. At such jumps, the function $(z, q) \mapsto V(z, q)$ decreases at least by δ. Since the definition of the flow set C_K in (7.14) includes points with μ_V equal to δ, solutions to the hybrid closed-loop system might be nonunique. In fact, for points (z, q) such that $\mu_V(z, q) = \delta(z, q)$, there exist solutions that jump and, if flows within the flow set are possible from such points, there would also exist solutions that flow.

The hybrid controller in (7.14) is well-posed as it satisfies the hybrid basic conditions.

Lemma 7.5 (Well-posedness of $\mathcal{H}_K$). *Let the set X be closed and the functions V, δ, and κ continuous on X. Then, the hybrid controller $\mathcal{H}_K$ in (7.14) satisfies the hybrid basic conditions.*[1]

Proof. On X, the function

$$(z,q) \mapsto \mu_V(z,q) = V(z,q) - \min_{p \in Q} V(z,p)$$

is continuous since V is continuous. Closedness of C_K and D_K is due to their construction and the continuity of μ_V and δ; see Exercise 100. The jump map G_K is nonempty for each $(z,q) \in D_K$ since $G_K(z,q)$ collects the points $p \in Q$ such that $V(z,p) = \min_{q \in Q} V(q,\eta)$, and there is always at least one such element $p \in Q$. Outer semicontinuity of G_K is a direct consequence of μ_V being continuous. $\square$

7.3 CLOSED-LOOP SYSTEM

Using the hybrid controller in (7.14) to control $\mathcal{H}_P$ as in (7.8), the hybrid closed-loop system has a state $x = (z, \eta) = (z, q) \in X$ that evolves according to

$$\dot{z} \in F_P(z, \kappa(z, q))$$
$$\dot{q} = 0$$

during flows, and at jumps, it is updated according to

$$z^+ = z$$
$$q^+ \in \left\{ p \in Q : \mu_V(z,p) = 0 \right\} = \left\{ p \in Q : V(z,p) = \min_{p' \in Q} V(z,p') \right\}$$

Following the construction in Definition 2.11, this closed loop can be written as $\mathcal{H} = (C, F, D, G)$ with state $x = (z, q)$, where

$$C := \{(z,q) \in X : (z,q) \in C_K\} \tag{7.15}$$

$$F(x) := \begin{bmatrix} F_P(z, \kappa_q(z)) \\ 0 \end{bmatrix} \qquad \forall x \in C \tag{7.16}$$

$$D := \{(z,q) \in X : (z,q) \in D_K\} \tag{7.17}$$

$$G(x) := \begin{bmatrix} z \\ G_K(z,q) \end{bmatrix} \qquad \forall x \in D \tag{7.18}$$

The feedbacks κ_q are given by the family (V, κ). The sets C_K and D_K, and the map G_K are defined in (7.14).

The next result establishes key properties of this hybrid closed-loop system.

---

[1] See Definition 2.25 and discussion below it.

Theorem 7.6 (Synergistic control). *Given*

$$\mathcal{H}_P \; : \; \begin{cases} (z,u) \in \mathbb{R}^{n_P} \times \mathbb{R}^{m_P} & \dot{z} \in F_P(z,u) \\ & y = z \end{cases}$$

satisfying the hybrid basic conditions, closed sets $X \subset \mathbb{R}^{n_P} \times Q$ and $\mathcal{M} \subset X$, and a compact set $\mathcal{A} \subset \mathcal{M}$ suppose that there exists a synergistic Lyapunov function and state-feedback pair candidate (V,κ) relative to $(\mathcal{A}, \mathcal{M})$ for $\mathcal{H}_P$, and that Assumption 7.4 holds. The following properties hold for the hybrid closed-loop system $\mathcal{H}$ with data (C, F, D, G) as in (7.15)-(7.18):

1. *The closed loop $\mathcal{H}$ satisfies the hybrid basic conditions.*

2. *The sets C, D, $\mathcal{M}$, and $\mathcal{A}$ satisfy*

   a) $C \subset \mathcal{M}$
   b) $\mathcal{A} \subset \mathcal{M}$
   c) $(C \setminus \mathcal{A}) \cap (\Psi \cup \overline{X \setminus \mathcal{M}}) = \emptyset$
   d) $C \cup D = X$

3. *Every maximal solution to $\mathcal{H}$ from $C \cup D$ is complete if*

$$\begin{bmatrix} F_P(z, \kappa_q(z)) \\ 0 \end{bmatrix} \cap T_C(z,q) \neq \emptyset \qquad \forall (z,q) \in C \setminus D \tag{7.19}$$

4. *The set $\mathcal{A}$ is globally pre-asymptotically stable for $\mathcal{H}$ and robust in the sense of Definition 3.16.*

Proof. To show item 1, note that since Assumption 7.4 holds, Lemma 7.5 implies that $\mathcal{H}_K$ satisfies the hybrid basic conditions. Then, $\mathcal{H}$ with data as in (7.15)-(7.18) satisfies the hybrid basic conditions by an application of Lemma 2.21.

Next, the properties in item 2 are established. By definition of the sets $C_{K,q}$ and $D_{K,q}$ in (7.14), the sets C and D satisfy $C = \overline{X \setminus D}$. Then, $C \cup D = X$. To show that $C \subset \mathcal{M}$, note that by definition of C

$$\mu_V(z,q) \leq \delta(z,q) \qquad \forall (z,q) \in C \tag{7.20}$$

and since the synergy gap has a lower bound given by δ, the condition in (7.13) holds. Then, with Ψ given as in Definition 7.3 and δ satisfying Assumption 7.4, the sets C and $(\Psi \cup \overline{X \setminus \mathcal{M}}) \setminus \mathcal{A}$ do not overlap – indeed, points in C have $\mu \leq \delta$ while at points in $(\Psi \cup \overline{X \setminus \mathcal{M}}) \setminus \mathcal{A}$, $\mu > \delta$; see item 4. Hence, item 2c holds. To show $C \subset \mathcal{M}$ in item 2a, note that both C and $\mathcal{M}$ are subsets of X and that $\mathcal{A} \subset \mathcal{M}$. Proceeding by contradiction, if $C \not\subset \mathcal{M}$ then there would exist $(z,q) \in C$ that is not in $\mathcal{M}$ and, necessarily, is in $\mathcal{A} \cup X$, which is impossible due to item 2c and the fact that $\mathcal{A} \subset \mathcal{M}$. To establish item 2b, note that since Definition 7.3 implies that points $(z,q) \in \mathcal{A}$ satisfy $V(z,q) = 0$, $\mu(z,q)$ is zero at those points. Then, (7.20) implies $\mathcal{A} \subset C$, which in turn, implies item 2b.

Stability of $\mathcal{A}$, as claimed in item 4, is shown using Theorem 3.19. According to Definition 3.17, the function V obtained from Assumption 7.4 is a Lyapunov function candidate on $\mathbb{R}^n$. Since $C \subset \mathcal{M}$, using (7.9), V satisfies[2]

$$\dot{V}(x) = \max_{\chi \in F(x)} \langle \nabla V(x), \chi \rangle = \max_{\chi' \in F_P(z, \kappa(z,q))} \langle \nabla_z V(z,q), \chi' \rangle \leq 0 \qquad (7.21)$$

for each $x = (z, q) \in C$. Furthermore, using the definition of the synergy gap in (7.11) and its lower bound, ΔV satisfies

$$\Delta V(x) = \max_{\chi \in G(x)} V(\chi) - V(x) = \min_{p \in Q} V(z, p) - V(z, q) = -\mu_V(z, q) \leq -\delta(z, q)$$
$$(7.22)$$

for each $(z, q) \in D$. Since δ is a nonnegative function, $\Delta V(x) \leq 0$ for each $(z, q) \in D$. Then, (3.18) and (3.19) in Theorem 3.19 hold and the set $\mathcal{A}$ is stable using item 1 therein. Furthermore, since from Assumption 7.4 the set $\{(z, q) \in \mathcal{M} : V(z, q) \leq c\}$ is compact for each $c \in \mathbb{R}_{\geq 0}$, every maximal solution is bounded.

To show global pre-attractivity of $\mathcal{A}$, Theorem 3.23 is applied. From (7.21), using the definition of $\mathcal{E}$,

$$\left\{ x \in X : \dot{V}(x) = 0 \right\} = \mathcal{E}$$

From (7.22), using positivity of δ on $X \setminus \mathcal{A}$,

$$\{ x \in D : \Delta V(x) = 0 \} = \mathcal{A}$$

Then, since it has already been shown that every maximal solution is bounded, item 1 in Theorem 3.23 implies that every maximal solution to $\mathcal{H}$ that is complete converges to

$$\{ x \in C : V(x) = r \} \cap (\mathcal{E} \cup \mathcal{A})$$

for some $r \in \mathbb{R}_{\geq 0}$. By definition of Ψ in Definition 7.3, this set is a subset of

$$\{ x \in X : V(x) = r \} \cap ((\Psi \cap C) \cup \mathcal{A})$$

But $\Psi \cap C \subset \mathcal{A}$ since, according to item 2c, every point in Ψ is not in $C \setminus \mathcal{A}$ – so, every point in $\Psi \cap C$ has to be in $\mathcal{A}$ due to Ψ being a weakly forward invariant set for the continuous-time system in (7.10). Then, $\mathcal{A}$ is globally pre-attractive for $\mathcal{H}$.

The last property to show is the one in item 3. Proposition 2.34 is applied to show that every maximal solution to $\mathcal{H}$ is complete, which, in turn, implies global (non-pre) asymptotic stability of $\mathcal{A}$ for $\mathcal{H}$. Since (7.19) holds, (VC) in Proposition 2.34 holds for each point in $C \setminus D$. Then, there exists a nontrivial solution to $\mathcal{H}$ from every initial point in $C \cup D$. The construction of the hybrid controller is such that $G(D) \subset C \cup D$. In fact, from the definition of G_K, for each $x \in D$, each $g \in G(x)$ satisfies $\mu_V(g) = 0$ for each $x \in D$. Then, since δ is nonnegative, each such point g belongs to C. Then, according to item 3 of Proposition 2.34, item c therein for maximal solutions does not hold. Since every maximal solution is bounded, then item b therein does not hold either. Then, every maximal solution to $\mathcal{H}$ is complete.

Finally, robustness of the asymptotic stability of $\mathcal{A}$ follows from Theorem 3.26, since the hybrid closed-loop system satisfies the hybrid basic conditions and $\mathcal{A}$ is a (globally) asymptotically stable compact set for the nominal closed-loop $\mathcal{H}$. $\qquad \square$

---

[2]Though it is avoided here to simplify notation, following Definition 3.18, the flow map can be intersected by the tangent cone of C at x.

Example 7.7 (Global stabilization of a two-point set on the line, revisited). *Using the individual feedback laws in (7.2) in Example 7.1, the data of the synergistic hybrid control strategy given in (7.14) is designed so as globally and robustly asymptotically stabilize the two-point set. Since the control objective is global stabilization on the state component z, then X is defined to collect all points in $\mathbb{R}$. Also, if V is quadratic for each q and gradient descent is used for each state-feedback law, then each Lyapunov function is nonincreasing on X. Then, $\mathcal{M}$ is taken to be equal to X. To define the synergistic hybrid controller, its data is designed as follows:*

$$q_{\max} := 1, \quad Q := \{0,1\}, \quad X := \mathbb{R} \times Q, \quad \mathcal{A} := (\{-z^*\} \times \{0\}) \cup (\{z^*\} \times \{1\})$$

$$V(z,0) := V_0(z) := \frac{1}{2}(z + z^*)^2, \quad V(z,1) := V_1(z) := \frac{1}{2}(z - z^*)^2 \qquad \forall z \in \mathbb{R}$$

$$\kappa(z,q) := -\nabla V_q(z) \qquad \forall (z,q) \in X, \quad \delta > 0$$

$$(7.23)$$

These choices are such that (V, κ) is a synergistic Lyapunov function and state-feedback pair relative to $(\mathcal{A}, X)$ with synergy gap lower bounded by the positive constant δ. More precisely, all of the items in Definition 7.3 hold:

1. *Item 1 in Definition 7.3 holds since $V(z,q)$ only vanishes when $z = -z^*$ and $q = 0$, or $z = z^*$ and $q = 1$.*

2. *For every $q \in Q$, $V(\cdot, q)$ has compact sublevel sets. Then, item 2 in Definition 7.3 holds.*

3. *For every $(z,q) \in \mathcal{M} = \mathbb{R} \times Q$,*

$$\langle \nabla_z V(z,q), -\nabla V_q(z) \rangle = -\nabla V_q(z)^\top \nabla V_q(z) = \begin{cases} -(z + z^*)^2 & \text{if } q = 0 \\ -(z - z^*)^2 & \text{if } q = 1 \end{cases} \quad (7.24)$$

which is nonpositive. Hence, item 3 therein holds.

4. *Using (7.24), the set $\mathcal{E}$ in item 4 is given by*

$$\mathcal{E} = \{(z,q) \subset \mathbb{R} \times Q : \langle \nabla_z V(z,q), -\nabla V_q(z) \rangle = 0\} = \mathcal{A}$$

Then, the synergy gap condition (7.13) holds for free since $\Psi = \mathcal{A}$ and $X = \mathcal{M}$.

Then, the synergistic hybrid controller $\mathcal{H}_K$ in (7.14) solves the problem of globally asymptotically stabilizing the two-point set $\mathcal{A}$. With the definition of μ_V in item 5, the resulting closed loop has state $x = (z,q) \in \mathbb{R} \times \{0,1\}$ and dynamics

$$\mathcal{H} : \begin{cases} V_q(z) \leq \delta + V_{1-q}(z) & \begin{bmatrix} \dot{z} \\ \dot{q} \end{bmatrix} = \begin{bmatrix} -\nabla V_q(z) \\ 0 \end{bmatrix} \\ V_q(z) \geq \delta + V_{1-q}(z) & \begin{bmatrix} z^+ \\ q^+ \end{bmatrix} = \begin{bmatrix} z \\ 1-q \end{bmatrix} \end{cases} \quad (7.25)$$

Global asymptotic stability of $\mathcal{A}$ follows directly from Theorem 7.6. Figure 7.7 depicts the flow and jump sets defined by V_0, V_1, and $\delta > 0$ for $\mathcal{H}$. The distance between the boundary of the jump set for $q = 1$ and the boundary of the jump set for $q = 0$ define the margin of robustness guaranteed by synergistic control.

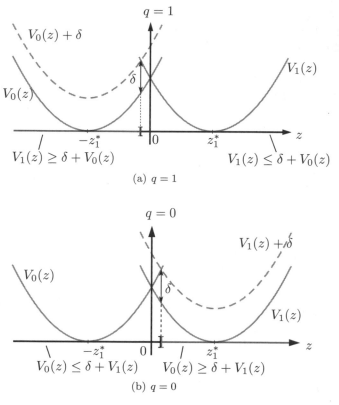

Figure 7.5: Flow and jump sets for the closed loop in Example 7.7.

Example 7.8 (Global stabilization of rigid body kinematics, revisited). *The control problem introduced in Example 7.2 is solved by constructing a synergistic Lyapunov function and state-feedback pair that uses potential functions. This class of functions is defined as follows.*

> **Definition 7.9** (Potential functions and critical points). *A continuously differentiable function $V : SO(3) \to \mathbb{R}_{\geq 0}$ is a potential function on $SO(3)$ with respect to the identity matrix I if $V(R) > 0$ for all $R \in SO(3) \backslash \{I\}$ and $V(I) = 0$. Its set of critical points is given by*
>
> $$\operatorname{Crit}(V) := \left\{ R \in SO(3) \,:\, \varphi(R^\top \nabla V(R)) = 0 \right\} \tag{7.26}$$
>
> *where $\varphi(A) := \widehat{S}^{-1}\left(A - A^\top\right)$ and $\nabla V(R)$ is the gradient of V evaluated at R, which is defined component-wise as*
>
> $$\nabla V(R)_{ij} := \frac{\partial V}{\partial R_{ij}}(R) \qquad \forall (i,j) \in \{1,2,3\} \times \{1,2,3\} \tag{7.27}$$
>
> *Recall that the map $\widehat{S}^{-1}$ is defined in (7.6).*

The following lemma provides conditions on potential functions V_q that, construc-tively, lead to a synergistic Lyapunov function and state-feedback pair for the plant in (7.7) modeling the attitude kinematics.

Lemma 7.10 (From potential to synergistic pair). *Given a family of potential functions $\{V_q\}_{q \in Q}$, $Q = \{0, 1, \ldots, q_{max}\}$, suppose there exists a constant $\delta > 0$ such that*

$$\min_{\substack{q \in Q \\ R \in \mathrm{Crit}(V_q) \backslash \{I\}}} \left\{ V_q(R) - \max_{p \in Q} V_p(R) \right\} > \delta \qquad (7.28)$$

Let $\mathcal{A} := \{I\} \times Q$ and $X := \mathcal{M} = SO(3) \times Q$. Then, the pair (V, κ) defined for each $(R, q) \in X$ as

$$V(R, q) := V_q(R), \quad \kappa(R, q) := \kappa_q(R) := -k_q \varphi(R^\top \nabla V_q(R)) \qquad k_q > 0$$

is a synergistic Lyapunov function and state-feedback pair relative to $(\mathcal{A}, X)$ for $\mathcal{H}_P$ as in (7.7) with synergy gap lower bounded by $\delta > 0$.

Proof. *It is shown first that (V, κ) defines a synergistic Lyapunov function and state-feedback pair candidate relative to $(\mathcal{A}, \mathcal{M})$ for $\mathcal{H}_P$ with data in (7.7) with $\mathcal{M} = X$. Note that by construction of X and the feedbacks κ_q, and the assumptions on the functions V_q, the sets X and $\mathcal{M}$ are closed, V and κ are defined on X, V is continuously differentiable, and κ is continuous. Following Definition 7.3, item 1 holds since by Definition 7.9, $V_q(R) = 0$ if and only if $R = I$, which implies that V only vanishes on the set $\mathcal{A} = \{I\} \times Q$. The compactness of the sublevel sets of V required in item 2 of Definition 7.3 follows directly from continuity of V, and the properties of X: V being continuous implies that, for each $c \geq 0$, $\{(R, q) \in X : V(R, q) \leq c\}$ is closed, and X being compact yields the desired property; see Exercise 100. Item 3 follows by direct computation: first, note that*

$$\langle \nabla_R V(R, q), F_P(R, u) \rangle = u^\top \varphi(R^\top \nabla V_q(R)) \qquad (7.29)$$

where $\nabla_R V$ denotes the gradient of V with respect to R, and that replacing u by $\kappa(R, q)$ yields

$$\langle \nabla_R V(R, q), F_P(R, \kappa(R, q)) \rangle = -k_q |\varphi(R^\top \nabla V_q(R))|^2 \qquad (7.30)$$

which is nonnegative for each $(R, q) \in X$ since $k_q > 0$ for each $q \in Q$. Hence, (V, κ) is a synergistic Lyapunov function and state-feedback pair candidate rel-ative to $(\mathcal{A}, \mathcal{M})$ for $\mathcal{H}_P$ with $\mathcal{M} = X$. Next, it is shown that items 4 and 5 hold. Note that for each $q \in Q$, the inner product $\langle \nabla_R V(R, q), F_P(R, \kappa(R, q)) \rangle$ is zero only at $R \in \mathrm{Crit}(V_q)$. Then, the set $\mathcal{E}$ in Definition 7.3 is given by $\mathcal{E} = \mathrm{Crit}(V_q) \times Q$. Using the definition of synergy gap in (7.12) and (7.28), it follows that (7.12) and (7.13) hold since $\mu_V(z, q)$ is positive and larger than δ at each $(z, q) \in \mathcal{E} \backslash \mathcal{A}$. ∎

By item 4 of Theorem 7.6, the resulting hybrid closed-loop system has the set $\mathcal{A}$ robustly and globally pre-asymptotically stable. Since, via Proposition 2.34, every maximal solution is complete, the set $\mathcal{A}$ is globally asymptotically stable.

The set $\mathcal{A}$ might contain points (z, q) such that, for some $p \in Q \setminus \{q\}$, (z, p) is not in $\mathcal{A}$. Such is the case in the control problem solved in Example 7.7: the point $(z^*, 1)$ belongs to $\mathcal{A}$, but the point $(z^*, 0)$ does not. Due to this, there would exist solutions to the hybrid closed-loop system that, from such points (z, p), move away from $\mathcal{A}$ before they converge to it. Such solutions can be avoided by making the compact set

$$\mathcal{A}_{\text{ext}} := \{(z, p) : \exists q \text{ s.t. } (z, q) \in \mathcal{A}\} \tag{7.31}$$

asymptotically stable as well. The following corollary of Theorem 7.6 establishes global pre-asymptotic stability of $\mathcal{A}_{\text{ext}}$.

Corollary 7.11 (Synergistic control, extended). *Suppose the conditions in Theorem 7.6 hold and that the synergistic Lyapunov function and state-feedback pair (V, κ) relative to $(\mathcal{A}, \mathcal{M})$ for $\mathcal{H}_P$ satisfies*

$$\mu_V(z, q) > \delta(z, q) > 0 \qquad \forall (z, q) \in \left(\Psi \cup \overline{X \setminus \mathcal{M}} \cup \mathcal{A}_{\text{ext}} \right) \setminus \mathcal{A} \tag{7.32}$$

Then, the set $\mathcal{A}_{\text{ext}}$ is globally pre-asymptotically stable for $\mathcal{H}$ with data as in (7.15)-(7.18) and robust in the sense of Definition 3.16.

Next, this result is illustrated in the problem of stabilizing a two-point set given in Example 7.1, which is revisited in Example 7.7.

Example 7.12 (Global stabilization of a two-point set on the line, revisited). *The hybrid controller designed in Example 7.7 is such that, if δ is large enough, solutions to the closed-loop system in (7.25) from $(-z^*, 1)$ would converge to $(z^*, 1) \in \mathcal{A}$, where $\mathcal{A} = (\{-z^*\} \times \{0\}) \cup (\{z^*\} \times \{1\})$. Though this behavior is perfectly fine and representative of global attractivity of $\mathcal{A}$ as defined above, one might want solutions from $(-z^*, 1)$ to jump immediately and then converge to $(-z^*, 0)$ by flowing. Such a property can be attained by stabilizing the set in (7.31), which results in $\mathcal{A}_{\text{ext}} = (\{-z^*\} \cup \{z^*\}) \times Q$. Such a design can be accomplished using the same data as in (7.23) but choosing $\delta \in (0, 2(z^*)^2)$. The only item in Definition 7.3 to check is the synergy gap condition, which, when $\mathcal{A}_{\text{ext}}$ is to be also stabilized, is given in (7.32). This condition reduces to*

$$\mu_V(z, q) > \delta > 0 \qquad \forall (z, q) \in \mathcal{A}_{\text{ext}} \setminus \mathcal{A} = (\{-z^*\} \times \{1\}) \cup (\{z^*\} \times \{0\}) \tag{7.33}$$

This condition holds since

$$\mu_V(-z^*, 1) = V(-z^*, 1) - V(-z^*, 0) = 2(z^*)^2$$
$$\mu_V(z^*, 0) = V(z^*, 0) - V(z^*, 1) = 2(z^*)^2$$

Hence, both $\mu_V(-z^*, 1)$ and $\mu_V(z^*, 0)$ are strictly larger than the chosen constant δ. Then, invoking Corollary 7.11, the synergistic hybrid controller $\mathcal{H}_K$ in (7.25) with $\delta \in (0, 2(z^*)^2)$ renders the sets $\mathcal{A}$ and $\mathcal{A}_{\text{ext}}$ globally pre-asymptotically stable. Robust, global asymptotic stability follows as in Example 7.8.

Simulation files implementing the synergistic control strategy to stabilize $\mathcal{A}$ and $\mathcal{A}_{\text{ext}}$ are at @BookSite/Simulation/TwoPointStabilization. The implementation includes disturbances that enable the user to validate the margin of robustness induced by synergistic control; see below (7.25) in Example 7.7.

7.4 DESIGN

7.4.1 The General Case

Given a desired region of operation X and a compact set $\mathcal{A}$ to asymptotically stabilize, the synthesis of the synergistic hybrid controller $\mathcal{H}_K$ in (7.14) requires the design of the synergistic Lyapunov function and state-feedback pair (V, κ) so that the induced synergy gap μ_V has a nonnegative lower bound δ that is positive on $X \setminus \mathcal{A}$. The cardinality of the set Q, which is q_{max}, is also a design parameter, which determines the dimension of the sets X and $\mathcal{A}$, as well as the number of functions V_q, κ_q leading to the pair (V, κ). The choice of q_{max} typically depends on the application. For the controller to be hybrid, q_{max} is to be larger than one.

The key step in the design of the synergistic hybrid controller is the design of the functions V and κ. The function $V : X \to \mathbb{R}_{\geq 0}$ can only vanish at points in $\mathcal{A}$, so as to satisfy item 1 of Definition 7.3, and have compact sublevel sets. The latter property is typically guaranteed by $z \mapsto V(z, q)$ being radially unbounded[3] for each $q \in Q$.

The combination of the properties required by items 3–5 in Definition 7.3 leads to the following design conditions:

($\star$) Find continuously differentiable functions $\{V_q\}_{q \in Q}$ with compact sublevel sets and such that, for each $q \in Q$, V_q is positive definite with respect to $\{z : (z, q) \in \mathcal{A}\}$, continuous functions $\{\kappa_q\}_{q \in Q}$, and a closed set $\mathcal{M} \subset X$ such that

$$\langle \nabla V_q(z), \chi \rangle \leq 0 \qquad \forall z, \, \forall \chi \in F_P(z, \kappa_q(z)) \, : \, (z, q) \in \mathcal{M} \tag{7.34}$$

$$\min_{p \in Q} V_p(z) - V_q(z) < 0 \qquad \forall z \, : \, (z, q) \in \left(\Psi \cup \overline{X \setminus \mathcal{M}} \right) \setminus \mathcal{A} \tag{7.35}$$

with $\Psi = \bigcup_{q \in Q} (\Psi_q \times \{q\})$, where

$$\Psi_q \subset \mathcal{E}_q := \{z : (z, q) \in \mathcal{M}, \exists \chi \in F_P(z, \kappa_q(z)) \text{ s.t. } \langle \nabla V_q(z), \chi \rangle = 0\}$$

is the largest weakly invariant set for the continuous-time system

$$\dot{z} \in F_P(z, \kappa_q(z)) \qquad z \in \mathcal{E}_q \tag{7.36}$$

Then, $V(z, q) := V_q(z)$ and $\kappa(z, q) := \kappa_q(z)$ define a synergistic Lyapunov function and state-feedback pair candidate (V, κ) relative to $(\mathcal{A}, \mathcal{M})$ with *synergy gap* μ_V.

The design condition (7.35) does not explicitly involve the positive lower bound δ in (7.13). The following result shows that a nonnegative lower bound satisfying (7.13) exists for free when (7.12) holds. It also provides a particular design of that lower bound.

---

[3]See Definition A.27.

Proposition 7.13 (Existence of lower bound). *For any given synergistic Lyapunov function and state-feedback pair (V, κ) relative to $(\mathcal{A}, \mathcal{M})$ such that the set Ψ for (7.36) is closed, there exists a continuous function $\delta : X \to \mathbb{R}_{\geq 0}$ that is positive and lower bounds the synergy gap on $(\Psi \cup \overline{X \setminus \mathcal{M}}) \setminus \mathcal{A}$. A particular such function is given as follows: for any $\epsilon \in (0, 1)$*

$$\delta(z, q) := \inf_{(\chi, p) \in \Psi \cup \overline{X \setminus \mathcal{M}}} \big(|(z, q) - (\chi, p)| + \epsilon \mu_V(\chi, p) \big) \tag{7.37}$$

Proof. The function δ in (7.37) is continuous and nonnegative by definition and the properties of μ_V coming from Definition 7.3. To show that it satisfies (7.13), note that for each $(z, q) \in \Psi \cup \overline{X \setminus \mathcal{M}}$, $\delta(z, q)$ is upper bounded by $\epsilon \mu_V(z, q)$ by picking $(\chi, p) = (z, q)$ inside the infimum – in fact, such a choice makes the first term in δ vanish. Then, $\delta(z, q) \leq \epsilon \mu_V(z, q) < \mu_V(z, q)$. This establishes that δ satisfies (7.13). To show that δ is positive on $(\Psi \cup \overline{X \setminus \mathcal{M}}) \setminus \mathcal{A}$ note that the invariant set Ψ being closed implies that $\Psi \cup \overline{X \setminus \mathcal{M}}$ is closed. Then, since μ_V is continuous and $\mu_V(z, q) > 0$ for all $(z, q) \in \big(\Psi \cup \overline{X \setminus \mathcal{M}} \big) \setminus \mathcal{A}$, and the first term in the definition of δ is nonnegative, it follows that $\delta(z, q) > 0$ for all $(z, q) \in (\Psi \cup \overline{X \setminus \mathcal{M}}) \setminus \mathcal{A}$. $\square$

Closedness of Ψ in Proposition 7.13 is guaranteed when the system in (7.36) satisfies the hybrid basic conditions. Under the assumptions of Theorem 7.6, Ψ is closed.

For the case when $\mathcal{M} = X$, condition (7.35) is satisfied when, in particular, at each point $z \notin \{z : (z, q) \in \mathcal{A}\}$ where V_q stops decreasing along flows, there exists a function V_p with smaller value than V_q. The following result captures this situation.

Proposition 7.14 (Decrease via jumps when $\dot{V}_q$ vanishes). *Suppose $\mathcal{M} = X$.*

1. *If for each $q \in Q$ and each*

$$z \in \widetilde{\Psi}_q := \{z : (z, q) \in X, \exists \chi \in F_P(z, \kappa_q(z)) \ \text{s.t.} \ \langle \nabla V_q(z), \chi \rangle = 0\}$$

   such that $(z, q) \notin \mathcal{A}$ there exists $p \in Q \setminus \{q\}$ such that

$$V_p(z) < V_q(z) \tag{7.38}$$

   then (7.35) holds.

2. *If for each $q \in Q$ and each $z \in \widetilde{\Psi}_q \cup \mathcal{A}_{\text{ext}, q}$ such that $(z, q) \notin \mathcal{A}$, where $\mathcal{A}_{\text{ext}, q} := \{z : \exists p \neq q \ \text{s.t.} \ (z, p) \in \mathcal{A}\}$, there exists $p \in Q \setminus \{q\}$ such that (7.38) holds, then (7.32) holds.*

Though condition (7.38) in Proposition 7.14 might be restrictive at times, the involvement of the change of V_q during flows through the set $\widetilde{\Psi}_q$ clearly shows the coupling between the properties of the synergistic pair during flows and jumps.

Example 7.15 (Global stabilization of a two-point set on the line, revisited). *The controller redesigned in Example 7.12 requires checking that the condition in (7.32) holds. Item 2 in Proposition 7.14 provides a way to check condition (7.32). With the choices in Example 7.12, $\widetilde{\Psi}_0 = \mathcal{A}_{\text{ext}, 0} = \{-z^*\}$ and $\widetilde{\Psi}_1 = \mathcal{A}_{\text{ext}, 1} = \{z^*\}$. For* *each* $z \in \widetilde{\Psi}_0 \cup \mathcal{A}_{\text{ext},0} = \{-z^*\}$ *and* $p = 1 \in Q \setminus \{0\}$, *it follows that*

$$V_0(z) = V_0(-z^*) = 0 < V_1(-z^*) = 2(z^*)^2$$

Similarly, for each $z \in \widetilde{\Psi}_1 \cup \mathcal{A}_{\text{ext},1} = \{z^*\}$ *and* $p = 0 \in Q \setminus \{1\}$, *it follows that*

$$V_1(z) = V_1(z^*) = 0 < V_0(z^*) = 2(z^*)^2$$

Then, by item 2 in Proposition 7.14, (7.32) holds.

Example 7.16 (Global stabilization of rigid body kinematics, revisited). *A particular construction for the functions* V_q *in Lemma 7.10 for the attitude control system in Example 7.8 is provided next. The idea is to design two potential functions,* V_0 *and* V_1, *and associated state-feedback laws, where both* V_0 *and* V_1 *vanish at* $R = I$ *but with critical points at different locations. A potential function on* $SO(3)$ *that has been used for gradient descent-type control is the modified trace function. Given a symmetric positive definite matrix* $M \in \mathbb{R}^3 \times \mathbb{R}^3$, *this function is given by*

$$P_M(R) := trace\,((I - R)M) \tag{7.39}$$

When M *has distinct eigenvalues, the number of critical points is equal to four, with one of them being at* $\{I\}$. *Though the modified trace function in (7.39) does not satisfy (7.28), the functions*

$$V_q(R) := P_{M_q}(R) := P_M(\mathcal{T}_q(R)), \qquad \mathcal{T}_q(R) := \exp(k_q P_M(R)\widehat{S}(s))R \tag{7.40}$$

defined for each $q \in Q := \{0,1\}$ *and each* $R \in SO(3)$ *can be designed to satisfy the synergy gap condition. The function* $\mathcal{T}_q$ *corresponds to a rotation of* $R \in SO(3)$ *by an amount* $k_q P_M(R) \in \mathbb{R}$ *around the axis* $s \in \mathbb{S}^2$. *The function* $\mathcal{T}_q : SO(3) \to SO(3)$ *is a diffeomorphism in* $SO(3)$ *as long as* $\sqrt{2}k_q \max \|\nabla P_M(R)\|_F < 1$. *The following result provides conditions guaranteeing that an angular warping of (7.39) provides a synergistic pair.*

> **Proposition 7.17** (Angular warping of potential functions on $SO(3)$). *Given a symmetric, positive definite matrix* $M \in \mathbb{R}^{3 \times 3}$ *with distinct eigenvalues and* $k > 0$ *satisfying*
>
> $$\sqrt{2}k \max_{R \in SO(3)} \|\nabla P_M(\mathcal{T}_q(R))\|_F < 1 \qquad \forall q \in Q = \{0,1\} \tag{7.41}$$
>
> *where* $\mathcal{T}_q$ *is defined in (7.40), there exist* $s \in \mathbb{S}^2$ *and* $\delta > 0$ *such that* $\{V_q\}_{q \in Q}$ *given in (7.40) with* $k_1 = k$ *and* $k_2 = -k$ *satisfies (7.28).*

Under the conditions in Proposition 7.17, the synergistic hybrid controller $\mathcal{H}_K$ *in (7.14) solves the problem of globally asymptotically stabilizing the attitude* R *to* I. *The resulting closed loop* $\mathcal{H}$ *has state* $x = (R,q) \in SO(3) \times Q$, $Q := \{0,1\}$, *and dynamics*

$$
\mathcal{H} \; : \; \begin{cases} P_M(\mathcal{T}_q(R)) \le \delta + P_M(\mathcal{T}_{1-q}(R)) & \begin{bmatrix} \dot{R} \\ \dot{q} \end{bmatrix} = \begin{bmatrix} R\widehat{S}\left(-k_q\varphi(R^\top \nabla V_q(R))\right) \\ 0 \end{bmatrix} \\[2ex] P_M(\mathcal{T}_q(R)) \ge \delta + P_M(\mathcal{T}_{1-q}(R)) & \begin{bmatrix} R^+ \\ q^+ \end{bmatrix} = \begin{bmatrix} R \\ 1-q \end{bmatrix} \end{cases}
$$

$$(7.42)$$

The following result establishes that the closed loop has the set $\mathcal{A}$ robustly globally asymptotically stable.

> **Proposition 7.18** (Global asymptotic stabilization of the attitude). *Given a symmetric, positive definite matrix $M \in \mathbb{R}^{3\times 3}$ with distinct eigenvalues and $k > 0$ satisfying (7.41) there exist $s \in \mathbb{S}^2$ and $\delta > 0$ defining V_q and $\mathcal{T}_q$ in (7.40) with $k_1 = k$ and $k_2 = -k$ such that the set $\mathcal{A} = \{I\} \times Q$ is globally asymptotically stable for the hybrid closed-loop system $\mathcal{H}$ in (7.42), with robustness in the sense of Definition 3.16.*

Simulation files validating the properties in Proposition 7.18 are at @BookSite/ Simulation/AttitudeControl. *This implementation also includes disturbances that enable the user to validate the margin of robustness induced by synergistic control.*

7.4.2   The Control Affine Case

The properties required by the synergistic Lyapunov function and state-feedback pair can be refined for the case of plants with a flow map that is affine in the control input. Such a refinement also enables a *backstepping-based design* for the synergistic hybrid controller. To this end, consider the special case of a plant $\mathcal{H}_P$ with state $z = (z_1, z_2) \in C_{P,1} \times \mathbb{R}^{m_P}$, input $u \in \mathbb{R}^{m_P}$, and dynamics given by

$$(z, u) \in C_{P,1} \times \mathbb{R}^{m_P} \times \mathbb{R}^{m_P} \qquad \dot{z} = F_P(z, u) := \phi_1(z) + \psi_1(z)u \qquad (7.43)$$

where

$$\phi_1(z) = \begin{bmatrix} \phi_0(z_1) + \psi_0(z_1)z_2 \\ 0 \end{bmatrix}, \qquad \psi_1(z) = \begin{bmatrix} 0 \\ I \end{bmatrix} \qquad (7.44)$$

It is possible to construct a synergistic Lyapunov function and state-feedback pair for this plant when a pair for the following *reduced plant* satisfying a weaker property than the one in Definition 7.3 exists:

$$(z_1, z_2) \in C_{P,0} \times \mathbb{R}^{m_P} \qquad \dot{z}_1 = \phi_0(z_1) + \psi_0(z_1)z_2 =: F_{P,0}(z_1, z_2) \qquad (7.45)$$

with z_2 as its (virtual) control input. Next, such weaker notion of synergistic pair is formally stated for the continuous-time plant in (7.43).

Definition 7.19 (Weak synergistic Lyapunov function and state-feedback pair). *Given a plant $\mathcal{H}_P$ as in (7.43), closed sets $X \subset \mathbb{R}^{n_P} \times Q$ and $\mathcal{M} \subset X$, and a compact set $\mathcal{A} \subset \mathcal{M}$, a continuously differentiable function $V : X \to \mathbb{R}_{\ge 0}$, and a continuous function $\kappa : X \to \mathbb{R}^{m_P}$ define a <u>weak synergistic Lyapunov function and state-feedback pair</u> candidate (V, κ) relative to $(\mathcal{A}, \mathcal{M})$ for $\mathcal{H}_P$ if items 1-3 in*

Definition 7.3 hold and, defining $\mathcal{W} := \{(z,q) \in X : \langle \nabla_z V(z,q), \psi_1(z) \rangle = 0\}$ *and with* $\mathcal{E}$ *as in Definition 7.3,*

4. *The set* $\Omega \subset \mathcal{E} \cap \mathcal{W}$ *denotes the largest weakly invariant set for the system*

$$\dot{z} = \phi_1(z) + \psi_1(z)\kappa(z,q), \quad \dot{q} = 0 \qquad (z,q) \in \mathcal{E} \cap \mathcal{W} \qquad (7.46)$$

and

5. *For each* $(z,q) \in X$, *define* $\mu_V(z,q)$ *as in (7.11) in Definition 7.3,*

the candidate pair (V, κ) *is called a weak synergistic Lyapunov function and state-feedback pair relative to* $(\mathcal{A}, \mathcal{M})$ *for* $\mathcal{H}_P$ *if*

$$\mu_V(z,q) > 0 \qquad\qquad \forall(z,q) \in \left(\Omega \cup \overline{X \setminus \mathcal{M}}\right) \setminus \mathcal{A} \qquad (7.47)$$

in which case μ_V *is called the weak synergy gap at* (z,q) *and any function* $\delta : X \to \mathbb{R}_{\geq 0}$ *such that*

$$\mu_V(z,q) > \delta(z,q) > 0 \qquad\qquad \forall(z,q) \in \left(\Omega \cup \overline{X \setminus \mathcal{M}}\right) \setminus \mathcal{A} \qquad (7.48)$$

is its positive lower bound.

For the class of continuous-time plants (7.43), a synergistic pair can be constructed from a weak synergistic pair (V, κ) with positive lower bound δ when there exist continuous functions $\sigma : \mathbb{R}^{m_P} \to \mathbb{R}_{\geq 0}$ and $\varrho : X \to \mathbb{R}_{\geq 0}$ with σ positive definite with respect to the origin such that

- For each $(z,q) \in X$ and each $p \in Q$

$$\sigma\big(\kappa(z,q) - \kappa(z,p)\big) \leq \varrho(z,q) \qquad (7.49)$$

- For each $(z,q) \in (\Omega \setminus \mathcal{A}) \cup \overline{X \setminus \mathcal{M}}$,

$$\mu_V(z,q) > \delta(z,q) + \varrho(z,q) \qquad (7.50)$$

The following result shows that a (nonweak) synergistic Lyapunov function and state-feedback pair with positive lower bound can be constructed for (7.43) from a weak synergistic Lyapunov function and state-feedback pair with positive lower bound for the reduced plant in (7.45).

Theorem 7.20 (From weak synergy pair to (nonweak) synergy pair). *Given a plant* $\mathcal{H}_P$ *as in (7.43), a closed set* $X_0 \subset C_{P,0} \times Q$, *a compact set* $\mathcal{A}_0 \subset X_0$, *and a symmetric positive definite matrix* Γ, *suppose that*

1. *There exist a pair* (V_0, κ_0) *defining a weak synergistic Lyapunov function and state-feedback pair relative to* $(\mathcal{A}_0, X_0)$ *for the system in (7.45) with synergy gap lower bounded by the continuous function* $\delta : X_0 \to \mathbb{R}_{\geq 0}$.

2. *The function* $\sigma(v) = v^\top \Gamma v$ *is such that* Γ *is invertible and there exists* ϱ *for which (7.49) and (7.50) hold on the set of points indicated therein.*

Then, with $X_1 = \{(z,q) : z = (z_1, z_2), (z_1, q) \in X_0, z_2 \in \mathbb{R}^{m_P}\}$, *the pair* (V_1, κ_1) *defined for each* $(z, q) \in X_1$ *as*

$$V_1(z, q) := V_0(z_1, q) + \sigma(z_2 - \kappa_0(z_1, q)) \tag{7.51}$$

$$\kappa_1(z, q) := \widetilde{\theta}(z_2 - \kappa_0(z_1, q)) - \frac{1}{2}\Gamma^{-1}\psi_0(z_1)^\top \nabla_{z_1} V_0(z_1, q) \tag{7.52}$$

$$+ \frac{\partial \kappa_0}{\partial z_1}(z_1, q)(\phi_0(z_1) + \psi_0(z_1)z_2)$$

is a (nonweak) synergistic Lyapunov function and state-feedback pair relative to $(\mathcal{A}_1, X_1)$ *for* $\mathcal{H}_P$ *as in* (7.43) *with synergy gap exceeding* δ, *where*

$$\mathcal{A}_1 := \{(z, q) \in X_1 : (z_1, q) \in \mathcal{A}_0, \ z_2 = \kappa_0(z_1, q)\} \tag{7.53}$$

and $\widetilde{\theta} : \mathbb{R}^m \to \mathbb{R}^m$ *is such that there exists a continuous function* $\theta : \mathbb{R}_{\geq 0} \to \mathbb{R}_{\geq 0}$ *that is positive definite with respect to the origin and satisfies*

$$v^\top \Gamma \widetilde{\theta}(v) + \widetilde{\theta}(v)^\top \Gamma v \leq -\theta(|v|) \quad \forall v \in \mathbb{R}^{m_P} \tag{7.54}$$

Proof. From the definition of (V_1, κ_1) in (7.51)-(7.52) and the properties of (V_0, κ_0), the change of V_1 along the flow of (7.43) with $u = \kappa_1(z)$ satisfies for each $(z, q) \in X_1$

$$\begin{aligned}
\langle \nabla_z V_1(z, q), \phi_1(z) + \psi_1(z)\kappa_1(z, q) \rangle \leq \ & \langle \nabla_{z_1} V_0(z_1, q), \phi_0(z_1) + \psi_0(z_1)z_2 \rangle \\
& - \theta(|z_2 - \kappa_0(z_1, q)|) \\
& - \langle \nabla_{z_1} V_0(z_1, q), \psi_0(z_1)(z_2 - \kappa_0(z_1, q)) \rangle \\
\leq \ & \langle \nabla_{z_1} V_0(z_1, q), \phi_0(z_1) + \psi_0(z_1)\kappa_0(z_1, q) \rangle \\
& - \theta(|z_2 - \kappa_0(z_1, q)|) \\
\leq \ & 0
\end{aligned}$$

$$\tag{7.55}$$

The last inequality follows from the fact that (V_0, κ_0) satisfy (7.9) on X_0. Define

$$\begin{aligned}
\mathcal{E}_1 &:= \{(z, q) \in X_1 : \langle \nabla_z V_1(z, q), \phi_1(z) + \psi_1(z)\kappa_1(z, q) \rangle = 0\} \\
\mathcal{W}_1 &:= \{(z, q) \in X_1 : \langle \nabla_z V_1(z, q), \psi_1(z) \rangle = 0\}.
\end{aligned} \tag{7.56}$$

Let $\mathcal{E}_0$, $\mathcal{W}_0$, and Ω_0 come from the pair (V_0, κ_0) defining a weak synergistic Lyapunov function and state-feedback pair relative to $(\mathcal{A}_0, X_0)$ for system (7.45) with synergy gap lower bounded by the continuous function $\delta : X_0 \to \mathbb{R}_{\geq 0}$. Using (7.55), the properties of θ, the definition of ψ_1 in (7.44), and the definition of V_1 it follows that

$$\mathcal{E}_1 = \{(z, q) \in X_1 : (z_1, q) \in \mathcal{E}_0, z_2 = \kappa_0(z_1, q)\} \subset \mathcal{W}_1. \tag{7.57}$$

Let $\Psi_1 \subset \mathcal{E}_1$ denote the largest weakly invariant set for the system

$$\dot{z} = \phi_1(z) + \psi_1(z)\kappa_1(z, q), \qquad \dot{q} = 0 \qquad (z, q) \in \mathcal{E}_1 \tag{7.58}$$

From the definition of κ_1 in (7.52), the fact that $\dot{z}_2 = \kappa_1(z, q)$, and the characterization of $\mathcal{E}_1$ in (7.57), the set Ψ_1 is given by

$$\Psi_1 = \{(z, q) \in X_1 : (z_1, q) \in \Omega_0, z_2 = \kappa_0(z_1, q)\} \tag{7.59}$$

Then, using (7.51), μ_{V_1} given in (7.11) satisfies

$$\mu_{V_1}(z, q) \geq V_0(z_1, q) - \min_{p \in Q} V_0(z_1, p) + \sigma(z_2 - \kappa_0(z_1, q)) - \min_{p \in Q} \sigma(z_2 - \kappa_0(z_1, p))$$

$$\geq \mu_{V_0}(z_1, q) + \sigma(z_2 - \kappa_0(z_1, q)) - \max_{p \in Q} \sigma(z_2 - \kappa_0(z_1, p))$$

Note that $X_1 \setminus \mathcal{M}_1 = \emptyset$. Moreover, $(z, q) \in \Psi_1 \setminus \mathcal{A}_1$ implies that $(z_1, q) \in \Omega_0 \setminus \mathcal{A}_0$. So, in particular, $z_2 = \kappa_0(z_1, q)$. Therefore, for each $(z, q) \in \left(\Psi_1 \cup \overline{X_1 \setminus \mathcal{M}_1} \right) \setminus \mathcal{A}_1$,

$$\mu_{V_1}(z, q) \geq \mu_{V_0}(z_1, q) - \max_{p \in Q} \sigma(\kappa_0(z_1, q) - \kappa_0(z_1, p))$$

$$\geq \mu_{V_0}(z_1, q) - \varrho(z_1, q) > \delta(z_1, q)$$

Thus, the pair (V_1, κ_1) is a synergistic Lyapunov function and state-feedback pair for $\mathcal{H}_P$ relative to $(\mathcal{A}_1, X_1)$ with synergy gap lower bounded by δ. $\qquad\square$

Example 7.21 (Quaternion-based attitude backstepping control). *Continuing from Example 3.3, consider the kinematic equations of a rigid body with attitude parameterized by unit quaternions, which are given by*

$$z = \mathfrak{q} = \begin{bmatrix} \mathfrak{n} \\ \mathfrak{e} \end{bmatrix} \in \mathbb{S}^3 \qquad \dot{z} = \frac{1}{2} \begin{bmatrix} -\mathfrak{e}^\top \omega \\ \mathfrak{n}I + \widehat{S}(\mathfrak{e})\omega \end{bmatrix} \tag{7.60}$$

where

$$\mathbb{S}^3 = \left\{ \mathfrak{q} \in \mathbb{R}^4 : \mathfrak{n}^2 + \mathfrak{e}^\top \mathfrak{e} = 1 \right\}$$

is the unit 3-sphere embedded in $\mathbb{R}^4$, $\mathfrak{q} \in \mathbb{S}^3$ is the unit quaternion representing attitude, $\omega \in \mathbb{R}^3$ is the angular velocity, and $\widehat{S}(\cdot)$ is defined in (7.6). A quaternion $\mathfrak{q} = (\mathfrak{n}, \mathfrak{e})$ is related to a rotation matrix through the map

$$\mathcal{R}(\mathfrak{q}) = I + 2\mathfrak{n}\widehat{S}(\mathfrak{e}) + 2\widehat{S}(\mathfrak{e})^2$$

which is known as Rodrigues formula – in other words, given $\mathfrak{q} \in \mathbb{S}^3$, $R = \mathcal{R}(\mathfrak{q})$ belongs to $SO(3)$. As mentioned in Example 3.3, for each $R \in SO(3)$ there exist exactly two (antipodal) unit quaternions, $\mathfrak{q}$ and $-\mathfrak{q}$, satisfying $\mathcal{R}(\mathfrak{q}) = \mathcal{R}(-\mathfrak{q}) = R$. Furthermore, since $\mathcal{R}(\mathfrak{q}) = I$ if and only if $\mathfrak{q} = \pm\mathbf{1} \in \mathbb{S}^3$, the objective is to render $\{\mathbf{1}\} \cup \{-\mathbf{1}\}$ globally asymptotically stable when the plant in (7.60) is controlled. This plant is denoted $\mathcal{H}_{P,0}$ and its data is given as

$$F_{P,0}(z_1, z_2) := \frac{1}{2} \begin{bmatrix} -\mathfrak{e}^\top \\ \mathfrak{n}I + \widehat{S}(\mathfrak{e}) \end{bmatrix} z_2 \qquad \forall (z_1, z_2) \in C_{P,0} := \mathbb{S}^3 \times \mathbb{R}^3 \tag{7.61}$$

and $z_1 = \mathfrak{q}$ and $z_2 = \omega$. Following (7.45), the state of $\mathcal{H}_{P,0}$ is z_1 and its input is z_2. In this way, as mentioned above Definition 7.19, ω is considered to be the "virtual" control input. Now, consider

- *$Q := \{0, 1\}$ and $X_0 := \mathbb{S}^3 \times Q$;*

- *$\mathcal{A}_0 := \{(z_1, q) \in X_0 : z_1 = \alpha(q)\mathbf{1}\}$, where α is a continuously differentiable function satisfying $\alpha(q) = 2q - 1$ for each $q \in Q$;*

- *$V_0(z_1, q) := 2k(1 - \alpha(q)\mathfrak{n})$ for each $(z_1, q) \in X_0$, where $k > 0$; and*

- $\kappa_0(z_1, q) := 0$ for each $(z_1, q) \in X_0$.

According to Definition 7.19, the pair (V_0, κ_0) is a weak synergistic Lyapunov function and state-feedback pair relative to $(\mathcal{A}_0, X_0)$ for $\mathcal{H}_{P,0}$ as in (7.61). In fact, V_0 and κ_0 are continuously differentiable. Moreover, since $\mathbb{S}^3$ is compact and V_0 is a continuous function that is positive definite with respect to $\mathcal{A}_0$, for each $c \geq 0$ the sublevel set $\{(z_1, q) \in X_0 : V_0(z_1, q) \leq c\}$ is compact. Furthermore, since $\kappa_0(z_1, q) = 0$ on X_0, it follows that $\langle \nabla_{z_1} V_0(z_1, q), F_{P,0}(z_1, \kappa_0(z_1, q)) \rangle = 0$ for each $(z_1, q) \in X_0$. Then, the conditions in items 1-3 in Definition 7.3 hold.

Denoting by $\mathcal{E}_0$ the set $\mathcal{E}$ in Definition 7.19 with V replaced by V_0, and with Ω_0 and $\mathcal{W}_0$ playing the role of Ω and $\mathcal{W}$ therein, respectively, it follows that $\mathcal{E}_0 = X_0$ and $\mathcal{W}_0 = \Omega_0 = \{(z_1, q) \in X_0 : z_1 = \pm \mathbf{1}\}$. Then,

$$\Omega_0 \setminus \mathcal{A}_0 = \{(z_1, q) \in X_0 : z_1 = -\alpha(q)\mathbf{1}\}$$

Since $\mathcal{M}_0 = X_0$, it follows that on

$$\left(\Omega_0 \cup \overline{X_0 \setminus \mathcal{M}_0} \right) \setminus \mathcal{A}_0 = \Omega_0 \setminus \mathcal{A}_0$$

the weak synergy gap satisfies

$$\begin{aligned}
\mu_{V_0}(z_1, q) &= V_0(z_1, q) - \min_{p \in Q} V_0(z_1, p) \\
&= V_0(-\alpha(q)\mathbf{1}, q) - \min_{p \in Q} V_0(-\alpha(q)\mathbf{1}, p) \\
&= 2k(1 + \alpha(q)^2) - \min_{p \in Q} 2k(1 + \alpha(q)\alpha(p)) \\
&= 4k > 0
\end{aligned}$$

The above steps establish that (V_0, κ_0) is a weak synergistic Lyapunov function and state-feedback pair for (7.60) relative to $(\mathcal{A}_0, X_0)$ for $\mathcal{H}_{P,0}$ as in (7.61), with gap exceeding $\delta \in (0, 4k)$.

Now, consider ω in $\mathcal{H}_{P,0}$ as a state denoting the angular velocity of the rigid body. The evolution of ω is governed by

$$J\dot{\omega} = \widehat{S}(J\omega)\omega + \widetilde{u} \tag{7.62}$$

where J is the inertia matrix and $\widetilde{u} \in \mathbb{R}^3$ is the torque applied to the rigid body. With the pre-feedback $\widetilde{u} = -\widehat{S}(J\omega)\omega + Ju$, the plant to stabilize is denoted $\mathcal{H}_P$ and its data is given as

$$F_P(z, u) := \begin{bmatrix} \frac{1}{2} \begin{bmatrix} -\mathfrak{e}^\top \\ \mathfrak{n}I + \widehat{S}(\mathfrak{e}) \end{bmatrix} z_2 \\ u \end{bmatrix} \qquad \forall (z, z_2) \in \mathbb{S}^3 \times \mathbb{R}^3 \times \mathbb{R}^3 \tag{7.63}$$

$z = (z_1, z_2)$, $z_1 = \mathfrak{q}$, and $z_2 = \omega$. Note that the functions ϕ_0 and ψ_0 defining ϕ_1 in (7.43) associated to this plant are

$$\phi_0(z_1) = 0, \qquad \psi_0(z_1) = \frac{1}{2} \begin{bmatrix} -\mathfrak{e}^\top \\ \mathfrak{n}I + \widehat{S}(\mathfrak{e}) \end{bmatrix} \qquad \forall z_1 \in \mathbb{S}^3$$

and the set $C_{P,1}$ is equal to $\mathbb{S}^3$. Furthermore, the weak synergistic Lyapunov function *and state-feedback pair* (V_0, κ_0) *given by*

$$V_0(z_1, q) = 2k(1 - \alpha(q)\mathfrak{n}), \qquad \kappa_0(z_1, q) = 0$$

for each $(z_1, q) \in X_0$ *is such that (7.49) holds (with* κ_0 *instead of* κ*) with* $\varrho \equiv 0$ *and, since* $\kappa_0 \equiv 0$*, with* σ *given as the zero function. Picking* $\Gamma = \frac{1}{2}J$*, the functions* V_1 *and* κ_1 *in Theorem 7.20 are defined on* $X_1 = \mathbb{S}^3 \times \mathbb{R}^3 \times Q$ *and are given by*

$$V_1(z, q) = V_0(z_1, q) + \sigma(z_2 - \kappa_0(z_1, q)) = 2k(1 - \alpha(q)\mathfrak{n}) + \tfrac{1}{2}z_2^\top J z_2$$

$$\kappa_1(z, q) = \widetilde{\theta}(z_2 - \kappa_0(z_1, q)) - \frac{1}{2}\Gamma^{-1}\psi_0(z_1)^\top \nabla_{z_1} V_0(z_1, q)$$

$$= \widetilde{\theta}(z_2) - J^{-1}\frac{1}{2}\begin{bmatrix} -\mathfrak{e}^\top \\ \mathfrak{n}I + \widehat{S}(\mathfrak{e}) \end{bmatrix}^\top \begin{bmatrix} \nabla_{\mathfrak{n}} V_0(z_1, q) & 0 \end{bmatrix} \qquad (7.64)$$

$$= \widetilde{\theta}(z_2) - k\alpha(q)J^{-1}\mathfrak{e}$$

where $\widetilde{\theta}$ *is to be designed to satisfy the conditions in Theorem 7.20. A possible choice for* $\widetilde{\theta}$ *is*

$$\widetilde{\theta}(z_2) = J^{-1}(\widehat{S}(Jz_2)z_2 - \Phi(z_2))$$

where $\Phi(0) = 0$ *and* $z_2^\top \Phi(z_2) \geq \theta(|z_2|)$ *for some positive definite function* $\theta : \mathbb{R}_{\geq 0} \to \mathbb{R}_{\geq 0}$*. Then, it follows that, with* $u = \kappa_1(z, q)$*, the actual input* $\widetilde{u}$ *applied to the plant is defined by the feedback*

$$\widetilde{\kappa}(z, q) = -\widehat{S}(Jz_2)z_2 + J\left(\widetilde{\theta}(z_2) - k\alpha(q)J^{-1}\mathfrak{e}\right) = -\Phi(z_2) - k\alpha(q)\mathfrak{e} \qquad (7.65)$$

Finally, according to Theorem 7.20, $(V_1, \widetilde{\kappa})$ *is a (nonweak) synergistic Lyapunov function and state-feedback pair for* $\mathcal{H}_P$ *with data in (7.63) relative to*

$$\mathcal{A}_1 = \left\{(z, q) \in \mathbb{S}^3 \times \mathbb{R}^3 \times Q : \mathfrak{q} = \alpha(q)\mathbf{1}, z_2 = 0\right\}$$

with gap exceeding $\delta \in (0, 4k)$*. Then, by item 4 of Theorem 7.6, the resulting hybrid closed-loop system has the set* $\mathcal{A}_1$ *globally pre-asymptotically stable.*

7.5 EXERCISES

Exercise 51 (Global stabilization of a two-point set on the plane). Consider the point-mass system $\dot{z} = u$ with z and u in $\mathbb{R}^2$, given $z^* \in \mathbb{R}^2 \setminus \{0\}$.

1. Design a quadratic function V_0 and a gradient-based state-feedback controller κ_0 that globally asymptotically stabilizes the point z^*.

2. Design a quadratic function V_1 and a gradient-based state-feedback controller κ_1 that globally asymptotically stabilizes the point $-z^*$.

3. Provide conditions under which the functions V_0, V_1 and the state-feedback laws κ_0, κ_1 define a synergistic Lyapunov function and state-feedback pair (V, κ) relative to $(\mathcal{A}, X)$ for the point-mass system with $\mathcal{A} = (\{-z^*\} \times \{0\}) \cup (\{z^*\} \times \{1\})$, $X = \mathbb{R}^2 \times Q$, and $Q = \{0, 1\}$, and with synergy gap lower bounded by (a constant or a function) δ that is positive on $X \setminus \mathcal{A}$.

4. Define the synergistic hybrid controller in (7.14) to solve the control problem.

5. Validate your design numerically.

6. Repeat items 3 and 5 for the case where $\mathcal{A} = (\{-z^*\} \cup \{z^*\}) \times \{0,1\}$.

Exercise 52 (Tracking control for rigid body rotational and translational kinematics). The kinematic equations describing translational and rotational motion of a rigid body with attached orthonormal (body) frame and evolving within an orthonormal inertial reference frame are given by

$$\dot{R} = R\widehat{S}(\omega), \qquad \dot{\xi} = \nu - \widehat{S}(\omega)\xi \qquad (7.66)$$

where, as in Example 7.2, $R \in SO(3)$ is the rotation matrix that maps vectors in body frame to inertial frame, $\omega \in \mathbb{R}^3$ denotes the angular velocity of the body frame with respect to the inertial frame expressed in body frame coordinates, $\xi \in \mathbb{R}^3$ denotes the position of the body frame with respect to the inertial frame, expressed in body frame, and $\nu \in \mathbb{R}^3$ denotes the linear velocity of the body frame with respect to the inertial frame (expressed in body frame).

1. Let $t \mapsto (R_d(t); \xi_d(t))$ denote a desired reference trajectory evolving on the manifold $SE(3) := SO(3) \times \mathbb{R}^3$ that is also a solution to (7.66) for some input $t \mapsto (\omega_d(t), \nu_d(t))$. Define the error quantities

$$R_e := R_d R^\top, \quad \xi_e := \xi - \xi_d, \quad \tilde{\omega} := \omega - \omega_d$$

and show that the dynamics of (R_e, ξ_e) are given by

$$\dot{R}_e = -\widehat{S}(R_d\tilde{\omega})R_e, \quad \dot{\xi}_e = \nu - \nu_d - \widehat{S}(\tilde{\omega})(\xi_e + \xi_d) - \widehat{S}(\omega_d)\xi_e \qquad (7.67)$$

where, for simplicity, the explicit dependency on t of some signals has been removed.

2. Use the linear and invertible transformation to the control input $\tilde{\omega}$ given by

$$\omega' := -R_e^\top R_d\tilde{\omega} \qquad (7.68)$$

in (7.67) to arrive to

$$\dot{R}_e = R_e\widehat{S}(\omega'), \quad \dot{\xi}_e = \nu - \nu_d + \widehat{S}(R_d^\top R_e\omega')(\xi_e + \xi_d) - \widehat{S}(\omega_d)\xi_e \qquad (7.69)$$

3. Define a compact set $\mathcal{A}^* \subset SE(3)$ that captures the goal of achieving global tracking of the given reference in error coordinates (R_e, ξ_e).

4. Use the functions V_0, V_1 in Example 7.8 and the state-feedback law κ that assigns the inputs (ω, ν) as

$$\kappa(R_e, \xi_e, R_d, \omega_d, \nu_d, q) := \begin{bmatrix} -k_\omega \varphi(R_e^\top \nabla V_q(R_e)) \\ \nu_d - k_e\xi_e - \widehat{S}(R_d^\top R_e\omega')(\xi_e + \xi_d) + \widehat{S}(\omega_d)\xi_e \end{bmatrix} \qquad (7.70)$$

for the synergistic hybrid controller in (7.14) with $k_\omega, k_e > 0$ and $q \in Q = \{0,1\}$, and show that the compact set $\mathcal{A} := \mathcal{A}^* \times Q$ is globally asymptotically stable for the resulting hybrid closed-loop system in error coordinates.

5. Rewrite the hybrid controller in item 4 in the original coordinates (R, ξ) and validate it numerically.

Exercise 53 (Stabilization of a point on the circle). For the system in the unit circle

$$\dot{z} = u \begin{bmatrix} 0 & -1 \\ 1 & 0 \end{bmatrix} z \qquad z \in \mathbb{S}^1, u \in \mathbb{R} \tag{7.71}$$

and the desired point to stabilize being $z^* = \mathbf{1} \in \mathbb{S}^1$

1. Show that, with $Q = \{0, 1\}$, $X = \mathbb{S}^1 \times Q$, $\mathcal{A} = \{z^*\} \times \{0\}$, and $\xi \in \mathbb{S}^1 \setminus (\{-z^*\} \cup \{z^*\})$, there exists $\alpha > 0$ such that the function

$$V(z, q) = (1 - q)(1 - z^{*\top} z) + q\alpha$$

and the state-feedback law

$$\kappa(z, q) = ((1 - q)z^{*\top} + q\xi) \begin{bmatrix} 0 & -1 \\ 1 & 0 \end{bmatrix} z$$

define a synergistic Lyapunov function and state-feedback pair relative to $(\mathcal{A}, X)$ for the plant in (7.71) with synergy gap lower bounded by a function δ that is positive on $X \setminus \mathcal{A}$.

2. Design a synergistic hybrid control algorithm to globally asymptotically stabilize $\mathcal{A}$.

3. Validate your design numerically.

Exercise 54 (Stabilization of a point on the circle as supervisory controller). Show that the hybrid controller in Exercise 15 (see also § 1.2.3) can be rewritten as a synergistic controller, similar to the one to design in Exercise 53.

Exercise 55 (Global asymptotic stabilization of the 3D pendulum). The differential equations capturing the reduced dynamics of a 3D pendulum are given by

$$\dot{\xi} = \widehat{S}(\xi)\omega, \qquad J\dot{\omega} = \widehat{S}(J\omega)\omega + m\gamma\widehat{S}(\nu)\xi + u \tag{7.72}$$

where $\xi \in \mathbb{S}^2$ is the direction of gravity in the body-fixed frame, $\omega \in \mathbb{R}^3$ is the angular velocity expressed in the body-fixed frame, $J \in \mathbb{R}^{3 \times 3}$ is the constant inertia matrix satisfying $J = J^\top > 0$, γ is the gravitational constant, ν is the vector from the pivot location to the center of mass, $u \in \mathbb{R}^3$ is a vector of input torques, and $\widehat{S}(\cdot)$ is defined in (7.6). The goal is to globally asymptotically stabilize (ξ, ω) to $(-\nu/|\nu|, 0)$, which represents an inverted configuration of the pendulum with zero angular velocity.

1. With Q a finite set, $X_0 = \mathbb{S}^2 \times Q$, $Q' \subset Q$ nonempty, and $\mathcal{A}_0 = \{-\nu/|\nu|\} \times Q'$, let $V_0 : X_0 \to \mathbb{R}$ be positive definite on X_0 relative to $\mathcal{A}_0$ and define $\kappa_0(\xi, q) = 0$ for each $(\xi, q) \in X_0 \times Q$. Show that if

$$\inf_{(\xi, q) \in \Omega_0 \setminus \mathcal{A}_0} \mu_{V_0}(\xi, q) > 0 \tag{7.73}$$

then (V_0, κ_0) is a weak synergistic Lyapunov function and state-feedback pair candidate relative to $(\mathcal{A}_0, X_0)$ for (7.72), where Ω_0 comes from Definition 7.19.

2. Suppose[4] that the weak synergy gap of (V_0, κ_0) is lower bounded by a constant $\delta > 0$. Consider the input transformation

$$u = -\widehat{S}(J\omega)\,\omega - m\gamma\widehat{S}(\nu)\,\xi + J\widetilde{u}$$

a) Show that the angular velocity dynamics of (7.72) reduces to $\dot{\omega} = \widetilde{u}$ with this input transformation.

b) Let $\sigma(\omega) = \frac{1}{2}\omega^\top J\omega$ and let $\widetilde{\theta}(\omega) = J^{-1}\left(\widehat{S}(J\omega)\,\omega - \widetilde{\Gamma}(\omega)\right)$, where $\widetilde{\Gamma} : \mathbb{R}^3 \to \mathbb{R}^3$ satisfies $\omega^\top\widetilde{\Gamma}(\omega) \geq \theta(|\omega|)$ for all $\omega \in \mathbb{R}^3$, and $\theta : \mathbb{R} \to \mathbb{R}$ is a continuous function that is positive definite with respect to the origin. Show that (7.52) leads to

$$\widetilde{u} = J^{-1}\left(\widehat{S}(J\omega)\,\omega - \widetilde{\Gamma}(\omega)\right) - J^{-1}\widehat{S}(\xi)^\top \nabla V_0(\xi, q),$$

which yields

$$u = \kappa_1(z, q) = -m\gamma\widehat{S}(\nu)\,\xi - \widetilde{\Gamma}(\omega) - \widehat{S}(\xi)^\top \nabla V_0(\xi, q). \qquad (7.74)$$

c) Show that (V_1, κ_1), with

$$V_1(z, q) = V_0(\xi, q) + \frac{1}{2}\omega^\top J\omega$$

is a synergistic Lyapunov function and state-feedback pair for (7.72) relative to $(\mathcal{A}_1, X_1)$ with synergy gap lower bounded by $\delta > 0$, where $X_1 = \mathbb{S}^2 \times \mathbb{R}^3 \times Q$ and

$$\mathcal{A}_1 = \{(\xi, \omega, q) \in X_1 : q \in Q', \xi = -\nu/|\nu|,\ \omega = 0\}$$

3. Design a synergistic hybrid controller that globally asymptotically stabilizes $\mathcal{A}_1$.

4. Validate your design numerically.

7.6 NOTES

Control strategies employing multiple state-feedback laws that are coordinated using logic have appeared in literature, many of which were already mentioned in §4.6 and §6.6. Early works where switching involves hysteresis appeared in [139] for the selection of identifier-based parameterized controllers – the algorithm therein is an extension of the one in [140]. Algorithms that select the mode of operation of the system based on the value of Lyapunov function associated to each mode have been widely used in the literature, with initial ideas in [141] and [65]; see also [142]. The scale-independent hysteresis switching strategy in [143] selects the current mode of operation of a switched system based on the relative value between performance

---

[4]A particular construction for δ is given in Proposition 7.13.

functions for each mode – in particular, the performance functions therein could
be Lyapunov-like functions. The switching strategy in [144] uses multiple state-
feedback controllers and picks one to apply to the plant based on the Lyapunov
functions for each controller. In [145], this strategy was applied to provide a solution
to the problem of swinging up a pendulum as well as for the control of a two-tank
system. A similar solution for the pendulum swing-up problem was proposed in
[146]; cf. the hybrid control strategy in Example 6.1. Using control Lyapunov functions and motivated by the impossibility of globally asymptotically stabilizing a
point on a manifold, [147, 148] proposes a minimum projection strategy to control
strict nonsmooth control Lyapunov functions on manifolds.

The notion of synergistic pairs introduced in this chapter and the main results
follow those in [149], [150], and [151]. The definition of synergistic pairs emerged
from hybrid control solutions proposed in [101, 152] to solve the problem of globally asymptotically stabilizing the attitude of a rigid body – see also [149, Section
VII]. Example 7.2 (and also Example 3.3) as well as its revisited versions in this
chapter compiles the main ideas in those references. The two-point set problem in
Example 7.1 was already considered in the literature, in particular, to illustrate that
discontinuous feedback laws are not robust and that logic-based (hybrid) feedback
is mandatory; see [1, Chapter 4] and [32]. The case when the Lyapunov functions
in the family are potential functions, which is considered in Example 7.8 in the
context of attitude control, is treated in detail in [153, 154], where applications to
control of the orientation on the 2-sphere and of the 3D pendulum are presented.
Other applications of synergistic feedback have been treated in the literature, including the global asymptotic stabilization of planar orientation appeared in [155],
of the attitude of a fully actuated rigid body in [156, 157] and [158], of the position
and attitude of an underactuated rigid body in [159, 160], and of a point on the
n-dimensional sphere in [161].

An initial version of the claims in Theorem 7.6 appeared in [149] for plants
modeled as continuous-time systems with a single-valued right-hand side that is
affine in the control input. The extension in Theorem 7.6 is enabled by the use of
the general tools in Chapter 3, as its proof indicates. The construction for the lower
bound on the synergy gap given in Proposition 7.13 is general and is expected to
be a good baseline design for most applications, as the revisited applications in
§ 7.4.1 suggest. The backstepping design approach in § 7.4.2 is a simplification of
the results presented in [150]. In addition, [150] introduces synergistic Lyapunov
function pairs for both *pure* and *ready-made* backstepping. In [150], a synergistic
Lyapunov function and state-feedback pair is pure when the Lyapunov function
is nonincreasing along solutions at every point in the state space when using the
feedback. A synergistic Lyapunov function and state-feedback pair is ready-made
when there is an appropriate relationship between the size of the jumps in the
feedback law and the synergy gap of the synergistic Lyapunov function and state-
feedback pair. Using such terminology, Theorem 7.20 corresponds to the *type I
ready-made* backstepping result in [150]. In addition, [150] provides a construction
that allows to remove the jumps in the feedback law, essentially smoothing out the
feedback law in situations where the feedback does not enter the system through
an integrator; see Section VI therein.

Chapter Eight

Supervisory Control

In this chapter, a hybrid control strategy that supervises multiple hybrid controllers is proposed. As illustrated by the strategies in Chapter 4, Chapter 6, and Chapter 7, control systems featuring multiple control laws employ a mechanism acting as a "supervisor" that selects the controller to be applied to the plant. This selection is performed in real time and involves the value of the states and outputs of the plant and of the individual hybrid controllers. The hybrid control strategy in this chapter is a supervisory algorithm that coordinates a family of hybrid controllers for asymptotic stabilization of a compact set $\mathcal{A}^*$. With knowledge of the operating regions for which each hybrid controller was designed for, the supervisor controls the value of a logic variable that determines the controller to be put in the loop. As depicted in Figure 8.1, the supervisor selects a different hybrid controller when the state reaches the set Ψ_p, which represents the said operating region of a hybrid controller, from where another controller can take over the control of the plant. Repeating this process, the plant state is steered towards the set $\mathcal{A}^*$.

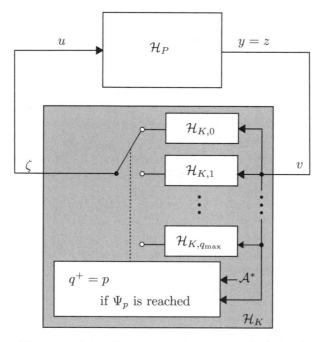

Figure 8.1: Closed-loop resulting from supervisory control. A family of hybrid controllers are coordinated to, sequentially, steer the state to the compact set $\mathcal{A}^*$.

8.1 OVERVIEW

Suppose that given a plant $\mathcal{H}_P$ the goal is to render a set-point (or a set), denoted $\mathcal{A}^*$, asymptotically stable by designing and coordinating a family of controllers whose combined region of operation covers a desired region of the state space. Similar to the strategy in Chapter 6, the idea is to exploit the "divide and conquer" advantage provided by hybrid feedback control – with the added value of robustness – and design the family of controllers and the associated supervisory algorithm to guarantee that solutions converge to the desired set-point, with stability. As a difference to the strategies in Chapter 4, Chapter 6, and Chapter 7, the individual controllers in the family are allowed to be hybrid. The supervisory control strategy is as follows:

> Given a family of hybrid controllers $\{\mathcal{H}_{K,q}\}_{q \in Q}$, where q indexes the controllers and Q is a finite set of the form $\{0, 1, \ldots, q_{\max}\}$, each individual controller is such that, when applied to the plant, solutions to the closed-loop system starting from a particular region of operation, denoted Ψ_q, either they converge to $\mathcal{A}^*$ or they converge to the region of operation of another hybrid controller in the family. Imposing this property for each one of the controllers in the family, a logic-based supervisory algorithm executing the following tasks is designed:
>
> - Apply controller $\mathcal{H}_{K,q}$ to the plant.
>
> - If the solution to the closed loop reaches a point in Ψ_p with p such that $p < q$, then switch to controller $\mathcal{H}_{K,p}$.
>
> Using the properties of the individual hybrid controllers in the family, $\mathcal{A}^*$ is attractive since, necessarily, there exists a controller that guarantees convergence to $\mathcal{A}^*$. Stability is guaranteed by the controller(s) whose region of operation includes points in $\mathcal{A}^*$.

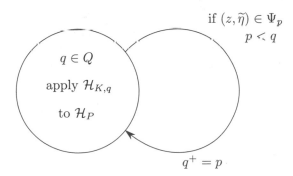

Figure 8.2: Logic implemented by supervisory control of a family of hybrid controllers $\{\mathcal{H}_{K,q}\}_{q \in Q}$. Switches among controllers in the family are triggered by a set-based condition involving the plant state z, the state $\widetilde{\eta}$ of the controllers in the family, and the logic state q of the supervisory algorithm.

Figure 8.2 depicts the logic implemented by this strategy.

The family of individual hybrid controllers $\{\mathcal{H}_{K,q}\}_{q \in Q}$ is designed so that the union of their region of operation is equal to the desired overall region from where attractivity is to hold; i.e., the desired region is a subset of the basin of attraction. Without loss of generality, and to simplify the design of the supervisory algorithm, a descending order on the application of the individual controllers is imposed: solutions starting in the region of operation of the controller associated with $q = 2$ either converge to the region of operation of the controller associated to $q = 1$ or they converge to $\mathcal{A}^*$, and so on.

With the above outline and main ingredients introduced, the formal control objective of the supervisory algorithm is follows:[1]

Given a closed set $\widetilde{X} \subset \mathbb{R}^{n_P} \times \mathbb{R}^{\widetilde{n}_K}$, a compact set $\mathcal{A}^* \subset \widetilde{X}$, where $\widetilde{X}$ is part of the definition of the region of operation for the closed-loop system, and a finite set $Q = \{0, 1, \ldots, q_{\max}\}$, and a family of hybrid controllers $\{\mathcal{H}_{K,q}\}_{q \in Q}$, the resulting hybrid closed-loop system $\mathcal{H} = (C, F, D, G)$ is such that

- The set $\mathcal{A}^* \times Q$ is pre-asymptotically stable for $\mathcal{H}$;

- The sets C, D, $\widetilde{X}$, and Q are such that $C \cup D = \widetilde{X} \times Q$;

- Every maximal solution starting from $\widetilde{X} \times Q$ is complete;[2]

- The basin of pre-attraction includes $C \cup D$.

When these objectives are satisfied, the compact set $\mathcal{A} = \mathcal{A}^* \times Q$ is stable and attractive from every point in $C \cup D$ for $\mathcal{H}$. Namely, the set $\mathcal{A}$ is *globally asymptotically stable* as the basin of attraction contains $C \cup D$ and every maximal solution from $C \cup D$ is complete.

Unlike the uniting control strategy in Chapter 4 or the throw-and-catch strategy in Chapter 6, more than one hybrid controller might be needed to render the compact set $\mathcal{A}^* \times Q$ locally asymptotically stable.

Before introducing the details of the hybrid controller accomplishing the stated control objective, examples illustrate the supervisory control strategy. The first example revisits a control problem already studied and solved in this book and puts it in the context of supervisory control.

Example 8.1 (Global stabilization on the unit circle). *Consider the problem of robustly and globally stabilizing a point on the unit circle introduced in § 1.2.3; see also Exercise 15 and Exercise 29. The point to stabilize is denoted $z^* = 1$ and the plant is given by*

$$\dot{z} = u \begin{bmatrix} 0 & -1 \\ 1 & 0 \end{bmatrix} z \qquad (z, u) \in \mathbb{S}^1 \times \mathbb{R} \tag{8.1}$$

where the state is z and the control input is u. This model also captures the evolution of the orientation of a rigid body on the plane as a function of the angular velocity u; see Example 6.1. Recall that in § 1.2.3 it is argued that robust and global stabilization

---

[1]Rather than using X, X_P, or X_K, the set $\widetilde{X}$ is used in this chapter to highlight that it involves plant states as well as controller states.

of z^* cannot be achieved using a static state-feedback law, in particular, due to such feedback laws having issues around another "antipodal point" in $\mathbb{S}^1$; e.g., $-z^*$; see Figure 8.3.

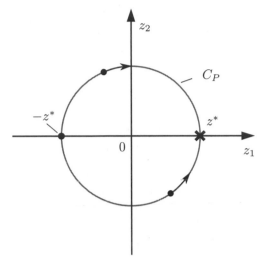

Figure 8.3: Topological obstruction to global asymptotic stabilization of a point on the unit circle.

As suggested in Exercise 15 and Exercise 29, robust and global asymptotic stabilization of z^* can be achieved with a hybrid control algorithm that combines two controllers. In the context of this chapter, such solutions lead to the following supervisory strategy:

- *A controller $\mathcal{H}_{K,1}$ steers, in finite flow time, the state z to a connected set $\Psi_0 \subset \mathbb{S}^1$ containing z^* in its interior. This property is guaranteed by $\mathcal{H}_{K,1}$ from $\mathbb{S}^1 \setminus \Psi_0$.*

- *From initial conditions in Ψ_0, a controller $\mathcal{H}_{K,0}$ asymptotically steers the state z to z^* while keeping the state in $\mathbb{S}^1$.*

Note that $\mathcal{H}_{K,1}$ renders $\mathcal{A} = \{z^*\} \times \{0\}$ asymptotically stable for the closed loop, but not globally. Then, as an alternative to the strategies in Chapter 4 and Chapter 6, the supervisory strategy proposed in this chapter solves the problem. Indeed, using the logic depicted in Figure 8.2, from initial conditions in $\mathbb{S}^1 \setminus \Psi_0$, the state reaches Ψ_0 in finite time using controller $\mathcal{H}_{K,1}$. Then, the supervisor switches to controller $\mathcal{H}_{K,0}$ and the state converges to z^*. The details of the resulting hybrid supervisory controller are in Example 8.7. To showcase the power of the supervisory controller, this strategy is extended to the tracking case in Example 8.10.

Example 8.2 (Stabilization of a mobile robot). *Consider the problem of globally stabilizing the position and orientation of a mobile robot modeled as the unicycle model*

$$\dot{\xi} = \begin{bmatrix} \cos(\theta) \\ \sin(\theta) \end{bmatrix} \vartheta, \qquad \dot{\theta} = \omega$$

where $\xi \in \mathbb{R}^2$ denotes its planar position and θ its orientation. In this model, θ

being zero corresponds to the mobile robot being oriented towards the right of the planar coordinate system and θ being positive corresponds to orientation of the mobile robot changing counterclockwise (from θ being zero). The inputs $\vartheta \in \mathbb{R}$ and $\omega \in \mathbb{R}$ denote the forward velocity and angular velocity inputs, respectively. Without loss of generality, the goal is to globally asymptotically stabilize the point $(\xi, \theta) = (0, 0) \in \mathbb{R}^2 \times \mathbb{R}$, namely, to steer the vehicle to zero position with orientation corresponding to that of the ξ_1 axis on the plane.

Unfortunately, this control system fails Brockett's well-known condition for robust local asymptotic stabilization by classical (even discontinuous) time-invariant feedback; see § 1.2.2. Nevertheless, robust and global asymptotic stability of the desired configuration can be achieved using the hybrid supervisory strategy proposed in this chapter. The details of the control strategy are given in Example 8.11. The idea behind the design of the individual controllers is as follows:

- *First, design a local hybrid controller that is able to locally asymptotically stabilize θ to zero in a way that the position converges to zero simultaneously. This design is possible with a hybrid controller that induces a persistency of excitation type of behavior on the orientation. This controller plays a role of $\mathcal{H}_{K,0}$.*

- *Then, design another hybrid controller – playing the role of $\mathcal{H}_{K,1}$ – capable of steering the vehicle to a neighborhood of the origin, both in position and orientation, contained in the region of operation of the first controller. The latter region plays the role of Ψ_0.*

8.2   HYBRID CONTROLLER

The supervisory algorithm coordinates a finite number of hybrid controllers for the asymptotic stabilization of the set $\mathcal{A} = \mathcal{A}^* \times Q$. For simplicity, the plant is assumed to only have continuous-time dynamics and an output equal to its state, resulting in a hybrid plant $\mathcal{H}_P$ given by[3]

$$\mathcal{H}_P : \quad \begin{cases} (z, u) \in \mathbb{R}^{n_P} \times \mathbb{R}^{m_P} & \dot{z} \in F_P(z, u) \\ & y = z \end{cases} \qquad (8.2)$$

To satisfy the stated objectives, it is assumed that a family of hybrid controllers $\{\mathcal{H}_{K,q}\}_{q \in Q}$ is available, none of which solves the stabilization problem. However, the assumption is such that the family of hybrid controllers have the ingredients necessary to construct a hybrid supervisor solving the problem. The assumption is as follows.

Assumption 8.3 (Family of controllers for supervisory control). *Given a plant $\mathcal{H}_P$ as in (7.8), a positive integer $\widetilde{n}_K$, a closed set $\widetilde{X} \subset \mathbb{R}^{n_P} \times \mathbb{R}^{\widetilde{n}_K}$, and a compact set[4] $\mathcal{A}^* \subset \mathbb{R}^{n_P + \widetilde{n}_K}$, there exist a nonnegative integer $q_{\max}$, a family of hybrid controllers $\{\mathcal{H}_{K,q}\}_{q \in Q}$, $Q := \{0, 1, \ldots, q_{\max}\}$, $\mathcal{H}_{K,q} = (C_{K,q}, F_{K,q}, D_{K,q}, G_{K,q}, \kappa_q)$ with sets*

---

[3]The case when the hybrid plant exhibits jumps can be dealt similarly, though it carries a larger notational burden.

[4]Note that in this formulation the set $\mathcal{A}^*$ involves plant and controller states.

$C_{K,q}$ and $D_{K,q}$ subsets of $\widetilde{X}$, and a collection of closed sets $\Psi_q \subset C_{K,q} \cup D_{K,q}$ such that the following conditions hold:

(SV-A1) $\displaystyle\bigcup_{q \in Q} \Psi_q = \widetilde{X}$.

(SV-A2) For each $q \in Q$, the hybrid closed-loop system obtained from the interconnection between $\mathcal{H}_P$ and the hybrid controller $\mathcal{H}_{K,q}$, which is denoted by $\mathcal{H}_q$, is such that

    a) The set $\mathcal{A}^*$ is globally pre-asymptotically stable;
    b) Each maximal solution is either complete or ends in

$$\Phi_q \cup \overline{\widetilde{X} \setminus (C_{K,q} \cup D_{K,q} \cup \Phi_q)} \tag{8.3}$$

    and
    c) No solution starting from Ψ_q reaches

$$\overline{\widetilde{X} \setminus (C_{K,q} \cup D_{K,q} \cup \Phi_q)} \setminus \mathcal{A}^* \tag{8.4}$$

    where

$$\Phi_q := \bigcup_{p \in Q, i < q} \Psi_p$$

Item (SV-A2)a does not assume a solution to the hybrid control problem treated in this chapter. For each $q \in Q$, item (SV-A2)a only guarantees convergence of maximal solutions to the set $\mathcal{A}^*$, which is different from the control objective. In fact, a controller that, when in the loop, only assures that all maximal solutions are bounded and have a bounded hybrid time domain would satisfy item (SV-A2)a. However, this property is not enough to solve the stated control problem. Certainly, the assumption guarantees that, with each controller, solutions that start close to $\mathcal{A}^*$ stay close to it (if they exist at all), but it does not guarantee that all maximal solutions are complete. For instance, if $C_{K,q} \cup D_{K,q}$ is disjoint from $\mathcal{A}^*$ then global pre-asymptotic stability of $\mathcal{A}^*$ for $\mathcal{H}_q$ as stated in item (SV-A2)a implies that $\mathcal{H}_q$ has no complete solutions.

Item (SV-A2)b, combined with item (SV-A2)a, asks that each maximal solution remains in $\widetilde{X}$ for all time, and either converges to $\mathcal{A}^*$,

- Ends at a point in Φ_q in finite hybrid time. This property enables the supervisor of the family of controllers to update q to the controller with a lower index p and adjust the state of the p-th controller, $\widetilde{\eta}$, to satisfy $(z, \widetilde{\eta}) \in \Psi_p$, from where solutions to the closed loop with $q = p$ can continue evolving.

or

- Ends at a point in $\overline{\widetilde{X} \setminus (C_{K,q} \cup D_{K,q} \cup \Phi_q)}$ in finite hybrid time. In this case, the supervisor can reset the controller state $\widetilde{\eta}$ and the mode q to satisfy $(z, \widetilde{\eta}) \in \Psi_q$.

Item (SV-A2)c requires that maximal solutions starting in Ψ_q never reach points outside of $\mathcal{A}^*$ that are in $\overline{\widetilde{X} \setminus (C_{K,q} \cup D_{K,q} \cup \Phi_q)}$. Hence, with item (SV-A2)b, such maximal solutions have to reach a point in Φ_q. Due to this property, once the state reaches Φ_q or is reset to a point in Ψ_q for some logic variable q, it is always possible to either keep q constant or decrease it, until the solution converges to the set $\mathcal{A}^*$. This eventual monotonicity property of the q component of the solution while outside of $\mathcal{A}^*$ prevents q from oscillating and thus leads to $(z, \widetilde{\eta})$ converging to $\mathcal{A}^*$. Finally, it should be also pointed out that the conditions in item (SV-A2) for the particular case of q equal to zero – the minimum value of q – implies that the maximal solutions to $\mathcal{H}_q$ with $q = 0$ starting in Ψ_0 converge to $\mathcal{A}^*$.

The hybrid controller implements the individual controllers as well as the strategy coordinating them. The state of the supervisor is defined as $\eta = (\widetilde{\eta}, q)$, where $\widetilde{\eta} \in \mathbb{R}^{\bar{n}_K}$ is the state of the individual controllers and the value of $q \in Q$ denotes the controller currently being used to control the plant. The input to the controller is $v = z$, its output is $\zeta \in \mathbb{R}^{m_P}$, and its data is defined as follows:

$$C_K := \bigcup_{q \in Q} (C_{K,q} \times \{q\}) \tag{8.5}$$

$$F_K(z, \eta) := \begin{bmatrix} F_{K,q}(z, \widetilde{\eta}) \\ 0 \end{bmatrix} \qquad \forall (z, \eta) \in C_K \tag{8.6}$$

$$D_K := \bigcup_{q \in Q} ((D_{K,q} \cup H_q) \times \{q\}) \tag{8.7}$$

$$G_K(z, \eta) := G'_{K,1}(z, \widetilde{\eta}) \cup G'_{K,2}(z, \widetilde{\eta}) \qquad \forall (z, \eta) \in D_K \tag{8.8}$$

$$\kappa(z, \eta) := \kappa_q(z, \widetilde{\eta}) \tag{8.9}$$

where

$$H_q := \Phi_q \cup \overline{\widetilde{X} \setminus (C_{K,q} \cup D_{K,q} \cup \Phi_q)} \qquad \forall q \in Q \tag{8.10}$$

and the maps $G'_{K,1}$ and $G'_{K,2}$ are defined as

$$G'_{K,1}(z, \widetilde{\eta}) := \begin{bmatrix} G_{K,q}(z, \widetilde{\eta}) \\ q \end{bmatrix} \qquad \forall (z, \eta) \in \bigcup_{q \in Q} (D_{K,q} \times \{q\})$$

$$G'_{K,2}(z, \widetilde{\eta}) := \begin{cases} \begin{bmatrix} \widetilde{\eta} \\ \{p \in Q : (z, \widetilde{\eta}) \in \Psi_p\} \end{bmatrix} \\ \qquad \qquad \text{if } (z, \widetilde{\eta}) \in \overline{\widetilde{X} \setminus (C_{K,q} \cup D_{K,q} \cup \Phi_q)}, \\ \begin{bmatrix} \widetilde{\eta} \\ \{p \in Q : p < q, (z, \widetilde{\eta}) \in \Psi_p\} \end{bmatrix} \\ \qquad \qquad \text{if } (z, \widetilde{\eta}) \in H_q \setminus \overline{\widetilde{X} \setminus (C_{K,q} \cup D_{K,q} \cup \Phi_q)} \\ \qquad \qquad \forall (z, \eta) \in H_q, \end{cases}$$

and empty elsewhere. The individual hybrid controllers $\mathcal{H}_{K,q}$ and associated data $(C_{K,q}, F_{K,q}, D_{K,q}, G_{K,q}, \kappa_q)$, and the sets Ψ_q and $\widetilde{X}$ come from Assumption 8.3.

As shown next, when the individual controllers satisfy the hybrid basic conditions, the proposed hybrid controller also satisfies the hybrid basic conditions.

Lemma 8.4 (Well-posedness of $\mathcal{H}_K$). *Suppose Assumption 8.3 holds and that, for each $q \in Q$, $\mathcal{H}_{K,q}$ satisfies the hybrid basic conditions. Then, the hybrid controller $\mathcal{H}_K$ with data in (8.5)-(8.9) satisfies the hybrid basic conditions.[5]*

Proof. By assumption, $\mathcal{H}_{K,q}$ satisfies the hybrid basic conditions for each $q \in Q$. The flow set C_K is closed since the sets $C_{K,q}$ are closed. Also, for each $q \in Q$, the set Φ_q is closed due to Ψ_q being closed by Assumption 8.3. The jump set D_K is closed since $D_{K,q}$ and H_q are closed. The flow map F_K satisfies the hybrid basic conditions by virtue of the properties of the maps $F_{K,q}$. The jump map is given by the union of two maps, $G'_{K,1}$ and $G'_{K,2}$. The domain of these maps is such that at least one of these maps is nonempty for each point in D_K. The map $G'_{K,1}$ is outer semicontinuous since $G_{K,q}$ satisfies hybrid basic conditions; see Lemma A.33. The map $G'_{K,2}$ is outer semicontinuous by construction due to closedness of the sets Ψ_q. Finally, κ is continuous due to the functions κ_q being continuous, which comes from $\mathcal{H}_{K,q}$ satisfying the hybrid basic conditions. $\qquad\square$

8.3 CLOSED-LOOP SYSTEM

The hybrid closed loop resulting from controlling a plant $\mathcal{H}_P$ as in (7.8) with the hybrid controller $\mathcal{H}_K$ with data $(C_K, F_K, D_K, G_K, \kappa)$ in (8.5)-(8.9) is defined by the assignment $u = \zeta$. It has state $x = (z, \eta) \in X := \widetilde{X} \times Q$ with $\eta = (\widetilde{\eta}, q)$, which changes according to

$$\begin{aligned}
\dot{z} &\in F_P(z, \kappa_q(z, \widetilde{\eta})) \\
\dot{\widetilde{\eta}} &\in F_{K,q}(z, \widetilde{\eta}) \\
\dot{q} &= 0
\end{aligned}$$

during flows, and at jumps, it is updated according to

$$\begin{aligned}
z^+ &= z \\
\begin{bmatrix} \widetilde{\eta}^+ \\ q^+ \end{bmatrix} &\in G_K(z, \eta)
\end{aligned}$$

The resulting hybrid closed-loop system $\mathcal{H}$ constructed according to Definition 2.11 has state $x = (z, \widetilde{\eta}, q) \in X$ and data (C, F, D, G) given by

$$F(x) := \begin{bmatrix} F_P(z, \kappa_q(z, \widetilde{\eta})) \\ F_{K,q}(z, \widetilde{\eta}) \\ 0 \end{bmatrix} \qquad \forall x \in C := C_K \qquad (8.11)$$

$$G(x) := \begin{bmatrix} z \\ G_K(z, \eta) \end{bmatrix} \qquad \forall x \in D := D_K \qquad (8.12)$$

The maps $F_{K,q}$ and G_K, and the sets C_K and D_K are defined in (8.5)-(8.9).

The next result establishes key properties of this hybrid closed-loop system.

---

[5]See Definition 2.25 and the discussion below it.

Theorem 8.5 (Supervisory control). *Given*

$$\mathcal{H}_P \; : \; \begin{cases} (z, u) \in \mathbb{R}^{n_P} \times \mathbb{R}^{m_P} & \dot{z} \in F_P(z, u) \\ & y = z \end{cases}$$

a positive integer $\widetilde{n}_K$, *a closed set* $\widetilde{X} \subset \mathbb{R}^{n_P} \times \mathbb{R}^{\widetilde{n}_K}$, *and a compact set* $\mathcal{A}^* \subset \mathbb{R}^{n_P + \widetilde{n}_K}$, *suppose Assumption 8.3 holds. Let the hybrid closed-loop system* $\mathcal{H}$ *have data as in* (8.11)-(8.12).

1. *If* $\mathcal{H}_P$ *and, for each* $q \in Q$, $\mathcal{H}_{K,q}$ *satisfy the hybrid basic conditions then* $\mathcal{H}$ *satisfies the hybrid basic conditions.*

2. *Every maximal solution to* $\mathcal{H}$ *is complete.*

3. *The set*

$$\mathcal{A} := \mathcal{A}^* \times Q \qquad\qquad (8.13)$$

   is globally asymptotically stable for $\mathcal{H}$. *Furthermore, when the conditions in item 1 hold, the asymptotic stability property of* $\mathcal{A}$ *is robust in the sense of Definition 3.16.*

The asymptotic stability property in item 1 in Theorem 8.5 is established by removing the "events" corresponding to changing the current controller $\mathcal{H}_{K,q}$ via an update of the logic variable q when the state is not in $\mathcal{A}$. The reason that removing such events is possible is due to the fact that, for each maximal solution to the closed loop $\mathcal{H}$, those events occur finitely many times. With such a property, (pre-)asymptotic stability of $\mathcal{A}$ for $\mathcal{H}$ can be established from (pre-)asymptotic stability of $\mathcal{A}$ for a hybrid system defined by a copy of $\mathcal{H}$ but with the said finite number of events removed. Before presenting a proof of Theorem 8.5, this relationship between the stability properties of $\mathcal{H}$ and its copy without the said events, which is denoted by $\mathcal{H}^0$, is presented.

A general event detector is defined as the set-valued map

$$\mathcal{E} : \mathbb{R}^n \times \mathbb{R}^n \rightrightarrows \{0\}$$

The argument of $\mathcal{E}$ is given by a pair (χ, x), where χ is the value of the state after a jump and x is the value of the state before the jump. These pairs take values from

$$\bigcup_{x \in D} (G(x) \times \{x\}) \qquad\qquad (8.14)$$

Events correspond to pairs (χ, x) in this set that satisfy $\mathcal{E}(\chi, x) = \emptyset$, namely, the pairs are not in the domain of $\mathcal{E}$. A special case of an event detector is the set-valued map that is empty on $\mathbb{R}^n \times \mathbb{R}^n$. In this case, any jump of $\mathcal{H}$ counts as an event. More relevant to supervisory control, for a hybrid closed-loop system $\mathcal{H} = (C, F, D, G)$ with state $x = (z, q) \in \mathbb{R}^{n_P} \times Q$, an event detector for the events that change the value of q is defined as follows. For each $x = (z, q) \in D$ and each $\chi = (z', q') \in G(x)$, such an event detector is defined as $\mathcal{E}(\chi, x) = \{0\}$ if $q = q'$, and empty otherwise. For the closed loop resulting from supervisory control in this chapter, the interest is on a detector for the events that change the controller being applied to the plant when the state is not in the compact set $\mathcal{A}$. As constructed in the proof of Theorem 8.5 given below, an event detector for such events leads to $\mathcal{E}$

being equal to zero at points $(\chi, x) = ((z', q'), (z', q))$ at which $q = q'$ or $x \in \mathcal{A}^* \times Q$, and empty otherwise. In this way, the jump set D^0 in (8.15) collects values of the state that do not change q when not in $\mathcal{A}$.

Given a general event detector $\mathcal{E}$ identifying a particular type of events in a hybrid closed-loop system $\mathcal{H} = (C, F, D, G)$, the hybrid system $\mathcal{H}^0$ with removed events detected by $\mathcal{E}$ is defined by the data (C, F, D^0, G^0), where D^0 and G^0 are given by

$$G^0(x) := G(x) \cap \{\chi : \mathcal{E}(\chi, x) = \{0\}\} \qquad \forall x \in D$$
$$D^0 := D \cap \{x : G^0(x) \neq \emptyset\} \tag{8.15}$$

When $\mathcal{H}$ satisfies the hybrid basic conditions and $\mathcal{E}$ is outer semicontinuous, the hybrid system $\mathcal{H}^0$ also satisfies the hybrid basic conditions and the solutions to $\mathcal{H}^0$ do not have jumps associated to the events that $\mathcal{E}$ detects.

In general, the number of events experienced by a solution x to $\mathcal{H}$ is given by the cardinality of the set

$$\{j : \exists t \in \mathbb{R}_{\geq 0} \text{ s.t. } (t, j), (t, j+1) \in \operatorname{dom} x, \ \mathcal{E}(x(t, j+1), x(t, j)) = \emptyset\} \tag{8.16}$$

The following general result links pre-asymptotic stability of $\mathcal{H}^0$ for a compact set $\mathcal{A}$ to that of $\mathcal{H}$. This result, which holds for general hybrid closed-loop systems $\mathcal{H}$, is used to show Theorem 8.5.

Theorem 8.6 (Stability under finite events). *Given a hybrid closed-loop system $\mathcal{H} = (C, F, D, G)$ satisfying the hybrid basic conditions, let the compact set $\mathcal{A} \subset \mathbb{R}^n$ satisfy $G(D \cap \mathcal{A}) \subset \mathcal{A}$. Suppose that for each compact set $K \subset \mathbb{R}^n$ there exists $N > 0$ such that each solution to the hybrid system $\mathcal{H} = (C, F, D, G)$ starting from K experiences no more than N events, i.e., for each such solution to $\mathcal{H}$ from K the cardinality of the set in (8.16) is less than or equal to N. If the set $\mathcal{A}$ is globally pre-asymptotically stable for $\mathcal{H}^0 = (C, F, D^0, G^0)$ then the set $\mathcal{A}$ is globally pre-asymptotically stable for $\mathcal{H}$.*

With this result in place, the proof of Theorem 8.5 is now presented.

Proof. To show item 1 in Theorem 8.5, note that since Assumption 8.3 holds, with the properties of $\{\mathcal{H}_{K,q}\}_{q \in Q}$, the hybrid controller $\mathcal{H}_K$ satisfies the hybrid basic conditions via Lemma 8.4. Then, since $\mathcal{H}_P$ satisfies the hybrid basic conditions by assumption, Lemma 2.21 implies that the closed loop $\mathcal{H}$ satisfies the hybrid basic conditions as well. In fact, the flow map F of $\mathcal{H}$ is outer semicontinuous and locally bounded relative to C since, due to $\mathcal{H}_P$ and the $\mathcal{H}_{K,q}$'s satisfying the hybrid basic conditions, F_P satisfies (A2$_P$) and the $F_{K,q}$'s satisfy (A2$_K$), respectively, and $(z, \widetilde{\eta}) \mapsto \kappa_q(z, \widetilde{\eta})$ is continuous for each $q \in Q$. Then, F satisfies (A1) in Definition 2.20. The jump map G satisfies (A2) therein by construction. Due to C_P, C_K, and $\widetilde{X}$ being closed, X being equal to $\widetilde{X} \times Q$, and κ being continuous on X, the flow set C is closed; see Exercise 100. Similar arguments show that the jump set D is closed. Then, item 1 holds.

To establish that $\mathcal{A} = \mathcal{A}^* \times Q$ is pre-asymptotically stable for the closed-loop system $\mathcal{H}$, Theorem 8.6 is applied. First, an appropriate event detector $\mathcal{E}$ is defined.

Let the jumps of the hybrid controller $\mathcal{H}_K$ corresponding to changes of mode q that occur at points not in $\mathcal{A}^* \times Q$ define the events to be removed. The event detector $\mathcal{E}$ is defined as follows: for each $(\chi, x) = ((z', \widetilde{\eta}', q'), (z, \widetilde{\eta}, q)) \in \cup_{x \in D} (G(x) \times \{x\})$, $\mathcal{E}$ is given by

$$\mathcal{E}(\chi, x) = \begin{cases} \{0\} & \text{if } x \in D \setminus \mathcal{A}, \ q' = q, \ \text{ or } \ x \in D \cap \mathcal{A} \\ \emptyset & \text{otherwise} \end{cases} \tag{8.17}$$

Note that $\mathcal{E}$ is equal to zero only at points in D at which, after a jump, q does not change or, regardless of the value of the state after jumps, at points in $D \cap \mathcal{A}$; hence, D^0 does not allow events that change the mode q and occur at points in $X \setminus \mathcal{A}$. Furthermore, the event detector is outer semicontinuous at each point it its domain. This property follows via Lemma A.31 since its graph, which is given by

$$\text{gph} \, \mathcal{E} = \{0\} \times (\{(\chi, x) : \chi = (z', \widetilde{\eta}', q'), q' = q, x \in D \setminus \mathcal{A}\} \cup (G(D \cap \mathcal{A}) \times D \cap \mathcal{A}))$$

is closed. With $\mathcal{E}$ specified, define the hybrid system $\mathcal{H}^0$ with flow map and flow set as that of $\mathcal{H}$, and with jump map and jump set given by (8.15). Next, the assumptions in Theorem 8.6 are checked one by one.

By construction of $\mathcal{H}_K$ and item (SV-A2)a of Assumption 8.3, $G(D \cap \mathcal{A}) \subset \mathcal{A}$. In fact, if there would exist $x \in D \cap \mathcal{A}$ such that $G(x) \cap \mathcal{A}$ is nonempty, then it would be a contradiction with the stability property of $\mathcal{A}^*$ in item (SV-A2)a.

By construction of C and D, each $x \in C \cup D$ with $x = (z, \widetilde{\eta}, q)$ satisfies $(z, \widetilde{\eta}) \in C_{K,q} \cup D_{K,q}$. By item (SV-A2)a, the interconnection between $\mathcal{H}_q$ and $\mathcal{H}_P$ is pre-asymptotically stable. Since $\mathcal{E}$ removed events changing the logic variable q at points not in $\mathcal{A}$, it follows that every maximal solution from x is either bounded or complete and converges to $\mathcal{A}$. Then, $\mathcal{A}$ is pre-asymptotically stable for $\mathcal{H}^0$.

The last property to check to be able to apply Theorem 8.6 and, in that way, assert the same pre-asymptotic stability property for $\mathcal{H}$, is that every solution x to $\mathcal{H}$ must have a finite number of events removed by $\mathcal{E}$ in (8.17). By item (SV-A2)b of Assumption 8.3, complete solutions to $\mathcal{H}_q$ that are also solutions to $\mathcal{H}$ do not experience any event, while solutions that end in the set in (8.3) have at least one event. After such an event the state $(z, \widetilde{\eta})$ is mapped to Ψ_q, where q denotes the new mode, from where solutions can only keep q constant or, by item (SV-A2)c, increase it. Using item (SV-A2)b once more, this argument can be repeated until $q = 0$, which occur with state $(z, \widetilde{\eta}) \in \Psi_0$. Then, q remains constant and the number of events until the solution reaches $\mathcal{A}$ is at most the cardinality of Q plus one, namely, $q_{\max} + 1$. Note that this property holds for solutions from every compact set $K \subset \mathbb{R}^{n_P} \times \mathbb{R}^{n_K}$. Then, with $N = q_{\max} + 1$, pre-asymptotic stability of $\mathcal{A}$ for $\mathcal{H}$ follows from Theorem 8.6.

Next, globality of pre-asymptotic stability of $\mathcal{A}$ for $\mathcal{H}$ is established by showing that $C \cup D = \widetilde{X} \times Q$. From Assumption 8.3, $C_{K,q} \subset \widetilde{X}$ and $D_{K,q} \subset \widetilde{X}$ for each $q \in Q$. Then, $C \cup D \subset \widetilde{X} \times Q$. Now, suppose there exists $(\xi, q) \in \widetilde{X} \times Q$ such that $(\xi, q) \notin C \cup D$. Then, $(\xi, q) \notin C$, which by construction of C_K in (8.5) implies $\xi \notin C_{K,q}$. It also implies that $(\xi, q) \notin D$, which by construction of D_K in (8.7) implies $\xi \notin D_{K,q} \cup \Phi_q \cup \overline{\widetilde{X} \setminus (C_{K,q} \cup D_{K,q} \cup \Phi_q)}$. Combining these properties, $\xi \notin C_{K,q} \cup D_{K,q} \cup \Phi_q \cup \overline{\widetilde{X} \setminus (C_{K,q} \cup D_{K,q} \cup \Phi_q)}$. In particular, $\xi \notin C_{K,q} \cup D_{K,q} \cup \overline{\widetilde{X} \setminus (C_{K,q} \cup D_{K,q})}$ which, due to closedness of the sets, implies that $\xi \notin \widetilde{X}$, which is a contradiction. Then, $C \cup D = \widetilde{X} \times Q$ and item 3 holds.

Finally, it is shown that item 2 holds. The pre-attractivity property of $\mathcal{A}$ for the closed loop $\mathcal{H}$ implies that every maximal solution to $\mathcal{H}$ is bounded and that the complete ones converge to $\mathcal{A}$. In particular, this holds for every solution starting from $C \cup D = \widetilde{X} \times Q$. To establish that all maximal solutions starting from $C \cup D$ are complete, since by construction $G(D) \subset C \cup D$, it is enough to show that, from each point $(z, \eta) \in X \backslash D = (\widetilde{X} \times Q) \backslash D$, there exists a solution that flows for some amount of time – see Proposition 2.34. For one such point, with $\eta = (\widetilde{\eta}, q)$, using the definition of D_K, note that $(z, \eta) \in (\widetilde{X} \times Q) \backslash D$ implies

$$(z, \widetilde{\eta}) \in \widetilde{X} \backslash \left(D_{K,q} \cup \Phi_q \cup \overline{\widetilde{X} \backslash (C_{K,q} \cup D_{K,q} \cup \Phi_q)} \right) \qquad (8.18)$$

In particular, this implies that

$$(z, \widetilde{\eta}) \in C_{K,q} \backslash \left(D_{K,q} \cup \Phi_q \cup \overline{\widetilde{X} \backslash (C_{K,q} \cup D_{K,q})} \right) \qquad (8.19)$$

By item (SV-A2)b of Assumption 8.3, the system $\mathcal{H}_q$ admits a flowing solution from such a point since the point is not in the jump set; in fact, if the solution is complete, it has to flow for some finite amount of time, while if it ends in the set (8.3) it also has to flow for some amount of time to reach (8.3). Since the sets Ψ_i are closed, this solution evolves within $C_{K,q} \backslash \Phi_q$ for some amount of time. From the definition of C, such solution defines a solution to $\mathcal{H}$ with constant logic component that belongs to C for some amount of time, and thus is a solution to $\mathcal{H}$ that flows from some time from a point in $X \backslash D$. $\qquad\qquad \square$

8.4   DESIGN

The synthesis of the hybrid controller $\mathcal{H}_K$ in (8.5)-(8.9) requires the design of the positive integer $\widetilde{n}_K$, which defines the dimension of the individual controllers, the set $\widetilde{X}$, which defines the region of operation for the states $(z, \widetilde{\eta})$ of the closed loop, as well as the individual hybrid controllers $\mathcal{H}_{K,q}$ and the sets Ψ_q for each $q \in Q$. Their designs have to satisfy Assumption 8.3.

The supervisory control strategy implemented by $\mathcal{H}_K$ assures that the individual hybrid controllers $\mathcal{H}_{K,q}$ are eventually applied with q monotonically nonincreasing. In the particular case that the controller with $q = 0$ is the only controller that locally asymptotically stabilizes $\mathcal{A}^*$, then q monotonically decreases to zero eventually. In such a case, the hybrid controller $\mathcal{H}_{K,0}$ is the one that would be used to steer the solution to $\mathcal{A}^*$ (with $q = 0$) and, hence, is to be designed to locally asymptotically stabilize that set. Such a design can be performed using the Lyapunov theorem in Theorem 3.19. Denoting the basin of attraction induced by $\mathcal{H}_{K,0}$ as $\mathcal{B}_{\mathcal{A}^*}$, the set Ψ_0 can be chosen as a subset of $\mathcal{B}_{\mathcal{A}^*}$. The controller $\mathcal{H}_{K,1}$ and the set Ψ_1 are to be designed to satisfy the following:

1. Every maximal solution to $\mathcal{H}_1$ is bounded and reaches the set in (8.3) with $q = 1$, namely,

$$\Psi_0 \cup \overline{\widetilde{X} \backslash (C_{K,1} \cup D_{K,1} \cup \Psi_0)} \qquad (8.20)$$

By virtue of the supervisory logic, solutions that end in Ψ_0 are steered by

$\mathcal{H}_{K,0}$ to the set $\mathcal{A}^*$ with $q = 0$. Solutions that reach $\overline{\widetilde{X} \backslash (C_{K,1} \cup D_{K,1} \cup \Psi_0)}$ are, after a jump triggered by the supervisory strategy, controlled by a hybrid controller $\mathcal{H}_q$ with q different than one.

2. For initial conditions in Ψ_1, solutions to $\mathcal{H}_1$ do not reach

$$\overline{\widetilde{X} \backslash (C_{K,1} \cup D_{K,1} \cup \Psi_0)} \backslash \mathcal{A}^* \qquad (8.21)$$

which is the set in (8.4) for $q = 1$. The set (8.21) corresponds to points that are neither in $\mathcal{A}^*$ nor in the closure of the region of operation of $\mathcal{H}_{K,1}$. This condition guarantees that when $\mathcal{H}_{K,1}$ is properly initialized, then the controller steers the solutions to a point in Ψ_0, from where $\mathcal{H}_{K,0}$ can take over and steer the state $(z, \widetilde{\eta})$ toward $\mathcal{A}^*$.

In simple terms, the set Ψ_1 should collect all points from where solutions to $\mathcal{H}_1$ are steered to the set where $\mathcal{H}_{K,0}$ is able to asymptotically stabilize $\mathcal{A}^*$.

Once $\mathcal{H}_{K,1}$ and Ψ_1 are designed to satisfy the above conditions, the controller $\mathcal{H}_2$ can be similarly designed so that every maximal solution to $\mathcal{H}_2$ is bounded and reaches $\Psi_0 \cup \Psi_1 \cup \overline{\widetilde{X} \backslash (C_{K,2} \cup D_{K,2} \cup \Psi_0 \cup \Psi_1)}$, and every solution to $\mathcal{H}_2$ from Ψ_2 does not reach $\overline{\widetilde{X} \backslash (C_{K,2} \cup D_{K,2} \cup \Psi_0 \cup \Psi_1)} \backslash \mathcal{A}^*$. The controllers $\mathcal{H}_{K,q}$ with $q > 2$ can be designed following the same steps.

The methodology outlined above leads to the following design conditions for the satisfaction of (SV-A1)-(SV-A2) in Assumption 8.3: given a compact set $\mathcal{A}^*$ and a closed set $\widetilde{X}$

(Step 1) Design $\mathcal{H}_{K,0}$ to locally asymptotically stabilize $\mathcal{A}^*$ with basin of attraction $\mathcal{B}_{\mathcal{A}^*}$. Pick Ψ_0 as a subset of $\mathcal{B}_{\mathcal{A}^*}$ that contains a neighborhood of $\mathcal{A}^*$. Set $k = 0$ and $\widetilde{n}_K = \widetilde{n}_{K_0}$, where $\widetilde{n}_{K_0}$ is the dimension of $\mathcal{H}_{K,0}$.

(Step 2) Let $q = k + 1$. Design $\mathcal{H}_{K,q}$ and Ψ_q such that

a) Each maximal solution to $\mathcal{H}_q$ is bounded and reaches

$$\Phi_q \cup \overline{\widetilde{X} \backslash (C_{K,q} \cup D_{K,q} \cup \Phi_q)} \qquad (8.22)$$

and

b) No maximal solution starting from Ψ_q reaches

$$\overline{\widetilde{X} \backslash (C_{K,q} \cup D_{K,q} \cup \Phi_q)} \backslash \mathcal{A}^* \qquad (8.23)$$

(Step 3) Increment k by one, set $\widetilde{n}_K = \max\{\widetilde{n}_{K_q}, \widetilde{n}_K\}$, where $\widetilde{n}_{K_q}$ is the dimension of $\mathcal{H}_{K,q}$, and repeat (Step 2) until

$$\widetilde{X} = \bigcup_{q \in Q} \Psi_q$$

The proposed design methods are exercised in examples.

Example 8.7 (Global stabilization on the unit circle, revisited). *In light of the impossibility of achieving robust and global asymptotic stability via discontinuous control for the system in Example 8.1, the proposed procedure is employed to design individual controllers and a supervisory algorithm yielding a hybrid controller $\mathcal{H}_K$ with data as in (8.5)-(8.9). The intuition behind the design procedure is as follows. Since global asymptotic stability is desired, the set $\widetilde{X}$ is chosen as $\mathbb{S}^1$. Then, a static continuous state-feedback law that asymptotically stabilizes $\mathcal{A}^* := \{z^*\} = \{\mathbf{1}\}$ with a basin of attraction that does not include the problematic point, namely, $-z^*$, is designed. This control law defines $\mathcal{H}_{K,0}$ outlined in Example 8.1. Then, another static continuous state-feedback law that asymptotically stabilizes a point in the interior of the basin of attraction induced by $\mathcal{H}_{K,0}$, with a basin of attraction that includes all of the points not included in the basin of $\mathcal{H}_{K,0}$, is designed. This second control law defines $\mathcal{H}_{K,1}$, which is also outlined in Example 8.1. These two controllers are designed next.*

Before designing the controllers, define the multiplication rule

$$y \odot x := \begin{bmatrix} y_1 x_1 - y_2 x_2 \\ y_2 x_1 + y_1 x_2 \end{bmatrix}$$

for generic vectors $x, y \in \mathbb{R}^2$ and note that the model of the plant in (8.1) can be rewritten as

$$\dot{z} = \begin{bmatrix} z_1 0 - z_2 u \\ z_2 0 + z_1 u \end{bmatrix} = z \odot \nu(u) \qquad (z, u) \in \mathbb{S}^1 \times \mathbb{R} \qquad (8.24)$$

where $u \mapsto \nu(u) \in \mathbb{R}^2$ is defined as

$$\nu(u) := \begin{bmatrix} 0 \\ u \end{bmatrix}$$

The set $\mathbb{S}^1$ is such that, regardless of the choice of u, any solution starting in it stays in it since $\langle z, z \odot \nu(u) \rangle = 0$ for each $z \in \mathbb{S}^1$ and for each $u \in \mathbb{R}$.

The design of the hybrid controller $\mathcal{H}_{K,0}$ is based on the observation in § 1.2.3, where a control law almost globally asymptotically stabilizing $\mathcal{A}^*$ is provided. The idea is to use the control law therein but restricted to a subset of $\mathbb{S}^1$, so as to preclude the existence of solutions that do not converge to $\mathcal{A}^*$. One such construction of $\mathcal{H}_{K,0}$ is given by the following constrained continuous state-feedback law:

$$C_{K,0} := \{z \in \mathbb{S}^1 : z_1 \geq c_0\} \qquad (8.25)$$
$$\kappa_0(z) := -z_2 \qquad (8.26)$$

with $c_0 \in (-1, 0)$. The analysis in § 1.2.3 restricted to $C_{K,0}$ leads to the following property.

> **Proposition 8.8** (Property of controller $\mathcal{H}_{K,0}$ for (8.24)). *The closed-loop system $\mathcal{H}_0$ resulting from controlling (8.24) with $\mathcal{H}_{K,0}$ is such that $\mathcal{A}^*$ is globally asymptotically stable.*

Note that "global" in Proposition 8.8 means that every maximal solution to the closed loop from $C_{K,0}$ converges to $\mathcal{A}^*$. However, $C_{K,0}$ does not include all of the *points in $\mathbb{S}^1$.*

To conclude (Step 1) of the design procedure, pick

$$\Psi_0 := C_{K,0}$$

Next, the controller $\mathcal{H}_{K,1}$ is designed. Following the design of $\mathcal{H}_{K,0}$, $\mathcal{H}_{K,1}$ is designed as the constrained continuous state-feedback law given by

$$C_{K,1} := \left\{ z \in \mathbb{S}^1 : z_1 \le c_1 \right\} \tag{8.27}$$

$$\kappa_1(z) := -z_1 \tag{8.28}$$

with $c_1 \in (c_0, 0)$. Due to the restriction to $C_{K,1}$, this controller renders the point $\{(0, -1)\} \in \mathbb{S}^1$ globally pre-asymptotically stabilizes. This point is in the interior of the basin of attraction induced by $\mathcal{H}_{K,0}$. However, note that due to the definition of $C_{K,1}$, no maximal solution to the closed-loop system resulting from using $\mathcal{H}_{K,1}$ converges to $(0, -1)$. In fact, every such maximal solution has a compact hybrid time domain and ends at the boundary of $C_{K,1}$. Conveniently, such a point is in the interior of $C_{K,0}$, so the supervisor is capable to extend the solutions through a jump that resets the value of q to zero.

The following property can be established using similar steps as those leading to Proposition 8.8. Its proof is also left as an exercise.[6]

> **Proposition 8.9** (Property of controller $\mathcal{H}_{K,1}$ for (8.24)). *The closed-loop system $\mathcal{H}_1$ resulting from controlling (8.24) with $\mathcal{H}_{K,1}$ is such that $\mathcal{A}^*$ is globally pre-asymptotically stable, and each[7] maximal solution to it is bounded and reaches Ψ_0 after a finite amount of flow time.*

To conclude (Step 2) of the design procedure, pick

$$\Psi_1 := \overline{\mathbb{S}^1 \backslash C_{K,1}}$$

This is (Step 2) with $k = 1$ of the design procedure. Note that since $\Psi_0 \cup \Psi_1 = \widetilde{X}$, Assumption 8.3 holds with $\rho \equiv 0$.

The closed-loop system resulting from using the two individual controllers designed above in the supervisory controller in (8.5)-(8.9) leads to the closed-loop system with data as in (8.11)-(8.12), but without a controller state $\widetilde{\eta}$, namely, $x = (z, \eta) = (z, q) =: X \in \mathbb{S}^1 \times \{0, 1\}$ and

$$C := \{ x \in X : z \in C_{K,q} \} \tag{8.29}$$

$$F(x) := \begin{bmatrix} z \odot \nu(\kappa_q(z)) \\ 0 \end{bmatrix} \quad \forall x \in C \tag{8.30}$$

$$D := \{ x \in X : z \in \Phi_q \} \tag{8.31}$$

$$G(x) := \begin{bmatrix} z \\ 1 - q \end{bmatrix} \quad \forall x \in D \tag{8.32}$$

where $\Phi_1 = \Psi_0$ and $\Phi_0 = \emptyset$. Figure 1.9 depicts the sets $C_{K,0} = \Psi_0 = \Phi_1$ and

---

[6]Hint: use $V(z) = 1 + z_2$.

[7]Each such maximal solution starts from Ψ_1.

$C_{K,1} = \mathbb{S}^1 \setminus \Psi_1$, *which are associated with this controller. Then, using* $\widetilde{n}_K = 0$, *by Theorem 8.5, the set* $\mathcal{A} = \mathcal{A}^* \times Q = \{z^*\} \times \{0,1\}$ *is globally asymptotically stable for* $\mathcal{H}$. *Furthermore, since* $\mathcal{H}_P$ *and, for each* $q \in Q$, $\mathcal{H}_{K,q}$ *satisfies the hybrid basic conditions, then* $\mathcal{H}$ *satisfies the hybrid basic conditions.*

Example 8.10 (Global stabilization on the unit circle). *The hybrid control algorithm proposed in Example 8.1 extends to the case of tracking a continuously differentiable signal* $\zeta : \mathbb{R}_{\geq 0} \to \mathbb{S}^1$. *By defining* $\xi = z \odot \zeta^c$, *where* $\zeta^c := (\zeta_1, -\zeta_2)$, *one has*

$$\dot{\xi} = \xi \odot \zeta \odot \left(\nu(u) - \zeta^c \odot \dot{\zeta} \right) \odot \zeta^c$$

Picking $\nu(u) = \zeta^c \odot \dot{\zeta} + \nu(\widetilde{u})$ *with* $\widetilde{u}$ *being an auxiliary input results in*

$$\dot{\xi} = \xi \odot \nu(\widetilde{u})$$

Then, the hybrid controller designed in Example 8.7 is applied to control the plant in the coordinates ξ *and through the auxiliary input* $\widetilde{u}$. *Using the definition of the data in* (8.29)-(8.32), *the resulting closed-loop system is*

$$\xi \in C_{K,q} \qquad \begin{bmatrix} \dot{\xi} \\ \dot{q} \end{bmatrix} = \begin{bmatrix} \xi \odot \nu(\kappa_q(\xi)) \\ 0 \end{bmatrix} \qquad (8.33)$$

$$\xi \in D_{K,q} \qquad \begin{bmatrix} \xi^+ \\ q^+ \end{bmatrix} = \begin{bmatrix} \xi \\ 1-q \end{bmatrix} \qquad (8.34)$$

which has the set $\{(\xi, q) \in \mathbb{S}^1 \times Q : \xi = \mathbf{1}\}$ *globally asymptotically stable. From the definition of* ξ, $\xi = \mathbf{1}$ *is equivalent to* $z = \zeta$. *Then, this hybrid controller assures that* ζ *is globally tracked, namely, for each initial condition for* z, *the tracking error* $z - \zeta$ *converges to zero. Note that in the original coordinates, the feedback law is* $\kappa_q(z \odot \zeta)$, *and that the flow and jump conditions in* (8.33) *and* (8.34) *are* $z \odot \zeta \in C_{K,q}$ *and* $z \odot \zeta \in C_{K,q}$, *respectively. Though this closed loop is time varying, when the signal* ζ *is generated by an exosystem, the closed-loop system in the original coordinates is time invariant.*

Similar stabilization and tracking results hold for the case when z *is a unit quaternion, i.e., when it lives in the hypersphere* $\mathbb{S}^3$. *This extension is left as an exercise.*

Example 8.11 (Stabilization of a mobile robot, revisited). *The problem of stabilizing the position and orientation of a mobile robot in Example 8.2 is revisited. The individual controllers outlined therein are now designed. Relabeling* ξ *as the first two components of* z, *which are denoted* $z_1 = (\xi_1, \xi_2) \in \mathbb{R}^2$, *and embedding the angle* θ *in the unit circle, which is denoted* $z_2 \in \mathbb{S}^1$, *the vehicle model in Example 8.2 can be rewritten as*

$$\dot{z}_1 = z_2 u_1, \qquad \dot{z}_2 = z_2 \odot \nu(u_2) \qquad (8.35)$$

whose right-hand side defines F_P. *In these coordinates, the goal is to globally asymptotically stabilize the plant to the point* $z = (0, \mathbf{1}) \in \mathbb{R}^2 \times \mathbb{S}^1$, *where* $\mathbf{1}$ *is the target orientation, with robustness. The proposed hybrid supervisory controller employs two individual controllers,* $\mathcal{H}_{K,0}$ *and* $\mathcal{H}_{K,1}$. *Following Example 8.2, these controllers are designed next.*

Both controllers use the same feedback law for the input u_1, which is given by

$$\kappa_a(z) := -\rho_0(z_1^\top z_2) \tag{8.36}$$

where ρ_0 is continuous and satisfies $s\rho_0(s) > 0$ for all $s \neq 0$. This design follows from the fact that, when the control law in (8.36) is used in (8.35), the time derivative of the quantity $|z_1|^2$ is equal to $-2z_1^\top z_2 \rho_0(z_1^\top z_2)$, which is negative except at points z such that $z_1^\top z_2 = 0$, for which also $\dot{z}_1 = 0$. In this way, in both controllers, the feedback for u_1 is used to control the position of the mobile robot.

The control input u_2 is assigned to different feedback control laws by each one of the individual controllers. The hybrid controller $\mathcal{H}_{K,0}$ locally asymptotically stabilizes $\mathcal{A}^*$ by steering z_2 to $\mathbf{1}$ with a persistency of excitation property, so as to allow z_1 to converge to zero. More precisely, $\mathcal{H}_{K,0}$ has state $\eta_0 := (q_0, \chi_0)$ where q_0 is a logic variable taking values in $Q := \{0, 1\}$ and $\chi_0 \in \mathbb{R}^2$ is a continuous state taking values from the compact set $\Omega_0 \subset \mathbb{R}^2$. The dynamics of q_0 correspond to those of the tracking controller on the unit circle given at the end of Example 8.1. The state χ_0 evolves according to the differential equation

$$\dot{\chi}_0 = \chi_0 \odot \nu(\omega)$$

where $\omega > 0$ is a constant. Since $\chi_0 \odot \nu(\omega) = \omega R(\pi/2)\chi_0$, the dynamics of χ_0 describe a linear oscillator with angular frequency ω. The signal to track is then given by $R(\sigma(z_1)\chi_{01})\mathbf{1}$, where χ_{01} is the first component of χ_0 and $\sigma : \mathbb{R}^2 \to \mathbb{R}$ is a continuously differentiable positive definite function. The definition of the signal is such that when $z_1 = 0$ and z_2 has reached a steady state value, which occurs in the limit, $z_2 = \mathbf{1}$. On the other hand, when $z_1 \neq 0$, the orientation rotates according to the periodic signal χ_{01} (modulated by $\sigma(z_1)$), which corresponds to a persistency of excitation property. Then, the hybrid controller $\mathcal{H}_{K,0}$ is given by

$$
\begin{aligned}
z \odot R(\sigma(z_1)\chi_{01})\mathbf{1} \in C_{K,q_0} \qquad &\begin{bmatrix} \dot{q}_0 \\ \dot{\chi}_0 \end{bmatrix} = \begin{bmatrix} 0 \\ \chi_0 \odot \nu(\omega) \end{bmatrix} \\
z \odot R(\sigma(z_1)\chi_{01})\mathbf{1} \in D_{K,q_0} \qquad &\begin{bmatrix} q_0^+ \\ \chi_0^+ \end{bmatrix} = \begin{bmatrix} 1 - q_0 \\ \chi_0 \end{bmatrix} \\
&\zeta_0 = \kappa_{q_0}(z \odot R(\sigma(z_1)\chi_{01})\mathbf{1})
\end{aligned}
$$

where C_{K,q_0} is given in (8.25) and (8.27), and $D_{K,q_0} = \Phi_{q_0}$ with Φ_{q_0} defined below (8.32).

The hybrid controller $\mathcal{H}_{K,1}$ is designed to steer the position of the mobile robot to nearby zero by making z_2 track $-z_1/|z_1|$ as long as $|z_1|$ is not too small. Figure 8.4(a) depicts the z_1 component of a solution with $\mathcal{H}_{K,1}$ while Figure 8.4(b) shows the attitude component (embedded in $\mathbb{S}^1$). For this purpose, $\mathcal{H}_{K,1}$ has state $\eta_1 = q_1$, with q_1 being a logic variable taking values from Q and dynamics very similar to those of the tracking controller on the unit circle given at the end of Example 8.1. The signal for this controller to track is given by $-z_1/|z_1|$ for each z_1 such that $|z_1| \geq \varepsilon_1$, $\varepsilon_1 > 0$, so as to avoid the singularity at $z_1 = 0$. The dashed circles in Figure 8.4 denote the points z_1 with norm equal to ε_1. Then, the hybrid *controller $\mathcal{H}_{K,1}$ is given by*

$$z \odot \frac{-z_1}{|z_1|} \in C_{K,q_1}, \ |z_1| \geq \varepsilon_1 \qquad \dot{q}_1 = 0$$

$$z \odot \frac{-z_1}{|z_1|} \in D_{K,q_1}, \ |z_1| \geq \varepsilon_1 \qquad q_1^+ = 1 - q_1$$

$$\zeta_1 = \kappa_{q_1}\left(z \odot \frac{-z_1}{|z_1|}\right)$$

where, as for $\mathcal{H}_{K,0}$, C_{K,q_1} is given in (8.25) and (8.27), and $D_{K,q_1} = \Phi_{q_1}$ with Φ_{q_1} defined below (8.32).

The design of the supervisor coordinating $\mathcal{H}_{K,0}$ and $\mathcal{H}_{K,1}$ requires the definition of the sets Ψ_0 and Ψ_1. As Figure 8.4 suggests, $\mathcal{H}_{K,0}$ is applied when the norm of z_1 is larger than or equal to ε_1. To capture this mechanism and to activate $\mathcal{H}_{K,1}$ when the norm of z_1 is smaller than ε_1, the sets Ψ_0 and Ψ_1 are chosen as

$$\Psi_0 := \overline{\mathbb{R}^2 \setminus \varepsilon_1 \mathbb{B}} \times \mathbb{S}^1 \times Q \times \Omega_0, \qquad \Psi_1 := \varepsilon_2 \mathbb{B} \times \mathbb{S}^1 \times Q \times \Omega_0$$

where $\varepsilon_2 > \varepsilon_1$. These definitions lead to

$$H_0 = \Psi_1, \qquad H_1 = \overline{\mathbb{R}^2 \setminus \varepsilon_2 \mathbb{B}} \times \mathbb{S}^1 \times Q \times \Omega_0, \qquad G'_{K,2}(z, \widetilde{\eta}) = 1 - q$$

Figure 8.4 depicts the sets associated with the proposed controller as well as a solution along the different stages using the hybrid controller. Figure 8.4(c) and Fig-

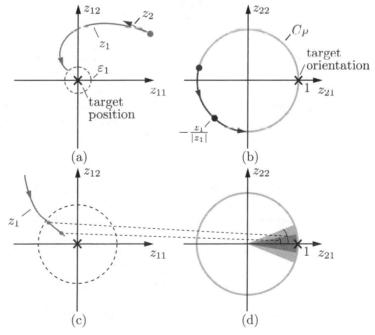

Figure 8.4: Pieces of a solution and sets associated with the supervisory controller for the nonholonomic vehicle in Example 8.11.

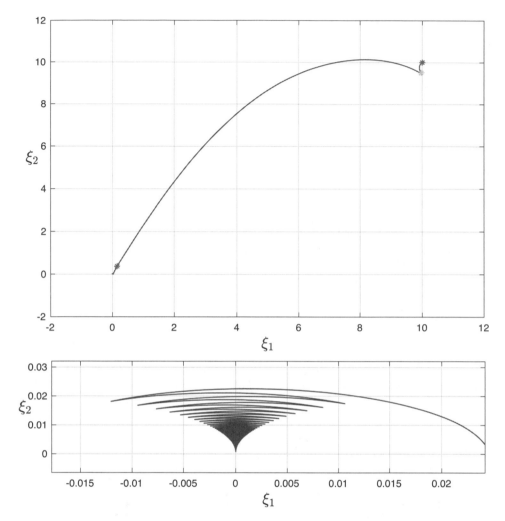

Figure 8.5: Position component of a closed-loop solution resulting from controlling the unicycle in Example 8.11 using the hybrid supervisory controller.

ure 8.4(d) illustrate how $\mathcal{H}_{K,0}$ drives the orientation to a cone centered at $\mathbf{1}$ whose aperture decreases with the norm of z_1. Figure 8.5 shows the position component of a closed-loop solution, showing convergence to the origin with the desired angle.

Simulation files implementing the supervisory control for the problems illustrated in examples in this chapter are available online:

- For Example 8.1 in @BookSite/Simulation/UnitCircle

- For Example 8.10 in @BookSite/Simulation/UnitCircleTracking

- For Example 8.11 in @BookSite/Simulation/Vehicle

8.5   EXERCISES

Exercise 56 (Properties of individual controllers in Example 8.1). For the stabilization problem in Example 8.1, show Proposition 8.8 and Proposition 8.9. Hint: to show the first result, use $V(z) = 1 - z_1$; to show the second result, use $V(z) = 1 + z_2$.

Exercise 57 (Global tracking on the unit circle). The goal of this exercise is to fill the details in some of the steps in the design of the supervisory controller for tracking in the unit circle in Example 8.10. Given a continuously differentiable signal $\zeta : \mathbb{R}_{\geq 0} \to \mathbb{S}^1$, show that ξ defined in Example 8.10 satisfies

$$\dot{\xi} = \xi \odot \zeta \odot \left(\nu(u) - \zeta^c \odot \dot{\zeta} \right) \odot \zeta^c$$

To arrive to such expression of $\dot{\xi}$, use the following properties:

- The multiplication rule, conjugate rule, and identity element satisfy

$$y \odot x := \begin{bmatrix} y_1 x_1 - y_2 x_2 \\ y_2 x_1 + y_1 x_2 \end{bmatrix}, \qquad x^c := \begin{bmatrix} x_1 \\ -x_2 \end{bmatrix}, \qquad \mathbf{1} := \begin{bmatrix} 1 \\ 0 \end{bmatrix}$$

  where x and y are vectors in $\mathbb{R}^2$. The multiplication rule is commutative, associative, and distributive. Note that $x = \mathbf{1} \odot x - x \odot \mathbf{1}$ and note that $x^c \odot x = x \odot x^c = |x|^2 \mathbf{1}$. Also, $(y \odot x)^c = x^c \odot y^c$.

- The signal ζ satisfies

$$\dot{\zeta}^c \odot \zeta = -\zeta^c \odot \dot{\zeta}$$
$$\zeta^c \odot \dot{\zeta} = \begin{bmatrix} 0 \\ \zeta_1 \dot{\zeta}_2 - \zeta_2 \dot{\zeta}_1 \end{bmatrix} \qquad (8.37)$$

  and, with the coordinate transformation $\xi = z \odot \zeta$, one has $z = \xi \odot \zeta^c$, $z \odot z^c = \xi \odot \xi^c = \mathbf{1}$, and $\xi = \zeta \iff z = \mathbf{1}$.

Then, treating $\zeta \odot \left(\nu(u) - \zeta^c \odot \dot{\zeta} \right) \odot \zeta^c$ as the control input, arrive to the controller leading to the closed loop in (8.33)-(8.34) following the steps in the design carried out in Example 8.7.

Exercise 58 (Global stabilization and tracking for unit quaternions). For the model of the plant (8.24) in Example 8.1 but with $z \in \mathbb{S}^3 = \{ z \in \mathbb{R}^4 : z^\top z = 1 \}$, perform the following tasks:

1. Design a hybrid feedback controller that globally asymptotically stabilizes the point $\mathbf{1} = (1,0,0,0)$. Hint: partition $z \in \mathbb{S}^3$ as $z = (z_1, \tilde{z})$ where $\tilde{z} = (z_2, z_3, z_4)$, and, following Example 8.1, design the first controller using $V(z) = 1 - z_1$ and the second controller to asymptotically stabilize a point in the interior of the basin of attraction of the first controller.

2. Following Exercise 57, extend the stabilizing hybrid controller to globally asymptotically track a given continuously differentiable signal $\zeta : \mathbb{R}_{\geq 0} \to \mathbb{S}^3$.

Exercise 59 (Nonholonomic integrator). Consider the model of the nonholonomic integrator given by

$$\dot{z}_1 = u_1, \qquad \dot{z}_2 = u_2, \qquad \dot{z}_3 = z_1 u_2 - z_2 u_1 \qquad z \in \mathbb{R}^3, u \in \mathbb{R}^2$$

Denoting the state of the Brockett integrator, define a nonnegative function r as

$$r(z) = \sqrt{z_1^2 + z_2^2}$$

for each $z \in \mathbb{R}^3$, and let positive constants c and ε satisfy $c(1 + \varepsilon) < 2$, design a supervisory controller that globally asymptotically stabilizes the origin with robustness using the following ideas:

1. For points z such that $r(z) \le \sqrt{c|z_3|}$, design a state-feedback control law that makes the nonnegative quantity

$$(1 + \varepsilon)\sqrt{c|z_3|} - z_1$$

   decrease to zero. Hint: consider a constant feedback law.

2. For points z such that $r(z) > \sqrt{c|z_3|}$, design a state-feedback control law that makes the nonnegative quantity

$$\frac{1}{2}\left(z_1^2 + z_2^2\right)$$

   decrease and simultaneously renders forward invariant the set of points z satisfying $r(z) > \sqrt{c|z_3|}$.

Validate your results numerically.

Exercise 60 (Global swing up of a pendulum on a cart, revisited). Consider the model of the angle and the angular velocity of a pendulum on a cart given in Exercise 18, where $\xi = (\xi_1, \xi_2)$ is a unit vector whose angle denotes the angle of the pendulum and ω corresponds to the angular velocity; see also Exercise 46. For this system, the control goal is to robustly and globally asymptotically stabilize the plant state $z = (\xi, \omega)$ to the point $z^* := (1, 0) \in \mathbb{S}^1 \times \mathbb{R}$.

1. Design a controller that steers the state z out of a neighborhood of $-z^*$.

2. Design a controller that steers the state to a neighborhood of z^*, except for solutions starting from $-z^*$.

3. Design a local stabilizer for z^*.

4. Design a hybrid controller that solves the control problem.

5. Validate your results numerically.

Exercise 61 (Rendezvous and docking control for spacecrafts, revisited). Given the relative motion model of two spacecrafts in (6.31) and positive constants m_c, m_t, $\tilde{\mu}$, r_o, u_{max}, $\rho_{max} > \rho_r > \rho_d$, $\overline{V}, V_{max}$, σ_1, σ_2, σ_3, σ_4, $t_f > t_e$, $\theta \in [0, \frac{\pi}{2})$, and $\overline{\dot{\xi}}_y > 0$, with $\xi = (\xi_x, \dot{\xi}_x, \xi_y, \dot{\xi}_y)$, design a hybrid feedback controller that measures

$$y = h(\eta)$$

and assigns u such that for every initial condition

$$\eta_0 \in \mathcal{M}_0 := \left\{ \eta \in \mathbb{R}^4 \ : \ \rho(\xi_x, \xi_y) \in [0, \rho_{max}], \rho(\dot{\xi}_x, \dot{\xi}_y) \in [0, \overline{V}] \right\}$$

with $\rho(\xi_x, \xi_y) := \sqrt{\xi_x^2 + \xi_y^2}$ and under the dynamics in (6.31) and constraints:

- The control signal $t \mapsto u(t) := (F_x(t), F_y(t))$ satisfies the "maximum thrust" constraint
$$\sup_{t \geq 0} \max\{|F_x(t)|, |F_y(t)|\} \leq u_{max}$$
namely, for each $t \geq 0$,
$$u(t) \in \mathbb{R}_P^{m_P} := \left\{ u \in \mathbb{R}^2 \ : \ \max\{|F_x|, |F_y|\} \leq u_{max} \right\} \qquad (8.38)$$

- For each $\eta \in \mathcal{M}_1 := \left\{ \eta \in \mathbb{R}^4 \ : \ \rho(\xi_x, \xi_y) \in [\rho_r, \infty) \right\}$, only angle measurements[8] are available, namely,
$$h(\eta) = \arctan\left(\frac{\xi_y}{\xi_x}\right)$$
where $\arctan : \mathbb{R} \to [-\pi, \pi]$ is the four-quadrant inverse tangent;

- For each $\eta \in \mathcal{M}_2 := \left\{ \eta \in \mathbb{R}^4 \ : \ \rho(\xi_x, \xi_y) \in [\rho_d, \rho_r) \right\}$, angle and range measurements are available, namely,
$$h(\eta) = \begin{bmatrix} \arctan\left(\frac{\xi_y}{\xi_x}\right) \\ \sqrt{\xi_x^2 + \xi_y^2} \end{bmatrix} \qquad (8.39)$$

- For each $\eta \in \mathcal{M}_3^a := \left\{ \eta \in \mathbb{R}^4 \ : \ \rho(\xi_x, \xi_y) \in [0, \rho_d) \right\}$, angle and range measurements are available, that is, h is given as in (8.39) while, in addition, if $\eta \in \mathcal{M}_3^a \cap \mathcal{M}_3^b$, where
$$\mathcal{M}_3^b(\theta) := \left\{ \eta \in \mathbb{R}^4 \ : \ \begin{bmatrix} \sin(\theta/2) & \cos(\theta/2) \\ \sin(\theta/2) & -\cos(\theta/2) \end{bmatrix} \begin{bmatrix} \xi_x \\ \xi_y \end{bmatrix} \leq \begin{bmatrix} 0 \\ 0 \end{bmatrix} \right\}$$
namely, the position state is in a cone with aperture θ centered about the ξ_x

---

[8]To overcome the discontinuities associated with angle calculations, embed the angle on a unit circle, in other words, consider line of sight (LOS) measurements given by $h(\eta) := \begin{bmatrix} \frac{\xi_x}{\rho(\xi_x, \xi_y)} & \frac{\xi_y}{\rho(\xi_x, \xi_y)} \end{bmatrix}^\top$.

axis, then the following constraint on closing/approaching velocity is satisfied:

$$\eta \in \mathcal{M}_3^c := \left\{ \eta \in \mathbb{R}^4 \ : \ \rho(\dot{\xi}_x, \dot{\xi}_y) \leq V_{\max} \right\}$$

where $\rho(\dot{\xi}_x, \dot{\xi}_y) := \sqrt{\dot{\xi}_x^2 + \dot{\xi}_y^2}$

and the following holds for the η-component $t \mapsto \eta(t)$ of each maximal solution to the closed-loop system: for some $t_{2f} < t_{3f} < t_{4f}$ such that $t_{3f} \leq t_e$, $t_{4f} \leq t_f$

1. $\eta(t_{2f}) \in \mathcal{M}_3^a \cap \mathcal{M}_3^b$ and $\rho(\xi_x(t_{2f}), \xi_y(t_{2f})) = \rho_d$; namely, the chaser reaches the cone first;

2. for some small $\epsilon > 0$, $\eta(t_{3f}) \in \mathcal{M}_3^c = \left\{ \eta \in \mathbb{R}^4 \ : \ |\eta| \leq \epsilon \right\}$; namely, the chaser practically docks on the target next, no later than t_{3f} time units;

3. $\eta(t_{4f}) \in \mathcal{M}_4$, where $\mathcal{M}_4 := \left\{ \eta \in \mathbb{R}^4 \ : \ \xi_x = 0, \xi_y = \bar{\xi}_y, \dot{\xi}_x = \dot{\xi}_y = 0 \right\}$; namely, the docked chaser (or chaser-target) reaches the desired final location no later than t_{4f} time units.

8.6 NOTES

Supervisory algorithms have found application in a wide range of systems, with a variety of different goals. One of the earlier articles proposing the use of supervisory-type algorithms is [162]; see also the references therein. The motivation of that work is that for a class of discrete-event systems – called therein *product systems* and consisting of discrete-event systems composed of a finite set of asynchronous components – the problem of synthesizing a controller is such that the size of the state space grows exponentially with the number of components in the discrete-event system. While this relationship between the number of systems states and components makes controller synthesis not computationally tractable, [162] shows that supervisory control makes certain control problems computationally tractable. A robust supervisory control algorithm for discrete-event systems appeared in [163]. The approach therein relies on "lower" and "upper" bounds on the model of the system, which adds a worst case flavor to the work. The main result is a set of sufficient conditions guaranteeing the existence of a supervisor realizing a desired behavior. Adaptive supervision, which is also considered in [163], is treated in depth in [164] and [165] for linear systems. The decision making strategy in the supervisory algorithm proposed therein selects the controller with smallest performance signal, which is defined as an error quantity (state dependent, output dependent, or state estimate dependent). The ideas therein were extended in [166] to the case of stabilizing controllers in the integral-input-to-state sense. A broad exposition and summary of supervisory algorithms with hybrid flavor was proposed in [167]. The definition of a hybrid system therein includes a continuous-time system with a discontinuous right-hand side or with a piecewise control input, as well as discrete-event systems. These mechanisms can be modeled by hybrid inclusions.

The hybrid supervisory control introduced in this chapter first appeared in [30]. It was later revisited in [31] and [11, Page 76]. A version of Theorem 8.6 appeared in [11, Theorem 31]. The version of the hybrid supervisory controller in this chapter extends those therein to the case of a continuous-time plant with state and input constraints. An extension to the case of a plant with hybrid dynamics follows similarly, but carries a much heavier notational burden. Indeed, in the same way that the results and ideas in Chapter 4, Chapter 6, and Chapter 7 extend to such a general setting, unavoidably, extensions of the supervisory control strategy presented in this chapter to the case of a hybrid plant require keeping track of flows and jumps of both the plant and the controller.

A hybrid supervisory approach to uniting two hybrid output feedback controllers was proposed in [113]. Based on estimates of the norm of the state and measurements of the output, the algorithm therein supervises two hybrid feedback controllers: one that guarantees stability of the set to stabilize, which is to be used locally (to that set), and another one that guarantees pre-attractivity of that set from points far from it. More details about uniting control are in Chapter 4; see also § 4.6. As pointed out in [11], supervisory control is more general than uniting control and can be captured using the control strategy presented in § 8.2.

The control problems in Example 8.1 and Example 8.2 were proposed and solved in [30]. In those references, extensions to the tracking are also proposed.

A hybrid feedback control strategy using *stabilizing patchy feedbacks* in the hybrid framework used in this book, with a discrete state, appeared in [168]. The construction therein, called *patchy feedback control*, follows closely that in [27], where a hybrid controller to robustly asymptotically stabilize a set is proposed. A precursor to the work in [27] is [169], where a family of discontinuous, piecewise smooth vector fields are patched together to show that an asymptotically controllable system can be stabilized by a piecewise constant patchy feedback control law. In [168], patchy feedback control is designed using a patchy control Lyapunov function, which consists of multiple sets, which are the patches, and Lyapunov-like functions, as defined below. The idea behind it can be summarized as follows. First, partition the state space $\mathbb{R}^{n_P}$ into subsets or patches on which a control Lyapunov function can be found. After that, from the inequality satisfied by the derivative of the control Lyapunov function, synthesize an asymptotically stabilizing static state-feedback law on each patch. Then, select the feedback law to apply at each point z according to the patch it belongs to. For the plant with continuous-time dynamics

$$\dot{z} = F_P(x, u) \qquad z \in \mathbb{R}^{n_P}, \ u \in \mathbb{R}^{m_P} \tag{8.40}$$

a smooth patchy control Lyapunov function with respect to a compact set $\mathcal{A}^* \subset \mathbb{R}^{n_P}$ consists of

- a finite set Q;

- collections $\{\Omega_q\}_{q \in Q}$ and $\{\Omega'_q\}_{q \in Q}$ of nonempty and open subsets of $\mathbb{R}^{n_P}$; and

- a collection of functions $\{V_q\}_{q \in Q}$

with the following properties:

(i) Each family of sets $\{\Omega_q\}_{q \in Q}$ and $\{\Omega'_q\}_{q \in Q}$ cover $\mathbb{R}^{n_P}$; i.e., $\mathbb{R}^{n_P} = \bigcup_{q \in Q} \Omega_q = \bigcup_{q \in Q} \Omega'_q$;

(ii) For each $q \in Q$, the unit (outward) normal vector to Ω_q is continuous on $\partial \Omega_q \setminus \bigcup_{i < q} \Omega'_i$, and the families of sets satisfy $\overline{\Omega'_q} \subset \Omega_q$;

(iii) Each V_q is smooth and defined on a neighborhood of $\overline{\Omega_q \setminus \bigcup_{i<q} \Omega_i'}$;

(iv) There exist a continuous, positive definite function ρ, and class-$\mathcal{K}_\infty$ functions α_1 and α_2 such that, for each $q \in Q$, the following hold: for each $z \in \Omega_q \setminus \bigcup_{i<q} \Omega_i'$ there exists $u_{q,z} \in \mathbb{R}^{m_P}$ such that

$$\alpha_1(|z|_{\mathcal{A}^*}) \le V_q(z) \le \alpha_2(|z|_{\mathcal{A}^*}) \qquad \forall z \in \Omega_q \setminus \bigcup_{i<q} \Omega_i'$$

$$\langle \nabla V_q(z), F_P(z, u_{q,z}) \rangle \le -\rho(|z|_{\mathcal{A}^*}) \qquad \forall z \in \Omega_q \setminus \bigcup_{i<q} \Omega_i'$$

$$\langle n_q(z), F_P(z, u_{q,z}) \rangle \le -\rho(|z|_{\mathcal{A}^*}) \qquad \forall z \in \partial\Omega_q \setminus \bigcup_{i<q} \Omega_i'$$

where $n_q(z)$ is the unit (outward) normal vector to Ω_q at z.

With these elements, the hybrid controller in (8.5)-(8.9) defined as follows implements the patchy feedback control strategy: for each $q \in Q$,

$$C_{K,q} = \overline{\Omega_q \setminus \bigcup_{i<q} \Omega_i'}, \quad D_{K,q} = \emptyset, \quad \Psi_q = \overline{\Omega_q' \setminus \bigcup_{i \in Q, i<q} \Omega_i'}$$

Now, suppose that, for each $z, \chi \in \mathbb{R}^{n_P}$ and $c \in \mathbb{R}$, the set

$$\{u \in \mathbb{R}^{m_P} \mid \langle \chi, F_P(z, u) \rangle \le c\}$$

is convex. (In particular, this property holds if $F_P(z, u)$ is affine in u and $\mathbb{R}^{m_P}$ is convex.) As argued in Chapter 10, for each $q \in Q$ there exists a continuous function $\kappa_q : C_{K,q} \to \mathbb{R}^{m_P}$ such that

$$\langle \nabla V_q(z), F_P(z, \kappa_q(z)) \rangle \le -\frac{\rho(|z|_{\mathcal{A}^*})}{2} \qquad \forall z \in C_{K,q}$$

all maximal solutions to

$$\dot{z} = F_P(z, \kappa_q(z))$$

are either complete or end in

$$\left(\bigcup_{i<q, i \in Q} \Psi_i \right) \cup \overline{\mathbb{R}^{n_P} \setminus C_{K,q}}$$

and no maximal solution starting in Ψ_q reaches

$$\overline{\mathbb{R}^{n_P} \setminus \left(C_{K,q} \cup \bigcup_{i \in Q, i<q} \Psi_i \right)}$$

This construction satisfies Assumption 8.3 with $\tilde{n}_K = 1$ and $\tilde{X} = \mathbb{R}^{n_P}$.

Chapter Nine

Passivity-Based Control

This chapter introduces passivity notions and tools for the design of passivity-based controllers for general hybrid plants $\mathcal{H}_P$. In simple terms, a system is passive if the energy it stores is no larger than the energy supplied. Hence, a passive system dissipates some of the energy that is supplied. Due to the combination of continuous and discrete dynamics, a hybrid system can dissipate energy during flows, at jumps, or both. The passivity notions introduced in this chapter include those that require "strict" passivity-type properties during flows and jumps, simultaneously. It also includes "weak" versions of these notions, in which the dissipation may not necessarily occur in both regimes. More precisely, a notion called *flow passivity*, in which dissipation happens along flows, and a notion called *jump passivity*, in which dissipation happens at jumps, are introduced. Through the aid of *detectability properties* needed to guarantee attractivity, these notions are linked to pre-asymptotic stability of a set-point (or set) $\mathcal{A}^*$. Conditions for the synthesis of passivity-based feedback laws are also provided and illustrated in applications. Figure 9.1 depicts the associated hybrid closed-loop system with such feedback law.

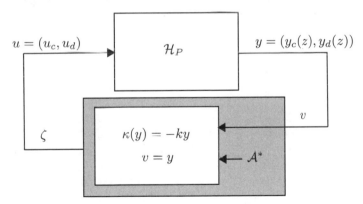

Figure 9.1: Hybrid closed-loop system resulting from passivity-based control.

9.1 OVERVIEW

Dissipativity and its special case, passivity, provide a useful physical interpretation of a feedback control system as they characterize the exchange of energy between the plant and its controller. For an open system, namely, a system with inputs not yet assigned, passivity (in its very pure form) is the property that the energy stored in the system is not larger than the energy it was supplied with. The energy stored in a system over a period of time is given by the difference between the initial and final energy. The function defining the energy stored in the system is the so-called *storage function*. Conveniently, passivity can be expressed in terms of the derivative of a storage function – that is, the rate of change of the internal energy – and the product between inputs and outputs – defining the system's power flow. Under further observability conditions, this power inequality can be employed as a control design tool, in particular, by assigning the inputs to a function depending on the outputs that makes the rate of change of the internal energy negative. Such function defines the control law and this method is called *passivity-based control design*.

The passivity-based control design method can be employed in the design of a controller for a "passive" hybrid plant $\mathcal{H}_P$, in which energy might be dissipated during flows, jumps, or both. Since the properties of the output function of the plant play a key role in asserting a passivity property, this property may not necessarily hold simultaneously during flows and at jumps, and may hold only for particular components of the inputs and of the outputs. As mentioned in Remark 2.16, in practice, the output of the plant is defined by the function $h = (h_c, h_d)$, where $y_c = h_c(z) \in \mathbb{R}^{m_{P,c}}$ is the output that is measured during flows and $y_d = h_d(z) \in \mathbb{R}^{m_{P,d}}$ is the output that is measured at jumps. Similarly, inputs u_c and u_d are defined as the inputs that affected the evolution of the state z during flows and at jumps, respectively, where $u = (u_c, u_d) \in \mathbb{R}^{m_{P,c}} \times \mathbb{R}^{m_{P,d}}$.

A storage function is a function of the state of the plant that satisfies the following properties. Roughly speaking, when F_P and G_P are single valued, a continuously differentiable function V certifies passivity with respect to a set $\mathcal{A}^*$ if

$$\langle \nabla V(z), F_P(z, u_c) \rangle \leq u_c^\top y_c$$

at every point in the flow set, and

$$V(G_P(z, u_d)) - V(z) \leq u_d^\top y_d$$

at every point in the jump set. Note that for these bounds to hold, the dimension of the space of the inputs u_c and u_d need to coincide with that of the outputs y_c and y_d, respectively, leading to $m_{P,c} = r_{P,c}$ and $m_{P,d} = r_{P,d}$. As formally defined below Definition 9.3, a function V with the properties above is said to be a *storage function*. In light of the above inequalities, the quantities introduced in Definition 3.18 can be naturally extended to

$$\dot{V}(z, u_c) = \langle \nabla V(z), F_P(z, u_c) \rangle$$

and

$$\Delta V(z, u_d) = V(G_P(z, u_d)) - V(z)$$

which now depend on the inputs, as those are not yet assigned.

To give an intuition behind the usefulness of such bounds, suppose that V is positive definite with respect to a set $\mathcal{A}^*$ and that there exists a state-feedback pair

$$z \mapsto \kappa(z) = (\kappa_c(z), \kappa_d(z))$$

such that when $u_c = \kappa_c(z)$ and $u_d = \kappa_d(z)$, the quantities

$$\dot{V}(z, u_c) \leq u_c^\top y_c = \kappa_c(z)^\top y_c = \kappa_c(z)^\top h_c(z) \leq 0$$

and

$$\Delta V(z, u_d) \leq u_d^\top y_d = \kappa_d(z)^\top y_d = \kappa_d(z)^\top h_d(z) \leq 0$$

on the respective sets. If V is positive definite with respect to $\mathcal{A}^*$, then the set $\mathcal{A}^*$ is stable. To further provide a decrease of V from points not in $\mathcal{A}^*$, a particular choice for the feedback pair (κ_c, κ_d) is as simple as "minus the output" (y_c, y_d), multiplied by a positive parameter that can be used to tune the decay rate. Figure 9.1 illustrates such a setting. Furthermore, if these quantities are negative definite with respect to $\mathcal{A}^*$, convergence of complete solutions to $\mathcal{A}^*$ ensues. When the said quantities are not negative definite with respect to $\mathcal{A}^*$, then a detectability-type property relative to the set of points where these quantities vanish can be exploited to conclude attractivity of $\mathcal{A}^*$ via the Hybrid Invariance Principle.

Before introducing notions and results, the following examples highlight hybrid plants for which passivity-based control is a suitable control design tool.

Example 9.1 (Robotic manipulator interacting with the environment). *Consider the mechanical system depicted in Figure 9.2, which represents a robotic manipulator interacting with an environment. The manipulator consists of a point-mass end effector driven by a controllable force input. The environment is defined as a contact surface at the origin of the coordinate system. The mass is constrained to move horizontally and, during its motion, it may come into contact with the surface. The position and the velocity of the mass have been denoted with z_1 and z_2, respectively.*

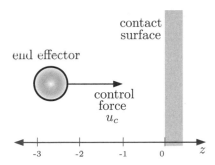

Figure 9.2: A robotic manipulator interacting with an environment in Example 9.1. The force input u_c affects the dynamics during flows only.

When the impact velocity is lower than or equal to a certain threshold, denoted as $\bar{z}_2 > 0$, a compliant impact model is used for the interaction between the end effector and the surface. Assuming unitary mass for the sake of simplicity, the *system is described by*

$$\dot{z}_1 = z_2, \qquad \dot{z}_2 = u_c - f_c(z) \tag{9.1}$$

where $u_c \in \mathbb{R}$ denotes the force input available for control and f_c is the contact force given by the (discontinuous) function

$$f_c(z) = \begin{cases} k_c z_1 + b_c z_2 & \text{if } z_1 > 0 \\ 0 & \text{if } z_1 \leq 0 \end{cases}$$

The positive constants k_c and b_c are, respectively, the elastic and damping coefficients of the compliant contact model. When a collision with the surface occurs with a velocity of the mass larger than or equal to $\bar{z}_2$, possible changes in the contact dynamics introduced, for example, by plastic deformations or other mechanical properties of the contact material, are captured by considering an impulsive impact model with uncertain restitution coefficient. The contact condition is modeled as

$$z_1 \geq 0 \text{ and } z_2 \geq \bar{z}_2 \tag{9.2}$$

while the new value of the state variables after the impact are described by the reset law

$$z_1^+ = z_1, \qquad z_2^+ = -\lambda z_2 \tag{9.3}$$

where $\lambda \in [0,1]$ represents the uncertain restitution coefficient. The expressions in (9.1)-(9.3) define the hybrid plant.

Now, suppose that the control goal is to stabilize this mechanical system to a fixed position in contact with the vertical surface, say, the origin. Consider the quadratic function

$$V(z) := \frac{1}{2}z_1^2 + \frac{1}{2}z_2^2 \qquad \forall z \in \mathbb{R}^2$$

and note that the following holds:

1. For each z such that (9.2) holds, since $\lambda \in [0,1]$, it follows that

$$V(z^+) = \frac{1}{2}z_1^2 + \frac{1}{2}\lambda^2 z_2^2 \leq V(z) \qquad \Rightarrow \qquad \Delta V(z) \leq 0$$

2. For each z not satisfying (9.2), if $z_1 \leq 0$

$$\dot{V}(z, u_c) = \left\langle \nabla V(z), \begin{bmatrix} z_2 \\ u_c - f_c(z) \end{bmatrix} \right\rangle = z_2(z_1 + u_c)$$

and, if $z_1 > 0$

$$\dot{V}(z, u_c) = \left\langle \nabla V(z), \begin{bmatrix} z_2 \\ u_c - f_c(z) \end{bmatrix} \right\rangle = z_2((1 - k_c)z_1 + u_c - b_c z_2)$$

Denoting a new (virtual) input as $\widetilde{u}_c$, the (discontinuous) choice

$$u_c = -z_1 + \widetilde{u}_c$$

for $z_1 \leq 0$ and

$$u_c = -(1 - k_c)z_1 + b_c z_2 + \widetilde{u}_c$$

for $z_1 > 0$ makes the right-hand side of the expressions in item 2 above to be equal to $z_2 \widetilde{u}_c$. The resulting expressions imply that the variation of V during flows is no larger than the product $z_2 \widetilde{u}_c$. This inequality can be interpreted as a passivity property[1] of the hybrid plant with input $\widetilde{u}_c$ and output $y_c = h_c(z) := z_2$. Moreover, the choice $\kappa_c(z) = -z_2 = -h_c(z)$ seems to be a suitable state-feedback control law emerging from the passivity inequality during flows, since z_2 identically zero corresponds to no motion. It is shown in Example 9.5 that this construction solves the control problem.

Note that a passivity property does not seem to hold at jumps for the system in Example 9.1, at least for the particular storage function chosen therein. The next example presents a hybrid system for which, opposite to the system in Example 9.1, a passivity property holds only at jumps.

Example 9.2 (One-degree-of-freedom juggling system). *Consider the model of the one-degree-of-freedom juggling system in Exercise 22, which is a simplification of the model introduced in Example 2.32. It consists of a ball bouncing on a fixed horizontal surface at zero height; see Exercise 11. The surface is equipped with a mechanical actuator that controls the speed of the ball resulting after impacts. From a physical standpoint, such control authority at impacts can be obtained by varying the viscoelastic properties of the surface and, in turn, the coefficient of restitution of the surface. As in Exercise 22, the position and the velocity of the ball are denoted, respectively, as z_1 and z_2; see Example 2.32. Between impacts, the motion of the ball is governed by*

$$\dot{z}_1 = z_2, \quad \dot{z}_2 = -\gamma \tag{9.4}$$

where $\gamma > 0$ is the gravity constant. Neglecting the input constraints in Exercise 22, impacts occur when the ball is in contact with the surface with nonpositive velocity, that is, when

$$z_1 = 0 \text{ and } z_2 \leq 0 \tag{9.5}$$

To further simplify the impact model, the new value of the state variables after each impact is given by

$$z_1^+ = z_1, \quad z_2^+ = u_d$$

in which u_d is the controlled velocity after the impact, capturing the effect of the actuator on the horizontal surface. Figure 9.3 depicts this plant, along with its state variables and input.

Now, suppose that the control goal is to stabilize the ball in contact with the horizontal surface, that is, $z = (z_1, z_2) = \{(0,0)\}$. Consider the energy of the system

$$V(z) := \gamma z_1 + \frac{1}{2}z_2^2 \qquad \forall z \in \mathbb{R}^2 \tag{9.6}$$

which is positive definite with respect $\{(0,0)\}$ in the region of operation given as $\{z \in \mathbb{R}^2 : z_1 \geq 0\}$. The following holds:

---

[1]This property is defined as flow passivity in Definition 9.4.

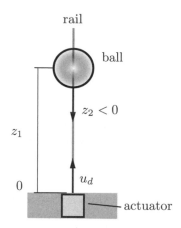

Figure 9.3: One-degree-of-freedom juggling system in Example 9.2: a ball controlled by an actuated surface. The position of the ball is denoted by z_1 and the velocity by z_2. The input u_d acts at jumps to control the energy of the ball.

1. *For each z such that (9.5) holds,*

$$\Delta V(z, u_d) = V((z_1, u_d)) - V(z) = \frac{1}{2}u_d^2 - \frac{1}{2}z_2^2$$

2. *For each $z \in \mathbb{R}^2$,*

$$\dot{V}(z) = \left\langle \nabla V(z), \begin{bmatrix} z_2 \\ -\gamma \end{bmatrix} \right\rangle = 0$$

Since $\dot{V}$ vanishes in the entire state space, then, V does not increase along flows. The choice for u_d given by

$$u_d = (c\lambda_1 + (1 - c)\lambda_2)|z_2|$$

with $0 < \lambda_1 < \lambda_2 < 1$ and c being a (potentially unknown) constant taking values from $[0, 1]$, makes the right-hand side of the expressions in item 1 above less than or equal to $\frac{1}{2}\frac{(1-\lambda_2^2)}{\lambda_2}z_2 u_d$. Indeed, the change of V at jumps is no larger than a function of the product of $z_2 u_d$. The associated inequality can be interpreted as a passivity property[2] of the system with input u_d, output

$$y_d = h_d(z) := \frac{1}{2}\frac{(1 - \lambda_2^2)}{\lambda_2}z_2$$

and storage function V. This property suggests that the feedback law

$$\kappa_d(z) = -z_2 = -h_d(z)$$

is a suitable choice to asymptotically stabilize the origin of the hybrid plant. The details behind the design of this control law are presented in Example 9.6.

---

[2]Defined as jump passivity in Definition 9.4.

9.2 PASSIVITY

Following the definition of candidate Lyapunov function for a hybrid closed-loop system $\mathcal{H}$ in Definition 3.17, the notion of storage function $z \mapsto V(z)$ for a hybrid plant $\mathcal{H}_P$ is introduced next. As a difference to the setting in previous chapters, the following general hybrid plant is considered:

$$\mathcal{H}_P \;:\; \begin{cases} (z, u_c) \in C_P & \dot{z} \;\in\; F_P(z, u_c) \\ (z, u_d) \in D_P & z^+ \;\in\; G_P(z, u_d) \\ & y \;=\; (y_c, y_d) = (h_c(z), h_d(z)) \end{cases} \tag{9.7}$$

For simplicity, the global case is treated – namely, the neighborhood $\mathcal{U}$ in Definition 3.17 is equal to $\mathbb{R}^{n_P}$. As stated in Remark 3.20, the local case follows similarly.

Definition 9.3 (Storage function candidate). *The set $\mathcal{A}^* \subset \mathbb{R}^{n_P}$ and the function $V : \operatorname{dom} V \to \mathbb{R}$ define a __storage function candidate__ with respect to $\mathcal{A}^*$ for the hybrid plant $\mathcal{H}_P = (C_P, F_P, \overline{D_P, G_P, h})$ as in (9.7) if the following conditions hold:*

1. $\Pi_c(C_P) \cup \Pi_d(D_P) \cup G_P(D_P) \subset \operatorname{dom} V$;

2. V is continuous, and locally Lipschitz on an open set containing $\overline{\Pi_c(C_P)}$;

3. V is positive definite on $\Pi_c(C_P) \cup \Pi_d(D_P) \cup G_P(D_P)$ with respect to $\mathcal{A}^*$.

Above, $\Pi_c(C_P)$ and $\Pi_d(D_P)$ are projections to the space for z; see List of Symbols.

According to § 9.1, the change of a storage function candidate V is to be upper bounded by a quantity involving the inputs and outputs of the system. In other words, given a solution to $\mathcal{H}_P$ for a given input, a similar property to the one for the change of a Lyapunov function along a closed-loop solution described in § 3.3.1 is desired. Following the arguments therein and assuming that $\mathcal{H}_P$ satisfies the hybrid basic conditions, when F_P is single valued and V is continuously differentiable on a open neighborhood containing $\overline{\Pi_c(C_P)}$, the quantity corresponding to the "continuous contribution" but with the input unassigned is given by

$$\dot{V}(z, u_c) := \langle \nabla V(z), F_P(z, u_c) \rangle \qquad \forall (z, u_c) \in C_P \tag{9.8}$$

When F_P is set valued or V is only locally Lipschitz, a construction that parallels (3.13) is given by

$$\dot{V}(z, u_c) := \max_{\chi \in F_P(z, u_c)} V^\circ(z, \chi) \qquad \forall (z, u_c) \in C_P \tag{9.9}$$

where, as defined in (3.7), $V^\circ(z, \chi)$ is the Clarke generalized directional derivative of V at z in the direction of χ. Similarly, the quantity paralleling (3.16), referred to in § 3.3.1 as the "discrete contribution" to the change of V, is given by

$$\Delta V(z, u_d) := \max_{\chi \in G_P(z, u_d)} V(\chi) - V(z) \qquad \forall (z, u_d) \in D_P \tag{9.10}$$

With these constructions, the following concept of passivity is considered for a hybrid plant $\mathcal{H}_P$.

Definition 9.4 (Storage function for passivity notions). *Given a closed set $\mathcal{A}^* \subset \mathbb{R}^{n_P}$, a storage function candidate V with respect to $\mathcal{A}^*$ for the hybrid plant $\mathcal{H}_P = (C_P, F_P, D_P, G_P, h)$ in (9.7) is a storage function with respect to $\mathcal{A}^*$ for $\mathcal{H}_P$ if there exist functions $\omega_c : \mathbb{R}^{m_{P_c}} \times \mathbb{R}^{n_P} \to \mathbb{R}$ and $\omega_d : \mathbb{R}^{m_{P_d}} \times \mathbb{R}^{n_P} \to \mathbb{R}$ such that*

$$\dot{V}(z, u_c) \leq \omega_c(u_c, z) \qquad \forall (z, u_c) \in C_P \qquad (9.11)$$

$$\Delta V(z, u_d) \leq \omega_d(u_d, z) \qquad \forall (z, u_d) \in D_P \qquad (9.12)$$

Furthermore, with $y_c = h_c(z)$ and $y_d = h_d(z)$ having the same dimension as u_c and u_d, respectively, a storage function V with respect to $\mathcal{A}^*$ for $\mathcal{H}_P$ certifies that $\mathcal{H}_P$ is

- *passive with respect to $\mathcal{A}^*$ if*

$$(u_c, z) \mapsto \omega_c(u_c, z) = u_c^\top y_c \qquad (9.13)$$

$$(u_d, z) \mapsto \omega_d(u_d, z) = u_d^\top y_d \qquad (9.14)$$

- *flow passive with respect to $\mathcal{A}^*$ if passive with $\omega_d \equiv 0$;*

- *jump passive with respect to $\mathcal{A}^*$ if passive with $\omega_c \equiv 0$;*

- *strictly passive with respect to $\mathcal{A}^*$ if*

$$(u_c, z) \mapsto \omega_c(u_c, z) = u_c^\top y_c - \rho_c(z) \qquad (9.15)$$

$$(u_d, z) \mapsto \omega_d(u_d, z) = u_d^\top y_d - \rho_d(z) \qquad (9.16)$$

where $\rho_c, \rho_d : \mathbb{R}^{n_P} \to \mathbb{R}_{\geq 0}$ are positive definite with respect to $\mathcal{A}^*$;

- *flow strictly passive with respect to $\mathcal{A}^*$ if strictly passive with $\omega_d \equiv 0$;*

- *jump strictly passive with respect to $\mathcal{A}^*$ if strictly passive with $\omega_c \equiv 0$;*

- *output strictly passive with respect to $\mathcal{A}^*$ if*

$$(u_c, z) \mapsto \omega_c(u_c, z) = u_c^\top y_c - y_c^\top \rho_c(y_c) \qquad (9.17)$$

$$(u_d, z) \mapsto \omega_d(u_d, z) = u_d^\top y_d - y_d^\top \rho_d(y_d) \qquad (9.18)$$

where $\rho_c : \mathbb{R}^{m_{P_c}} \to \mathbb{R}^{m_{P_c}}$, $\rho_d : \mathbb{R}^{m_{P_d}} \to \mathbb{R}^{m_{P_d}}$ are functions such that $y_c^\top \rho_c(y_c) > 0$ for all $y_c \neq 0$ and such that $y_d^\top \rho_d(y_d) > 0$ for all $y_d \neq 0$, respectively;

- *flow output strictly passive with respect to $\mathcal{A}^*$ if it is output strictly passive with $\omega_d \equiv 0$;*

- *jump output strictly passive with respect to $\mathcal{A}^*$ if it is output strictly passive with $\omega_c \equiv 0$.*

The storage function and the passivity notions introduced in Definition 9.4 cover the counterparts for continuous-time and discrete-time systems. The hybrid specific cases introduced therein, termed as *flow passivity* and *jump passivity*, are motivated by the applications introduced in Examples 9.1 and 9.2 in which energy dissipation happens along flows or jumps, but not necessarily along both regime. It is shown in § 9.3 that these passivity notions can be employed to establish pre-asymptotic stability of the set $\mathcal{A}^*$ for $\mathcal{H}_P$ with zero input. In turn, in § 9.4, passivity-based control techniques for such hybrid specific cases are introduced.

Example 9.5 (Robotic manipulator interacting with the environment, revisited). *Consider the mechanical system with impacts introduced in Example 9.1. First, note that the Filippov regularization of the discontinuous contact force f_c is given by*

$$f_c^r(z) = \begin{cases} k_c z_1 + b_c z_2 & \text{if } z_1 > 0 \\ \overline{\text{con}}\{0, b_c z_2\} & \text{if } z_1 = 0 \\ 0 & \text{if } z_1 < 0 \end{cases} \tag{9.19}$$

For more details, see Example 2.3. Then, the mechanical system of interest is described by means of the following (regularized) hybrid plant:

$$\mathcal{H}_P : \begin{cases} (z, u_c) \in C_P & \dot{z} \in F_P(z, u_c) := \begin{bmatrix} z_2 \\ u_c - f_c^r(z) \end{bmatrix} \\ z \in D_P & z^+ = G_P(z) := \begin{bmatrix} z_1 \\ -\lambda z_2 \end{bmatrix} \end{cases} \tag{9.20}$$

with state $z = (z_1, z_2) \in \mathbb{R}^2$, input $u_c \in \mathbb{R}$, and sets C_P and D_P given by

$$\begin{aligned} C_P &:= \{(z, u_c) \in \mathbb{R}^2 \times \mathbb{R} : z_1 \leq 0\} \cup \{(z, u_c) \in \mathbb{R}^2 \times \mathbb{R} : z_1 \geq 0, z_2 \leq \bar{z}_2\} \\ D_P &:= \{z \in \mathbb{R}^2 : z_1 \geq 0, z_2 \geq \bar{z}_2\} \end{aligned}$$
$$\tag{9.21}$$

Next, it is shown that the control input u_c can be designed so that the resulting system is such that, for a suitable choice of the output y_c of $\mathcal{H}_P$, is flow passive with respect to the set $\mathcal{A}^* = \{(z_1^\star, 0)\}$. The constant $z_1^\star \geq 0$ denotes the desired set-point position for the mass. The choice $z_1^\star \geq 0$ requires the mass to maintain contact with the vertical surface. The idea is to design the control input following an energy shaping approach: define a storage function that is based on the energy of the hybrid closed-loop system.

Consider the hybrid plant $\mathcal{H}_P$ in (9.20) with input assigned via the continuous (pre-)feedback

$$u_c = \tilde{\kappa}_c(z_1, \tilde{u}_c) := \begin{cases} k_c z_1 - k_p(z_1 - z_1^\star) + \tilde{u}_c & \text{if } z_1 > 0 \\ -k_p(z_1 - z_1^\star) + \tilde{u}_c & \text{if } z_1 \leq 0 \end{cases} \tag{9.22}$$

where $k_p > 0$ and $\tilde{u}_c \in \mathbb{R}$ is a new (virtual) input. Let the storage function be defined as

$$V(z) = \frac{1}{2} k_p (z_1 - z_1^\star)^2 + \frac{1}{2} z_2^2 \qquad \forall z \in \mathbb{R}^2 \tag{9.23}$$

Observe that V represents the mechanical energy of the closed-loop system in the case where contact forces are neglected. When u_c is assigned via (9.22), the resulting *model of the plant in (9.20) is such that the flow map is given by*

$$
F_P(z, \widetilde{\kappa}_c(z_1, \widetilde{u}_c)) = \begin{cases} \begin{bmatrix} z_2 \\ k_c z_1 - k_p(z_1 - z_1^\star) + \widetilde{u}_c \end{bmatrix} & \text{if } z_1 > 0 \\ \left\{ \begin{bmatrix} z_2 \\ k_p z_1^\star - \chi_2 + \widetilde{u}_c \end{bmatrix} : \chi_2 \in \overline{\text{con}}\{0, b_c z_2\} \right\} & \text{if } z_1 = 0 \\ \begin{bmatrix} z_2 \\ -k_p(z_1 - z_1^\star) + \widetilde{u}_c \end{bmatrix} & \text{if } z_1 < 0 \end{cases}
$$

$$(9.24)$$

With this flow map for the plant and with V as in (9.23), the following holds for each $(z, \widetilde{u}_c) \in C_P$:

$$
\left\langle \nabla V(z), \begin{bmatrix} z_2 \\ -k_p(z_1 - z_1^\star) - b_c z_2 + \widetilde{u}_c \end{bmatrix} \right\rangle = \widetilde{u}_c z_2 - b_c z_2^2 \qquad \text{if } z_1 > 0
$$

$$
\left\langle \nabla V(z), \begin{bmatrix} z_2 \\ k_p z_1^\star - \chi_2 + \widetilde{u}_c \end{bmatrix} \right\rangle \leq \widetilde{u}_c z_2 \qquad \forall \chi_2 \in \overline{\text{con}}\{0, b_c z_2\}, \ \text{if } z_1 = 0
$$

$$
\left\langle \nabla V(z), \begin{bmatrix} z_2 \\ -k_p(z_1 - z_1^\star) + \widetilde{u}_c \end{bmatrix} \right\rangle = \widetilde{u}_c z_2 \qquad \text{if } z_1 < 0
$$

Then, it follows that

$$
\max_{\chi \in F_P(z, \widetilde{u}_c)} \langle \nabla V(z), \chi \rangle \leq \widetilde{u}_c y_c \qquad \forall (z, \widetilde{u}_c) \in C_P \tag{9.25}
$$

Now, for each $z \in D_P$,

$$
V(G_P(z)) - V(z) \leq -\frac{1}{2}(1 - \lambda^2) z_2^2 \leq 0 \tag{9.26}
$$

since $\lambda \in [0, 1]$. Then, according to Definition 9.4, the hybrid plant $\mathcal{H}_P$ in (9.20) with input assigned via (9.22) is flow passive with respect to $\mathcal{A}^* = \{(z_1^\star, 0)\}$ with output $y_c = h_c(z) := z_2$, input $\widetilde{u}_c$, and $\omega_c(\widetilde{u}_c, z) := \widetilde{u}_c y_c$.

The flow passivity property established above can be strengthened to flow output strict passivity. To this end, consider the hybrid plant $\mathcal{H}_P$ in (9.20) with input assigned via

$$
u_c = \widetilde{\kappa}_c(z_1, \widetilde{u}_c) - k_1 z_2 \tag{9.27}
$$

where $\widetilde{\kappa}_c$ is defined in (9.22) and $k_1 > 0$ is the damping injection gain. Using (9.23), and with the choice for u_c given in (9.27), it follows that

$$
\max_{\chi \in F_P(z, \widetilde{u}_c)} \langle \nabla V(z), \chi \rangle \leq \widetilde{u}_c y_c - k_1 y_c^2 \qquad \forall (z, \widetilde{u}_c) \in C_P
$$

With (9.26), the hybrid plant $\mathcal{H}_P$ in (9.20) with input assigned via (9.27) is flow output strictly passive with respect to the set $\mathcal{A}^* = \{(z_1^\star, 0)\}$ with storage function given in (9.23), input $\widetilde{u}_c$, output $y_c = h_c(z) = z_2$, and functions $\omega_c(\widetilde{u}_c, z) := \widetilde{u}_c y_c$ and $\rho_c(y_c) := k_1 y_c$.

Note that since ΔV is only nonpositive, the storage function V in (9.23) cannot be used to establish a passivity property at jumps. Nevertheless, as clarified in the forthcoming § 9.3, the passivity properties during flows can already be exploited to show that $\mathcal{A}^*$ is attractive for an appropriately designed feedback law assigning $\widetilde{u}_c$.

Example 9.6 (One-degree-of-freedom juggling system, revisited). *Consider the actuated bouncing ball system in Example 9.2. The system therein can be written as the hybrid plant $\mathcal{H}_P$ given by*

$$
\mathcal{H}_P \ : \ \begin{cases} z \in C_P & \dot{z} = F_P(z) := \begin{bmatrix} z_2 \\ -\gamma \end{bmatrix} \\[2mm] (z, u_d) \in D_P & z^+ = G_P(z, u_d) := \begin{bmatrix} z_1 \\ u_d \end{bmatrix} \end{cases} \tag{9.28}
$$

where

$$
C_P := \{ z \in \mathbb{R}^2 \ : \ z_1 \geq 0 \}
$$
$$
D_P := \{ (z, u_d) \in \mathbb{R}^2 \times \mathbb{R} \ : \ z_1 = 0, z_2 \leq 0, u_d \in U(z) \}
$$

The set-valued map U defines a constraint on the input u_d. Its definition is motivated by the analysis in Example 9.2. It is defined to force impacts to be inelastic, that is, totally elastic (unitary restitution coefficient) and plastic (zero restitution coefficient) impacts are not possible. To this end, given λ_1 and λ_2 such that $0 < \lambda_1 < \lambda_2 < 1$, the set-valued map U defining the constraint for u_d is given by

$$
U(z) := [\lambda_1 |z_2|, \lambda_2 |z_2|] \qquad \forall z \in \mathbb{R}^2 \tag{9.29}
$$

In this way, the choice $u_d = -\lambda z_2$ with $\lambda \in [\lambda_1, \lambda_2]$ is a feasible choice. Note that the parameter λ plays the role of the coefficient of restitution for the impacts between the ball and the controlled surface. This hybrid plant with output

$$
y_d = h_d(z) := \frac{1}{2} \frac{(1 - \lambda_2^2)}{\lambda_2} z_2 \tag{9.30}
$$

and input u_d is jump passive with respect to the set $\mathcal{A}^* := \{(0,0)\}$. This property can be certified using the storage function candidate V in (9.6). The function V is positive definite on $C_P \cup \Pi_d(D_P) \cup G_P(D_P)$ with respect to $\mathcal{A}^*$ and h_d vanishes on $\mathcal{A}^*$. With this storage function candidate, for each $z \in C_P$,

$$
\langle \nabla V(z), F_P(z) \rangle = 0
$$

Using the definition of U above, for each $(z, u_d) \in D_P$, since $(z, u_d) \in D_P$ implies that $z_2 \leq 0$,

$$
\begin{aligned}
V(G_P(z, u_d)) - V(z) = \frac{1}{2}(u_d^2 - z_2^2) &\leq -\frac{1}{2}(1 - \lambda_2^2) z_2^2 \\
&\leq \frac{1}{2}(1 - \lambda_2^2)|z_2| z_2 \\
&\leq \frac{1}{2} \frac{(1 - \lambda_2^2)}{\lambda_2} u_d z_2 \\
&\leq u_d y_d
\end{aligned}
$$

With these properties for V, the hybrid plant in (9.28) with output y_d in (9.30), input u_d taking values from the set-valued map U in (9.29), and $\omega_d(u_d, z) := u_d y_d$ is jump passive with respect to the set $\mathcal{A}^*$.

Since $\dot{V}$ is only nonpositive on C_P, the storage function V in (9.6) does not assure a passivity property during flows. However, the input u_d can be designed to establish jump output strict passivity for $\mathcal{H}_P$. Indeed, let the control input u_d be given by

$$u_d = -\lambda z_2 - \widetilde{u}_d \tag{9.31}$$

where $\lambda \in [\lambda_1, \lambda_2]$ and $\widetilde{u}_d$ is a new (virtual) control input. This new input is constrained to take values from $\widetilde{U}$, which is a set-valued map defined as

$$\widetilde{U}(z) := \{\widetilde{u}_d \in \mathbb{R} : |\widetilde{u}_d| \leq \bar{u}|z_2|\} \qquad \forall z \in \mathbb{R}^2$$

where $\bar{u} := \min\{\lambda - \lambda_1, \lambda_2 - \lambda\}$. From the definition of $\widetilde{U}$ and the choice of λ, it follows that $u_d \in U(z)$ for all z such that $z_2 \leq 0$. In fact, for each $(z, u_d) \in D_P$, $u_d = -\lambda z_2 - \widetilde{u}_d = \lambda|z_2| - \widetilde{u}_d$, and, from the definition of $\widetilde{U}$, $(\lambda - \bar{u})|z_2| \leq u_d \leq (\lambda + \bar{u})|z_2|$. Since $\min\{\lambda - \lambda_1, \lambda_2 - \lambda\} \leq \lambda - \lambda_1$, it follows that $\lambda - \bar{u} \geq \lambda_1$, and, since $\min\{\lambda - \lambda_1, \lambda_2 - \lambda\} \leq \lambda_2 - \lambda$, it follows that $\lambda + \bar{u} \leq \lambda_2$, from which $u_d \in U(z)$. Then, for each $(z, \widetilde{u}_d) \in D_P$ such that $\widetilde{u}_d \in \widetilde{U}(z)$, since $\bar{u} \leq \lambda_2 - \lambda < 1 - \lambda$, it follows that

$$
\begin{aligned}
V(G_P(z, \widetilde{u}_d)) - V(z) &= -\frac{1}{2}(1 - \lambda^2)z_2^2 + \frac{1}{2}\widetilde{u}_d^2 + \lambda z_2 \widetilde{u}_d \\
&\leq -\frac{1}{2}(1 - \lambda^2)z_2^2 + \frac{1}{2}|\bar{w}|^2 z_2^2 + \lambda z_2 \widetilde{u}_d \\
&\leq -\frac{1}{2}(1 - \lambda^2)z_2^2 + \frac{1}{2}(1 - \lambda)^2 z_2^2 + \lambda z_2 \widetilde{u}_d \\
&\leq -\lambda(1 - \lambda)z_2^2 + \lambda z_2 \widetilde{u}_d \\
&\leq -\frac{1 - \lambda}{\lambda}y_d^2 + \widetilde{u}_d y_d
\end{aligned}
$$

Then, the hybrid plant in (9.28) is jump output strictly passive with respect to the compact set $\mathcal{A}^* = \{(0, 0)\}$ with input $\widetilde{u}_d$, output $y_d = \lambda z_2$, and functions $\omega_d(z, \widetilde{u}_d) := \widetilde{u}_d y_d$ and $\rho_d(y_d) := \dfrac{1 - \lambda}{\lambda}y_d$.

9.3 PRE-ASYMPTOTIC STABILITY FROM PASSIVITY

The passivity notions introduced in Definition 9.3 provide a bound on the change of energy of the system that depends on a function of its inputs and outputs. Conveniently, as outlined in §9.1, when its input is properly assigned, such a bound can be exploited to assure stability and pre-attractivity of the set $\mathcal{A}^*$ for the resulting hybrid closed-loop system. In this section, results that establish such properties when the input is set to zero are provided. These results are used in the next section for the design of pre-asymptotically stabilizing state-feedback laws that only depend on the output.

The main challenge in asserting pre-asymptotic stability via storage functions is that their change along flows or jumps is typically *weak*, in the sense that the storage function does not strictly decrease along both regimes. For instance, the storage function proposed in Example 9.1 for the problem of controlling a robotic manipulator interacting with its environment does not decrease at jumps – see item 1 therein – but it can be made to decrease during flows. Similarly, the storage function introduced in Example 9.2 for the one-degree-of-freedom juggling system does not decrease during flow – see item 2 therein – but it can be made to decrease at jumps. Due to the nonstrict nature of the change of the storage function along flows and jumps in general, for pre-asymptotic stability to hold, the Hybrid Invariance Principle suggests that solutions that remain in the set where the change of the storage function is zero have to converge to $\mathcal{A}^*$. The following detectability notion is introduced to capture such a property.

Definition 9.7 (Detectability). *Given sets $\mathcal{A}^*, K \subset \mathbb{R}^{n_P}$, the distance to $\mathcal{A}^*$ is 0-input detectable relative to K for the hybrid plant $\mathcal{H}_P$ if every complete solution z to $\mathcal{H}_P$ with zero input is such that*

$$z(t,j) \in K \quad \vee (t,j) \in \operatorname{dom} z \tag{9.32}$$

implies

$$\lim_{(t,j) \in \operatorname{dom} z,\ t+j \to \infty} |z(t,j)|_{\mathcal{A}^*} = 0 \tag{9.33}$$

Given sets $\mathcal{A}, K \subset \mathbb{R}^n$, the distance to $\mathcal{A}$ is detectable relative to K for $\mathcal{H}$ if every complete solution x to the hybrid closed-loop system $\mathcal{H}$ such that (9.32) holds with x instead of z, then (9.33) holds with x and $\mathcal{A}$ instead of z and $\mathcal{A}^*$, respectively.

When K is given by the set of points z such that $h(z) = 0$, the condition $z(t,j) \in K$ for all $(t,j) \in \operatorname{dom} z$ is equivalent to holding the output of the hybrid plant (with zero input) to zero. In such a case, Definition 9.7 reduces to the classical notion of detectability.

To relate different forms of passivity to asymptotic stability with zero input, the hybrid plant $\mathcal{H}_P$ in (9.7) with its input u held to zero is defined as

$$\mathcal{H}_{P,0} : \begin{cases} (z,0) \in C_P & \dot{z} \in F_P(z,0) \\ (z,0) \in D_P & z^+ \in G_P(z,0) \\ & y = (y_c, y_d) = (h_c(z), h_d(z)) \end{cases} \tag{9.34}$$

When a set is stable, attractive, or pre-asymptotically stable for $\mathcal{H}_{P,0}$, it is said that it is 0-input stable, 0-input attractive, or 0-input pre-asymptotically stable for $\mathcal{H}_P$. In the remainder of this chapter, the set X is defined as[3] $X := \Pi_{c,0}(C_P) \cup \Pi_{d,0}(D_P) \cup G_P(\Pi_{d,0}(D_P), 0)$.

For the next result to hold, the data of $\mathcal{H}_{P,0}$ has to satisfy the hybrid basic conditions, that is, the following properties (cf. Definition 2.18 and Definition 2.20):

(A1$_0$) $\Pi_{c,0}(C_P)$ and $\Pi_{d,0}(D_P)$ are closed in $\mathbb{R}^{n_P}$;

(A2$_0$) $(z,0) \mapsto F_P(z,0)$ is outer semicontinuous and locally bounded relative to $\mathbb{R}^{n_P} \times \{0\}$, and for each $z \in \Pi_{c,0}(C_P)$, $F_P(z,0)$ is nonempty and convex.

(A3$_0$) $(z,0) \mapsto G_P(z,0)$ is outer semicontinuous and locally bounded relative to

---

[3]The maps $\Pi_{c,0}$ and $\Pi_{d,0}$ are the projection to the space for z with $u_c = 0$ and $u_d = 0$, respectively; see List of Symbols.

$\mathbb{R}^{n_P} \times \{0\}$, and for each $z \in \Pi_{d,0}(D_P)$, $G(x, 0)$ is nonempty.

Observe that property ($A1_0$) simply requires that the sets C_P and D_P are closed for the case in which $u = 0$.

Theorem 9.8 (0-input pre-asymptotic stability from passivity). *Given a hybrid plant $\mathcal{H}_P = (C_P, F_P, D_P, G_P, h)$ as in (9.7) satisfying ($A1_0$)-($A3_0$) and a compact set $\mathcal{A}^* \subset \mathbb{R}^{n_P}$, if the hybrid plant $\mathcal{H}_P$ is*

1. *passive with respect to $\mathcal{A}^*$ then $\mathcal{A}^*$ is 0-input stable for $\mathcal{H}_P$;*

2. *output strictly passive with respect to $\mathcal{A}^*$ and the distance to $\mathcal{A}^*$ is detectable relative to[4]*

$$\left\{z \in \Pi_{c,0}(C_P) : h_c(z)^\top \rho_c(h_c(z)) = 0\right\} \cup \\ \left\{z \in \Pi_{d,0}(D_P) : h_d(z)^\top \rho_d(h_d(z)) = 0\right\} \qquad (9.35)$$

   for $\mathcal{H}_{P,0}$, then $\mathcal{A}^*$ is 0-input pre-asymptotically stable for $\mathcal{H}_P$;

3. *strictly passive with respect to $\mathcal{A}^*$ then $\mathcal{A}^*$ is 0-input pre-asymptotically stable for $\mathcal{H}_P$.*

Proof. According to Definition 9.4 with $u \equiv 0$, the passivity property in item 1 implies that

$$\dot{V}(z, 0) \leq 0 \qquad \forall (z, 0) \in C_P \qquad (9.36)$$
$$\Delta V(z, 0) \leq 0 \qquad \forall (z, 0) \in D_P \qquad (9.37)$$

Since $\mathcal{H}_{P,0}$ satisfies ($A1_0$)-($A3_0$), $\mathcal{A}^*$ is compact, and (9.36)-(9.38) imply that (3.18)-(3.19) hold with $\mathcal{U} = \mathbb{R}^{n_P}$, $\mathcal{A}^*$ is stable for $\mathcal{H}_{P,0}$ via item 1 of Theorem 3.19. This proves item 1. The 0-input stability property in the other items follow similarly.

To show pre-attractivity of $\mathcal{A}^*$ for $\mathcal{H}_{P,0}$ under the conditions in item 2, note that from the output strict passivity property, using Definition 9.4 with $u \equiv 0$, it follows that

$$\dot{V}(z, u_c) \leq -h_c(z)^\top \rho_c(h_c(z)) \qquad \forall (z, 0) \in C_P \qquad (9.38)$$
$$\Delta V(z, u_d) \leq -h_d(z)^\top \rho_d(h_d(z)) \qquad \forall (z, 0) \in D_P \qquad (9.39)$$

Now consider complete solutions $(z, 0)$ to $\mathcal{H}_P$ starting nearby $\mathcal{A}^*$. By stability and compactness of $\mathcal{A}^*$, these solutions are bounded. Note that $\mathcal{H}_{P,0}$ satisfying ($A1_0$)-($A3_0$) implies that the Hybrid Invariance Principle in Theorem 3.23 applies. Using the property that $h_c(z)^\top \rho_c(h_c(z)) > 0$ for all z such that $h_c(z) \neq 0$ and that $h_d(z)^\top \rho_d(h_d(z)) > 0$ for all $h_d(z) \neq 0$, item 1 of Theorem 3.23 with $\dot{V}(z) = -h_c(z)^\top \rho_c(h_c(z))$ for each $z \in \Pi_{c,0}(C_P)$ and $\Delta V(z) = -h_d(z)^\top \rho_d(h_d(z))$ for each $z \in \Pi_{d,0}(D_P)$ implies that each such complete solution converges to the largest weakly invariant set contained in (3.32) for some $r \geq 0$. In this case, the set in (3.32) is given by

$$V^{-1}(r) \cap \left(\left\{z \in \Pi_{c,0}(C_P) : \dot{V}(z) = 0\right\} \cup \{z \in \Pi_{c,0}(D_P) : \Delta V(z) = 0\}\right) \qquad (9.40)$$

---

[4]See Definition 9.7.

or, equivalently, by

$$V^{-1}(r) \cap \left(\left\{ z \in \Pi_{c,0}(C_P) \; : \; h_c(z)^\top \rho_c(h_c(z)) = 0 \right\} \cup \right. \\ \left. \left\{ z \in \Pi_{d,0}(D_P) \; : \; h_d(z)^\top \rho_d(h_d(z)) = 0 \right\} \right) \tag{9.41}$$

for some $r \geq 0$. Due to detectability relative to the set (9.35), every maximal and complete solution starting from and staying in (9.41) converges to $\mathcal{A}^*$. Then, since V is positive definite with respect to $\mathcal{A}^*$, the only invariant set contained in (9.41) is for $r = 0$. It follows that $\mathcal{A}^*$ is pre-attractive for $\mathcal{H}_{P,0}$. This establishes item 2.

Using Definition 9.4 with $u \equiv 0$, item 3 implies that

$$\dot{V}(z,0) \leq -\rho_c(z) \qquad \forall (z,0) \in C_P \tag{9.42}$$
$$\Delta V(z,0) \leq -\rho_d(z) \qquad \forall (z,0) \in D_P \tag{9.43}$$

Then, 0-input pre-asymptotic stability of the set $\mathcal{A}^*$ follows from item 2a in Theorem 3.19 with $\mathcal{U} = \mathbb{R}^{n_P}$, $\dot{V} = -\rho_c$, and $\Delta V = -\rho_d$. □

Remark 9.9 (On assumptions and the global case). The conditions in item 2 of Theorem 9.8 require finding a storage function V, and functions ρ_c and ρ_d such that, using detectability, pre-attractivity of the solutions to $\mathcal{H}_P$ with zero input can be established. Such a property follows from the Hybrid Invariance Principle in Theorem 3.23. Finding such a combination of functions and properties requires insight on solutions to the system, in particular, their behavior for "large" hybrid time. The pre-asymptotic stability property obtained in Theorem 9.8 is global when, in particular, the storage function V is radially unbounded. △

The following result is a version of Theorem 9.8 that is tailored to hybrid plants for which the storage function only guarantees passivity during flows or jumps. Its proof follows from similar steps to those in the proof of Theorem 9.8.

Theorem 9.10 (0-input pre-asymptotic stability from flow passivity and jump passivity). *Given a hybrid plant $\mathcal{H}_P = (C_P, F_P, D_P, G_P, h)$ as in (9.7) satisfying $(A1_0)$-$(A3_0)$ and a compact set $\mathcal{A}^* \subset \mathbb{R}^{n_P}$, if the hybrid plant $\mathcal{H}_P$ is*

1. *flow passive or jump passive with respect to $\mathcal{A}^*$ with a storage function V then $\mathcal{A}^*$ is 0-input stable for $\mathcal{H}_P$;*

2. *flow output strictly passive with respect to $\mathcal{A}^*$ with a storage function V,*

   a) *the distance to $\mathcal{A}^*$ is detectable relative to*

   $$\left\{ z \in \Pi_{c,0}(C_P) \; : \; h_c(z)^\top \rho_c(h_c(z)) = 0 \right\} \tag{9.44}$$

   for $\mathcal{H}_{P,0}$, and
   b) *every complete solution z to $\mathcal{H}_{P,0}$ is such that for some $\delta > 0$ and some $J \in \mathbb{N}$ the jump times t_j of z satisfy $t_{j+1} - t_j \geq \delta$ for all $j \geq J$,*

   then $\mathcal{A}^*$ is 0-input pre-asymptotically stable for $\mathcal{H}_P$;
3. *jump output strictly passive with respect to $\mathcal{A}^*$ with a storage function V,*

   a) *the distance to $\mathcal{A}^*$ is detectable relative to*

   $$\left\{ z \in \Pi_{d,0}(D_P) \; : \; h_d(z)^\top \rho_d(h_d(z)) = 0 \right\} \tag{9.45}$$

   for $\mathcal{H}_{P,0}$, and

b) every complete solution z to $\mathcal{H}_{P,0}$ is Zeno,

then $\mathcal{A}^*$ is 0-input pre-asymptotically stable for $\mathcal{H}_P$;

4. *flow strictly passive with respect to $\mathcal{A}^*$ and 2.2b holds, then $\mathcal{A}^*$ is 0-input pre-asymptotically stable for $\mathcal{H}_P$;*

5. *jump strictly passive with respect to $\mathcal{A}^*$ and 3.3b holds, then $\mathcal{A}^*$ is 0-input pre-asymptotically stable for $\mathcal{H}_P$.*

Remark 9.11 (About detectability properties). The detectability properties required by Theorem 9.10 can be conveniently interpreted as checking if every complete solution to $\mathcal{H}_P$ with zero input and with flow and jump sets intersected by the set (9.44) or by the set (9.45) converge to $\mathcal{A}^*$. A key difference between the detectability properties imposed by Theorem 9.10 and Theorem 9.8 is that the former result requires checking if solutions converge to $\mathcal{A}^*$ from a potentially smaller set. In fact, the detectability requirements in Theorem 9.10 are relative to (9.44) or to (9.45), while Theorem 9.8 requires detectability relative to their union; see (9.35). To illustrate the benefits of the conditions in Theorem 9.10, consider the one-degree-of-freedom juggling system in Example 9.6. According to the derivations therein, the choice u_d in (9.31) yields a hybrid system that is jump output strictly passive with respect to the set $\mathcal{A}^* = \{(0,0)\}$ with storage function as in (9.6), output $y_d = \kappa_d(z) := \lambda z_2$, input $\tilde{u}_d$, and function $\rho_d(y_d) = ((1-\lambda)/\lambda)y_d$. According to item 2 in Theorem 9.8, pre-asymptotic stability of the set $\mathcal{A}^*$ for the hybrid closed-loop system with zero input can be proved by verifying the detectability property with respect to the set in (9.35), which results in

$$C_P \cup \left\{z \in \Pi_{d,0}(D_P) \,:\, \lambda(1-\lambda)z_2^2 = 0\right\} = \left\{z \in \mathbb{R}^2 \,:\, z_1 \geq 0\right\} \qquad (9.46)$$

Since this set includes all of the admissible solutions to the hybrid plant with zero input, asserting that the detectability in Theorem 9.8 holds requires checking every possible solution – in fact, the flow and jump sets are not changed when intersected by (9.46). On the other hand, with the knowledge that every maximal solution to the hybrid plant with zero input is Zeno, which is a property that can be asserted without computing solutions (see item 1i in Proposition 2.34), item 3 in Theorem 9.10 requires the distance to $\mathcal{A}^*$ to be detectable relative to the set in (9.45), which results to be simply the origin. Then, since both the flow and jump sets are intersected by (9.45), which is the origin, establishing such a detectability property reduces to checking that solutions from the origin stay at the origin, which for this particular system is easy to check.   △

9.4 DESIGN

With the results relating passivity to pre-asymptotic stability in § 9.3, conditions for the design of static output-feedback control laws for hybrid plants are proposed in this section. As outlined in § 9.1 and depicted in Figure 9.1, the static output-feedback law designed in this section assigns the plant input to a function that depends only on the plant output, as follows.

Given a hybrid plant $\mathcal{H}_P$ and a compact set $\mathcal{A}^*$ to stabilize, the passivity-based output-feedback control law formulated in this section assigns the input $u = (u_c, u_d)$ via

$$u_c = -\kappa_c(y_c), \qquad u_d = -\kappa_d(y_d)$$

where, in particular, the pair (κ_c, κ_d) is such that the quantities $y_c^\top \kappa_d(y_c)$ and $y_d^\top \kappa_d(y_d)$ are nonnegative, vanishing only when the output is zero.

When the passivity property holds both for flows and jumps, then Theorem 9.8 can be applied to show that such "minus the output" static output-feedback law can be synthesized and that it induces pre-asymptotic stability of the desired set. As motivated by the previous examples, the more interesting cases are when the passivity property holds in one regime only, namely, during flows or at jumps only. In such cases, the input corresponding to the regime for which the passivity property holds is designed as an output-feedback law and the other input is assigned to zero. The following design result pertains to such cases.

Theorem 9.12 (Passivity-based control). *Given a hybrid plant $\mathcal{H}_P$ with data (C_P, F_P, D_P, G_P, h) as in (9.7) satisfying the hybrid basic conditions in Definition 2.25 and a compact set $\mathcal{A}^* \subset \mathbb{R}^{n_P}$, the following hold:*

1. *If $\mathcal{H}_P$ is flow passive with respect to $\mathcal{A}^*$ with a storage function V with respect to $\mathcal{A}^*$ for $\mathcal{H}_P$ and there exists a continuous function $\kappa_c : \mathbb{R}^{m_{Pc}} \to \mathbb{R}^{m_{Pc}}$ such that $y_c^\top \kappa_c(y_c) > 0$ for all $y_c \neq 0$ and the resulting hybrid closed-loop system $\mathcal{H}$ with state $x = z$, $u_c = -\kappa_c(y_c)$, and $u_d \equiv 0$ has the following properties:*

   a) *The distance to $\mathcal{A}^*$ is detectable relative to*

$$\left\{ z \in \Pi_c(C_P) \cup \Pi_d(D_P) \cup G_P(D_P) : h_c(z)^\top \kappa_c(h_c(z)) = 0, \ (z, -\kappa_c(h_c(z))) \in C_P \right\}$$

   for $\mathcal{H}$;

   b) *Every complete solution x to $\mathcal{H}$ is such that, for some $\delta > 0$ and some $J \in \mathbb{N}$, the jump times t_j of x satisfy $t_{j+1} - t_j \geq \delta$ for all $j \geq J$;*

   then the output-feedback control law $u_c = -\kappa_c(y_c)$, $u_d \equiv 0$ renders the set $\mathcal{A} := \mathcal{A}^*$ pre-asymptotically stable for $\mathcal{H}$;

2. *If $\mathcal{H}_P$ is jump passive with respect to $\mathcal{A}^*$ with a storage function V and there exists a continuous function $\kappa_d : \mathbb{R}^{m_{Pd}} \to \mathbb{R}^{m_{Pd}}$ such that $y_d^\top \kappa_d(y_d) > 0$ for all $y_d \neq 0$ and the resulting hybrid closed-loop system $\mathcal{H}$ with state $x = z$, $u_c \equiv 0$, and $u_d = -\kappa_d(y_d)$ has the following properties:*

   a) *The distance to $\mathcal{A}^*$ is detectable relative to*

$$\left\{ z \in \Pi_c(C_P) \cup \Pi_d(D_P) \cup G_P(D_P) : h_d(z)^\top \kappa_d(h_d(z)) = 0, \ (z, -\kappa_d(h_d(z))) \in D_P \right\}$$

   for $\mathcal{H}$;

   b) *Every complete solution x to $\mathcal{H}$ is Zeno;*

   then the output-feedback control law $u_c \equiv 0$, $u_d = -\kappa_d(y_d)$ renders $\mathcal{A} := \mathcal{A}^*$ pre-asymptotically stable for $\mathcal{H}$.

Furthermore, the pre-asymptotic stability property of $\mathcal{A}$ is robust in the sense of Definition 3.16.

Theorem 9.12 is for the case when the hybrid plant has a passivity property that is not strict. A result for cases when the plant has a strict passivity property is immediate. In particular, strict passivity and output strict passivity can be employed to assert pre-asymptotic stability with zero inputs.

The two examples introduced in §9.1 are revisited and Theorem 9.12 is exercised in them. For both applications, the passivity-based control law proposed in this section renders the set of interest globally asymptotically stable.

Example 9.13 (Robotic manipulator interacting with the environment, revisited). *Consider the hybrid plant given in Example 9.1 and later revisited in Example 9.5. As in those examples, the goal is to robustly and globally asymptotically stabilize the point-mass to a position in contact with the vertical surface, in other words, to render $\mathcal{A}^* = \{(z_1^*, 0)\}$, with $z_1^* \geq 0$, globally asymptotically stable for the resulting hybrid closed-loop system. The approach to achieve this is as follows. When the energy-based control law given in (9.22) is used, the hybrid plant has a passivity property that involves the (virtual) control input $\widetilde{u}_c$ – this approach is known as passivation via energy shaping. With this passivity property on hand, the control law assigning the input $\widetilde{u}_c'$ is synthesized as a damping injection. Finally, Theorem 9.12 establishes pre-asymptotic stability. The details are given next.*

Consider the hybrid plant given in (9.20) with control input u_c chosen as in (9.22). As shown in Example 9.5, the choice (9.22) and the storage function V in (9.23) lead to flow passivity of $\mathcal{H}_P$ with respect to $\mathcal{A}^* = \{(z_1^*, 0)\}$ with output $y_c = h_c(z) = z_2$ and input $\widetilde{u}_c$. Following the construction of the control law in Theorem 9.12, pick[5] $\widetilde{u}_c = \kappa_c(z) := -k_1 y_c$, with $k_1 > 0$. The resulting hybrid closed-loop system has state $x = z$ and dynamics given by

$$\mathcal{H} \; : \quad \begin{cases} x \in C := \Pi_c(C_P) & \dot{x} \;\in\; F(x) := F_P(z, -k_1 y_c) \\ x \in D := D_P & x^+ = G(x) := G_P(z) \end{cases} \tag{9.47}$$

where C_P and D_P are given in (9.21), F_P is given in (9.24), and G_P is given in (9.20). Next, the conditions in item 1 of Theorem 9.12 are checked.

- *Item 1.1a requires that the distance to $\mathcal{A}^*$ is 0-input detectable relative to the set on which $y_c \top \kappa_c(y_c)$ vanishes. With $x = z$, this set is given by $K := \{x \in C : y_c = 0\} = \{x \in \mathbb{R}^2 : x_2 = 0\}$. Now, pick a solution x to the closed loop (9.47) such that $x(t,j) \in K$ for all $(t,j) \in \operatorname{dom} x$. Then, from the definitions of C and D, the solution x is such that $x(t,j)$ belongs to $C \backslash \bar{D}$ for all $(t,j) \in \operatorname{dom} x$. Then, from the definition of the flow map F, $x(t,j) \in \mathcal{A}^*$ for all $(t,j) \in \operatorname{dom} x$. Hence, the distance to $\mathcal{A}$ is 0-input detectable relative to K for $\mathcal{H}$ – in fact, the distance is observable.*

- *To show item 1.1b holds, observe that the time between consecutive jumps is lower bounded by a finite amount of time, uniformly. In fact, this property follows by the fact that $|G(x)|_D \geq (1 + \lambda)\bar{z}_2 > 0$ for all $x \in D$ – that is, G maps points in D to points in $C \setminus D$ – and the flow map is locally bounded; see item 1ii in Proposition 2.34.*

---

[5]Note that, once convergence to $\mathcal{A}^*$ is achieved, this feedback is equal to $k_c z_1^*$. From a physical standpoint, this condition corresponds to the point-mass applying a force to the vertical surface that can be varied according to the choice of the set-point position z_1^*.

Then, item 1 of Theorem 9.12 establishes that the set $\mathcal{A}^*$ is pre-asymptotically stable for $\mathcal{H}$.

Asymptotic stability follows from the fact that all maximal solutions to (9.47) are also complete. Furthermore, nontrivial solutions exist from every point in $C \cup D$. These properties are asserted using Proposition 2.34. In fact, to establish the existence of nontrivial solutions from each point in $C \cup D$, according to Proposition 2.34, it is enough to show that $F(x, -k_1 x_2) \subset T_C(x)$ for each $x \in C \backslash D$. From the definitions of the sets C and D, it follows that $T_C(x)$ at points $x \in C \backslash D$ is given by $\mathbb{R}^2$; hence, $F(x, -k_1 x_2) \subset T_C(x)$ trivially holds. Moreover, $G(D) \subset C \cup D$. Furthermore, the storage function V in (9.23) is such that

$$\alpha_1 |x|_{\mathcal{A}^*}^2 \leq V(x) \leq \alpha_2 |x|_{\mathcal{A}^*}^2 \qquad \forall x \in \mathbb{R}^2$$

with $\alpha_1 = 1/2 \min\{k_p, 1\}$ and $\alpha_2 = 1/2 \max\{k_p, 1\}$. This fact, stability of $\mathcal{A}^*$, and compactness of $\mathcal{A}^*$ imply that each solution x to $\mathcal{H}$ in (9.47) is bounded. In fact, each solution x to $\mathcal{H}$ satisfies

$$|x(t, j)|_{\mathcal{A}^*}^2 \leq \frac{\alpha_2}{\alpha_1} |x(0, 0)|_{\mathcal{A}^*}^2 \qquad \forall (t, j) \in \operatorname{dom} x$$

Then, since every maximal solution to $\mathcal{H}$ is bounded, Proposition 2.34 implies that every maximal solution to $\mathcal{H}$ is complete. Hence, $\mathcal{A}^*$ is asymptotically stable for $\mathcal{H}$. Finally, since V is radially unbounded, $\mathcal{A}^*$ is globally asymptotically stable for $\mathcal{H}$. Robustness is a direct consequence of Theorem 3.26.

Example 9.14 (One-degree-of-freedom juggling system, revisited). *Consider the hybrid plant $\mathcal{H}_P$ given in Example 9.2 and Example 9.6. The goal is to robustly and globally asymptotically stabilize the ball to the origin, in contact with the horizontal surface; in other words, to render the set $\mathcal{A}^* = \{(0, 0)\}$ robustly and globally asymptotically stable. By taking advantage of the passivity property of the system shown in Example 9.6, this goal can be obtained by designing a passivity-based control law using the construction in Theorem 9.12. To that end, let $\bar{\lambda} > 0$ be such that*

$$\bar{\lambda} \in \left(\frac{2\lambda_1 \lambda_2}{1 - \lambda_2^2}, \frac{2\lambda_2^2}{1 - \lambda_2^2} \right)$$

and define the feedback κ_d as

$$\kappa_d(y_d) := \bar{\lambda} y_d \qquad (9.48)$$

with y_d given by (9.30). Then, since from Example 9.6 the hybrid plant $\mathcal{H}_P$ is jump passive with respect to the set $\mathcal{A}^*$ with input u_d, output y_d and storage function V given in (9.6), following Theorem 9.12, assign u_d to $-\kappa_d(y_d)$. This choice for u_d is such that $y_d^\top \bar{\lambda} y_d > 0$ for all $y_d \neq 0$ and such that $(z, -\bar{\lambda} y_d) \in D_P$ for all $z \in \Pi_d(D_P)$. In fact, from the definition of y_d in (9.30), it follows that $\bar{\lambda} y_d = \frac{1}{2} \frac{\bar{\lambda}(1 - \lambda_2^2)}{\lambda_2} z_2$. From $2\lambda_1 \lambda_2 / (1 - \lambda_2^2) < \bar{\lambda} < 2\lambda_2^2 / (1 - \lambda_2^2)$, it follows that $\lambda_1 < \frac{1}{2} \bar{\lambda} \frac{(1 - \lambda_2^2)}{\lambda_2} < \lambda_2$, and then $\lambda_1 |z_2| < -\bar{\lambda} y_d < \lambda_2 |z_2|$ for all $z \in \Pi_d(D_P)$. To establish pre-asymptotic stability of $\mathcal{A}^*$, items 2.2a and 2.2b of Theorem 9.12 are verified. First, note that the hybrid closed-loop system $\mathcal{H}$ with the chosen input has *state $x = z$ and dynamics given by*

$$\mathcal{H} \; : \; \begin{cases} \quad x \in C := C_P & \dot{x} \;\; = F(x) := F_P(z) \\ (x, -\bar{\lambda}y_d) \in D_P & x^+ := G(x) := G_P(x, -\bar{\lambda}y_d) \end{cases} \qquad (9.49)$$

where C_P, F_P, D_P, and G_P are given in (9.28). Note that the condition $(x, -\bar{\lambda}y_d) \in D_P$ leads to the jump set $D = \big\{ x \in \mathbb{R}^2 : x_1 = 0, x_2 \leq 0 \big\}$. For this system, the distance to $\mathcal{A}^*$ is 0-input detectable relative to the set

$$K := \big\{ x \in \mathbb{R}^2 : y_d = 0, (x, -\bar{\lambda}y_d) \in D_P \big\} = \big\{ x \in \mathbb{R}^2 : x_1 = 0, x_2 = 0 \big\}$$

since every solution that stays in K is also in $\mathcal{A}^*$. Then, item 2.2a holds. Item 2.2b holds since $\bar{\lambda}$ is less than one, leading to every maximal solution to the closed-loop system being Zeno. Then, by item 2 in Theorem 9.12, $\mathcal{A}^*$ is pre-asymptotically stable for $\mathcal{H}$.

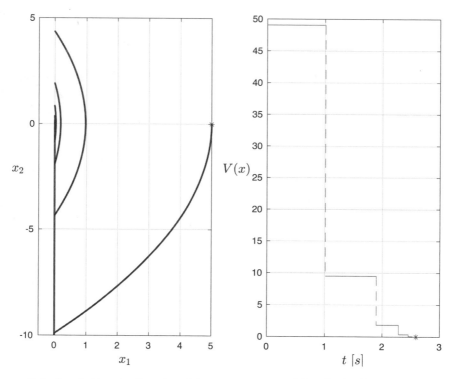

Figure 9.4: A solution to the closed-loop system resulting from the passivity-based controller in Example 9.14, with parameters $\lambda_1 = 0.1$, $\lambda_2 = 0.8$, and $\bar{\lambda} = 1.96$.

To establish asymptotic stability, completeness of maximal solutions is shown next, by applying Proposition 2.34. Existence of nontrivial solutions from each point in $C \cup D$ follows due to the following three properties:

- *For each $x = (x_1, x_2) \in C$ such that $x_1 > 0$, $T_C(x) = \mathbb{R}^2$ and, as a consequence, $F(x) \in T_C(x)$ holds.*
- *For each $x \in C$ such that $x_1 = 0$, either $x \in D$ or $x_2 > 0$. Since jumps from D are always possible, it suffices to show that $F(x) \in T_C(x)$ for each x such that $x_1 = 0$ and $x_2 > 0$. At such points, $T_C(x) = \mathbb{R}_{\geq 0} \times \mathbb{R}$ and, by definition, the first component of F being positive implies $F(x) \in T_C(x)$.*

- *For each $x \in D$, $G(x) \in \partial C \subset C$; hence, solutions cannot leave $C \cup D$ via jumps.*

Then, for every point in $C \cup D$ there exists a nontrivial solution. Now, note that on $C \cup \Pi_c(D) \subset C$ and that since $x_1 \geq 0$, V defined in (9.6) is equal to $V(x) = \frac{1}{2}x_2^2 + \gamma|x_1|$. Then, if $|x_1| \geq |x_2|$ it follows that $\gamma|x_1| = \gamma \max\{|x_1|, |x_2|\}$. Since $|x| \leq \sqrt{2}\max\{|x_1|, |x_2|\}$ then $V(x) \geq \gamma|x_1| \geq (\gamma/\sqrt{2})|x|$. If $|x_1| < |x_2|$ then $|x_2|^2 = \max\{|x_1|, |x_2|\}^2 \geq 1/2|x|^2$. Hence, $V(x) \geq (1/2)x_2^2 \geq 1/4|x|^2$. Both α_1 and α_2 are of class-$\mathcal{K}_\infty$. Accordingly $\alpha_1(|x|) \leq V(x) \leq \alpha_2(|x|)$ for each x, where $\alpha_1(s) := \min\{1/4|s|^2, (\gamma/\sqrt{2})|s|\}$ and $\alpha_2(s) := (1/2)|s|^2 + \gamma|s|$ for each $s \geq 0$. This fact and stability of $\mathcal{A}^*$ imply that every solution x to (9.49) is bounded. From the above arguments and since solutions are bounded, it follows that each maximal solution is complete. Consequently, the set $\mathcal{A}^*$ is asymptotically stable for $\mathcal{H}$. Finally, global asymptotic stability of $\mathcal{A}^*$ follows from the fact that V is radially unbounded.

Figure 9.4 shows a solution to the closed-loop system resulting from controlling the juggling system with the passivity-based controller designed above, and the change of the storage function along the solution. As the simulation shows, through impacts with the controlled surface, the position and velocity of the ball approach rest. Associated simulation files are at @BookSite/Simulation/Juggling2.

9.5 EXERCISES

Exercise 62 (Fully actuated one-degree-of-freedom juggling system). Consider the fully actuated version of the model of the one-degree-of-freedom juggling system in Example 9.6. This system is modeled as in (9.28) but with the dynamics during flows given by

$$\dot{z}_1 = z_2, \quad \dot{z}_2 = -\gamma + u_c \tag{9.50}$$

where $u_c \in \mathbb{R}$ is a new control input.

1. Derive a hybrid plant model $\mathcal{H}_P$ capturing the dynamics of the system.

2. Show that the hybrid plant is passive with respect to the compact set $\mathcal{A}^* = \{0\}$.

3. Define the output to the plant to require the least information possible during flows and at jumps for which the set $\mathcal{A}^*$ can be asymptotically stabilized.

4. Design a feedback controller rendering globally asymptotically stable $\mathcal{A}^*$ for the resulting closed-loop system.

5. Validate your results numerically using $\gamma = 9.8 \ m/s^2$.

Exercise 63 (Point-mass interacting with the environment). Consider the model of a point-mass interacting with the environment in Example 9.5 but, instead, use the so-called Hunt-Crossley contact model for the contact force. Nonlinear Hunt-Crossley contact models are preferred to linear Kelvin-Voigt models, such as the one employed to model the contact force f_c in Example 9.1, since they can more accurately describe the behavior of viscous materials. In the Hunt-Crossley contact model the force f_c is given by

$$\hat{f}_c(z) = \begin{cases} k_c z_1^a + b_c z_1^a z_2 & \text{if } z_1 > 0 \\ 0 & \text{if } z_1 \leq 0 \end{cases}$$

where $a \geq 1$. For this purpose, replace $\widetilde{\kappa}_c$ in (9.22) by

$$\widetilde{\kappa}_c(z_1, \widetilde{u}_c) := \begin{cases} k_c z_1^a - k_p(z_1 - z_1^\star) + \widetilde{u}_c & \text{if } z_1 > 0 \\ -k_p(z_1 - z_1^\star) + \widetilde{u}_c & \text{if } z_1 \leq 0 \end{cases}$$

and show the following:

1. The resulting hybrid plant with $u_c = \widetilde{\kappa}_c(z_1, \widetilde{u}_c)$ is flow passive with respect to $\mathcal{A}^* = \{(z_1^\star, 0)\}$.

2. The resulting hybrid plant with $u_c = \widetilde{\kappa}_c(z_1, \widetilde{u}_c)$ and $\widetilde{u}_c$ as in (9.27) is flow output strictly passive with respect to $\mathcal{A}^* = \{(z_1^\star, 0)\}$.

3. Validate your results numerically using $k_c = 100$, $b_c = 0.1$, $\bar{z}_2 = 0.1\ m/s$, $a = \frac{3}{2}$, $\lambda = 0.8$, and $z_1^\star = 0.01\ m$.

Exercise 64 (Interconnection of single integrators). Consider the hybrid plant with dynamics

$$\mathcal{H}_P : \begin{cases} (z, u_c) \in C_P & \dot{z} = u_c \\ (z, u_d) \in D_P & z^+ = -u_d \end{cases} \tag{9.51}$$

where the state is $z \in \mathbb{R}$, input $u_c, u_d \in \mathbb{R}$, and sets $C_P = \{(z, u_c) \in \mathbb{R} \times \mathbb{R} : z \leq 0\}$, $D_P = \{(z, u_d) \in \mathbb{R} \times \mathbb{R} : z \geq 0, u_d \in U(z)\}$ with $U : \mathbb{R} \to \mathbb{R}_{\leq 0}$ given by $z \mapsto U(z) := \{u_d \in \mathbb{R}_{\leq 0} : u_d = -(c\lambda_1 + (1 - c)\lambda_2)|z|\}$ for $c \in [0, 1]$ and $0 < \lambda_1 < \lambda_2 < 1$. With the outputs defined as

$$y_c = h_c(x) := z, \qquad y_d = h_d(z) := -\lambda_2 z \tag{9.52}$$

where $\lambda_2 := \frac{1}{2}\frac{(1 - \lambda_2^2)}{\lambda_2}$:

1. Show that the hybrid plant is passive with respect to the compact sets $\mathcal{A}^* = \{0\}$ with quadratic storage function V.

2. Design a feedback controller rendering globally asymptotically stable $\mathcal{A}^*$ for the resulting closed-loop system.

3. Validate your results numerically using $c = \frac{1}{2}$, $\lambda_1 = \frac{1}{4}$, and $\lambda_2 = \frac{3}{4}$.

Exercise 65 (Passivity of interconnections). Consider the model of one-degree-of-freedom juggling system given by

$$\mathcal{H}_P : \begin{cases} (z, u_c) \in C_P & \dot{z} = F_P(z, u_c) := \begin{bmatrix} z_2 \\ -\gamma + u_c \end{bmatrix} \\ (z, u_d) \in D_P & z^+ = G_P(z, u_d) := \begin{bmatrix} z_1 \\ -\lambda z_2 - u_d \end{bmatrix} \end{cases} \tag{9.53}$$

with state $z = (z_1, z_2) \in \mathbb{R}^2$, inputs $u_c, u_d \in \mathbb{R}$, fixed restitution coefficient $\lambda \in [0, 1)$, gravity constant $\gamma > 0$ and sets C_P and D_P given by

$$C_P := \{(z, u_c) \in \mathbb{R}^2 \times \mathbb{R} : z_1 \geq 0\}$$
$$D_P := \{(z, u_d) \in \mathbb{R}^2 \times \mathbb{R} : z_1 = 0, z_2 \leq 0, u_d \in U(z_2)\} \quad (9.54)$$

where $U : \mathbb{R} \to \mathbb{R}$ defines the constraint set for the input $u_{d,1}$ which is given by $z_2 \mapsto U(z_2) := \{u_d \in \mathbb{R} : u_d^2 \leq (1 - \lambda^2)z_2^2\}$; namely, the applied input at impacts is upper bounded by the ball's velocity; cf. the models in Exercise 62 and in Example 9.2.

1. Show that the hybrid plant (9.53) with inputs u_c and u_d, and outputs

   $$y_c = h_c(z) := z_2, \qquad y_d = h_d(z) := \lambda z_2 \quad (9.55)$$

   is passive with respect to set $\mathcal{A}^* = \{(0,0)\}$ with storage function $V : \mathbb{R}^2 \to \mathbb{R}_{\geq 0}$ given in (9.6).

2. Consider the negative feedback interconnection of two hybrid systems, each modeled as in (9.53)-(9.55), denoted $\mathcal{H}_{P,1}$ and $\mathcal{H}_{P,2}$, respectively, shown in Figure 9.5.

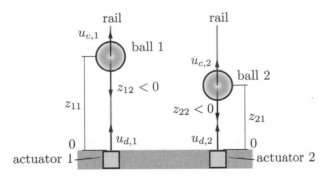

Figure 9.5: Interconnected one-degree-of-freedom juggling systems in Exercise 65.

   a) Derive the negative feedback interconnection between $\mathcal{H}_{P,1}$ and $\mathcal{H}_{P,2}$ by choosing $u_{c,1} = \tilde{u}_{c1} - y_{c2}$, $u_{c,2} = \tilde{u}_{c2} + y_{c1}$, $u_{d,1} = \tilde{u}_{d1} - y_{d2}$ and $u_{d,2} = \tilde{u}_{d2} + y_{d1}$ where $\tilde{u}_{c1}, \tilde{u}_{c2}, \tilde{u}_{d1}, \tilde{u}_{d2}$ are additional inputs. Denote it as a new hybrid system $\mathcal{H}_{P,12}$.
   b) Let V_1 and V_2 be the individual storage function for each system. Is $\mathcal{H}_{P,12}$ passive with respect to the origin with storage function $V = V_1 + V_2$?

Exercise 66 (Interconnection of single integrators). As defined in Exercise 65, consider the negative feedback interconnection between two hybrid systems with dynamics as those in Example 64.

1. Show that the interconnection is flow passive with respect to the origin.

2. Show that the origin for the resulting interconnection is 0-input stable.

3. Design a feedback controller that globally asymptotically stabilizes the origin.

4. Validate your results numerically using $c = \frac{1}{2}$, $\lambda_1 = \frac{1}{4}$, and $\lambda_2 = \frac{3}{4}$.

Exercise 67 (Interconnection of hybrid plants with linear maps). For each $i \in \{1, 2\}$, consider the hybrid plant

$$\mathcal{H}_{P,i} : \quad \begin{cases} (z_i, u_{c,i}) \in C_{P,i} & \dot{z}_i = Az_i + Bu_{c,i} \\ z_i \in D_{P,i} & z_i^+ = Rz_i \end{cases} \qquad (9.56)$$

with state $z_i = (p_i, v_i) \in \mathbb{R}^2$, input $u_{c,i} \in \mathbb{R}$,

$$A = \begin{bmatrix} 0 & 1 \\ -a_1 & -a_2 \end{bmatrix}, \quad B = \begin{bmatrix} 0 \\ 1 \end{bmatrix}, \quad R = \begin{bmatrix} 1 & 0 \\ 0 & -\lambda \end{bmatrix},$$

sets

$$C_{P,i} = \left(\left\{ z_i \in \mathbb{R}^2 : p_i \leq 0 \right\} \cup \left\{ z_i \in \mathbb{R}^2 : p_i \geq 0, \ v_i \leq \bar{v} \right\} \right) \times \mathbb{R}$$

$$D_{P,i} = \left\{ z_i \in \mathbb{R}^2 : p_i \geq 0, \ v_i \geq \bar{v} \right\} \cap \left\{ z_i \in \mathbb{R}^2 : v_i(\beta v_i - \alpha p_i) \leq 0 \right\}$$

where $\bar{v}$, a_1, and a_2 are positive parameters, $\lambda \in [0, 1]$, $\alpha = \frac{\lambda}{a_2}$, and $\beta = \frac{(\lambda^2 - 1)(1 + a_1)}{2a_1 a_2}$.

1. Let $P = P^\top > 0$ be the solution to the Lyapunov equation $A^\top P + PA = -I$. Show that, for each $i \in \{1, 2, \}$, system (9.56) is flow strictly passive with respect to the compact set $\mathcal{A}_i = \{(0, 0)\}$ with storage function

$$V_i(z_i) := \frac{1}{2} z_i^\top P z_i \qquad \forall z_i \in \mathbb{R}^2 \qquad (9.57)$$

   input $u_{c,i}$, output $y_{c,i} = h_c(z_i) := 2B^\top P z_i$, and function $\rho_c(z_i) := z_i^\top z_i$.

2. Consider the negative feedback interconnection obtained by considering the assignments $u_{c,i} = -y_{c,i}$ for each $i \in \{1, 2\}$ and show that the origin is 0-input pre-asymptotically stable for the resulting interconnection.

9.6   NOTES

Several textbooks [170, 171, 95, 172] and seminal papers [173, 174, 175, 176] document dissipativity and passivity concepts, sufficient conditions, and passivity-based feedback control designs; for a detailed survey on the latter see [177]. For passive systems, the passivity-based control design technique has been shown to be particularly useful in designing controllers that can be well understood from an energetic perspective. The problem of stabilizing a system to a given equilibrium point, in particular, can be addressed using passivity techniques by designing a feedback controller such that the storage function has the desired form and minimum. With such a function, convergence is obtained by selecting the input so that the energy of the system is dissipated. Modifications of the storage function and of the dissipation rate are often referred to as *energy shaping* and *damping injection*; see, e.g., [175].

Dissipativity and passivity have been studied for several types of hybrid systems. Passivity of switching systems was investigated in [178]. Motivated by haptic and teleoperation applications, a notion of passivity for systems in which the controller switches between different operative modes was proposed in [179]. Results about dissipativity of switching systems appeared also in [180], where multiple storage functions were considered. Passivity and passivity-based control for systems undertaking impacts and unilateral constraints have been investigated in [181] by first extending the Lagrange-Dirichlet theorem to a class of nonsmooth Lagrangian systems. The results therein are applied to mechanical systems including robotic manipulators with rigid or flexible joints. In [182], passivity-based control techniques are employed to regulate walking for a class of bipedal robots (see also [183]). In that reference, impact Poincaré maps are considered as a tool to investigate stability of periodic orbits characterizing the desired walking behavior. In [19], the authors consider dissipativity theory for a class of impulsive dynamical systems. In particular, the proposed framework considers different input and output maps for, respectively, the continuous-time evolution and the instantaneous changes, and results linking observability to asymptotic stability for the design of feedback controllers are presented. A general notion of dissipativity for a class of hybrid systems was linked to detectability and used to establish asymptotic stability for large-scale interconnections of hybrid systems in [184].

The results in this chapter and some of the illustrations in examples appeared in R. Naldi, and R. G. Sanfelice "Passivity-based Control for Hybrid Systems with Applications to Mechanical Systems Exhibiting Impacts," *Automatica*, vol. 49, no. 5, pp. 1104–1116, May, 2013 [78]. These results have been reproduced here by permission of Elsevier.

The main purpose of Theorem 9.8 and Theorem 9.10 is to enable the passivity-based control design tool in Theorem 9.12 for hybrid plants that are only flow passive and jump passive. For the case $\mathcal{A}^* = \{0\}$, the property that standard passivity implies 0-input stability (as in item 1 of Theorem 9.8) was established in [19, Proposition 12.3] (see also [185]) for left-continuous hybrid systems. Additional dissipativity and observability conditions leading to asymptotic stability of the origin were introduced therein. The 0-input stability property of $\mathcal{A}^*$ in items 1 and 2 of Theorem 9.8 can be established without insisting on conditions $(A1_0)$-$(A3_0)$ and, instead, proceeding as in [184, Theorem 2] – in fact, item 2 of Theorem 9.8 follows from [184, Theorem 2] when specializing the general dissipativity concept therein to passivity. A complete proof of Theorem 9.10 is in [78]. It should be pointed out that the application of [184, Theorem 2] to the one-degree-of-freedom juggling system in Example 9.2, which is also considered as an example in [184], only intersects the jump set with $\{(0,0)\}$ and leaves the flow set unchanged. As a consequence, establishing the detectability property in [184, Theorem 2] requires checking solutions from points in C. For the particular case of the bouncing ball, asymptotic stability of $\mathcal{A}^*$ is established in [184] using the fact that the only complete solutions to the system start and stay at the origin.

A sufficient condition for item 2b to hold for the hybrid plant $\mathcal{H}_{P,0}$ is given in [9, Lemma 2.7]; see also item 1(ii) in Proposition 2.34. Accordingly, item 2b holds if $\mathcal{H}_{P,0}$ satisfies $(A1_0)$-$(A3_0)$ and the jump set does not map points back to D_P (for zero input). Observe that this condition does not require checking solutions to $\mathcal{H}_{P,0}$. Similarly, a sufficient condition to assert item 3b without checking solutions to $\mathcal{H}_{P,0}$ is given in [186, Theorem 1], which holds for a class of Lagrangian hybrid systems modeling mechanical systems exhibiting impacts. Results therein link Zeno behavior and stability of Zeno equilibria to properties of the coefficient of restitution and the system's unilateral constraints.

The systems featured in Example 9.1 and Example 9.2 appeared in [78]. More details can be found therein. The compliant impact model used in Example 9.1 is an adaptation of the one in [187]; see also Exercise 63.

Chapter Ten

Feedback Design via Control Lyapunov Functions

In this chapter, control Lyapunov functions for hybrid plants $\mathcal{H}_P$ are introduced and employed to systematically design a state-feedback law that asymptotically stabilizes a given set-point (or set) $\mathcal{A}^*$. In simple terms, a control Lyapunov function (CLF) is a regular enough scalar function that decreases along solutions to the system for some values of the input. When a CLF exists, it is very tempting to exploit its properties to construct an asymptotically stabilizing state-feedback law that takes input values for which the CLF decreases. The main challenge in doing so is the regularity properties of the resulting feedback – in particular, continuity with respect to the state. This chapter provides a CLF-based construction of a pre-asymptotically stabilizing static control law that, as a function of the state z, is continuous. Figure 10.1 depicts the hybrid closed-loop system resulting from using such a state-feedback law. A continuous static state-feedback law enables the resulting hybrid closed-loop system to satisfy the hybrid basic conditions. Conditions for the existence of such a feedback revealing the basic properties needed from the CLF and from the data of the plant, relative to $\mathcal{A}^*$, are given. These conditions are shown to lead to a "universal" construction of a control law that is asymptotically stabilizing and has minimum pointwise norm.

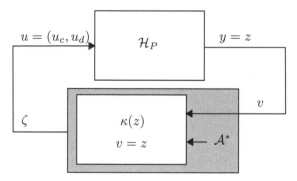

Figure 10.1: Hybrid closed-loop system resulting from CLF-based control.

10.1 OVERVIEW

Regardless of whether the plant is a continuous-time system, a discrete-time system, or a hybrid system, the existence of a CLF leads to a natural feedback design methodology capable of asymptotically driving the plant state to a set $\mathcal{A}^*$:

- Construct the feedback control law assigning the input to the plant so that the CLF decreases when the plant state is not in $\mathcal{A}^*$.

Such a choice guarantees convergence of the state to $\mathcal{A}^*$ since, as formally defined in the next section, a CLF takes positive values at points not in $\mathcal{A}^*$ and only vanishes on $\mathcal{A}^*$. The existence of such a feedback control law for each value of the plant state relies on the fact that the "minimum" of the change along solutions of the CLF is negative – more precisely, the (pointwise, in terms of the state) "minimum" is taken over all possible inputs allowed for each value of the state.

To outline the CLF-based strategy formulated in this chapter, for simplicity, consider a hybrid plant $\mathcal{H}_P$ with single-valued maps F_P and G_P and, with u_c and u_d as the input components affecting the change of z during flows and at jumps, respectively. Then, the function

$$z \mapsto V(z)$$

is a *control Lyapunov function* if it is positive definite with respect to the set $\mathcal{A}^*$ and can be made to decrease along solutions, both when they flow and when they jump, by properly choosing the control inputs u_c and u_d. More precisely, assuming that V is continuously differentiable, the approach is to design u_c so that during flows

$$\langle \nabla V(z), F_P(z, u_c) \rangle < 0$$

and to design u_d so that at jumps

$$V(G_P(z, u_d)) - V(z) < 0$$

where "< 0" means negative definiteness with respect to the set $\mathcal{A}^*$. The map

$$(z, u_c) \mapsto \langle \nabla V(z), F_P(z, u_c) \rangle$$

can be interpreted as an input-dependent version of the infinitesimal quantity $\dot{V}$ defined in (3.13), which, as in Chapter 9, can be conveniently denoted as $\dot{V}(z, u_c)$ for each $(z, u_c) \in C_P$. Similarly, the map

$$(z, u_d) \mapsto V(G_P(z, u_d)) - V(z)$$

can be treated as an input-dependent version of ΔV in (3.16), which can be denoted $\Delta V(z, u_d)$ for each $(z, u_d) \in D_P$.

For each $z \in \Pi_c(C_P)$, the construction

$$\kappa_c(z) = u_c^*$$

where

$$u_c^* \in \left\{ u_c : \dot{V}(z, u_c) < 0 \right\} \qquad \text{if } z \notin \mathcal{A}^*$$
$$u_c^* \in \left\{ u_c : \dot{V}(z, u_c) = 0 \right\} \qquad \text{if } z \in \mathcal{A}^*$$

(10.1)

defines a static state-feedback law that, when assigned to u_c, during flows, makes V decreases along solutions that are not in $\mathcal{A}^*$ and remains constant when solutions that are in $\mathcal{A}^*$.

Similarly, for each $z \in \Pi_d(D_P)$, the construction

$$\kappa_d(z) = u_d^*$$

where

$$u_d^* \in \{u_d : \Delta V(z, u_d) < 0\} \qquad \text{if } z \notin \mathcal{A}^*$$
$$u_d^* \in \{u_d : \Delta V(z, u_d) = 0\} \qquad \text{if } z \in \mathcal{A}^* \tag{10.2}$$

defines a static state-feedback law that, when assigned to u_d, at jumps, makes V decrease along solutions that are not in $\mathcal{A}^*$ and remains constant for solutions that are in $\mathcal{A}^*$.

As shown in § 10.3, under further assumptions, these feedback laws can also be constructed to be continuous and to have pointwise minimum norm.

10.2 CONTROL LYAPUNOV FUNCTIONS

Similar to the notion of a candidate Lyapunov function for a hybrid closed-loop system $\mathcal{H}$ in Definition 3.17, a control Lyapunov function for a hybrid plant $\mathcal{H}_P$ is given by a function V with large enough domain of definition, sufficient regularity on the flow set so that its change during flow can be determined infinitesimally, and positive definiteness with respect to the set $\mathcal{A}^*$. The following definition introduces such a candidate. In this chapter, a general hybrid plant is considered, namely

$$\mathcal{H}_P : \begin{cases} (z, u_c) \in C_P & \dot{z} \in F_P(z, u_c) \\ (z, u_d) \in D_P & z^+ \in G_P(z, u_d) \\ & y = z \end{cases} \tag{10.3}$$

For simplicity, the case when the neighborhood $\mathcal{U}$ of interest is large enough or equal to $\mathbb{R}^{n_P}$ is considered.[1]

Definition 10.1 (Control Lyapunov function candidate). *The set $\mathcal{A}^* \subset \mathbb{R}^{n_P}$ and the function $V : \mathrm{dom}\, V \to \mathbb{R}$ define a control Lyapunov function candidate with respect to $\mathcal{A}^*$ for the hybrid plant $\mathcal{H}_P = \overline{(C_P, F_P, D_P, G_P, Id)}$ as in (10.3) if the following conditions hold:*

1. $\Pi_c(C_P) \cup \Pi_d(D_P) \cup G_P(D_P) \subset \mathrm{dom}\, V$;

2. *V is continuous, and locally Lipschitz on an open set containing $\overline{\Pi_c(C_P)}$;*

3. *There exist $\alpha_1, \alpha_2 \in \mathcal{K}_\infty$ satisfying*

$$\alpha_1(|z|_{\mathcal{A}^*}) \le V(z) \le \alpha_2(|z|_{\mathcal{A}^*}) \qquad \forall z \in \Pi_c(C_P) \cup \Pi_d(D_P) \cup G_P(D_P) \tag{10.4}$$

Above, $\Pi_c(C_P)$ and $\Pi_d(D_P)$ are projections to the space for z; see List of Symbols.

---

[1]The case when $\mathcal{U}$ is a neighborhood of $\mathcal{A}^*$ follows similarly – in fact, it just requires intersecting C_P and D_P by $\mathcal{U} \times \mathbb{R}^{m_{P_c}}$ and $\mathcal{U} \times \mathbb{R}^{m_{P_d}}$, respectively. See Remark 3.20.

Given a solution to $\mathcal{H}_P$ and associated input, a control Lyapunov function candidate V changes along the state component of the solution as explained in §9.2, in the discussion below Definition 9.3. In particular, when F_P is single valued and V is continuously differentiable on a open neighborhood containing $\overline{\Pi_c(C_P)}$, the quantity governing the change of V during flow is

$$\dot{V}(z, u_c) := \langle \nabla V(z), F_P(z, u_c) \rangle \qquad \forall (z, u_c) \in C_P \qquad (10.5)$$

This expression is given in (9.8) and repeated here for convenience. As stated in §9.2 when F_P is set valued or V is only locally Lipschitz, a construction that extends (3.13) to the case when the input is not assigned is given by

$$\dot{V}(z, u_c) := \sup_{\chi \in F_P(z, u_c)} V^\circ(z, \chi) \qquad \forall (z, u_c) \in C_P \qquad (10.6)$$

where $V^\circ(z, \chi)$ is the Clarke generalized directional derivative of V at z in the direction of χ; see (3.7) for a definition.

As given in (9.10), the quantity paralleling the "discrete contribution" to the change of V in (3.16) is given by

$$\Delta V(z, u_d) := \sup_{\chi \in G_P(z, u_d)} V(\chi) - V(z) \qquad \forall (z, u_d) \in D_P \qquad (10.7)$$

Using these constructions, the definition of CLF for a hybrid plant $\mathcal{H}_P$ is given next. Note that sup's are used since the hybrid basic conditions may not hold.

Definition 10.2 (Control Lyapunov function). *Given a closed set $\mathcal{A}^* \subset \mathbb{R}^{n_P}$ and $r^* \geq 0$, a control Lyapunov function candidate V with respect to $\mathcal{A}^*$ for the hybrid plant $\mathcal{H}_P = (C_P, F_P, D_P, G_P, \mathrm{Id})$ given in (10.3) is a control Lyapunov function (CLF) with respect to $(\mathcal{A}^*, r^*)$ for $\mathcal{H}_P$ if there exist continuous and positive definite functions $\rho_c : \mathbb{R}^{n_P} \to \mathbb{R}_{\geq 0}$ and $\rho_d : \mathbb{R}^{n_P} \to \mathbb{R}_{\geq 0}$ such that[2]*

$$\inf_{u_c \in \Psi_c^u(z)} \dot{V}(z, u_c) \leq -\rho_c(|z|_{\mathcal{A}^*}) \qquad \forall z \in \Pi_c(C_P) \cap \mathcal{I}(r^*) \qquad (10.8)$$

$$\inf_{u_d \in \Psi_d^u(z)} \Delta V(z, u_d) \leq -\rho_d(|z|_{\mathcal{A}^*}) \qquad \forall z \in \Pi_d(D_P) \cap \mathcal{I}(r^*) \qquad (10.9)$$

The map Ψ_c^u evaluated at z collects all u_c's such that $(z, u_c) \in C_P$, Ψ_d^u evaluated at z collects all u_d's such that $(z, u_d) \in D_P$, and $\mathcal{I}(r^*)$ collects all z's such that $V(z) \geq r^*$. See List of Symbols.

Remark 10.3 (On nonstrict CLFs). The CLF in Definition 10.2 is "strict" in the sense that, for points not in $\mathcal{A}^*$, V can be made decreasing by the inputs in both the continuous and discrete regime of the hybrid plant. Following Theorem 3.19, weaker conditions, for instance, for the case when one of these functions is identically zero (or even positive), can be handled via the conditions in the Hybrid Lyapunov Theorem allowing the Lyapunov function to be weak or via the Hybrid Invariance Principle. The price to pay is the need of information about solutions. $\triangle$

The CLF definition is exercised in examples.

---

[2]Since ρ_c is defined on $\mathbb{R}^{n_P}$ and the infimum of the empty set is assumed to be ∞, (10.8) implies that $\dot{V}(z, u_c)$ is defined for each $z \in \Pi_c(C_P) \cap \mathcal{I}(r^*)$, for some $u_c \in \Psi_c^u(z)$. A similar comment applies for $\Delta V(z, u_d)$.

Example 10.4 (Pendulum with impacts). *Consider the hybrid plant given by*

$$
\mathcal{H}_P \ : \
\begin{cases}
(z, u_c) \in C_P & \begin{bmatrix} \dot{z}_1 \\ \dot{z}_2 \end{bmatrix} = \begin{bmatrix} z_2 \\ -a \sin z_1 - b z_2 + u_c \end{bmatrix} =: F_P(z, u_c) \\[12pt]
z \in D_P & \begin{bmatrix} z_1^+ \\ z_2^+ \end{bmatrix} = \begin{bmatrix} z_1 \\ -\lambda z_2 \end{bmatrix} =: G_P(z)
\end{cases}
\tag{10.10}
$$

where $a, b \in \mathbb{R}$, and $\lambda \in [0, 1)$ are constants, $u_c \in \mathbb{R}$ is the control input,

$$
C_P := \{(z, u_c) \in X_P \times \mathbb{R} \ : \ z_1 \leq 0\}, \quad D_P := \{z \in X_P \ : \ z_1 = 0, z_2 \geq 0\}
$$

and $X_P := [-\pi/2, 0] \times \mathbb{R}$. This hybrid plant models the motion of a pendulum with impacts. The state z_1 denotes the angle between the pendulum and the vertical, which, for simplicity, is restricted to evolve in the third quadrant; i.e., $z_1 \in [-\pi/2, 0]$. The state z_2 denotes the angular velocity, which is assumed to be positive when the pendulum moves counterclockwise. The constants a and b represent parameters such as the length, mass, and viscous friction of the pendulum as well as gravity. The parameter λ represents the coefficient of restitution at impacts. For this hybrid plant, consider the control Lyapunov function candidate with respect to $\mathcal{A}^* = \{0\} \subset \mathbb{R}^2$ given by the quadratic function

$$
V(z) = z^\top P z, \qquad P = \begin{bmatrix} 2 & 1 \\ 1 & 1 \end{bmatrix}
\tag{10.11}
$$

defined on $\mathbb{R}^2$. Condition 1 in Definition 10.1 holds since $\operatorname{dom} V = \mathbb{R}^2$. Since V is continuously differentiable, positive definite with respect to the origin, and radially unbounded, conditions 2 and 3 hold. Now, during flows, namely, for each $(z, u_c) \in C_P$,

$$
\begin{aligned}
\dot{V}(z, u_c) &= \langle \nabla V(z), F_P(z, u_c) \rangle \\
&= 4 z_1 z_2 + 2 z_2^2 + 2(-a \sin z_1 - b z_2 + u_c)(z_2 + z_1)
\end{aligned}
\tag{10.12}
$$

This equality implies that condition (10.8) is satisfied with $\rho_c(s) := s^2$ for each $s \geq 0$. In fact, since $\Psi_c^u(z) = \mathbb{R}$ for each $z \in X_P$ and $\Pi_c(C_P) = X_P$,

$$
\inf_{u_c \in \Psi_c^u(z)} \dot{V}(z, u_c) = \inf_{u_c \in \mathbb{R}} \langle \nabla V(z), F_P(z, u_c) \rangle = -z^\top z
$$

for all $z \in \Pi_c(C_P)$ such that $z_1 + z_2 = 0$, while, for each $z \in \Pi_c(C_P)$ satisfying $z_1 + z_2 \neq 0$,

$$
\inf_{u_c \in \Psi_c^u(z)} \dot{V}(z, u_c) = \inf_{u_c \in \mathbb{R}} \langle \nabla V(z), F_P(z, u_c) \rangle = -\infty
$$

At jumps, since no inputs are involved, it follows that

$$
V(G_P(z)) - V(z) = -(1 - \lambda^2) z_2^2 = -(1 - \lambda^2) z^\top z
\tag{10.13}
$$

for each $z \in \Pi_d(D_P) = D_P$. The property that each $z \in D_P$ satisfies $z_1 = 0$ was used to arrive to (10.13). Then, (10.9) is satisfied with $\rho_d(s) := (1 - \lambda^2) s^2$ for each $s \geq 0$. Hence, V in (10.11) is a CLF with respect to $(\mathcal{A}^*, r^*) = (\{0\}, 0)$ for $\mathcal{H}_P$ in (10.10).

The following example introduces a hybrid plant with inputs affecting both the flows and jumps, and with a set-valued jump map.

Example 10.5 (Planar system with jumps). *The hybrid plant considered in this example has a state z taking values from a subset of $\mathbb{R}^2$ and an input $u = (u_c, u_d)$. Similar to the model of the plant in the stabilization on a circle problem in § 1.2.3 and Example 8.1, during flows the state z describes a circle in a direction defined by u_c, but with angular velocity $\omega > 0$. In addition, the state exhibits jumps when at the boundary of a triangular region in the bottom half plane, shown in Figure 10.2. To define these regions, given vectors $w_1, w_2 \in \mathbb{R}^2$, define $\mathcal{W}(w_1, w_2) := \{z \in \mathbb{R}^2 : z = r(\lambda w_1 + (1 - \lambda)w_2), r \geq 0, \lambda \in [0,1]\}$. Then, the hybrid plant capturing the behavior described above is defined as*

$$\mathcal{H}_P \begin{cases} (z, u_c) \in C_P & \dot{z} = F_P(z, u_c) := u_c \begin{bmatrix} 0 & -\omega \\ \omega & 0 \end{bmatrix} z \\ (z, u_d) \in D_P & z^+ \in G_P(z, u_d) \end{cases} \tag{10.14}$$

where

$$C_P := \left\{ (z, u_c) \in X_P \times \mathbb{R} : u_c \in \{-1, 1\}, z \in \widehat{C}_P \right\}$$
$$\widehat{C}_P := \overline{X_P \setminus (\mathcal{W}(v_1^1, v_2^1) \cup \mathcal{W}(v_1^2, v_2^2))}$$
$$D_P := \left\{ (z, u_d) \in X_P \times \mathbb{R}_{\geq 0} : u_d \geq \gamma |z|_{\mathcal{A}^*} + \delta, z \in \partial \mathcal{W}(v_1^2, v_2^2) \right\}$$
$$X_P := \left\{ z \in \mathbb{R}^2 : |z| \geq \delta \right\}, \qquad \mathcal{A}^* := \left\{ z \in \mathbb{R}^2 : |z| = \delta \right\}$$

for each $(z, u_d) \in X_P \times \mathbb{R}_{\geq 0}$ the set-valued map G_P is given by

$$G_P(z, u_d) := \left\{ R(\pi/4) \begin{bmatrix} 0 \\ u_d \end{bmatrix}, R(-\pi/4) \begin{bmatrix} 0 \\ u_d \end{bmatrix} \right\},$$

$\gamma > 0$ is such that $\exp(\pi/(2\omega))\gamma < 1$, $\delta > 0$, $v_1^1 := [1\ 1]^\top$, $v_2^1 := [-1\ 1]^\top$, $v_1^2 := [1\ -1]^\top$, and $v_2^2 := [-1\ -1]^\top$. Figure 10.2 depicts the flow and jump sets projected to the z plane, and also the set $\mathcal{A}^*$.

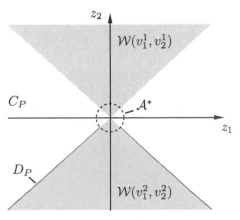

Figure 10.2: Sets for Example 10.5. The shaded region (and its boundary) corresponds to the set $\mathcal{W}(v_1^1, v_2^1) \cup \mathcal{W}(v_1^2, v_2^2)$ and the closure of its complement (restricted to X_P) is the flow set projected onto the z plane. The gray solid line represents D_P. The dashed line represents $\mathcal{A}^*$.

Consider a control Lyapunov function candidate with respect to $\mathcal{A}^*$ given by a function $V : \mathbb{R}^2 \to \mathbb{R}$ that is continuously differentiable on an open set containing $\widehat{C}_P$ and, for each $z \in \Pi_c(C_P) \cup \Pi_d(D_P) \cup G_P(D_P)(= \widehat{C}_P)$, is defined as

$$V(z) = \exp(t_\mathcal{W}(z))(|z| - \delta), \qquad t_\mathcal{W}(z) = \frac{1}{\omega}\arcsin\left(\frac{\sqrt{2}}{2}\frac{|z_1| + z_2}{|z|}\right) \tag{10.15}$$

At such points, $|z|_{\mathcal{A}^*} = |z - \delta\, z/|z|| = |z| - \delta$. The function $t_\mathcal{W}$ denotes the minimum time to reach the set $\mathcal{W}(v_1^2, v_2^2)$ with the continuous dynamics of (10.14), with u_c chosen from the set $\{-1, 1\}$. Note that $\Pi_c(C_P) = \widehat{C}_P$ and $\Pi_d(D_P) = \partial\mathcal{W}(v_1^2, v_2^2) \cap X_P$. For each $z \in \mathbb{R}^2$,

$$\Psi_c^u(z) = \begin{cases} \{-1, 1\} & \text{if } z \in \widehat{C}_P \\ \emptyset & \text{otherwise} \end{cases}$$

$$\Psi_d^u(z) = \begin{cases} \{u_d \in \mathbb{R}_{\geq 0} : u_d \geq \gamma|z|_{\mathcal{A}^*} + \delta\} & \text{if } z \in \partial\mathcal{W}(v_1^2, v_2^2) \cap X_P \\ \emptyset & \text{otherwise} \end{cases}$$

Since the quantity $|z|$ remains constant during flows, for each $(z, u_c) \in C_P$

$$\langle \nabla V(z), F_P(z, u_c)\rangle = \langle \nabla t_\mathcal{W}(z), F_P(z, u_c)\rangle V(z)$$

Now, for each $z \in \widehat{C}_P$ such that $z_1 > 0$,

$$\langle \nabla t_\mathcal{W}(z), F_P(z, u_c)\rangle = -\frac{\sqrt{2}}{2\omega}\frac{1}{\sqrt{1 - \frac{1}{2}\left(\frac{|z_1|+z_2}{|z|}\right)^2}}\left\langle \nabla\frac{|z_1| + z_2}{|z|}, F_P(z, u_c)\right\rangle$$

$$= -\frac{u_c}{\omega}\begin{bmatrix} \frac{z_2}{|z|^2} & -\frac{z_1}{|z|^2}\end{bmatrix}\begin{bmatrix} 0 & -\omega \\ \omega & 0\end{bmatrix}z$$

which is equal to -1 when $u_c = -1$. Similarly, for each $z \in \widehat{C}_P$ such that $z_1 < 0$,

$$\langle \nabla t_\mathcal{W}(z), F_P(z, u_c)\rangle = -1$$

when $u_c = 1$. Then,

$$\inf_{u_c \in \Psi_c^u(z)} \langle \nabla V(z), F_P(z, u_c)\rangle \leq -|z|_{\mathcal{A}^*} \tag{10.16}$$

for all $z \in \Pi_c(C_P)$. At jumps, for each $(z, u_d) \in D_P$ and each $\chi \in G_P(z, u_d)$

$$V(\chi) = \exp(t_\mathcal{W}(\chi))|\chi|_{\mathcal{A}^*} = \exp\left(\frac{\pi}{2\omega}\right)(u_d - \delta)$$

It follows that

$$\inf_{u_d \in \Psi_d^u(z)} \sup_{\chi \in G_P(z, u_d)} V(\chi) - V(z) \tag{10.17}$$

$$= \inf_{u_d \in \Psi_d^u(z)} \exp\left(\frac{\pi}{2\omega}\right)(u_d - \delta) - \exp(t_\mathcal{W}(z))|z|_{\mathcal{A}^*} \leq -\left(1 - \exp\left(\frac{\pi}{2\omega}\right)\gamma\right)|z|_{\mathcal{A}^*}$$

for each $z \in \Pi_d(D_P)$. Finally, both (10.8) and (10.9) hold with $s \mapsto \rho_c(s) = \rho_d(s) := \left(1 - \exp\left(\frac{\pi}{2\omega}\right)\gamma\right)s$. Then, V in (10.15) is a CLF with respect to $(\mathcal{A}^*, r^*)$ for $\mathcal{H}_P$ in (10.14), where $\mathcal{A}^* = \{z \in \mathbb{R}^2 : |z| = \delta\}$ and $r^* = 0$.

The CLF notion in Definition 10.2 is for a hybrid plant without disturbances. Its extension to the case with disturbances, termed *robust CLF*, along with versions of Example 10.4 including disturbances is given in §10.3.2.

10.3 DESIGN

10.3.1 Nominal Design

Given a closed set $\mathcal{A}^*$, a constant $r^* \geq 0$, and a control Lyapunov function V for $\mathcal{H}_P$ with respect to $(\mathcal{A}^*, r^*)$ satisfying Definition 10.2 with positive definite functions ρ_c and ρ_d, using (10.6) and (10.7), define the function

$$\Gamma_c(z, u_c, r) := \begin{cases} \dot{V}(z, u_c) + \sigma \rho_c(|z|_{\mathcal{A}^*}) & \text{if } (z, u_c) \in C_P, z \in \mathcal{I}(r) \\ -\infty & \text{otherwise} \end{cases} \quad (10.18)$$

for each $(z, u_c) \in \mathbb{R}^{n_P} \times \mathbb{R}^{m_P}$ and each $r \geq r^*$ and the function

$$\Gamma_d(z, u_d, r) := \begin{cases} \Delta V(z, u_d) + \sigma \rho_d(|z|_{\mathcal{A}^*}) & \text{if } (z, u_d) \in D_P, z \in \mathcal{I}(r) \\ -\infty & \text{otherwise} \end{cases} \quad (10.19)$$

for each $(z, u_d) \in \mathbb{R}^{n_P} \times \mathbb{R}^{m_P}$ and $r \geq r^*$, where $\sigma \in (0, 1)$ is an extra degree of freedom. Evaluating these functions at points (z, u_c, r) and (z, u_d, r) such that z and r satisfy $r = V(z)$ leads to functions with nonnegative values that depend on the state and input pairs only. The set-valued map $\mathcal{T}_c$ defined as

$$\mathcal{T}_c(z) := \{u_c \in \Psi_c^u(z) : \Gamma_c(z, u_c, V(z)) \leq 0\} \qquad \forall z \in \mathbb{R}^{n_P} \quad (10.20)$$

collects all allowed values of u_c that guarantee that $\dot{V}$ is decreasing on the flow set. In fact, following the discussion in §10.1, for each (z, u_c) satisfying $u_c \in \mathcal{T}_c(z)$ and $z \in \mathcal{I}(r^*)$, by definition of $\mathcal{T}_c$ it follows that

$$\dot{V}(z, u_c) < 0 \quad \text{if } z \notin \mathcal{A}^*, \qquad \dot{V}(z, u_c) \leq 0 \quad \text{if } z \in \mathcal{A}^* \quad (10.21)$$

Similarly, the set-valued map

$$\mathcal{T}_d(z) := \{u_d \in \Psi_d^u(z) : \Gamma_d(z, u_d, V(z)) \leq 0\} \qquad \forall z \in \mathbb{R}^{n_P} \quad (10.22)$$

is such that for each (z, u_d) satisfying $u_d \in \mathcal{T}_d(z)$ and $z \in \mathcal{I}(r^*)$,

$$\Delta V(z, u_d) < 0 \quad \text{if } z \notin \mathcal{A}^*, \qquad \Delta V(z, u_d) \leq 0 \quad \text{if } z \in \mathcal{A}^* \quad (10.23)$$

In simple words, for each $z \in \mathbb{R}^{n_P}$, the set-valued maps $\mathcal{T}_c$ and $\mathcal{T}_d$ collect all of the input values that guarantee a decrease of V for points not in $\mathcal{A}^*$ and a nonincrease of V on $\mathcal{A}^*$, both along flows and at jumps. These constructions suggest that when the inputs u_c and u_d are selected from $\mathcal{T}_c$ and $\mathcal{T}_d$, respectively, the function V decreases along solutions that are not in $\mathcal{A}^*$.

The CLF-based static state-feedback construction outlined in §10.1 is formalized in the following result.

Theorem 10.6 (CLF-based feedback control). *Given a hybrid plant $\mathcal{H}_P$ as in (10.3) with data (C_P, F_P, D_P, G_P, Id), a closed set $\mathcal{A}^* \subset \mathbb{R}^{n_P}$, $r^* \geq 0$, and a control Lyapunov function V for $\mathcal{H}_P$ with respect to $(\mathcal{A}^*, r^*)$ as in Definition 10.2, define the state-feedback law pair (κ_c, κ_d) as follows:*

- *For each $z \in \Pi_c(C_P) \cap \mathcal{I}(r^*)$,*

$$\kappa_c(z) := u_c^* \qquad \text{where } u_c^* \in \mathcal{T}_c(z) \tag{10.24}$$

- *For each $z \in \Pi_d(D_P) \cap \mathcal{I}(r^*)$,*

$$\kappa_d(z) := u_d^* \qquad \text{where } u_d^* \in \mathcal{T}_d(z) \tag{10.25}$$

The following hold for each $r \geq r^*$:

1. *The hybrid closed-loop system resulting from using the state-feedback law pair (κ_c, κ_d) restricted to $\mathcal{I}(r^*)$, namely, $\mathcal{H}_{r^*} = (C \cap \mathcal{I}(r^*), F, D \cap \mathcal{I}(r^*), G)$, where (C, F, D, G) is given in (2.24), has the set*

$$\mathcal{A} := \{z \in \mathbb{R}^{n_P} : V(z) \leq r\} \tag{10.26}$$

uniformly globally pre-asymptotically stable.

2. *If (10.9) holds for each $z \in \Pi_d(D_P)$, then with κ_c extended to $\Pi_c(C_P)$ arbitrarily and κ_d extended to $\Pi_d(D_P)$ as in (10.25), the set $\mathcal{A}$ in (10.26) is uniformly globally pre-asymptotically stable for the hybrid closed-loop system $\mathcal{H} = (C, F, D, G)$ given in (2.24).*

Proof. By definition of $\mathcal{T}_c$ in (10.20) and the maps involved therein, for each $z \in \Pi_c(C_P) \cap \mathcal{I}(r^*)$, $\kappa_c(z)$ satisfies

$$\kappa_c(z) \in \Psi_c^u(z), \qquad \Gamma_c(z, \kappa_c(z), V(z)) \leq 0$$

or, equivalently,

$$(z, \kappa_c(z)) \in C_P, \qquad \dot{V}(z, \kappa_c(z)) \leq -\sigma\rho_c(|z|_{\mathcal{A}^*})$$

Similarly, by definition of $\mathcal{T}_d$ in (10.22) and the maps involved therein, for each $z \in \Pi_d(D_P) \cap \mathcal{I}(r^*)$, $\kappa_d(z)$ satisfies

$$\kappa_d(z) \in \Psi_d^u(z), \qquad \Gamma_d(z, \kappa_d(z), V(z)) \leq 0$$

or, equivalently,

$$(z, \kappa_d(z)) \in D_P, \qquad \Delta V(z, \kappa_d(z)) \leq -\sigma\rho_d(|z|_{\mathcal{A}^*})$$

Combining these properties of V for $\mathcal{H}_P$ leads to the following property for the closed-loop system $\mathcal{H}_{r^*} = (C \cap \mathcal{I}(r^*), F, D \cap \mathcal{I}(r^*), G)$ where (C, F, D, G) is the data of (2.24):

$$\dot{V}(z) \leq -\sigma\rho_c(|z|_{\mathcal{A}^*}) \qquad \forall z \in C \cap \mathcal{I}(r^*) \tag{10.27}$$

$$\Delta V(z) \leq -\sigma\rho_d(|z|_{\mathcal{A}^*}) \qquad \forall z \in D \cap \mathcal{I}(r^*) \tag{10.28}$$

Since, according to item 3 in Definition 10.1, V is positive definite with respect to $\mathcal{A}^*$, for each $r \geq r^*$, conditions (10.27) and (10.28) imply that the set $\mathcal{A}$ is uniformly globally pre-asymptotically stable for $\mathcal{H}_{r^*}$, by an application of item 3a in Theorem 3.19. Note that if $r^* = 0$, then picking $r = r^*$ implies that $\mathcal{A}^*$ is uniformly globally pre-asymptotically stable for $\mathcal{H} = (C, F, D, G)$. When $r^* > 0$, for $\mathcal{A}$ with $r \geq r^*$ to have such property it suffices to show that solutions to $\mathcal{H}$ from $\mathcal{A}$ do not leave $\mathcal{A}$. Since $r > 0$, $\dot{V}(z) < 0$ for points $z \in C$ such that $V(z) = r^*$; hence, solutions to $\mathcal{H}$ cannot flow out of $\mathcal{A}$. The only way that solutions can leave $\mathcal{A}$ is through a jump. Since, by the assumption item 2, (10.9) holds for each $r \in [0, r^*]$, κ_d can be extended to $\Pi_d(D_P)$ as in (10.25) to guarantee that solutions do not jump outside of $\mathcal{A}$. Then, uniform global pre-asymptotic stability of $\mathcal{A}$ for $\mathcal{H}$ follows from an application of item 3a in Theorem 3.19. $\qquad\square$

The (static) state-feedback laws κ_c and κ_d constructed in Theorem 10.6 are not guaranteed to be continuous functions on $\Pi_c(C_P) \cap \mathcal{I}(r^*)$ and $\Pi_d(D_P) \cap \mathcal{I}(r^*)$, respectively. Furthermore, the conditions in Theorem 10.6 implicitly require the existence of values u_c^* and u_d^* from the sets $\mathcal{T}_c$ and $\mathcal{T}_d$, respectively, but such existence does not come for free from just having a CLF. When the functions

$$u_c \mapsto \Gamma_c(z, u_c, V(z)), \qquad u_d \mapsto \Gamma_d(z, u_d, V(z))$$

are convex and the set-valued maps Ψ_c^u and Ψ_d^u are inner semicontinuous with nonempty closed convex values, the set-valued maps $\mathcal{T}_c$ and $\mathcal{T}_d$ have nonempty convex closed values and unique elements with minimum norm; that is, the state-feedback laws

$$\kappa_c(z) := \arg \min_{u_c} \left\{ |u_c| : u_c \in \mathcal{T}_c(z) \right\} \tag{10.29}$$

$$\kappa_d(z) := \arg \min_{u_d} \left\{ |u_d| : u_d \in \mathcal{T}_d(z) \right\} \tag{10.30}$$

are well defined, on $\Pi_c(C_P) \cap \mathcal{I}(r^*)$ and $\Pi_d(D_P) \cap \mathcal{I}(r^*)$, respectively.

These conditions are collected in the next assumption.

Assumption 10.7 (Conditions for CLF-based control with continuity). *Given a hybrid plant $\mathcal{H}_P = (C_P, F_P, D_P, G_P, \mathrm{Id})$ as in (10.3), a closed set $\mathcal{A}^* \subset \mathbb{R}^{n_P}$, $r^* \geq 0$ and a control Lyapunov function V for $\mathcal{H}_P$ with respect to $(\mathcal{A}^*, r^*)$, suppose*

(CLF-A1) The set-valued maps Ψ_c^u and Ψ_d^u are inner semicontinuous with convex values.[3]

(CLF-A2) For $r > r^*$, the function $u_c \mapsto \Gamma_c(z, u_c, r)$ is convex on $\Psi_c^u(z)$ for each $z \in \Pi_c(C_P) \cap \mathcal{I}(r)$ and the function $u_d \mapsto \Gamma_d(z, u_d, r)$ is convex on $\Psi_d^u(z)$ for each $z \in \Pi_d(D_P) \cap \mathcal{I}(r)$.

The following result provides an explicit construction of a practically stabilizing state-feedback law that is continuous on the desired sets.

---

[3]See Definition A.36.

Theorem 10.8 (CLF-based control with continuity and minimum norm). *Given a hybrid plant $\mathcal{H}_P = (C_P, F_P, D_P, G_P, Id)$ as in (10.3) satisfying the hybrid basic conditions, a closed set $\mathcal{A}^* \subset \mathbb{R}^{n_P}$, $r^* \geq 0$, and a control Lyapunov function V for $\mathcal{H}_P$ with respect to $(\mathcal{A}^*, r^*)$, suppose that Assumption 10.7 is satisfied and let $r > r^*$ be given in (CLF-A2). The following hold:*

1. *The hybrid closed-loop system resulting from using the state-feedback law pair (κ_c, κ_d) restricted to $\mathcal{I}(r)$, namely, $\mathcal{H}_r = (C \cap \mathcal{I}(r), F, D \cap \mathcal{I}(r), G)$, where (C, F, D, G) is given in (2.24), with κ_c defined in (10.29) for each $z \in \Pi_c(C_P) \cap \mathcal{I}(r)$ and κ_d defined in (10.30) for each $z \in \Pi_d(D_P) \cap \mathcal{I}(r)$, has the set $\mathcal{A}$ in (10.26) uniformly globally pre-asymptotically stable. Furthermore, if the set-valued maps Ψ_c^u and Ψ_d^u are outer semicontinuous[4] then κ_c and κ_d are continuous and $\mathcal{H}_r$ satisfies the hybrid basic conditions.*

2. *If (10.9) holds for each $z \in \Pi_d(D_P)$ then the set $\mathcal{A}$ in (10.26) with r as in item 1 is uniformly globally pre-asymptotically stable for the resulting hybrid closed-loop system $\mathcal{H} = (C, F, D, G)$ given in (2.24) with κ_c defined in (10.29) for each $z \in \Pi_c(C_P) \cap \mathcal{I}(r)$ and κ_d defined in (10.30) for each $z \in \Pi_d(D_P) \cap \mathcal{I}(r)$, that, respectively, are extended to $\Pi_c(C_P)$ arbitrarily and to $\Pi_d(D_P)$ as in (10.25).*

Proof. The given hybrid plant $\mathcal{H}_P$, the pair (V, r^*), the constant $r > r^*$, and the functions ρ_c and ρ_d satisfy

$$\inf_{u_c \in \Psi_c^u(z)} \dot{V}(z, u_c) + \sigma \rho_c(|z|_{\mathcal{A}^*}) \leq -(1-\sigma)\rho_c(|z|_{\mathcal{A}^*}) \quad \forall z \in \Pi_c(C_P) \cap \mathcal{I}(r) \tag{10.31}$$

$$\inf_{u_d \in \Psi_d^u(z)} \Delta V(z, u_d) + \sigma \rho_d(|z|_{\mathcal{A}^*}) \leq -(1-\sigma)\rho_d(|z|_{\mathcal{A}^*}) \quad \forall z \in \Pi_d(D_P) \cap \mathcal{I}(r) \tag{10.32}$$

From these properties, since $r > 0$, the functions Γ_c and Γ_d defined in (10.18) and (10.19), respectively, satisfy

$$\inf_{u_c \in \Psi_c^u(z)} \Gamma_c(z, u_c, V(z)) < 0 \quad \forall z \in \Pi_c(C_P) \cap \mathcal{I}(r)$$

$$\inf_{u_d \in \Psi_d^u(z)} \Gamma_d(z, u_c, V(z)) < 0 \quad \forall z \in \Pi_d(D_P) \cap \mathcal{I}(r)$$

Furthermore, the following properties hold:

1. The functions Γ_c and Γ_d are upper semicontinuous. In fact, since V is continuous and locally Lipschitz on a neighborhood of $\Pi_c(C_P)$, and since F_P and G_P satisfy (A2$_P$) and (A3$_P$), respectively, using Lemma A.24, $(z, u) \mapsto \dot{V}(z, u)$ and $(z, u) \mapsto \Delta V(z, u)$ are upper semicontinuous. Then, with the functions ρ_c and ρ_d being continuous, upper semicontinuity of Γ_c and Γ_d is established.

2. The sets $C_P \cap (\mathcal{I}(r) \times \mathbb{R}^{m_P})$ and $D_P \cap (\mathcal{I}(r) \times \mathbb{R}^{m_P})$ are closed. This property follows since C_P and D_P are closed by (A1$_P$) in Definition 2.25 and V is continuous by Definition 10.1.

---

[4]Since the set-valued maps Ψ_c^u and Ψ_d^u are already required to be inner semicontinuous (via (CLF-A1)), as a whole, these maps are required to be continuous.

3. The set-valued maps Ψ_c^u and Ψ_d^u have nonempty closed convex values on $\Pi_c(C_P) \cap \mathcal{I}(r)$ and $\Pi_d(D_P) \cap \mathcal{I}(r)$, respectively. This property follows from the closedness property of C_P and D_P along with item (CLF-A1) of Assumption 10.7; see Exercise 100.

4. The functions $u_c \mapsto \Gamma_c(z, u_c, V(z))$ and $u_d \mapsto \Gamma_c(z, u_c, V(z))$ defined in (10.18) and (10.19) are convex on $\Pi_c(C_P) \cap \mathcal{I}(r)$ and $\Pi_d(D_P) \cap \mathcal{I}(r)$, respectively. This property follows directly from (CLF-A2).

The set-valued maps $\mathcal{T}_c$ and $\mathcal{T}_d$ in (10.20) and (10.22), respectively, are lower semicontinuous with nonempty closed convex values on $\Pi_c(C_P) \cap \mathcal{I}(r)$ and on $\Pi_d(D_P) \cap \mathcal{I}(r)$, respectively. To establish this property, define

$$\mathcal{T}_c^\circ(z) := \{u_c \in \Psi_c^u(z) : \Gamma_c(z, u_c, V(z)) < 0\} \qquad \forall z \in \Pi_c(C_P) \cap \mathcal{I}(r) \qquad (10.33)$$

Recall that Ψ_c^u is inner semicontinuous and has nonempty closed convex values on $\Pi_c(C_P) \cap \mathcal{I}(r)$, and $(z, u_c) \mapsto \Gamma_c(z, u_c, V(z))$ is upper semicontinuous. Then, using Corollary A.38 with $W = \Psi_c^u$ and $w = \Gamma_c(\cdot, \cdot, V(\cdot))$, the set-valued map $\mathcal{T}_c^\circ$ is inner semicontinuous on $\Pi_c(C_P) \cap \mathcal{I}(r)$. A definition of and inner semicontinuity of $\mathcal{T}_d^\circ$ on $\Pi_d(D_P) \cap \mathcal{I}(r)$ follows similarly. The last sentence of Corollary A.38 establishes the inner semicontinuity property of $\mathcal{T}_c$ and $\mathcal{T}_d$. Moreover, via Proposition A.40, $\mathcal{T}_c$ and $\mathcal{T}_d$ have unique elements of minimum norm, and their minimal selections κ_c and κ_d are, by definition, given by (10.29) and (10.30) on $\Pi_c(C_P) \cap \mathcal{I}(r)$ and on $\Pi_d(D_P) \cap \mathcal{I}(r)$, respectively. Now, note that by construction, these selections satisfy

$$\kappa_c(z) \in \Psi_c^u(z), \quad \Gamma_c(z, \kappa_c(z), V(z)) \le 0 \qquad \forall z \in \Pi_c(C_P) \cap \mathcal{I}(r)$$

$$\kappa_d(z) \in \Psi_d^u(z), \quad \Gamma_d(z, \kappa_d(z), V(z)) \le 0 \qquad \forall z \in \Pi_d(D_P) \cap \mathcal{I}(r)$$

Using the definitions of Ψ_c^u, Ψ_d^u and (10.31)-(10.32),

$$\dot{V}(z, \kappa_c(z)) \le -\sigma \rho_c(|z|_{\mathcal{A}^*}) \quad \forall z \in \mathcal{I}(r) : (z, \kappa_c(z)) \in C_P \qquad (10.34)$$

$$\Delta V(z, \kappa_d(z)) \le -\sigma \rho_d(|z|_{\mathcal{A}^*}) \quad \forall z \in \mathcal{I}(r) : (z, \kappa_d(z)) \in D_P \qquad (10.35)$$

Then, the state-feedback pair (κ_c, κ_d) renders the closed set $\mathcal{A}$ is uniformly (locally) pre-asymptotically stable for $\mathcal{H}_r$. Establishing this property follows as in the proof of Theorem 10.6.

If Ψ_c^u and Ψ_d^u are outer semicontinuous, then they have closed graphs. It follows that the graphs of $\mathcal{T}_c$ and $\mathcal{T}_d$ are closed since, for $\star = c, d$,

$$\mathrm{graph}(\mathcal{T}_\star) = \mathrm{graph}(\Psi_\star^u) \cap \mathrm{graph}(\{u_\star \in \mathbb{R}^{m_P} : \Gamma_\star(\cdot, u_\star, V(\cdot)) \le 0\})$$

where the first graph is closed by assumption while the second one is closed by closedness of $\mathbb{R}^{m_P}$ and the continuity properties of $\Gamma_\star$ and V. Then, according to Proposition A.40, the feedback laws κ_c and κ_d in (10.29) and (10.30), respectively, are minimal selections that are continuous on $\Pi_c(C_P) \cap \mathcal{I}(r)$ and on $\Pi_d(D_P) \cap \mathcal{I}(r)$, respectively. The fact that $\mathcal{H}_P$ satisfies (A1$_P$)-(A3$_P$) in Definition 2.25, combined with continuity of these feedback laws implies, via Lemma 2.21, that the closed loop $\mathcal{H}_r$ satisfies the hybrid basic conditions. The proof of the last item follows as in the proof of Theorem 10.6. $\qquad \square$

Theorem 10.8 assures pre-asymptotic stability of $\mathcal{A}$ for chosen $r > r^*$, which comes from item (CLF-A2) in Assumption 10.7. Note that on $\mathcal{I}(r)$, the state-feedback law proposed therein is continuous. To achieve pre-asymptotic stability with $r = 0$ via a continuous feedback pair, in which case $\mathcal{A}$ reduces to $\mathcal{A}^*$, extra conditions are required to hold nearby the set. In the literature of continuous-time systems, such conditions are referred to as *small control properties*. In simple terms, to obtain a state-feedback pair that is not only continuous for $r > r^*$ (as in Theorem 10.8), but also for $r = r^*$ when $r^* = 0$, a continuous extension of the feedback pair to the set $\mathcal{A}^*$ has to be possible. The following assumptions guarantee such extension.

Assumption 10.9 (Conditions for CLF-based control on $\mathcal{A}^*$). *Given a hybrid plant $\mathcal{H}_P = (C_P, F_P, D_P, G_P, Id)$ as in (10.3), a closed set $\mathcal{A}^* \subset \mathbb{R}^{n_P}$, $r^* \geq 0$ and a control Lyapunov function V for $\mathcal{H}_P$ with respect to $(\mathcal{A}^*, r^*)$, suppose there exist continuous functions $\kappa_{c,0}$ and $\kappa_{d,0}$ such that*

(CLF-A3) Every maximal solution $t \mapsto z(t, 0)$ to

$$(z, \kappa_{c,0}(z)) \in C_P \qquad \dot{z} \in F_P(z, \kappa_{c,0}(z))$$

with $z(0, 0) \in \mathcal{A}^*$ satisfies $z(t, 0) \in \mathcal{A}^*$ for all $(t, 0) \in \operatorname{dom} z$.

(CLF-A4) Every maximal solution $j \mapsto z(0, j)$ to

$$(z, \kappa_{d,0}(z)) \in D_P \qquad z^+ \in G_P(z, \kappa_{d,0}(z))$$

with $z(0, 0) \in \mathcal{A}^*$ satisfies $z(0, j) \in \mathcal{A}^*$ for all $(0, j) \in \operatorname{dom} z$.

(CLF-A5) The set-valued map $\mathcal{T}_c'$ defined for each $z \in \mathbb{R}^{n_P}$ as

$$\mathcal{T}_c'(z) := \begin{cases} \mathcal{T}_c(z) & \text{if } z \in \Pi_c(C_P) \setminus \mathcal{A}^* \\ \kappa_{c,0}(z) & \text{if } z \in \Pi_c(C_P) \cap \mathcal{A}^* \\ \mathbb{R}^{m_P} & \text{otherwise} \end{cases} \qquad (10.36)$$

is inner semicontinuous at each $z \in \Pi_c(C_P) \cap \mathcal{A}^*$.

(CLF-A6) The set-valued map $\mathcal{T}_d'$ defined for each $z \in \mathbb{R}^{n_P}$ as

$$\mathcal{T}_d'(z) := \begin{cases} \mathcal{T}_d(z) & \text{if } z \in \Pi_d(D_P) \setminus \mathcal{A}^* \\ \kappa_{d,0}(z) & \text{if } z \in \Pi_d(D_P) \cap \mathcal{A}^* \\ \mathbb{R}^{m_P} & \text{otherwise} \end{cases} \qquad (10.37)$$

is inner semicontinuous at each $z \in \Pi_d(D_P) \cap \mathcal{A}^*$.

Items (CLF-A3) and (CLF-A4) guarantee that the state-feedback laws $\kappa_{c,0}$ and $\kappa_{d,0}$ induce forward (pre-)invariance of $\mathcal{A}^*$. Under the conditions in Assumption 10.7, the set-valued maps $\mathcal{T}_c$ and $\mathcal{T}_d$ are lower semicontinuous for every z such that $V(z) > 0$; that is, for points z in $\mathcal{I}(r)$ with $r > 0$. To be able to make continuous selections at points in $\mathcal{A}^*$, these maps are further required to be inner semicontinuous on $\mathcal{A}^*$ (for points z such that $V(z) = r$ with $r = 0$) in items (CLF-A5) and (CLF-A6). These conditions resemble those already in the literature of continuous-time systems. Below, $X_P := \Pi_c(C_P) \cup \Pi_d(D_P) \cup G_P(D_P)$.

Theorem 10.10 (Global CLF-based control with continuity and minimum norm). *Given a hybrid plant $\mathcal{H}_P = (C_P, F_P, D_P, G_P, Id)$ as in (10.3) satisfying the hybrid basic conditions, a closed set $\mathcal{A}^* \subset \mathbb{R}^{n_P}$, and a control Lyapunov function V for $\mathcal{H}_P$ with respect to $(\mathcal{A}^*, r^*)$ with $r^* = 0$, suppose that Assumption 10.7 and Assumption 10.9 hold. The following hold:*

1. *The hybrid closed-loop system resulting from using the state-feedback law pair (κ_c, κ_d) – namely, $\mathcal{H}$ as in (2.24) – with κ_c defined in (10.29) for each $z \in \Pi_c(C_P)$ with $\mathcal{T}_c$ replaced by $\mathcal{T}_c'$ and with κ_d defined in (10.30) for each $z \in \Pi_d(D_P)$ with $\mathcal{T}_d$ replaced by $\mathcal{T}_d'$, has the set $\mathcal{A} = \mathcal{A}^*$ uniformly globally preasymptotically stable.*

2. *If the set-valued maps Ψ_c^u and Ψ_d^u are outer semicontinuous, and $\kappa_{c,0}$ and $\kappa_{d,0}$ are such that $\kappa_{c,0}(\mathcal{A}^*) = \kappa_{d,0}(\mathcal{A}^*) = \{0\}$, then κ_c and κ_d as defined in item 1 are continuous and $\mathcal{H}$ therein satisfies the hybrid basic conditions. Furthermore, when $\mathcal{A}^*$ is compact, the uniform global pre-asymptotic stability property of $\mathcal{A}$ in item 1 is robust in the sense of Definition 3.16.*

Proof. Following similar steps as in the proof of Theorem 10.8, using items (CLF-A5) and (CLF-A6) in Assumption 10.9, and Lemma A.37, $\mathcal{T}_c'$ and $\mathcal{T}_d'$ are inner semicontinuous with nonempty closed values on $\Pi_c(C_P)$ and $\Pi_c(D_P)$, respectively. In fact, from the proof of Theorem 10.8, $\mathcal{T}_c'$ and $\mathcal{T}_d'$ are inner semicontinuous with nonempty closed values on $\Pi_c(C_P) \setminus \mathcal{A}^*$ and $\Pi_c(D_P) \setminus \mathcal{A}^*$, respectively. Then, by (CLF-A5) and (CLF-A6), the same property for $\mathcal{T}_c'$ and $\mathcal{T}_d'$ holds on $\Pi_c(C_P)$ and $\Pi_c(D_P)$, respectively. Inner semicontinuity of $\mathcal{T}_c'$ on $\mathbb{R}^{n_P}$ follows from Lemma A.37 with F_1, F_2, and K therein given by $F_1 = \mathcal{T}_c'$ on $\Pi_c(C_P)$, $K = \Pi_c(C_P)$, and $F_2 = \mathbb{R}^{m_P}$ – inner semicontinuity of $\mathcal{T}_d'$ on $\mathbb{R}^{n_P}$ follows similarly. Then, according to Proposition A.40, $\mathcal{T}_c'$ and $\mathcal{T}_d'$ have unique elements of minimum norm, and their minimal selections are κ_c and κ_d on $\Pi_c(C_P)$ and $\Pi_d(D_P)$, respectively. The feedbacks κ_c and κ_d are defined in (10.29) for each $z \in \Pi_c(C_P)$ with $\mathcal{T}_c$ replaced by $\mathcal{T}_c'$ and in (10.30) for each $z \in \Pi_d(D_P)$ with $\mathcal{T}_d$ replaced by $\mathcal{T}_d'$, respectively. Then, (10.34) and (10.35) hold for $r = 0$: from the proof of Theorem 10.8, these conditions hold for each $r > 0$, while for $r = 0$, they hold due to (CLF-A3) and (CLF-A4). Using similar steps as those in the proof of Theorem 10.8, pre-asymptotic stability of $\mathcal{A}^*$ for the closed-loop system $\mathcal{H}$ follows.

When the set-valued maps Ψ_c^u and Ψ_d^u are outer semicontinuous, as established in the proof of Theorem 10.8, κ_c and κ_d are continuous on $\Pi_c(C_P) \setminus \mathcal{A}^*$ and on $\Pi_d(D_P) \setminus \mathcal{A}^*$, respectively. Now, since $(\kappa_{c,0}, \kappa_{d,0})(\mathcal{A}^*) = 0$, Theorem A.41 implies that there exists a continuous feedback pair $(\widetilde{\kappa}_c, \widetilde{\kappa}_d)$ (not necessarily of pointwise minimum norm) pre-asymptotically stabilizing the closed set $\mathcal{A}^*$ and with the property $(\widetilde{\kappa}_c, \widetilde{\kappa}_d)(\mathcal{A}^*) = \{0\}$ – the pair $(\widetilde{\kappa}_c, \widetilde{\kappa}_d)$ vanishes on $\mathcal{A}^*$ due to the fact that the only possible selection for $r = 0$ is the pair $(\kappa_{c,0}, \kappa_{d,0})$, which, by assumption, vanishes at such points. Since κ_c and κ_d have pointwise minimum norm, we have

$$0 \leq |\kappa_c(z)| \leq |\widetilde{\kappa}_c(z)| \qquad \forall z \in \Pi_c(C_P) \tag{10.38}$$

$$0 \leq |\kappa_d(z)| \leq |\widetilde{\kappa}_d(z)| \qquad \forall z \in \Pi_d(D_P) \tag{10.39}$$

Then, since $\widetilde{\kappa}_c$ and $\widetilde{\kappa}_d$ are continuous and vanish at points in $\mathcal{A}^*$, the feedbacks κ_c and κ_d are continuous on $\Pi_c(C_P)$ and $\Pi_d(D_P)$, respectively. $\qquad\square$

Remark 10.11 (Special cases). Theorem 10.8 and Theorem 10.10 do not explicitly require both u_c and u_d to affect the continuous and discrete dynamics of $\mathcal{H}_P$. In fact, when only one of the inputs is effectively present in the plant, the results therein apply with obvious modifications. On the other hand, the case when u_c and u_d are the same physical signal (or they have common components) requires the design of a feedback law that is uniquely defined at points z that belong to both the flow and jump sets. In the extreme case that the input for flows and jumps are the same, i.e., $u_c = u_d =: u$, to be able to synthesize a stabilizing state-feedback law, common input values u such that $(z, u) \in C_P$ and $(z, u) \in D_P$ need to exist. More precisely, it is required that

$$\mathcal{T}'_c(z) \cap \mathcal{T}'_d(z) \neq \emptyset \qquad \forall z \in \Pi_c(C_P) \cap \Pi_d(D_P) \cap \mathcal{I}(r) \tag{10.40}$$

for the desired $r \geq r^*$. A result paralleling Theorem 10.10 follows using

$$\mathcal{T}'(z) := \begin{cases} \mathcal{T}'_c(z) & \text{if } z \in (\Pi_c(C_P) \setminus \Pi_d(D_P)) \cap \mathcal{I}(r) \\ \mathcal{T}'_c(z) \cap \mathcal{T}'_d(z) & \text{if } z \in \Pi_c(C_P) \cap \Pi_d(D_P) \cap \mathcal{I}(r) \\ \mathcal{T}'_d(z) & \text{if } z \in (\Pi_d(D_P) \setminus \Pi_c(C_P)) \cap \mathcal{I}(r) \\ \mathbb{R}^{m_P} & \text{otherwise,} \end{cases}$$

which, when further assuming (10.40), is lower semicontinuous and has nonempty, convex values. $\triangle$

The nominal design results are illustrated in the examples introduced in §10.2.

Example 10.12 (Pendulum with impacts, revisited). *Consider the pendulum system in Example 10.4 but now exhibiting impacts on a controlled slanted surface. As in Example 10.4, the angle of the pendulum is denoted z_1 and its velocity by z_2. In the model considered in this example, when $z_1 \geq \mu$ with μ denoting the angle of the controlled surface, its continuous evolution is given by the flow of the hybrid plant in (10.10). For simplicity, to avoid embedding[5] the angle in the unit circle, it is assumed that $(z_1, z_2) \in [-\pi/2, \pi] \times \mathbb{R} =: X_P$ and $\mu \in [-\pi/2, 0] =: U_2$. Impacts between the pendulum and the surface occur when*

$$z_1 \leq \mu, \quad z_2 \leq 0 \tag{10.41}$$

At such events, the jump map takes the form

$$z_1^+ = z_1 + \widetilde{\rho}(\mu)z_1, \qquad z_2^+ = -\widetilde{\lambda}(\mu)z_2,$$

where the functions $\widetilde{\rho} : U_2 \to (-1, 0)$ and $\widetilde{\lambda} : U_2 \to [0, 1)$ are continuous and capture the effect of pendulum compression and restitution at impacts, respectively, as a function of μ.

The function $\widetilde{\rho}$ captures rapid displacements of the pendulum at collisions while $\widetilde{\lambda}$ models the effect of the angle μ on energy dissipation at impacts. When the surface is in the vertical position ($\mu = 0$), $\widetilde{\rho}$ is chosen such that $\widetilde{\rho}(0) \in (-1, 0)$ and $\widetilde{\lambda}$ is chosen to satisfy $\widetilde{\lambda}(0) = \lambda_0$, where $\lambda_0 \in (0, 1)$ is the nominal (no gravity effect) restitution coefficient. For slanted surfaces ($\mu \in [-\frac{\pi}{2}, 0)$), when conditions (10.41) hold, $\widetilde{\rho}$ is chosen as $z_1 + \widetilde{\rho}(\mu)z_1 > z_1$, $\widetilde{\rho}(\mu) \in (-1, 0)$, so that, after the impacts,

---

[5]An embedding of the angle in the unit circle as in Example 6.1 can also be performed for this system.

the pendulum is pushed away from the contact condition, while the function $\widetilde{\lambda}$ is chosen as a nondecreasing function of μ satisfying $\lambda_0 \leq \widetilde{\lambda}(\mu) < 1$ at such angles so that, due to the effect of the gravity force at impacts, less energy is dissipated as $|\mu|$ increases.

The model above can be captured by the hybrid plant $\mathcal{H}_P$ given by

$$\mathcal{H}_P \ : \ \begin{cases} (z, u_c) \in C_P & \begin{bmatrix} \dot{z}_1 \\ \dot{z}_2 \end{bmatrix} = \begin{bmatrix} z_2 \\ -a\sin z_1 - bz_2 + u_{c,1} \end{bmatrix} =: F_P(z, u_c) \\ (z, u_d) \in D_P & \begin{bmatrix} z_1^+ \\ z_2^+ \end{bmatrix} = \begin{bmatrix} z_1 + \widetilde{\rho}(u_d)z_1 \\ -\widetilde{\lambda}(u_d)z_2 \end{bmatrix} =: G_P(z, u_d) \end{cases} \tag{10.42}$$

where, now, inputs play a role in F_P, C_P, G_P, and D_P, $u_c = (u_{c,1}, u_{c,2}) \in U_1 \times U_2$, $U_1 := \mathbb{R}$, with $u_{c,2} = \mu$, $u_d = \mu \in U_2$,

$$C_P := \{(z, u_c) \in X_P \times U : z_1 \geq u_{c,2}\}$$
$$D_P := \{(z, u_d) \in X_P \times U_2 : z_1 \leq u_d, z_2 \leq 0\}$$

where $U := U_1 \times U_2$. Note that the definitions of C_P and D_P impose state constraints on the inputs.

As in Example 10.4, let $\mathcal{A}^* = \{(0,0)\}$, $r^* = 0$, and consider the candidate control Lyapunov function for $\mathcal{H}_P$ given in (10.11). For each $(z, u) \in C_P$, $\dot{V}$ satisfies (10.12) with u_c replaced by $u_{c,1}$. For each $z \in \mathbb{R}^2$,

$$\Psi_c^u(z) = \begin{cases} \{u_c \in U : z_1 \geq u_{c,2}\} = \mathbb{R} \times [-\frac{\pi}{2}, \min\{z_1, 0\}] & \text{if } z_1 \in [-\frac{\pi}{2}, \pi] \\ \emptyset & \text{if } z_1 \notin [-\frac{\pi}{2}, \pi] \end{cases}$$

Furthermore, $\Pi_c(C_P) = [-\frac{\pi}{2}, \pi] \times \mathbb{R}$. Then

$$\inf_{u_c \in \Psi_c^u(z)} \dot{V}(z, u_c) = -z^\top z$$

for all $z \in \Pi_c(C_P)$ such that $z_1 + z_2 = 0$, while when $z_1 + z_2 \neq 0$,

$$\inf_{u_c \in \Psi_c^u(z)} \dot{V}(z, u_c) = -\infty$$

Then, condition (10.8) holds with $\rho_c(s) := s^2$ for all $s \geq 0$.

For each $(z, u) \in D_P$,

$$V(G_P(z, u_d)) - V(z) = (2(\chi_1)^2 + 2(\chi_1)(\chi_2) + (\chi_2)^2) - V(z) \tag{10.43}$$

where $(\chi_1, \chi_2) = G_P(z, u_d)$. For each $z \in \mathbb{R}^2$,

$$\Psi_d^u(z) = \begin{cases} \{u_d \in U_2 : z_1 \leq u_d\} = [z_1, 0] & \text{if } z_1 \in [-\frac{\pi}{2}, 0], z_2 \leq 0 \\ \emptyset & \text{otherwise} \end{cases}$$

and that $\Pi_d(D_P) = [-\frac{\pi}{2}, 0] \times (-\infty, 0]$. Then, at jumps, we have

$$\inf_{u_d \in \Psi_d^u(z)} V(G_P(z, u_d)) - V(z) = V(G_P(z, z_1)) - V(z)$$

$$\leq -\min\{2(1 - (1 + \widetilde{\rho}(z_1))^2), 1 - \widetilde{\lambda}(z_1)^2\}z^\top z$$

for all $z \in \Pi_c(D_P)$. Then, condition (10.9) is satisfied with $\rho_d(s) := \lambda s^2$ for all $s \geq 0$, with $\lambda := \min_{z_1 \in U_2}\{2(1 - (1 + \widetilde{\rho}(z_1))^2), 1 - \widetilde{\lambda}(z_1)^2\}$.

The definition of Γ_c in (10.18) with $\sigma \in (0, 1)$ gives

$$\Gamma_c(z, u_c, r) = \begin{cases} 4z_1 z_2 + 2z_2^2 + 2(-a \sin z_1 - bz_2 + u_{c,1})(z_2 + z_1) + \sigma\rho_c(|z|_{\mathcal{A}^*}) \\ \qquad\qquad\qquad\qquad\qquad\qquad\qquad\quad if\ (z, u_c) \in C_P, z \in \mathcal{I}(r), \\ -\infty \qquad\qquad\qquad\qquad\qquad\qquad\qquad otherwise \end{cases}$$

Then, for each $r > 0$, and for each $z \in \mathcal{I}(r)$ such that $(z, u_c) \in C_P$, the set-valued map $\mathcal{T}_c$ in (10.20) is given by

$$\mathcal{T}_c(z) = \left\{ u_c \in \mathbb{R} \times \left[-\frac{\pi}{2}, \min\{z_1, 0\} \right] \; : \; 4z_1 z_2 + 2z_2^2 \right.$$
$$\left. + 2(-a \sin z_1 - bz_2 + u_{c,1})(z_2 + z_1) + \sigma z^\top z \leq 0 \right\}$$

Defining $\psi_0(z) := 4z_1 z_2 + 2z_2^2 + 2(-a \sin z_1 - bz_2)(z_2 + z_1) + \sigma z^\top z$, and $\psi_1(z) := 2(z_1 + z_2)$, $\mathcal{T}_c$ can be rewritten as

$$\mathcal{T}_c(z) = \left\{ u_c \in \mathbb{R} \times \left[-\frac{\pi}{2}, \min\{z_1, 0\} \right] \; : \; \psi_0(z) + \psi_1(z)u_{c,1} \leq 0 \right\}$$

for each $z \in \Pi_c(C_P)$. Next, the pointwise minimum norm control selection in (10.29) is obtained. Note that the conditions on $u_{c,1}$ and $u_{c,2}$ are decoupled, which permits to compute the expressions of $\kappa_{c,1}$ and $\kappa_{c,2}$ assigning those inputs independently. Since $u_{c,2}$ is constrained to U_2, the optimal choice of the feedback $\kappa_{c,2}$ is the zero function. To determine $\kappa_{c,1}$, note that at points z such that $\psi_0(z) \leq 0$, the pointwise minimum norm control selection is $\kappa_{c,1}(z) = 0$ and that at z's such that $\psi_0(z) > 0$, by[6]

$$\kappa_{c,1}(z) = -\frac{\psi_0(z)\psi_1(z)}{\psi_1^2(z)} = -\frac{\psi_0(z)}{\psi_1(z)}$$

Then, $\kappa_c(z) = (\kappa_{c,1}(z), \kappa_{c,2}(z))$ is given by

$$\kappa_{c,1}(z) := \begin{cases} -\frac{\psi_0(z)}{\psi_1(z)} & if\ \psi_0(z) > 0 \\ 0 & if\ \psi_0(z) \leq 0 \end{cases} \qquad \kappa_{c,2}(z) := 0$$

on $\Pi_c(C_P)$ such that $V(z) > 0$. Note that there is no division by zero in the construction of $\kappa_{c,1}$ since, when $\psi_1(z) = 0$ the definition of $\mathcal{T}_c$ implies that $\psi_0(z) \leq 0$, in which case, $\rho_{c,1}$ is defined as zero.

Proceeding in the same way, the definition of Γ_d in (10.18) gives

$$\Gamma_d(z, u_d, r) = \begin{cases} -2(1 - (1 + \widetilde{\rho}(u_d))^2)z_1^2 - 2(1 + (1 + \widetilde{\rho}(u_d))\widetilde{\lambda}(u_d))z_1 z_2 \\ \qquad\qquad\qquad\qquad\quad -(1 - \widetilde{\lambda}(u_d)^2)z_2^2 + \sigma\lambda z^\top z \\ \qquad\qquad\qquad\qquad\qquad\qquad if\ (z, u_d) \in D_P, z \in \mathcal{I}(r), \\ -\infty \qquad\qquad\qquad\qquad\qquad\qquad otherwise \end{cases}$$

Then, for each $r > 0$, and for each $(z, u_d) \in D_P$ such that $z \in \mathcal{I}(r)$, the set-valued

---

[6]This choice leads to $\psi_0(z) + \psi_1(z)\kappa_{c,1}(z) = 0$.

map $\mathcal{T}_d$ in (10.22) is given by

$$\mathcal{T}_d(z) = \Big\{ u_d \in [z_1, 0] \ : \ -2(1 - (1 + \widetilde{\rho}(u_d))^2)z_1^2 - 2(1 + (1 + \widetilde{\rho}(u_d))\widetilde{\lambda}(u_d))z_1 z_2$$
$$-(1 - \widetilde{\lambda}(u_d)^2)z_2^2 + \sigma\lambda z^\top z \leq 0 \Big\}$$

According to (10.30), from the expression of $\mathcal{T}_d$, since $\widetilde{\rho}$ maps to $(-1, 0)$ and $\widetilde{\lambda}$ to $(0, 1)$, for each $z \in \Pi_d(D_P)$ such that $V(z) > 0$, the pointwise minimum norm control selection for κ_d is given by the zero function. Since $\kappa_d = \kappa_{c,2}$, the selection above can be implemented.

Figure 10.3(a) depicts a closed-loop trajectory on the plane with the control selections above when the region of operation is restricted to the region of points z such that $V(z) \geq r$ with $r = 0.0015$. Figure 10.3(b) shows the position and velocity trajectories projected on the t axis. The functions $\widetilde{\rho}$ and $\widetilde{\lambda}$ used in the simulations are defined as $\widetilde{\rho}(s) = 0.5s - 0.1$ and $\widetilde{\lambda}(s) = -0.28s + 0.5$ for each $s \in [-\pi/2, 0]$. Associated simulation files are at @BookSite/Simulation/CLFpendulum.

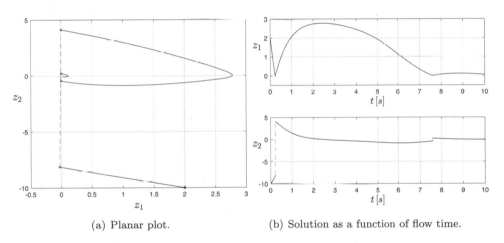

(a) Planar plot. (b) Solution as a function of flow time.

Figure 10.3: Closed-loop solution to the system in Example 10.4 starting from $z(0,0) = (2, -10)$ and evolving on the region $V(z) \geq r$ with $r = 0.0015$.

Example 10.13 (Planar system with jumps, revisited). *Consider the hybrid plant in Example 10.5. To construct a state-feedback law for (10.14), consider the candidate control Lyapunov function V given in (10.15) and compute the map $\mathcal{T}_c$ in (10.20) and $\mathcal{T}_d$ in (10.22). To this end, the definition of Γ_c gives*

$$\Gamma_c(z, u_c, r) = \begin{cases} \dfrac{u_c}{\omega} \begin{bmatrix} \frac{z_2}{|z|^2} & -\frac{z_1}{|z|^2} \end{bmatrix} \begin{bmatrix} 0 & \omega \\ -\omega & 0 \end{bmatrix} zV(z) + \sigma|z|_{\mathcal{A}^*} \\ \qquad\qquad \text{if } (z, u_c) \in C_P, z \in \mathcal{I}(r), \\[2ex] -\infty \qquad\qquad\qquad\qquad\qquad otherwise \end{cases}$$

Then, for each $r > 0$, and for each $z \in \Pi_c(C_P) \cap \mathcal{I}(r)$, the map $\mathcal{T}_c$ is given by[7]

$$\mathcal{T}_c(z) = \{u_c \in \{-1, 1\} : \Gamma_c(z, u_c, V(z)) \leq 0\} = \begin{cases} -1 & \text{if } z_1 > 0 \\ 1 & \text{if } z_1 < 0 \end{cases}$$

Proceeding in the same way, the definition of Γ_d gives

$$\Gamma_d(z, u_d, r) = \begin{cases} \exp\left(\dfrac{\pi}{2\omega}\right)(u_d - \delta) - V(z) + \sigma\left(1 - \exp\left(\dfrac{\pi}{2\omega}\right)\gamma\right)|z|_{\mathcal{A}^*} \\ \qquad \text{if } (z, u_d) \in D_P, z \in \mathcal{I}(r), \\ -\infty \qquad \text{otherwise} \end{cases}$$

Then, for each $r > 0$, and each $z \in \Pi_d(D_P) \cap \mathcal{I}(r)$, the map $\mathcal{T}_d$ is given by

$$\begin{aligned} \mathcal{T}_d(z) &= \left\{u_d : u_d - \delta \geq \gamma|z|_{\mathcal{A}^*}, \exp\left(\frac{\pi}{2\omega}\right)(u_d - \delta) - |z|_{\mathcal{A}^*} \right. \\ &\qquad\qquad \left. +\sigma\left(1 - \exp\left(\frac{\pi}{2\omega}\right)\gamma\right)|z|_{\mathcal{A}^*} \leq 0\right\} \\ &= \left\{u_d : u_d - \delta \geq \gamma|z|_{\mathcal{A}^*}, 0 \leq u_d - \delta \leq \sigma\left(\gamma + \frac{1}{\exp\left(\frac{\pi}{2\omega}\right)}\right)|z|_{\mathcal{A}^*}\right\} \\ &= \gamma|z|_{\mathcal{A}^*} + \delta \end{aligned}$$

where the fact that $\exp(\pi/(2\omega))\gamma < 1$ was used.

Using (10.29), the pointwise minimum norm control selection from $\mathcal{T}_c$ is

$$\kappa_c(z) := \begin{cases} -1 & \text{if } z_1 > 0 \\ 1 & \text{if } z_1 < 0 \end{cases}$$

According to (10.30), the minimum norm feedback law to use at jumps is

$$\kappa_d(z) := \gamma|z|_{\mathcal{A}^*} + \delta$$

Figure 10.4 depicts a closed-loop trajectory with the control selections above when the state z evolves in $V(z) \geq r$. Associated simulation files are available at @BookSite/Simulation/HybridOscillator.

10.3.2 Robust Design

Theorem 10.10 provides a design method of a CLF-based controller that renders a compact set pre-asymptotically stable. In addition, when the set is bounded, the pre-asymptotic stability property is robust in the sense of Definition 3.16. As argued in § 2.3.5, such a robustness property may require the disturbances to be small in size. In this section, the design methodology proposed in § 10.3.1 is extended to the case when the hybrid plant is affected by potentially large disturbances.

Following the model in §2.3.5, a hybrid plant $\mathcal{H}_P$ under the effect of disturbances w can be modeled by appending w to the data of $\mathcal{H}_P$ in (2.7), resulting in $\mathcal{H}_{P,w}$ given, for example, as in (2.43). With some abuse of notation, the model of $\mathcal{H}_{P,w}$

---

[7]Since $r > 0$, then $z \in \mathcal{I}(r)$ implies that z_1 is nonzero.

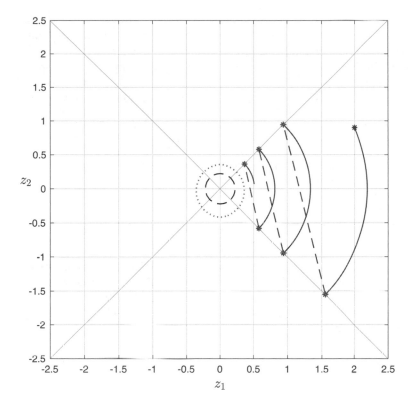

Figure 10.4: Closed-loop trajectory to the system in Example 10.13 starting from $z(0,0) = (2, 0.9)$ evolving within $V(z) \geq r$. The parameters used are $\omega = 5$, $\gamma = 0.6$, $\delta = 0.05$, and $r = 0.15$. The lines at ± 45 deg define the boundary of the flow and jump sets projected onto the z plane. The r-contour plot of V is shown in dotted line and $\mathcal{A}^*$ in dashed line.

considered in this chapter is given by

$$\mathcal{H}_{P,w} \ : \ \begin{cases} (z, u_c, w_c) \in C_P & \dot{z} \in F_P(z, u_c, w_c) \\ (z, u_d, w_d) \in D_P & z^+ \in G_P(z, u_d, w_d) \end{cases} \tag{10.44}$$

where $w_c \in \mathbb{R}^{s_{P_c}}$ denotes the disturbance affecting the flows and $w_d \in \mathbb{R}^{s_{P_d}}$ the disturbance affecting the jumps, with output given by $y = z$. The control Lyapunov function in Definition 10.2 can be seamlessly extended to the case of a hybrid plant with disturbances. When disturbances are present, the change during flow and at jumps of a control Lyapunov function V – which, in § 10.2 is characterized by $\dot{V}$ and ΔV – also depend on the value of the disturbances. For the particular case of a continuously differentiable CLF and single-valued maps F_P and G_P, these quantities are given as

$$\dot{V}(z, u_c, w_c) := \langle \nabla V(z), F_P(z, u_c, w_c) \rangle \qquad \forall (z, u_c, w_c) \in C_P$$

$$\Delta V(z, u_d, w_d) := V(G_P(z, u_d, w_d)) - V(z) \qquad \forall (z, u_d, w_d) \in D_P$$

(Cf. (10.5) and (10.7)).

To be able to synthesize a state-feedback law that guarantees stabilization under the presence of disturbances, an appropriate notion of CLF in such a setting should guarantee the existence (in the infimum sense) of an input value for all possible disturbances, in particular, for the worst case disturbance. For the hybrid plant with disturbances in (10.44), the quantities $\dot{V}$ and ΔV in Definition 10.2 are replaced by

$$\sup_{w_c \in \Psi_c^w(z)} \dot{V}(z, u_c, w_c), \qquad \sup_{w_d \in \Psi_d^w(z)} \Delta V(z, u_d, w_d) \qquad (10.45)$$

respectively, where Ψ_c^w and Ψ_d^w collect all disturbances w_c and w_d allowed for a given z. See List of Symbols.

Following Definition 10.2, a robust control Lyapunov function for the hybrid plant $\mathcal{H}_{P,w}$ as in (10.44) is given as follows.

Definition 10.14 (Robust control Lyapunov function). *Given a closed set $\mathcal{A}^* \subset \mathbb{R}^{n_P}$ and $r^* \geq 0$, a control Lyapunov function candidate V with respect to $\mathcal{A}^*$ for the hybrid plant $\mathcal{H}_P = (C_P, F_P, D_P, G_P, Id)$ as in in (10.3) is a robust control Lyapunov function (RCLF) with respect to $(\mathcal{A}^*, r^*)$ for $\mathcal{H}_P$ if there exist positive definite functions ρ_c and ρ_d such that*

$$\inf_{u_c \in \Psi_c^u(z)} \sup_{w_c \in \Psi_c^w(z)} \dot{V}(z, u_c, w_c) \leq -\rho_c(|z|_{\mathcal{A}^*}) \qquad \forall z \in \Pi_c(C_P) \cap \mathcal{I}(r^*) \ (10.46)$$

$$\inf_{u_d \in \Psi_d^u(z)} \sup_{w_d \in \Psi_d^w(z)} \Delta V(z, u_d, w_d) \leq -\rho_d(|z|_{\mathcal{A}^*}) \qquad \forall z \in \Pi_d(D_P) \cap \mathcal{I}(r^*) \ (10.47)$$

The same notion of control Lyapunov function candidate introduced in Definition 10.1 applies for the case of a hybrid plant with disturbances as in (10.44).

The following example illustrates this definition on the controlled pendulum system with added disturbances.

Example 10.15 (Pendulum with impacts, revisited). *Consider the hybrid plant in Example 10.4 with additive actuator noise, and uncertainty in the friction and restitution coefficients. Such a model is given as $\mathcal{H}_P$ in (10.42) but with these disturbances added, as follows. Denoting these disturbances as w_{c1}, w_{c2}, and w_d, respectively, the resulting hybrid plant is given by*

$$\mathcal{H}_P : \begin{cases} (z, u_c, w_c) \in C_P & \begin{bmatrix} \dot{z}_1 \\ \dot{z}_2 \end{bmatrix} = \begin{bmatrix} z_2 \\ -a \sin z_1 - (b + w_{c2}) z_2 + u_c + w_{c1} \end{bmatrix} \\ & \qquad =: F_P(z, u_c, w_c) \\[2em] (z, w_d) \in D_P & \begin{bmatrix} z_1^+ \\ z_2^+ \end{bmatrix} = \begin{bmatrix} z_1 \\ -(\lambda + w_d) z_2 \end{bmatrix} =: G_P(z, w_d) \end{cases}$$

$$(10.48)$$

where

$$C_P := \left\{ (z, u_c, w_c) \in X_P \times \mathbb{R} \times \mathbb{R}^2 : z_1 \leq 0 \right\}$$

$$D_P := \left\{ (z, w_d) \in X_P \times [\underline{w_d}, \overline{w_d}] : z_1 = 0, z_2 \geq 0 \right\}$$

with $-\lambda \leq \underline{w_d} \leq \overline{w_d} < 1 - \lambda$, $w_c = (w_{c1}, w_{c2})$, and $X_P = [-\pi/2, 0] \times \mathbb{R}$. Now, let $\mathcal{A}^* = \{(0,0)\}$ and consider the control Lyapunov function candidate given in (10.11). For each $(z, u_c, w_c) \in C_P$

$$\langle \nabla V(z), F_P(z, u_c, w_c) \rangle = 4z_1 z_2 + 2z_2^2$$
$$+ 2(-a \sin z_1 - (b + w_{c2})z_2 + u_c + w_{c1})(z_2 + z_1)$$

It follows that (10.46) is satisfied with $\rho_c(s) := s^2$ for all $s \geq 0$. In fact, note that, for each $z \in \mathbb{R}^2$,

$$\Psi_c^u(z) = \begin{cases} \mathbb{R}^2 & \text{if } z \in X_P \\ \emptyset & \text{otherwise} \end{cases} , \quad \Psi_c^w(z) = \begin{cases} \mathbb{R}^2 & \text{if } z \in X_P \\ \emptyset & \text{otherwise} \end{cases}$$

and that $\Pi_c(C_P) = [-\frac{\pi}{2}, 0] \times \mathbb{R}$. Then, it follows that

$$\inf_{u_c \in \Psi_c^u(z)} \sup_{w_c \in \Psi_c^w(z)} \dot{V}(z, u_c, w_c) = -z^\top z$$

for each $z \in \Pi_c(C_P)$ such that $z_1 + z_2 = 0$, while when $z_1 + z_2 \neq 0$, we have

$$\inf_{u_c \in \Psi_c^u(z)} \sup_{w_c \in \Psi_c^w(z)} \dot{V}(z, u_c, w_c) = -\infty$$

For each $z \in \mathbb{R}^2$, we have

$$\Psi_d^w(z) = \begin{cases} [\underline{w_d}, \overline{w_d}] & \text{if } z \in \Pi_d(D_P) \\ \emptyset & \text{otherwise} \end{cases}$$

and that $\Pi_d(D_P) = \{0\} \times (-\infty, 0]$. Then, at each $(z, w_d) \in D_P$,

$$\sup_{w_d \in \Psi_d^w(z)} V(G_P(z, w_d)) - V(z) = -\bar{\lambda} z^\top z$$

with $\bar{\lambda} := 1 - (\lambda + \overline{w_d})^2 > 0$ since, using the property that z_1 is reset (by the identity) to zero and z_2 to $-(\lambda + w_d)z_2$,

$$\sup_{w_d \in \Psi_d^w(z)} V(G_P(z, w_d)) = \max_{w_d \in [\underline{w_d}, \overline{w_d}]} \left[2z_1^2 - 2z_1(\lambda + w_d)z_2 + (\lambda + w_d)^2 z_2^2\right]$$
$$= (\lambda + \overline{w_d})^2 z_2^2$$

Since $\overline{w_d} < 1 - \lambda$, $(\lambda + \overline{w_d})^2 \in [0, 1)$. Then, condition (10.47) is satisfied with $\rho_d(s) := \bar{\lambda} s^2$ for all $s \geq 0$. Then, with $r^* = 0$, V is a RCLF with respect to $(\mathcal{A}^*, r^*)$.

When a RCLF is available for a hybrid plant $\mathcal{H}_{P,w}$ as in (10.44), the design method for the nominal case given in §10.3.1 can be applied by using the quantities in (10.45) instead of $\dot{V}$ and ΔV in (10.6) and (10.7), respectively.[8] In fact, replacing $\dot{V}(z, u_c)$ by $\sup_{w_c \in \Psi_c^w(z)} \dot{V}(z, u_c, w_c)$ in the definition of Γ_c given in (10.18), the set-valued map $\mathcal{T}_c$ from where the state-feedback law κ_c is designed can still be con-

---

[8]For this reason, the constructions in (10.46) and (10.47) use values of u_c and of u_d that are independent of w_c and w_d, respectively. Defining (10.46) and (10.47) using fixed values for (z, w_c) and (z, w_d) in the inf is also possible.

structed as in (10.20). Similarly, replacing $\Delta V(z, u_d)$ by $\sup_{w_d \in \Psi_d^w(z)} \Delta V(z, u_d, w_d)$ in Γ_d given in (10.19), the construction of $\mathcal{T}_d$ in (10.22) can still be employed to design κ_d. This already permits the use of Theorem 10.6 for control design under disturbances. To employ Theorem 10.8 and Theorem 10.10, in particular, so as to guarantee continuity of the feedback pair, Ψ_c^w and Ψ_d^w need to satisfy the following regularity property so that Γ_c and Γ_d are upper semicontinuous:

(CLF-A7) The set-valued maps Ψ_c^w and Ψ_d^w are outer semicontinuous, locally bounded, and nonempty for each $z \in \Pi_c(C_P) \cap \mathcal{I}(r^*)$ and each $z \in \Pi_d(D_P) \cap \mathcal{I}(r^*)$, respectively.

In fact, according to the proof of Theorem 10.8, when Γ_c and Γ_d satisfy (CLF-A2) (with the disturbance-dependent quantities) and (CLF-A7) holds, then upper semicontinuity follows from an application of Lemma A.24.

Using the ideas outlined above, an extension of Theorem 10.8 to the case with disturbances is presented. The hybrid closed-loop system resulting from state-feedback control includes the disturbance $w = (w_c, w_d)$ and is given by

$$
\mathcal{H}_w \; : \; \begin{cases} (x, w_c) \in C & \dot{x} \in F(x, w_c) \\ (x, w_d) \in D & x^+ \in G(x, w_d) \end{cases} \tag{10.49}
$$

where $w_c \in \mathbb{R}^{s_c}$ denotes the disturbance affecting the flows and $w_d \in \mathbb{R}^{s_d}$ the disturbance affecting the jumps; cf. (2.48). The notion of pre-asymptotic stability of a set introduced in Definition 3.1 extended to the hybrid system with disturbances $\mathcal{H}_w$ requires that stability and pre-attractivity hold for all disturbances. In the next result, this property is referred to as w-robust pre-asymptotic stability.

Theorem 10.16 (Robust CLF-based control with continuity and minimum norm). *Given a hybrid plant $\mathcal{H}_{P,w}$ as in (10.44) satisfying, for each w_c and each w_d, (A1$_P$)-(A3$_P$) in Definition 2.25, a closed set $\mathcal{A}^* \subset \mathbb{R}^{n_P}$, $r^* \geq 0$, and a robust control Lyapunov function V for $\mathcal{H}_P$ with respect to $(\mathcal{A}^*, r^*)$, suppose that Assumption 10.7 and (CLF-A7) are satisfied, and let $r > r^*$ be given in (CLF-A2). The following hold:*

1. *The hybrid closed-loop system resulting from using the state-feedback law pair (κ_c, κ_d) restricted to $\mathcal{I}(r)$ – namely, $\mathcal{H}_{w,r} = (C \cap (\mathcal{I}(r) \times \mathbb{R}^{s_{P_c}}), F, D \cap (\mathcal{I}(r) \times \mathbb{R}^{s_{P_d}}), G)$ with (C, F, D, G) being the data of (2.24) including disturbances w_c and w_d – with κ_c defined in (10.29) for each $z \in \Pi_c(C_P) \cap \mathcal{I}(r)$ and κ_d defined in (10.30) for each $z \in \Pi_d(D_P) \cap \mathcal{I}(r)$, has the set $\mathcal{A}$ in (10.26) w-robustly globally pre-asymptotically stable. Furthermore, if the set-valued maps Ψ_c^u and Ψ_d^u are outer semicontinuous then κ_c and κ_d are continuous and $\mathcal{H}_{w,r}$ satisfies, for each w_c and each w_d, the hybrid basic conditions.*

2. *If (10.9) holds for each $z \in \Pi_d(D_P)$ then the closed set $\mathcal{A}$ is w-robustly globally pre-asymptotically stable for the closed-loop system $\mathcal{H} = (C, F, D, G)$ given in (2.24) including disturbances w_c and w_d and with κ_c defined in (10.29) for each $z \in \Pi_c(C_P) \cap \mathcal{I}(r)$ and κ_d defined in (10.30) for each $z \in \Pi_d(D_P) \cap \mathcal{I}(r)$, and, respectively extended to $\Pi_c(C_P)$ arbitrarily and to $\Pi_d(D_P)$ as in (10.25).*

The next example applies the robust CLF-based design in Theorem 10.16 to the problem of controlling a pendulum with impacts, now with disturbances.

Example 10.17 (Controlled pendulum with impacts, revisited). *Consider the model of the controlled pendulum with impacts in Examples 10.15 with disturbances. The functions Γ_c and Γ_d can be directly constructed from the computations therein, leading to*

$$\mathcal{T}_c(x) = \left\{ u_c \in \mathbb{R} \times \left[-\frac{\pi}{2}, \min\{z_1, 0\} \right] \; : \; 4z_1 z_2 + 2z_2^2 + 2(z_2 + z_1) \times \right.$$

$$\left. (-a \sin z_1 - bz_2 + u_c) - \sigma z^\top z - 2(z_2 + z_1) \sup_{w_c \in \Psi_c^w(z)} (-w_{c2} z_2 + w_{c1}) \right\}$$

$$= \left\{ u_c \in \mathbb{R} \times \left[-\frac{\pi}{2}, \min\{z_1, 0\} \right] \; : \; 4z_1 z_2 + 2z_2^2 + 2(z_2 + z_1) \times \right.$$

$$\left. (-a \sin z_1 - bz_2 + u_c) - \sigma z^\top z + 2|z_2 + z_1| \left(\overline{w_{c2}}|z_2| + \overline{w_{c1}} \right) \right\}$$

for each $z \in \Pi_c(C_P)$ such that $V(z) > 0$. To synthesize a state-feedback law with minimum pointwise norm assigning u_c, define $\psi_0^{w_c}(z) := 2|z_2 + z_1| \left(\overline{w_{c2}}|z_2| + \overline{w_{c1}} \right)$ and using the definitions of ψ_0 and ψ_1 in Example 10.12, $\mathcal{T}_c$ can be rewritten as

$$\mathcal{T}_c(z) = \left\{ u_c \in \mathbb{R} \times \left[-\frac{\pi}{2}, \min\{z_1, 0\} \right] \; : \; \psi_0(z) + \psi_0^{w_c}(z) + \psi_1(z)u_c \leq 0 \right\}$$

Proceeding as in Example 10.12 to determine the feedback according to (10.29), note that, when $\psi_0(z) + \psi_0^{w_c}(z) \leq 0$, the pointwise minimum norm control selection is $\kappa_c(z) = 0$ and that, when $\psi_0(z) + \psi_0^w(z) > 0$, is given by

$$\kappa_c(z) = -\frac{(\psi_0(z) + \psi_0^{w_c}(z))\psi_1(z)}{\psi_1^2(z)} = -\frac{\psi_0(z) + \psi_0^{w_c}(z)}{\psi_1(z)}$$

Then, the feedback law is

$$\kappa_c(z) := \begin{cases} -\frac{\psi_0(z) + \psi_0^{w_c}(z)}{\psi_1(z)} & \text{if } \psi_0(z) + \psi_0^{w_c}(z) > 0 \\ 0 & \text{if } \psi_0(z) + \psi_0^{w_c}(z) \leq 0 \end{cases}$$

Figure 10.5 shows closed-loop trajectories using the designed pointwise minimum norm state-feedback law κ_c. The restitution function $\tilde{\lambda}$ used is linear, with lower and upper bounds given by $\frac{1}{3}$ and $\frac{2}{3}$, respectively, and the function $\tilde{\rho}$ is constant and equal to $-\frac{1}{20}$. The simulation results show convergence to nearby the set $\mathcal{A}^* = \{(0,0)\}$, even under the presence of disturbances. For simplicity, the simulations are performed under constant disturbances (w_c, w_d), for different values of w_c and w_d. The disturbances used in the simulations are constant.

The plots in Figure 10.5(a) and Figure 10.5(b) correspond to solutions for different values of w_c and with $w_d = 0$. For each $i \in \{1, 2\}$, $w_{c,i} = 0$, $w_{c,i} = 0.1$, $w_{c,i} = 0.5$; and $w_d \equiv 0$ (all simulations). The velocity component jumps at the impact time and then rapidly gets close to nearby zero. The larger the disturbance, the longer it takes for the solutions to converge. While not being part of the design procedure, the state-feedback law κ_c steers the solutions to the origin from within the flow set. In fact, as the solutions converge to a neighborhood of $\mathcal{A}^*$, they evolve nearby the manifold $z_1 + z_2 = 0$, which leads to large input values.

The plots in Figure 10.5(c) and Figure 10.5(d) correspond to solutions for different values of w_d and with $w_c = 0$. For each $i \in \{1, 2\}$, $w_{c,i} \equiv 0$ (all simulations); $w_d = 0$, $w_d = 0.4$, and $w_d = 0.8$. Since the disturbance w_d is positive and captures

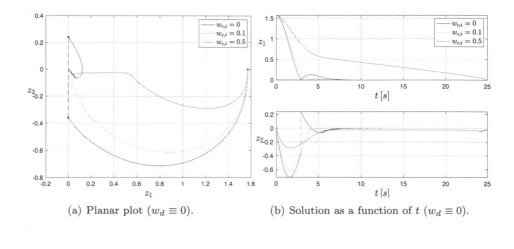

(a) Planar plot ($w_d \equiv 0$).   (b) Solution as a function of t ($w_d \equiv 0$).

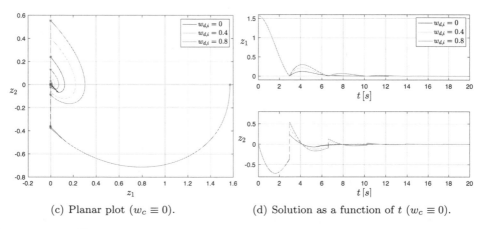

(c) Planar plot ($w_c \equiv 0$).   (d) Solution as a function of t ($w_c \equiv 0$).

Figure 10.5: Closed-loop solutions to the system in Example 10.17 starting from $z(0,0) = (1.5707, 0)$ (marked with $\star$). The $\star$'s after the initial interval of flow in the plot of the solutions denote the values of the solution before and after the jump.

the uncertainty in the restitution coefficient function, large values of the disturbance cause large peaks after every jump as well as more jumps during the transient, when compared to the results in Figure 10.5(b) and Figure 10.5(a). After a few jumps, the solutions converge to a neighborhood of $\mathcal{A}^*$ along the manifold $z_1 + z_2 = 0$.

The simulation files used to generate the plots above are at @BookSite/Simulation/RCLFpendulum.

10.4   EXERCISES

Exercise 68 (Planar hybrid plant with linear flow and jump maps). Consider the planar hybrid plant given by

$$\mathcal{H}_P : \quad \begin{cases} (z, u_c) \in C_P & \dot{z} = A_c z + B_c u_c =: F_P(z, u_c) \\ (z, u_d) \in D_P & z^+ = A_d z + B_d u_d =: G_P(z, u_d) \end{cases}$$

where $C_P := \left\{ (z, u_c) \in \mathbb{R}^2 \times \mathbb{R} : z_1 \geq 0, z_2 \geq 0 \right\}$, $D_P := (D_P^a \cup D_P^b) \times \mathbb{R}$, $D_P^a = \left\{ z \in \mathbb{R}^2 : z_1 = 0, z_2 \geq 0 \right\}$, and $D_P^b = \left\{ z \in \mathbb{R}^2 : z_1 \geq 0, z_2 = 0 \right\}$.

1. Using the control Lyapunov function candidate $V(z) = z^\top P z$ with $P = P^\top > 0$, determine conditions on A_c, B_c, A_d, B_d, and P such that V is a control Lyapunov function with respect to $(\mathcal{A}^*, r^*)$ with $\mathcal{A}^* = \{0\}$ and $r^* = 0$.

2. For each $r > r^*$, design a state-feedback law that globally asymptotically stabilizes the set $\mathcal{A}$ in (10.26) and is such that the resulting hybrid closed-loop system satisfies the hybrid basic conditions.

3. Redesign the controller to globally asymptotically stabilize $\mathcal{A}^*$.

4. Validate your designs numerically.

Exercise 69 (Fully actuated bouncing ball). Consider the one-degree-of-freedom juggling system in Example 9.2 but with $u_d \subset [u_{d\min}, \infty)$, $u_{d\min} > 0$, and dynamics in between impacts given by

$$\dot{z}_1 = z_2, \quad \dot{z}_2 = -\gamma + u_c$$

where $\gamma > 0$ is the gravity constant and $u_c \in \mathbb{R}$ is a control input.

1. Derive a hybrid plant model $\mathcal{H}_P$ capturing its dynamics.

2. Let V be the total energy of the system. Compute $\dot{V}$ and ΔV. Is V a control Lyapunov function with respect to $(\mathcal{A}^*, r^*)$ for $\mathcal{H}_P$ with $\mathcal{A}^* = \{0\}$ and for some $r^* \geq 0$?

3. Design a state-feedback law that globally asymptotically stabilizes $\mathcal{A}$ in (10.26) for the smallest possible $r \geq 0$ such that it has pointwise minimum norm on the region $\mathcal{I}(r)$. Hint: employ the Hybrid Invariance Principle in Theorem 3.23.

4. Validate your design numerically.

5. Building from the design for this particular system, the special *flow* and *jump* passivity notions in Chapter 9, and Remark 10.11, propose a relaxation of the CLF-approach in this chapter for the case when $\dot{V}$ can be made strictly negative during flows and ΔV is nonincreasing at jumps.

Exercise 70 (Point-mass interacting with the environment). Consider the model of a point-mass interacting with a surface in Example 9.1.

1. Derive a hybrid plant model $\mathcal{H}_P$ capturing its dynamics.

2. Let $V(z) = z^\top P z$ with $P = P^\top > 0$. Compute $\dot{V}$ and ΔV. Is there a matrix P such that V is a control Lyapunov function with respect to $(\mathcal{A}^*, r^*)$ for $\mathcal{H}_P$ with $\mathcal{A}^* = \{0\}$ and for some $r^* \geq 0$?

3. Design a state-feedback law that globally asymptotically stabilizes $\mathcal{A}$ in (10.26) for the smallest possible $r \geq 0$ such that it has pointwise minimum norm on the region $\mathcal{I}(r)$. Hint: employ the Hybrid Invariance Principle in Theorem 3.23.

4. Validate your design numerically.

5. Building from the design for this particular system, the special *flow* and *jump* passivity notions in Chapter 9, and Remark 10.11, propose a relaxation of the CLF-approach in this chapter for the case when neither $\dot{V}$ nor ΔV are strictly decreasing, but rather nonincreasing.

Exercise 71 (Pendulum with impacts and disturbances). Consider the pendulum system in Example 10.12 but with additive actuator noise, and uncertainty in the friction and restitution coefficients. Following the model in Example 10.15, the flow and jump maps with such disturbances are given by

$$F_P(z, u_c, w_c) := \begin{bmatrix} z_2 \\ -a \sin z_1 - (b + w_{c2})z_2 + u_{c,1} + w_{c1} \end{bmatrix}$$

$$G_P(z, u_d, w_d) := \begin{bmatrix} z_1 + \widetilde{\rho}(u_d)z_1 \\ -(\widetilde{\lambda}(u_d) + w_d)z_2 \end{bmatrix}$$

and the flow and jump sets remain unchanged. Suppose that $w_c = (w_{c1}, w_{c2}) \in W_c := [0, \overline{w}_1] \times [0, \overline{w}_2]$ with $\overline{w}_1, \overline{w}_2 \in \mathbb{R}_{\geq 0}$, $w_d \in W_d := [0, \lambda_1 - \lambda_0]$, where $0 < \lambda_0 \leq \widetilde{\lambda}(u_d) \leq \lambda_1$ for all $u_d \in U_2$; see Example 10.12.

1. Using the control Lyapunov function candidate V given in (10.11), determine conditions on the parameters $\overline{w}_1$, $\overline{w}_2$, λ_0, and λ_1 such that V is a robust control Lyapunov function with respect to $(\mathcal{A}^*, r^*)$ with $\mathcal{A}^* = \{0\}$ and some $r^* \geq 0$.

2. For each $r > r^*$, design a state-feedback law that globally asymptotically stabilizes the set $\mathcal{A}$ in (10.26) and is such that the resulting closed-loop system satisfies the hybrid basic conditions.

3. Validate your design numerically.

Exercise 72 (Planar system with jumps). Consider the planar system with jumps in Example 10.5 under the effect of the following disturbances:

- Additive actuator disturbance w_{c1} on u_c taking values from the set $W_{c,1} := [\overline{w}_{c1,\min}, \overline{w}_{c1,\max}]$, where $\overline{w}_{c1,\min} \leq \overline{w}_{c1,\max}$.

- Uncertainty on the parameter ω modeled as an additive disturbance w_{c2} taking values from $W_{c,2} := [\overline{w}_{c2,\min}, \overline{w}_{c2,\max}]$, where $\overline{w}_{c2,\min} \leq \overline{w}_{c2,\max}$.

- Additive actuator disturbance w_d on u_d taking values from the set $W_d := [\overline{w}_{d\min}, \overline{w}_{d\max}]$, where $\overline{w}_{d\min} \leq \overline{w}_{d\max}$.

The disturbances are unknown but their bounds are known.

1. Given γ, δ, v_1^1, v_2^1, v_1^2, and v_2^2 as in Example 10.5, use the control Lyapunov function candidate V given in (10.15) to determine conditions on the disturbance parameters $\overline{w}_{c1,\min}$, $\overline{w}_{c1,\max}$, $\overline{w}_{c2,\min}$, $\overline{w}_{c2,\max}$, and $\overline{w}_d$ such that V is a robust control Lyapunov function with respect to $(\mathcal{A}^*, r^*)$ where $\mathcal{A}^* = \{0\}$ and $r^* \geq 0$.

2. For each $r > r^*$, design a state-feedback law that globally asymptotically stabilizes the set $\mathcal{A}$ in (10.26) and is such that the resulting closed-loop system satisfies the hybrid basic conditions.

3. Validate your design numerically.

Exercise 73 (Planar system with jumps with different CLF). Repeat the tasks in Exercise 72 for the system therein using $V(z) = \exp(t_W(z))z^\top z$ and $\delta = 0$ instead.

Exercise 74 (CLF-based control with common input). Show that the construction in Remark 10.11 guarantees that a pre-asymptotically stabilizing state-feedback law can be synthesized, with minimum pointwise norm.

Exercise 75 (Boost converter). Consider the Boost converter in Exercise 7 and the hybrid plant $\mathcal{H}_P$ derived from solving that exercise. Consider the function V defined as

$$V(z) = (z - z^*)^\top P(z - z^*), \quad P = \begin{bmatrix} p_{11} & 0 \\ 0 & p_{22} \end{bmatrix} > 0, \quad z^* = (v_c^*, i_L^*)$$

with $i_L^* = \frac{v_c^*}{R V_{DC}}$, $v_c^* > V_{DC}$.

1. Show that V is a candidate control Lyapunov function with respect to $(\mathcal{A}^*, r^*)$ for $\mathcal{H}_P$ with $\mathcal{A}^* = \{z^*\}$ and $r^* \geq 0$.

2. Following item 5 in Example 69, show that V suffices to design a stabilizing state-feedback law and synthesize it.

10.5 NOTES

Control Lyapunov functions have been shown to be very useful in constructively designing feedback control algorithms [188, 189, 190, 191]. The concept of control Lyapunov function has been extended to different classes of hybrid systems without disturbances. In particular, the article [192] extends the CLF notion to the hybrid inclusions setting and provides conditions guaranteeing the existence of a stabilizing state-feedback law. In [193], a CLF notion is defined for discrete-time systems with continuous and discrete states. Due to the combination of such mixed-valued states, this class of systems is treated as hybrid; see § 2.2 for a discussion about what, according to this book, is a hybrid system (and what is not).

The nominal CLF notion in Definition 10.2 follows those in [192] and [194], which are inspired by those for continuous-time systems in [188, 189, 190, 191]. The construction of the functions Γ_c and Γ_d in (10.18) and (10.19), respectively, and of the maps $\mathcal{T}_c$ and $\mathcal{T}_d$ in (10.20) and (10.22), respectively, follow the ideas in [191] and [195] for continuous-time systems. In those references, the maps $\mathcal{T}_c$ and $\mathcal{T}_d$ are referred to as "regulation maps." The state-feedback law constructed in Theorem 10.6 is not necessarily a continuous function of the state. The literature on CLF-based control for continuous-time systems already indicates that regularity on the data of the system and the CLF is needed to guarantee a continuous selection – such conditions for hybrid equations are given in [192] – leading to the assumptions imposed in Theorem 10.8 and Theorem 10.10, and the preliminary version of the latter result in [196]. The regularity properties imposed in Assumption 10.7 and in Assumption 10.9 are extensions of those in [188, 191, 197] – in particular, the conditions imposed in Assumption 10.9 are referred to as a *small control property* as they guarantee the existence of a feedback nearby the attractor of interest, which in the references mentioned is the origin.

Example 10.5 is a slight variation of Example 2.2 in R. Sanfelice "On the Existence of Control Lyapunov Functions and State-Feedback Laws for Hybrid Systems" in *IEEE Transactions on Automatic Control*, Volume: 58, Issue: 12, pp. 3242-3248, Dec. 2013 [192], which has been reproduced here by permission of IEEE.

The book [191] introduces robust control Lyapunov functions for the design of robustly stabilizing feedback controllers for continuous-time systems. A salient feature of using robust control Lyapunov functions is that, even under the presence of large disturbances, an asymptotic stability of a set, typically defined by a residual neighborhood around the desired equilibrium, can be guaranteed. Following [191], Definition 10.14 provides a robust CLF notion that builds from the construction in [191, Definition 3.8] for continuous-time systems. The design result in Theorem 10.16 provides a CLF-based feedback law that guarantees robust pre-asymptotic stability of a given set. More details are in [198]. It is important to remember that existence of solutions with inputs – in this case, disturbances – is not guaranteed by the constructions of the state-feedback laws proposed in this chapter. As already pointed out in Chapter 9, such existence and completeness of maximal solutions needs to be checked separately, for instance, using Proposition 2.34.

The implementation of the pointwise minimum norm state-feedback law proposed in (10.29)-(10.30) requires the computation of the minimum of a set along the solution to the plant. When a minimizer can be explicitly computed as a function of the state z, as in the examples in this chapter, then its implementation is straightforward. In general, a computationally tractable implementation consists of computing the minimizer frequently enough (say, periodically or upon events) and updating the input to the plant using a zero-order hold strategy. Such an implementation would guarantee the asymptotic stability but only semiglobally and practically, in this way trading the computation demands by degradation of the property guaranteed for the hybrid closed-loop system. A computationally tractable implementation of the feedback in (10.29)-(10.30) that uses memory states and timers is proposed in [199].

Chapter Eleven

Invariants and Invariance-Based Control

Given a dynamical system and a set K, the property that every solution to the system from K stays in K is known as *flow invariance, positive invariance, viability*, or, as referred in this chapter, *forward invariance*. Asymptotically stabilizing feedback control algorithms typically induce this property for the point or set that they stabilize. Forward invariance-type properties are also instrumental in the design of algorithms that guarantee *safety* for a system. For many applications, safety can be recast as a forward invariance property of the set that excludes all points that make the system unsafe. For example, the set K to render forward invariant can be defined as the complement – perhaps after some inflation to account for disturbances – of the set of unsafe sets. Techniques to verify forward invariance, both in nominal conditions and under disturbances are vital for the design of algorithms that not only guarantee convergence of solutions to a desired set but also assure that such convergence occurs without reaching unsafe state values. A challenge in generating analysis and design tools guaranteeing forward invariance is the formulation of sufficient conditions that are not overly conservative and that do not require to solve for the solutions forward in time. In this chapter, tools to certify forward invariance of a set and to design feedback controllers that render a given set K forward invariant are presented. Similar to the conditions to certify pre-asymptotic stability in Chapter 3, the conditions formulated in this chapter are infinitesimal, and depend only on the data of the system, the set to render invariant, and a properly defined certificate for invariance. Figure 11.1 depicts a set being invariant for a hybrid closed-loop system. The nominal case is considered first, and the perturbed case follows it.

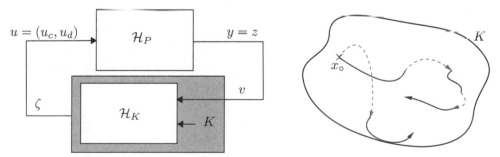

Figure 11.1: Hybrid closed-loop system with invariance-based controller inducing invariance of the set K.

11.1   OVERVIEW

Suppose that given a plant $\mathcal{H}_P$ and a set K the goal is to design a feedback control algorithm that achieves the following property for the hybrid closed-loop system:

For each initial condition in the set K, every solution that starts in K stays in K.

The set K collects the points in the state space of the closed loop where solutions need to remain. Interest in this property arises in many situations, including the following:

1. When a controller is to be designed to guarantee safety, then it has to be the case that there exists a forward invariant set that does not intersect with the set of points considered unsafe. For example, in the context of Example 4.13, suppose the safety set is given by every point in the state space except those in the set $\mathcal{O}$, which represents an obstacle. A hybrid controller that renders the set $K = \mathbb{R}^2 \setminus \mathcal{O}$ forward invariant assures safety.

2. When a set is to be rendered asymptotically stable, then the control algorithm should also guarantee forward invariance of that set. For example, in the context of Example 6.2, the design of a state-feedback controller for a vehicle that renders a neighborhood of a target set asymptotically stable requires the neighborhood itself (or a subset of it) to be forward invariant.

The main idea employed in this chapter to guarantee forward invariance of a set stems from the observations in §3.1, mainly, those pertaining to Lyapunov functions and their change along solutions when the infinitesimal inequalities therein hold. More precisely, given a closed set K to render forward invariant, suppose that there exists a function $B : \mathbb{R}^n \to \mathbb{R}$ that assumes positive values outside of K and nonpositive values on K. In other words, suppose that B and K are such that

$$B(x) > 0 \qquad \forall x \in \mathbb{R}^n \setminus K, \qquad B(x) \leq 0 \qquad \forall x \in K$$

In simple words, the function B is an indicator for the set K: when $B(x)$ is positive, then the state x is not in K, and when $B(x)$ is less than or equal to zero, then the state x is in K. Certainly, the indicator property of B for K alone does not prevent solutions that start from K to leave K. To assure that the set K is forward invariant for the system under study, conditions on the variation of B relative to the system dynamics are required.

To outline conditions guaranteeing forward invariance of a closed set K, consider the autonomous continuous-time system $\dot{x} = f(x)$ with the state x taking values from $\mathbb{R}^n$. It is evident that extra care needs to be paid for solutions $t \mapsto x(t)$ that start right on the boundary of the set K. Certainly, if from such points the direction of the vector field f points outside of K, then there could exist solutions that leave the set K. When K has a nonempty interior, solutions that start from and stay in that interior are benign and do not challenge invariance of the set K. These observations suggest that the following property is crucial:

Solutions from the boundary of K do not leave K.

A direct consequence of this property is the following fact:

> Forward pre-invariance of a closed set K for $\dot{x} = f(x)$ is guaranteed when for each solution $t \mapsto x(t)$ starting from points in ∂K, the function
>
> $$t \mapsto B(x(t))$$
>
> is nonincreasing, where ∂K is the boundary of K.

Indeed, if $t \mapsto B(x(t))$ is allowed to increase, then it would mean that the solution $t \mapsto x(t)$, which starts from a point $x_\circ \in \partial K$ – hence, $B(x_\circ) \leq 0$ – is such that $B(x(t')) > 0$ for some $t' > 0$ in the domain of the solution x. Since, by definition, B is nonpositive on K, then it has to be the case that $x(t') \notin K$. Therefore, the nonincreasing property for $t \mapsto B(x(t))$ is critical. Fortunately, as it should already be clear from the conditions in Theorem 3.19 given in terms of Lyapunov functions, this monotonicity property can be assured without explicitly computing the solutions to the system. In fact, as in Chapter 3, considering points x rather than solutions, and with the functions B and f being smooth enough, the said monotonicity property holds when

$$\dot{B}(x) \leq 0$$

Since the objective is to prevent solutions to flow out of K, this condition should only be enforced at points x nearby the boundary of K. Imposing such a condition for all points $x \in \mathbb{R}^n$ could be restrictive as, in particular, it would require the monotonicity property to hold on the entirety of K. However, as argued above, monotonicity is not really needed at points in the interior of the set K.

Now consider the autonomous discrete-time system $x^+ = g(x)$. In contrast to the continuous-time case discussed above, solutions can now evolve from points in the interior of K to points outside of K, in one step. In fact, if there exists $x_\circ \in K$ such that $g(x_\circ) \notin K$, then there would exist a solution $j \mapsto x(j)$ to $x^+ = g(x)$ with $x(0) = x_\circ$ and $x(1) \notin K$. This observation leads to the following fact:

> Forward invariance of a closed set K for $x^+ = g(x)$ is guaranteed when $B(g(x)) \leq 0$ for each $x \in K$.

As argued for the continuous-time case, if B were to increase, then it would mean that the solution $j \mapsto x(j)$ from $x_\circ \in K$ – hence, $B(x_\circ) \leq 0$ – is such that $B(g(x(j))) > 0$ for some j in the domain of the solution x. Then, by definition of B, it has to be the case that $g(x(j)) \notin K$. Therefore, the solution would leave the set K. Due to this, the natural condition that B, g, and K should satisfy is

$$B(g(x)) \leq 0 \qquad \forall x \in K$$

Note that as a difference to the continuous-time case, this condition needs to hold for all points in K.

Similar arguments to the ones above apply to nonautonomous systems $\dot{x} = f(x, u)$ and $x^+ = g(x, u)$ with their control input u assigned by a controller. This invariance property is typically referred to as *controlled invariance* since it is under the effect of a controller.

Controlled forward pre-invariance of a closed set K for $\dot{x} = f(x, u)$ is guaranteed via a state-feedback law κ_c when, for each closed-loop solution $t \mapsto x(t)$ starting from points in ∂K, the function

$$t \mapsto B(x(t))$$

is nonincreasing. Similarly, controlled forward invariance of a closed set K for $x^+ = g(x, u)$ via a state-feedback law κ_d is guaranteed when $B(g(x, \kappa_d(x))) \leq 0$ for each $x \in K$.

The insight provided in the particular settings above is exploited in this chapter to formulate sufficient conditions for forward invariance of a given closed set. The conditions are in terms of the function B and the data of the hybrid system. In related literature, the function B is referred to as *barrier function, positive function, potential function,* among others. Examples showcasing systems for which forward invariance is important in the context of feedback control already appeared in previous chapters. Example 1.1 – along with Example 1.4 and Example 2.37 – pertain to rendering a set forward invariant for a hybrid closed-loop system with a state that includes a logic variable. Example 4.13 provides an example of a set to be rendered invariant for the purposes of safety, which therein and in this chapter are recast as forward invariance specifications.

11.2   NOMINAL AND ROBUST FORWARD INVARIANCE

11.2.1   Forward Invariance

Given a hybrid closed-loop system $\mathcal{H}$ as in (2.19), the forward invariance notions introduced in Definition 3.13 – namely, *forward pre-invariance* and *forward invariance* – are guaranteed using a barrier function B satisfying conditions similar to those outlined in § 11.1. Inspired by the notion of Lyapunov function candidate in Definition 3.17, the following notion is introduced for the purposes of forward invariance of a set.

Definition 11.1 (Barrier function candidate). *The set $K \subset \mathbb{R}^n$ and the function $B : \operatorname{dom} B \to \mathbb{R}$ define a barrier function candidate with respect to K for the hybrid closed-loop system $\mathcal{H} = \overline{(C, F, D, G)}$ if the following conditions hold:*

1. *$\overline{C} \cup D \cup G(D) \subset \operatorname{dom} B$ and $K \subset C \cup D$;*

2. *B is continuous and, for some open neighborhood $\mathcal{U}$ of ∂K, locally Lipschitz on $(\mathcal{U} \setminus K) \cap \overline{C}$;*

3. *$B(x) > 0$ for all $x \in (\overline{C} \cup D) \backslash K$;*

4. *$B(x) \leq 0$ for all $x \in K$.*

The first property in item 1 of Definition 11.1 assures that B is defined where solutions to $\mathcal{H}$ can evolve. The second property therein ensures that K is a subset of the region of the state space from where flows and jumps are allowed. Item 2 assumes mild regularity on the barrier function. In particular, the local Lipschitz property allows the use of the Clarke generalized gradient to characterize the change of B during flows. Note that local Lipschitzness of B is only required at points in C that are nearby and outside of K. As argued in §11.1 and shown in Theorem 11.4 below, these conditions are instrumental in formulating infinitesimal conditions guaranteeing that the monotonicity property mentioned in §11.1 holds for solutions starting from points in the boundary of K. Items 3 and 4 in Definition 11.1 formalize the indicator property of B for the set K, in turn, establishing a relationship between K, B, and the data (C, F, D, G) of $\mathcal{H}$.[1] Furthermore, these conditions imply that K can be written as

$$K = \{x \in C \cup D : B(x) \leq 0\} \tag{11.1}$$

that is, K is the zero-sublevel set of B restricted to $C \cup D$. Showing this property is left as an exercise to the reader; see Exercise 76.

Figure 11.2 depicts the sets associated with a barrier function candidate. The set K therein (with a solid fill) is the subset of $C \cup D$ on which B is nonpositive. The set $\mathcal{U}$ is an open neighborhood of ∂K. The set of points in $\mathbb{R}^n$ where B is nonpositive and where B is positive are also depicted.

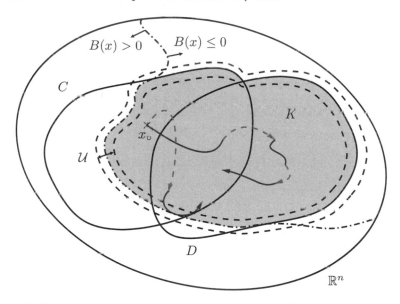

Figure 11.2: Sets associated with a barrier function candidate as in Definition 11.1.

Before getting technical with the conditions on B, K, and the data of $\mathcal{H}$ guaranteeing forward invariance, the definition of barrier function candidate is illustrated in an example.

---

[1]Due to the fact that K, B, and (C, F, D, G) are linked by the conditions in Definition 11.1, "with respect to K" is written when defining the barrier function candidate notion.

Example 11.2 (Forward invariance for a simplified juggling system). *Consider the simplification in Exercise 22 of the one-degree-of-freedom juggling system defined in Example 2.32 (see also Example 2.35). For the case when $u \equiv 0$, the resulting hybrid system is the well-known bouncing ball. Its data (C, F, D, G) is given by*

$$F(x) := \begin{bmatrix} x_2 \\ -\gamma \end{bmatrix} \qquad \forall x \in C := \{ x \in \mathbb{R}^2 : x_1 \geq 0 \} \qquad (11.2)$$

$$G(x) := \begin{bmatrix} x_1 \\ -\lambda x_2 \end{bmatrix} \qquad \forall x \in D := \{ x \in \mathbb{R}^2 : x_1 = 0, x_2 \leq 0 \} \qquad (11.3)$$

where x_1 and x_2 denote the vertical position and velocity of the ball, respectively. The constants $\gamma \geq 0$ and $\lambda \in [0, 1]$ are the gravity constant and the restitution coefficient, respectively. Since the restitution coefficient is no larger than one, the total energy of the system does not increase along solutions and every maximal solution to the system is complete; hence, any sublevel set of the total energy is forward invariant. To establish such a property, note that the total energy of the one-degree-of-freedom juggling system is given by

$$V(x) = \gamma x_1 + \frac{1}{2} x_2^2$$

for each $x \in C \cup D$. With $c \geq 0$ representing the level of energy, the set K is defined as the c-sublevel set of V:

$$K := \{ x \in \mathbb{R}^2 : 2\gamma x_1 + x_2^2 - 2c \leq 0, x_1 \geq 0 \}$$

Then, a barrier function candidate with respect to this set as in Definition 11.1 is given by

$$B(x) := 2\gamma x_1 + x_2^2 - 2c \qquad \forall x \in C \cup D$$

In fact, all of the items in Definition 11.1 hold:

- *Item 1 holds since B is defined on $C \cup D$, $G(D) \subset C$, and, by definition of K, $K \subset C \cup D$;*

- *Item 2 holds since B is continuously differentiable;*

- *Item 3 holds since, due to $B(x) = 2(V(x) - c)$, $B(x)$ is positive for points $x \in (C \cup D) \setminus K$;*

- *Item 4 follows directly from the definition of K and B.*

Later in this chapter, this example is revisited and forward invariance of K is established using appropriate sufficient conditions.

Following the expressions of $\dot{V}$ and ΔV in Definition 3.18, and the discussion right above it, similar quantities are introduced for a barrier function candidate B. For simplicity, and to motivate the need for nonsmooth barrier functions, a definition of $\dot{B}$ for the case when B is continuously differentiable is introduced first. The case when B is locally Lipschitz is treated in Definition 11.13.

Definition 11.3 ($\dot{B}$ and ΔB with continuously differentiable B). *Suppose a hybrid closed-loop system* $\mathcal{H} = (C, F, D, G)$ *satisfies the hybrid basic conditions, and that a set* $K \subset \mathbb{R}^n$ *and a function* $B : \operatorname{dom} B \to \mathbb{R}$ *defining a barrier function candidate with respect to* K *for* $\mathcal{H}$ *are given. Let* $\mathcal{U}$ *be an open neighborhood of* ∂K *as in Definition 11.1 and suppose that* B *is continuously differentiable on* $(\mathcal{U} \setminus K) \cap \overline{C}$.

- *The change of* B *along flows is given by*

$$\dot{B}(x) := \max_{\chi \in F(x) \cap T_C(x)} \langle \nabla B(x), \chi \rangle \qquad \forall x \in (\mathcal{U} \setminus K) \cap C \tag{11.4}$$

Furthermore, if F *is single valued, then* $\dot{B}$ *is given by*

$$\dot{B}(x) := \langle \nabla B(x), F(x) \cap T_C(x) \rangle \qquad \forall x \in (\mathcal{U} \setminus K) \cap C \tag{11.5}$$

- *The largest value of* B *after jumps is given by*

$$B^+(x) := \max_{\chi \in G(x)} B(\chi) \qquad \forall x \in K \cap D \tag{11.6}$$

Furthermore, if G *is single valued, then* B^+ *is given by*

$$B^+(x) := B(G(x)) \qquad \forall x \in K \cap D \tag{11.7}$$

The intersection by T_C in the definition of $\dot{B}$ removes directions in F at points x from which solutions cannot initially flow; see Lemma A.22. For points x in the interior of C, such intersection is equal to F itself.

The following result provides sufficient conditions for pre-forward invariance and forward invariance of a set. It requires that the barrier function candidate is continuously differentiable on a neighborhood of ∂K. This assumption is relaxed later in this chapter.

Theorem 11.4 (Forward pre-invariance and forward invariance using a C^1 scalar barrier function). *Given* $\mathcal{H} = (C, F, D, G)$ *as in (2.19) satisfying the hybrid basic conditions and a closed set* $K \subset \mathbb{R}^n$, *suppose* B *is a barrier function candidate for* $\mathcal{H} = (C, F, D, G)$ *with respect to* K. *Let* $\mathcal{U}$ *be an open neighborhood of* ∂K *as in Definition 11.1 and suppose that* B *is continuously differentiable on an open neighborhood of* $(\mathcal{U} \setminus K) \cap C$.

1. *The set* K *is forward pre-invariant for* $\mathcal{H}$ *if*

$$\dot{B}(x) \leq 0 \qquad \forall x \in (\mathcal{U} \setminus K) \cap C \tag{11.8}$$

$$B^+(x) \leq 0 \qquad \forall x \in K \cap D \tag{11.9}$$

$$G(K \cap D) \subset C \cup D \tag{11.10}$$

When these properties hold, B *is said to be a* <u>*barrier function for forward pre-invariance*</u> *of* K *for* $\mathcal{H}$.

2. *The set K is forward invariant for $\mathcal{H}$ if (11.8)-(11.10) hold, each maximal solution to $\mathcal{H}$ starting from the set $(K \cap \partial C) \setminus D$ is nontrivial, and at least one of the following conditions holds:*

   a) $K \cap C$ is compact;
   b) F has linear growth on $K \cap C$; or
   c) The constrained differential inclusion

   $$\dot{x} \in F(x) \quad x \in K \cap C$$

   does not have a maximal solution with finite escape time; i.e., there is no solution x such that $\lim_{t+j \nearrow \sup \operatorname{dom} x} |x(t,j)| = \infty$.

   When these properties hold, B is said to be a <u>barrier function for forward invariance</u> of K for $\mathcal{H}$.

Proof. Proceeding by contradiction, suppose there exists a maximal solution x starting from $x_\circ \in K$ that leaves the set K. The following cases are possible:

- The solution x leaves K after a jump from K: there exists $(t,j) \in \operatorname{dom} x$ such that $x(t,j) \in K \cap D$ and $x(t, j+1) \notin K$. Using the definition of B, $B(x(t, j+1)) > 0$. On the other hand, from the definition of solution to $\mathcal{H}$ in Definition 2.33, $x(t, j+1) \in G(x(t,j))$. Since (11.9) implies that $B(x(t, j+1)) \leq 0$ and (11.10) implies that $x(t, j+1) \in C \cup D$, then $x(t, j+1) \in K$.

- The solution x leaves the set K by flow from K: there exists $(t', j'), (t'', j') \in \operatorname{dom} x$ such that $x(t,j) \in K$ for all $(t,j) \in \operatorname{dom} x$, $t+j \leq t'+j'$, and $x(t, j') \in (\mathcal{U} \setminus K) \cap C$ for all $(t, j') \in \operatorname{dom} x$, $t' < t \leq t''$. Using continuous differentiability of B and absolute continuity[2] of $t \mapsto x(t, j')$ on $[t', t'']$, $t \mapsto B(x(t, j'))$ is also absolutely continuous on $[t', t'']$ and, via integration, satisfies

$$B(x(t'', j')) - B(x(t', j')) = \int_{t'}^{t''} \langle \nabla B(x(t, j')), \dot{x}(t, j') \rangle dt \tag{11.11}$$

Since $B(x(t, j')) > 0$ for all $t \in [t', t'']$ and $B(x(t', j')) = 0$, the expression in (11.11) is positive. On the other hand, since $x((t', t''], j') \subset (\mathcal{U} \setminus K) \cap C$, using Lemma A.22, $\dot{x}(t, j') \in T_C(x(t, j'))$ for almost all $t \in [t', t'']$. In turn, (11.8) implies that $\langle \nabla B(x(t, j')), \chi \rangle \leq 0$ for all $\chi \in F(x(t', j)) \cap T_C(x(t', j))$ for almost all $t \in (t', t'')$. Hence, via integration again, the expression in (11.11) is less than or equal to zero.

Since a contradiction is reached in both cases, item 1 is established.

Item 2 requires showing that every maximal solution from K is complete. The analysis at jumps in the proof of item 1 above already shows that solutions from K cannot flow or jump to points outside of K. Moreover, since for each $x \in K \setminus C$, $x \in D$ and $G(x) \subset K$, then solutions starting from K cannot end at points $K \cap D$. Then, according to Proposition 2.34, for a maximal solution from K to not be complete it has to either: i') reach a point in K that is in the boundary of C, ii') reach a point in ∂K that is in the interior of C, from where neither flow nor

---

[2]See Definition A.20 and the discussion following it.

jump are possible, or iii') have a finite escape time within the set $K \cap C$. Case i') cannot happen by assumption. Case ii') cannot happen since the point belongs to the interior of C, from where flow is possible. Finally, conditions 2a-2c rule out case iii'). □

Remark 11.5 (About conditions in Theorem 11.4). Condition (11.8) in Theorem 11.4 enforces that each solution x from ∂K that initially flows does not make the barrier function B positive. Note that the size of the neighborhood $\mathcal{U}$ can be arbitrarily small. Figure 11.2 depicts this set. Condition (11.9) assures that every solution that jumps from a point in K does not make B positive. This condition alone does not guarantee that solutions that jump from K remain in K since B could be nonnegative at points not in $C \cup D$ – recall that $K \subset C \cup D$. When, in addition, condition (11.10) holds then solutions that jump from K remain in K. △

Remark 11.6 (About using different barrier functions along flows and at jumps). The results established by Theorem 11.4 hold when, given $\mathcal{H}$ and a closed set K, a barrier function candidate satisfies the flow condition in (11.8) and a different barrier function candidate satisfies the condition at jumps in (11.9). In fact, with a pair of such functions denoted (B_c, B_d), respectively, the invariance properties established therein hold with B in (11.8) replaced by B_c and B in (11.9) replaced by B_d. △

The following example illustrates Theorem 11.4.

Example 11.7 (Invariance for the simplified juggling system, revisited). *Using the barrier function candidate B defined in Example 11.2 for the system therein, forward invariance of K is established using Theorem 11.4. To this end, pick any $\varepsilon > 0$ and define the open neighborhood $\mathcal{U}$ of K of size ε as the inflation of K given by[3]*

$$\mathcal{U} := K + \varepsilon \mathbb{B}^\circ$$

Since C collects all points (x_1, x_2) in $\mathbb{R}^2$ with $x_1 \geq 0$, the set in (11.8) results in

$$(\mathcal{U} \setminus K) \cap C = \left\{ x \in \mathbb{R}^2 : 2\gamma x_1 + x_2^2 - 2c \in (0, 2\varepsilon), x_1 \geq 0 \right\}$$

Note that, for each $x \in C$, the function B and the flow map F satisfy

$$\langle \nabla B(x), F(x) \rangle = \left\langle \begin{bmatrix} 2\gamma \\ 2x_2 \end{bmatrix}, \begin{bmatrix} x_2 \\ -\gamma \end{bmatrix} \right\rangle = 0$$

Hence, (11.8) holds for any neighborhood $\mathcal{U}$ of ∂K, in particular, it holds for the chosen ε. Now, the largest value of B after jumps from $D \cap K$ satisfies

$$B^+(x) = B(G(x)) = 2\gamma x_1 + \lambda^2 x_2^2 - 2c \leq 2\gamma x_1 + x_2^2 - 2c = B(x)$$

Since by definition of B and K, B is nonpositive at points in $D \cap K$, it follows that $B^+(x) \leq 0$. Then, (11.9) in Theorem 11.4 holds. Consequently, by item 1 in Theorem 11.4, the set K is forward pre-invariant for $\mathcal{H}$ with data as in (11.2)-(11.3), and B is a barrier function for forward pre-invariance of K for $\mathcal{H}$. Forward invariance of K for $\mathcal{H}$ follows from Theorem 11.4, after an application of Proposition 2.34.

---

[3]Recall that $\mathbb{B}^\circ$ is the open unit ball centered at the origin; see List of Symbols.

Theorem 11.4 provides conditions for forward invariance for the case that, given a hybrid closed-loop system, a single barrier function defines the set K as in (11.1). At times, it is convenient to use more than one barrier function to establish forward invariance of a set. In this way, one barrier function can assure that solutions do not leave the set from certain points in the boundary of the set to be rendered forward invariant, while other barrier functions can be used to assure that solutions do not leave the set from other boundary points. For instance, consider a planar system and suppose K is the upper half of the unit disk, which is given by

$$K = \left\{x \in \mathbb{R}^2 : |x| \leq 1, x_2 \geq 0\right\} \tag{11.12}$$

Certainly, the distance to this set[4] is a reasonable barrier function candidate. However, such a barrier function would only be locally Lipschitz and Theorem 11.4 would not apply. Though the forthcoming Theorem 11.14 applies, it is possible to construct two barrier functions, one that is only zero on the boundary of the unit disk and negative in its interior, and another one that is only zero on the x_1-axis and negative in the upper half plane. With these two independent barrier functions, the set of points where both of them are simultaneously nonpositive is equal to the set K defined above. In fact, the functions

$$B_1(x) = x_1^2 + x_2^2 - 1, \qquad B_2(x) = -x_2 \tag{11.13}$$

are such that the set of points on which both functions are nonpositive is equal to the set K defined above. In such a case, forward pre-invariance and forward invariance can be established using an extension of Theorem 11.4 to the case of multiple barrier functions.

To establish such an extension, the concept of *multiple barrier function* is introduced next. Similar to the scalar case in Definition 11.1, a vector-valued function denoted B, mapping from $\mathbb{R}^n$ to $\mathbb{R}^N$ with $N \in \mathbb{N} \setminus \{0\}$, defines a multiple barrier function if, for each $i \in \{1, 2, \ldots, N\}$, each of its components B_i is an indicator of a set K_i and satisfies infinitesimal conditions that are similar to those in (11.8)-(11.10), and the set K to render forward invariant is the intersection of all of the sets K_i.

Definition 11.8 (Multiple barrier function candidate)**.** *The set $K \subset \mathbb{R}^n$ and the function $B : \operatorname{dom} B \to \mathbb{R}^N$, $B = (B_1, B_2, \ldots, B_N)$, define a* multiple barrier function candidate *with respect to K for the hybrid closed-loop system $\mathcal{H}$ with data (C, F, D, G) if the following conditions hold:*

1. $\overline{C} \cup D \cup G(D) \subset \operatorname{dom} B$ *and* $K \subset C \cup D$;

2. *For each $i \in \{1, 2, \ldots, N\}$, B_i is continuous and, for some open neighborhood $\mathcal{U}_i$ of M_i, locally Lipschitz on $(\mathcal{U}_i \setminus K_{ei}) \cap \overline{C}$, where*

$$M_i := \{x \in \partial K : B_i(x) = 0\}, \qquad K_{ei} := \{x \in \mathbb{R}^n : B_i(x) \leq 0\} \tag{11.14}$$

3. *For each $i \in \{1, 2, \ldots, N\}$, $B_i(x) > 0$ for all $x \in (\overline{C} \cup D) \setminus K_i$, where*

$$K_i := \{x \in C \cup D : B_i(x) \leq 0\}$$

---

[4]Using the square of the distance is an alternative, which is actually continuously differentiable.

4. *For each $i \in \{1, 2, \ldots, N\}$, $B_i(x) \leq 0$ for all $x \in K_i$;*

5. $K = \displaystyle\bigcap_{i \in \{1,2,\ldots,N\}} K_i.$

When $N = 1$, a multiple barrier function candidate reduces to a scalar barrier function candidate as introduced in Definition 11.1. In fact, in that case, the set K_1 is equal to K and items 3-4 in Definition 11.8 coincide with those in Definition 11.1. The set M_1 is the collection of points x in ∂K such that $B_1(x) = 0$ – note that ∂K includes points that might be in the boundary of $C \cup D$ but not in the boundary of K_{ei}. However, since the requirement in item 2 is only at points in $\overline{C}$, item 2 in Definition 11.8 implies the condition in the respective item in Definition 11.1.

Theorem 11.9 (Forward pre-invariance and forward invariance using a C^1 multiple barrier function). *Given $\mathcal{H} = (C, F, D, G)$ as in (2.19) satisfying the hybrid basic conditions and a closed set $K \subset \mathbb{R}^n$, suppose B is a multiple barrier function candidate for $\mathcal{H}$ with respect to K. For each $i \in \{1, 2, \ldots, N\}$, let $\mathcal{U}_i$ be an open neighborhood of M_i as in Definition 11.8 and suppose that B_i is continuously differentiable on an open neighborhood of $(\mathcal{U}_i \setminus K_{ei}) \cap C$, where M_i and K_{ei} are given in (11.14).*

1. *The set K is forward pre-invariant if, for each $i \in \{1, 2, \ldots, N\}$,*

$$\dot{B}_i(x) \leq 0 \qquad \forall x \in (\mathcal{U}_i \setminus K_{ei}) \cap C \qquad (11.15)$$

$$B_i^+(x) \leq 0 \qquad \forall x \in K_i \cap D \qquad (11.16)$$

$$G(K_i \cap D) \subset C \cup D \qquad (11.17)$$

   When these properties hold, B is said to be a <u>multiple barrier function for forward pre-invariance</u> of K for $\mathcal{H}$.

2. *The set K is forward invariant for $\mathcal{H}$ if (11.15)-(11.17) hold, each maximal solution to $\mathcal{H}$ starting from the set $(K \cap \partial C) \setminus D$ is nontrivial, and at least one of the conditions in items 2a-2c in Theorem 11.4 holds. When these properties hold, B is said to be a <u>multiple barrier function for forward invariance</u> of K for $\mathcal{H}$.*

Proof. The proof follows using similar arguments to those in the proof of Theorem 11.4. Under conditions (11.15)-(11.17), solutions starting from K_i stay in K_i for all time. As a difference to the proof of Theorem 11.4, for the case that a solution x leaves K via a jump, the arguments therein imply the existence of $k \in \{1, 2, \ldots, N\}$ and $(t, j) \in \operatorname{dom} x$ such that $(t, j + 1) \in \operatorname{dom} x$ and $B_k(x(t, j + 1)) > 0$ with $x(t, j + 1) \in G(x(t, j))$. A contradiction follows using the same arguments as in the proof of Theorem 11.4. For the case of a solution x leaving K via flow, the arguments therein also imply the existence of $k \in \{1, 2, \ldots, N\}$ such that $x((t', t''], j') \subset (\mathcal{U}_k \setminus K_{ek}) \cap C$. A contradiction follows also by integration, leading to (11.11) with B replaced by B_k being strictly positive. The rest of the proof follows from the steps of the proof of Theorem 11.4. $\qquad\square$

Following the discussion motivating the use of multiple barrier functions above Definition 11.8, the next example illustrates the use of Theorem 11.9 to establish forward invariance of K as in (11.12).

Example 11.10 (Forward invariance for a linear oscillator with jumps). *Consider the hybrid closed-loop system* $\mathcal{H} = (C, F, D, G)$ *in* $\mathbb{R}^2$ *given by*

$$F(x) := \begin{cases} \begin{bmatrix} x_2 \\ -x_1 \end{bmatrix} & \text{if } x_2 > 0 \\ \text{con} \left\{ \begin{bmatrix} 0 \\ -x_1 \end{bmatrix}, \begin{bmatrix} -x_1 \\ 0 \end{bmatrix} \right\} & \text{if } x_2 = 0 \\ \begin{bmatrix} -x_1 \\ 0 \end{bmatrix} & \text{if } x_2 < 0 \end{cases} \quad \begin{aligned} \forall x \in C &:= \left\{ x \in \mathbb{R}^2 : x_2 \geq 0 \right\} \\ &\cup \left\{ x \in \mathbb{R}^2 : x_1 \leq 0 \right\} \end{aligned}$$

$$G(x) := \begin{bmatrix} [-x_1, x_1] \\ x_2 \end{bmatrix} \qquad\qquad \forall x \in D := \left\{ x \in \mathbb{R}^2 : x_2 = 0 \right\}$$

Following the discussion above Definition 11.8, suppose that the goal is to show that the upper closed half of the unit disk is forward invariant for this system, that is, the set K in (11.12). For this purpose, consider the functions B_1 and B_2 given in (11.13). Theorem 11.9 is applied with $B = (B_1, B_2)$.

First, note that B is a multiple barrier function candidate for $\mathcal{H} = (C, F, D, G)$ with respect to K. In fact, item 1 and item 2 in Definition 11.8 hold $N = 2$ using B_1 and B_2 defined on $\mathbb{R}^2$, and by the fact that $K \subset C$. It is easy to check that B_1 and B_2 have the correct sign on K_1 and K_2, and on their complements, from where item 3 and item 4 follow. Confirming that item 5 therein holds is also immediate. Next, the conditions in Theorem 11.9 are checked.

- *The condition in (11.15) in item 1 holds due to the fact that, for any open neighborhood $\mathcal{U}_1$ of $M_1 = \left\{ x \in \mathbb{R}^2 : |x| = 1 \right\}$, $\dot{B}_1(x) \leq 0$ for all $x \in (\mathcal{U}_1 \setminus K_{e1}) \cap C$, and that for any open neighborhood $\mathcal{U}_2$ of $M_2 = \mathbb{R} \times \{0\}$, $\dot{B}_2(x) \leq 0$ for all $x \in (\mathcal{U}_2 \setminus K_{e2}) \cap C$ – note that the sets K_{e1} and K_{e2} are given as $K_{e1} = \left\{ x \in \mathbb{R}^2 : x_1^2 + x_2^2 \leq 1 \right\}$ and $K_{e2} = \left\{ x \in \mathbb{R}^2 : x_2 \geq 0 \right\}$, respectively.*

- *Condition (11.16) in item 1 holds since jumps are only possible from $x = (x_1, x_2)$ with $x_2 = 0$, at which x_1 can be mapped to a point in $[-x_1, x_1]$. This fact implies that $B_1^+(x) \leq 0$ and $B_2^+(x) = 0$. It also shows that (11.17) holds.*

Since B_1 and B_2 are continuously differentiable, item 1 in Theorem 11.9 implies that K is forward pre-invariant, and B is a barrier function for forward pre-invariance of K for $\mathcal{H}$. Forward invariance of K follow from item 2 therein since $K \cap C$ is compact, and for points x in $K \cap \partial C$, flows are always possible if $x_2 > 0$, while if $x_2 = 0$ then jumps are always possible.

Example 11.10 showcases the advantages of using more than one barrier function to certify invariance of a set for a planar hybrid closed-loop system. The next example further illustrates the advantages of multiple barrier functions for a hybrid closed-loop system with a logic state, which, in particular, can be modeled as a hybrid automaton.

Example 11.11 (Invariance in the thermostat). *Consider the problem of controlling the temperature of a room using a heating device that can only be turned on or off. Denote the temperature of the room by z. Its change over time as a function of the external temperature z_{out} and the heater capacity z_{heat} is given by*

$$\dot{z} = -z + z_{out} + z_{heat}u$$

where the input u can be either equal to zero – meaning that the heater is off – or equal to one – indicating that the heater is on. Suppose that the goal is to design a controller that regulates the temperature z to the range $[z_{min}, z_{max}]$ by turning the heater on and off, as needed, where $z_{min} \leq z_{max}$ are the minimum and maximum desired temperature values.

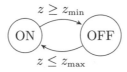

Figure 11.3: Logic proposed to control the temperature of a room.

A hybrid controller that accomplishes such a goal is as follows:

- *While the heater is on, if $z \geq z_{max}$ then turn the heater off to prevent the temperature from going above the desired range;*

- *While the heater is off, if $z \leq z_{min}$ then turn the heater on to prevent the temperature from going below the range.*

Figure 11.3 shows this logic. A hybrid controller $\mathcal{H}_K$ capturing this logic has state $\eta = q \in Q := \{0, 1\}$ and, with the assignments $u = q$ and $v = z$, leads to a hybrid closed-loop system $\mathcal{H}$ with state $x = (z, q) \in \mathbb{R} \times Q =: X$ and data given by

$$C := \bigcup_{q \in Q} (C_q \times \{q\}), \quad F(x) := \begin{bmatrix} -z + z_{out} + z_{heat} q \\ 0 \end{bmatrix} \quad \forall x \in C$$

$$D := \bigcup_{q \in Q} (D_q \times \{q\}), \quad G(x) := \begin{bmatrix} z \\ 1 - q \end{bmatrix} \quad \forall x \in D \tag{11.18}$$

where

$$C_0 := \{z \in \mathbb{R} : z \geq z_{min}\}, \quad C_1 := \{z \in \mathbb{R} : z \leq z_{max}\}$$
$$D_0 := \{z \in \mathbb{R} : z \leq z_{min}\}, \quad D_1 := \{z \in \mathbb{R} : z \geq z_{max}\}$$

For this system, the set to render forward invariant is

$$K := \{x = (z, q) \in C \cup D : z \in [z_{min}, z_{max}]\} \tag{11.19}$$

To certify its invariance, consider the multiple barrier function $B = (B_1, B_2)$ with

$$B_1(x) := z - z_{max}, \qquad B_2(x) = z_{min} - z$$

when the parameters satisfy $z_{out} \leq z_{max}$ and $z_{out} + z_{heat} \geq z_{min}$. These are natural constructions since each one of these functions vanish only on one of the boundaries of K. It is immediate to check that the conditions in Definition 11.8 hold for these functions, the given set K, and the data of $\mathcal{H}$. In particular, K satisfies item 5 therein since $K_1 = \{x \in X : z \leq z_{max}\}$ and $K_2 = \{x \in X : z \geq z_{min}\}$. Hence, $B = (B_1, B_2)$ is a barrier function candidate for forward pre-invariance of K for $\mathcal{H}$.

Next, the conditions in each of the items in Theorem 11.9 are verified.

- *To show that the condition in (11.15) in item 1 holds, note that for this system, $K_{e1} = \{x \in \mathbb{R}^2 : z \leq z_{max}\}$ and $K_{e2} = \{x \in \mathbb{R}^2 : z \geq z_{min}\}$, and that, for each $i \in \{1,2\}$, $M_i = \partial K_{ei} \cap (C \cup D)$. Next, pick open neighborhoods $\mathcal{U}_1$ and $\mathcal{U}_2$ of M_1 and M_2, such that, for some $\varepsilon_1 > 0$ and $\varepsilon_2 > 0$,*

$$(\mathcal{U}_1 \setminus K_{e1}) \cap C = \{x \in \mathbb{R}^2 : z \in (z_{max}, z_{max} + \varepsilon_1), q = 0\}$$

  and

$$(\mathcal{U}_2 \setminus K_{e2}) \cap C = \{x \in \mathbb{R}^2 : z \in (z_{min} - \varepsilon_2, z_{min}), q = 1\}$$

  Since

$$\langle \nabla B_1(x), F(x) \rangle = z_{out} - z$$

  and $z_{out} \leq z_{max}$, this inner product is less than or equal to zero for each x in $(\mathcal{U}_1 \setminus K_{e1}) \cap C$. Similarly,

$$\langle \nabla B_1(x), F(x) \rangle = z - z_{out} - z_{heat}$$

  and $z_{out} + z_{heat} \geq z_{min}$, this inner product is also less than or equal to zero for each x in $(\mathcal{U}_2 \setminus K_{e2}) \cap C$.

- *Condition (11.16) in item 1 holds since B_1 and B_2 depend only on z, which at jumps does not change, and B_1 and B_2 take nonnegative values on K; hence, on $K \cap D$, B_1^+ and B_2^+ are nonpositive.*

- *Condition (11.17) holds since $C \cup D = X$.*

Then, item 1 of Theorem 11.9 implies that K is forward pre-invariant for $\mathcal{H}$, and that B is a multiple barrier function for forward pre-invariance of K for $\mathcal{H}$. Forward invariance of K follows from item 2 of the same result since $K \cap C$ is compact and points x in $K \cap \partial C$ are either in the boundary of C and flow into K is possible, or are in D and a jump resetting q leads to flow afterward. According to the definition near the end of § 11.1, the proposed hybrid controller renders the set K controlled forward invariant.

The following example employs Theorem 11.9 to show that the set K of interest for the DC/AC inverter problem in Example 1.1 is forward invariant.

Example 11.12 (Invariance-based control for the DC/AC inverter). *Consider the controlled single-phase DC/AC inverter shown in Figure 1.3 and introduced in Example 1.1; see also Example 1.4 and Example 2.37. With the properties in Lemma 2.38, the function $B = (B_1, B_2)$ defined as*

$$B_1(x) := V(z) - c_o, \qquad B_2(x) := c_i - V(z)$$

is a multiple barrier function candidate with respect to K for the resulting hybrid closed-loop system in (2.41)-(2.42) in Example 2.37. It is straightforward to check that the set K in (2.40), the multiple barrier function B, and the hybrid closed-loop system $\mathcal{H}$ satisfy the conditions in Definition 11.8. Due to M_1 and M_2 being the boundary of $K_{e1} = \{z \in \mathbb{R}^2 : V(x) \leq c_o\} \times Q$ and of $K_{e2} = \{z \in \mathbb{R}^2 : c_i = V(x)\} \times Q$, respectively, condition (11.15) holds for each $i \in \{1,2\}$. Furthermore, since z

remains constant at jumps and B depends on z only, conditions (11.16) and (11.17) also hold for each $i \in \{1,2\}$. Then, Theorem 11.9 implies forward pre-invariance of K for the closed loop. Forward invariance of K follows from item 2 in Theorem 11.9. Furthermore, according to the definition near the end of § 11.1, the proposed hybrid controller renders the set K controlled forward invariant.

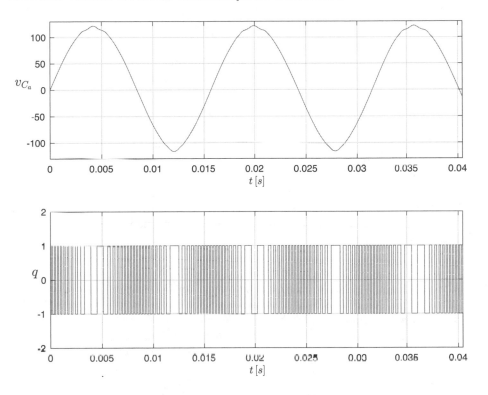

Figure 11.4: The voltage output v_{C_a} and logic variable for a solution to the hybrid closed-loop system in Example 11.12.

Numerical simulations are performed to verify forward invariance of K. The following parameters are used in the simulation: $R = 1\,\Omega$, $L = 0.1\,H$, $C_a = 66.6\,\mu F$, $V_{DC} = 220\,V$, $b = 120\,V$, $\omega = 120\pi$, $c_i = 0.9$, and $c_o = 1.1$. The plot shown in Chapter 1, in Figure 1.5, shows three simulated solutions to the hybrid closed-loop system $\mathcal{H}$ with initial condition for z equal to $(bC_a\omega, 0) = (3.013, 0)$ and initial q as -1, 0, and 1. The three solutions stay within the projection of K onto the $z = (i_L, v_{C_a})$ plane. Figure 11.4 shows that the output voltage v_{C_a} behaves very close to a sinusoidal signal. Associated simulation files are at @BookSite/Simulation/DCAC.

As the examples above illustrate, Theorem 11.9 provides extra flexibility in certifying forward invariance of a set for a hybrid closed-loop system. Indeed, it is a "compositional" type of tool as it permits rendering a set forward invariant by decomposing it into multiple sets for which an independent (scalar) barrier function can be found. Such a decomposition of the set and the construction of the individual barrier functions is typically intuitive and does not add extra difficulty, as the examples show.

A reinterpretation of Definition 11.8 is that a multiple barrier function $B = (B_1, B_2, \ldots, B_N)$ is a certificate for the simultaneous satisfaction of multiple constraints of the form $B_i(x) \leq 0$, where x evolves according to the dynamics of the hybrid closed-loop system $\mathcal{H}$. An alternative way to assure that these constraints are simultaneously satisfied is by defining the scalar function

$$\widetilde{B}(x) := \max_{i \in \{1, 2, \ldots, N\}} B_i(x) \tag{11.20}$$

and then enforcing that the solutions to the system remain in the set of points x satisfying $\widetilde{B}(x) \leq 0$. In fact, when such a property holds, then, along solutions and for each $i \in \{1, 2, \ldots, N\}$, B_i is nonpositive. However, the function $\widetilde{B}$ may not be continuously differentiable at points x at which more than one function B_i takes the same value. For instance, even though the functions B_1 and B_2 in Example 11.10 given in (11.13) are continuous differentiable, the function $\widetilde{B}$ resulting from (11.20) is only locally Lipschitz. In fact, B_1 and B_2 are zero at $x = (-1, 0)$, and their gradients at that point are $\nabla B_1(x) = (-2, 0)$ and $\nabla B_2(x) = (1, 0)$, respectively. This issue with nonsmoothness motivates the use of locally Lipschitz barrier functions, which requires extending Definition 11.3. With this motivation, and following the expression of $\dot{V}$ in Definition 3.18 for the case of V being locally Lipschitz, the definition of the change of B in (11.3) is revisited.

Definition 11.13 ($\dot{B}$ with locally Lipschitz B). *Suppose a hybrid closed-loop system $\mathcal{H} = (C, F, D, G)$ satisfies the hybrid basic conditions, and that a set $K \subset \mathbb{R}^n$ and a function $B : \mathrm{dom}\, B \to \mathbb{R}$ defining a barrier function candidate with respect to K for $\mathcal{H}$ are given. Let $\mathcal{U}$ be an open neighborhood of ∂K as in Definition 11.1.*

- *The change of B along flows is given by*

$$\dot{B}(x) := \max_{\chi \in F(x) \cap T_C(x)} B^\circ(x, \chi) \qquad \forall x \in (\mathcal{U} \setminus K) \cap C \tag{11.21}$$

Furthermore, if F is single valued, then $\dot{B}$ is given by

$$\dot{B}(x) := B^\circ(x, F(x) \cap T_C(x)) \qquad \forall x \in (\mathcal{U} \setminus K) \cap C \tag{11.22}$$

Theorem 11.14 (Forward pre-invariance and forward invariance using a scalar barrier function). *Given $\mathcal{H} = (C, F, D, G)$ as in (2.19) satisfying the hybrid basic conditions and a closed set $K \subset \mathbb{R}^n$, suppose B is a barrier function candidate for $\mathcal{H} = (C, F, D, G)$ with respect to K. Let $\mathcal{U}$ be an open neighborhood of ∂K as in Definition 11.1.*

1. *The set K is forward pre-invariant if*

$$\dot{B}(x) \leq 0 \qquad \forall x \in (\mathcal{U} \backslash K) \cap C \tag{11.23}$$
$$B^+(x) \leq 0 \qquad \forall x \in D \cap K \tag{11.24}$$
$$G(D \cap K) \subset C \cup D \tag{11.25}$$

Furthermore, B is a <u>barrier function for forward pre-invariance</u> of K for $\mathcal{H}$.

2. *The set K is forward invariant for $\mathcal{H}$ if (11.23)-(11.25) hold, each maximal solution to $\mathcal{H}$ starting from the set $(K \cap \partial C) \setminus D$ is nontrivial, and at least one of the following conditions holds:*

   a) *$K \cap C$ is compact;*
   b) *F has linear growth on $K \cap C$; or*
   c) *The constrained differential inclusion*

   $$\dot{x} \in F(x) \quad x \in K \cap C$$

   does not have a maximal solution with finite escape time; i.e., there is no solution x such that $\lim_{t+j \nearrow \sup \operatorname{dom} x} |x(t,j)| = \infty$.

Furthermore, B is a <u>barrier function for forward invariance</u> of K for $\mathcal{H}$.

Remarkably, even though B is not necessarily continuously differentiable on a neighborhood of ∂K, the conditions for invariance required in Theorem 11.14 are the same as those in Theorem 11.4. The main difference in its proof compared to that of Theorem 11.4 is that the generalized derivative of B is used in (11.23) instead. Note that $t \mapsto B(x(t,j'))$ is still locally Lipschitz on $[t',t'']$, where t' and t'' are in the proof of Theorem 11.4. Similar to the steps therein, due to (3.9), the change of B during flows is upper bounded by $\dot{B}$ given as in Definition 11.3.

Theorem 11.14 is illustrated in the system considered in Example 11.10.

Example 11.15 (Forward invariance for a linear oscillator with jumps, revisited). *A multiple barrier function is used in Example 11.10 to show that the set K in (11.12) is forward invariant for the hybrid closed-loop system therein. Now, following the construction in (11.20), define the function B as $B(x) := \max\{B_1(x), B_2(x)\}$ with functions B_1, B_2 given as in (11.13). This function is not continuously differentiable at points $x = (x_1, x_2) \in \partial K \cap C \cap D$, which are points such that $|x_1| = 1$ and $x_2 = 0$. Similarly, B is not continuously differentiable at points x in a neighborhood of ∂K intersected by C, which, according to (11.23), are points at which $\dot{B}$ needs to be computed. Then, to check (11.23), the expression of $\dot{B}$ in (11.21) is employed. Let $\mathcal{U}$ be an open neighborhood of ∂K and let $x = (x_1, x_2) \in (\mathcal{U} \setminus K) \cap C$. Then, the following cases are possible:*

- *If $x_2 > 0$, then $B(x) = B_1(x)$ and $\dot{B}(x) = 0$.*

- *If $x_2 = 0$, then B is not differentiable and x_1 is in a neighborhood of $\{-1, 1\}$. The Clarke generalized gradient is given by*

$$\partial B(x) = \overline{\operatorname{con}} \{\nabla B_1(x), \nabla B_2(x)\} = \overline{\operatorname{con}} \left\{ \begin{bmatrix} 2x_1 \\ 0 \end{bmatrix}^\top, \begin{bmatrix} 0 \\ -1 \end{bmatrix}^\top \right\}$$

Then, since $x_2 = 0$, due to x_1 being nonpositive by definition of C,

$$\dot{B}(x) = \max_{\chi \in F(x)} \max_{\chi' \in \partial B(x)} \langle \chi', \chi \rangle = \max_{(\lambda, \lambda') \in [0,1] \times [0,1]} -2(1-\lambda)\lambda' x_1^2 + \lambda(1-\lambda') x_1 \leq 0$$

- *If $x_2 < 0$, then $B(x) = B_2(x)$ and $\dot{B}(x) = 0$.*

The other conditions in Theorem 11.14 needed to establish forward invariance of K already hold from the analysis in Example 11.10.

It is left as an exercise for the reader to check that the function $B(x) = \max\{B_1(x), B_2(x)\}$ is a barrier function for forward invariance of the set K in (11.19) for the thermostat system in Example 11.11, with B_1 and B_2 given therein. Note that although the function B is not smooth on $\mathbb{R}^2$, it is sufficiently smooth on small enough neighborhoods of ∂K.

11.2.2 Weak Forward Invariance

Weak forward invariance of a set has already been introduced in Chapter 3 as it plays a key role in the Hybrid Invariance Principle in Theorem 3.23. A result like Theorem 11.4 providing sufficient conditions for weak forward pre-invariance of a set can also be formulated, as outlined next.

When K is a closed subset of the interior of $C \cup D$, instead of requiring $\dot{B}$ to be nonpositive for all elements χ in the flow map, such a result requires ∇B to be nonzero on $(\mathcal{U}\backslash K) \cap C$ and the existence of a direction of flow allowing solutions to continuously evolve on K, that is, it requires the existence of χ in $F(x)$ as follows:

$$\langle \nabla B(x), \chi \rangle \leq 0 \qquad \forall x \in (\mathcal{U}\backslash K) \cap C, \ \exists \chi \in F(x) \tag{11.26}$$

At points in $K \cap D$ it requires the existence of an element in the jump map that belongs to K. More precisely, the following condition is required:

$$B(\chi) \leq 0 \qquad \forall x \in D \cap K, \ \exists \chi \in G(x) \tag{11.27}$$

The following example illustrates these conditions.

Example 11.16 (A planar system with finite escape time). *Consider the planar hybrid closed-loop system $\mathcal{H} = (C, F, D, G)$ with data given by*

$$F(x) := \begin{bmatrix} x_2 x_1^2 \\ [-1, 1] \end{bmatrix} \qquad \forall x \in C := \left\{ x \in \mathbb{R}^2 : x_2 \in \overline{[-1, 1] \backslash \frac{c}{4}\mathbb{B}} \right\}$$

$$G(x) := \begin{bmatrix} x_1 \\ x_2 + \mathbb{B} \end{bmatrix} \qquad \forall x \in D := \left\{ x \in \mathbb{R}^2 : |x_2| = c/4 \right\}$$

where $c \in (0, 1/2)$. For this system, the set

$$K = \left\{ x \in \mathbb{R}^2 : x_2 \in \overline{\left([-1, 1] \backslash \frac{c}{2}\mathbb{B}\right)} \cap \left(1 - \frac{c}{2}\right)\mathbb{B} \right\}$$

is weakly forward pre-invariant for $\mathcal{H}$:

- *From each point in K, there exists a maximal solution that stays within K for all hybrid time. However, such solutions are not complete since x_2 is never allowed to be zero, leading to solutions with finite escape times due to the fact that its x_1 components are lower bounded by the solutions to $\dot{x}_1 = \frac{c}{2}x_1^2$.*

- *From points in D, there exist maximal solutions that leave K and are not complete. Such solutions end after a jump because their x_2 component is mapped outside of K.*

Weak forward pre-invariance of K for $\mathcal{H}$ is certified by the barrier function candidate $B(x) = x_2^2 - (1 - c/2)$. Let $\mathcal{U}$ be a neighborhood of ∂K such that $(\mathcal{U} \setminus K) \cap C = (K + \varepsilon \mathbb{B}) \setminus K$ with $\varepsilon \in (0, c/4)$. Note that for each $x \in (\mathcal{U} \setminus K) \cap C$,

$$\langle \nabla B(x), \chi \rangle = 2 x_2 \chi_2$$

where $\chi = (\chi_1, \chi_2) \in F(x)$. Since $\chi_2 \in [-1, 1]$, there exists $\chi \in F(x)$ such that $\langle \nabla B(x), \chi \rangle = 0$. Now, for each $x \in K \cap D$,

$$B(\chi) = \chi_2^2 - (1 - c/2)$$

where χ_2 belongs to $x_2 + \mathbb{B}$. Due to the fact that $x_2 \in K$, $\chi_2 = x_2$ is a possible choice from $G(x)$ for which $B(\chi) \leq 0$. Note that due to maximal solutions exhibiting finite escape time, K is not weakly forward invariant.

11.2.3   Robust Forward Invariance

Thus far, the forward invariance notions considered in this chapter are for nominal hybrid closed-loop systems, namely, systems without disturbances. Unfortunately, the invariance properties shown in the earlier examples may not be robust to disturbances, no matter what their size is. For instance, even if the disturbances are arbitrarily small, a solution can land outside K after a jump. This is the case for the invariance property shown in Example 11.7 when the restitution coefficient λ is unitary: from $x = (0, -\sqrt{2c}) \in (\partial K) \cap D$, the additive disturbance $w_d = (0, \varepsilon)$ to the jump map leads to a solution that jumps to $(0, \sqrt{2c} + \varepsilon)$, which is not in K no matter how small $\varepsilon > 0$ might be. In fact, unlike asymptotic stability of a compact set, to guarantee robustness of forward invariance, disturbances have to be taken into account at the design stage.

In this section, robust versions of the nominal forward invariance notions in Definition 3.13 are introduced for $\mathcal{H}$ with disturbances, namely, for $\mathcal{H}_w$ in (10.49). In simple terms, robust forward invariance of a set K is the property that solutions to the perturbed system that start in the set K remain in the set K for all possible allowed disturbances.

A general model of a hybrid plant and of a hybrid closed-loop system under disturbances are given in (2.7) and in (10.49), respectively. In Chapter 10, a model of a hybrid plant with disturbances is proposed for robust CLF-based control is given in (10.44). When this plant is controlled by a static-state feedback pair, the resulting hybrid closed-loop system is denoted $\mathcal{H}_w$ and given in (10.49). The disturbances affecting the flows are denoted w_c and those affecting the jumps by w_d. The notions introduced in the following definition extend the nominal notions in Definition 3.13 to the case with disturbances. When no disturbances are present, they reduce to the nominal ones.

Definition 11.17 (Robust forward invariance). *Given a hybrid closed-loop system $\mathcal{H}_w$ with disturbances as in (10.49), a nonempty set $K \subset \mathbb{R}^n$ is said to be*

  1. *w-robustly forward pre-invariant for $\mathcal{H}_w$ if each maximal solution (x, w) to $\mathcal{H}_w$ with $x(0, 0) \in K$ satisfies $x(t, j) \in K$ for all $(t, j) \in \operatorname{dom} x$;*

2. *w-robustly forward invariant* for $\mathcal{H}_w$ if each maximal solution (x, w) to $\mathcal{H}_w$ with $x(0, 0) \in K$ is complete and satisfies $x(t, j) \in K$ for all $(t, j) \in \operatorname{dom} x$.

Similar weak forward invariance notions to the ones in Definition 3.12 can be formulated for the case with disturbances.

The notion of barrier function candidate in Definition 11.1 immediately extends to the case with disturbances. The only difference from the nominal case is that, in the presence of disturbances, the indicator properties of B in Definition 11.1 need to hold for each x such that there exist a disturbance $w = (w_c, w_d)$ for which (x, w_c) belongs to C or for which (x, w_d) belongs to D. Indeed, the projections of the sets C and D of $\mathcal{H}_w$ to the state space need to be used in Definition 11.1. The conditions for forward invariance in Theorem 11.4 can also be extended to the case with disturbances. In such a case, under the effect of disturbances, $\dot{B}$ and B^+ depend on the disturbances. For the case when the barrier function candidate is smooth enough and the maps F and G are single valued, given an open neighborhood $\mathcal{U}$ of ∂K as in Definition 11.1, $\dot{B}$ and B^+ are, respectively, defined as

$$\dot{B}(x, w_c) := \langle \nabla B(x), F(x, w_c) \rangle \quad \forall (x, w_c) \in C \;:\; x \in (\mathcal{U} \setminus K) \cap \Pi_c^w(C)$$

and

$$B^+(x, w_d) := B(G(x, w_d)) \qquad \forall (x, w_d) \in D$$

Using these constructions, a result extending Theorem 11.4 to the case of robust forward pre-invariance is given next. Only the case when F and G are single-valued maps and when B is C^1 is considered. The general case when F and G are set valued and when B is locally Lipschitz follows directly from the results in § 11.2.1.

Theorem 11.18 (Robust forward pre-invariance using a C^1 scalar barrier function). *Given $\mathcal{H}_w = (C, F, D, G)$ as in (10.49) satisfying the hybrid basic conditions and a closed set $K \subset \mathbb{R}^n$, suppose B is a barrier function candidate[5] for $\mathcal{H}_w$ with respect to K. Let $\mathcal{U}$ be an open neighborhood of ∂K as in Definition 11.1 and suppose that B is continuously differentiable on an open neighborhood of $(\mathcal{U} \setminus K) \cap \Pi_c^w(C)$. Then, the set K is w-robustly forward pre-invariant for $\mathcal{H}_w$ if*

$$\dot{B}(x, w_c) \leq 0 \qquad \forall (x, w_c) \in C \;:\; x \in \mathcal{U} \backslash K \tag{11.28}$$

$$B^+(x, w_d) \leq 0 \qquad \forall (x, w_d) \in D \;:\; x \in K \tag{11.29}$$

$$G((K \times \mathbb{R}^{s_d}) \cap D) \subset \Pi_c^w(C) \cup \Pi_d^w(D) \tag{11.30}$$

Furthermore, B is said to be a barrier function for w-robust forward pre-invariance of K for $\mathcal{H}_w$.

---

[5] This barrier function candidate satisfies the conditions in Definition 11.1 with $\Pi_c^w(C)$ and $\Pi_d^w(D)$ instead of C and D therein, respectively.

11.3   DESIGN

Given a hybrid plant $\mathcal{H}_P$ and a set K, a hybrid controller $\mathcal{H}_K$ that renders the set K forward invariant in the sense of the notions introduced in Definition 3.13 and Definition 11.17 can be designed using the sufficient conditions given in § 11.2.

For the case when $\mathcal{H}_K$ is given by a static state-feedback law $\kappa = (\kappa_c, \kappa_d)$, the goal is to design the functions κ_c and κ_d such that the resulting hybrid closed-loop system $\mathcal{H}$ with data given in (2.25), which, since $x = z$, is given by

$$
\mathcal{H} : \begin{cases} x \in C := \{x \in \mathbb{R}^n : (x, \kappa_c(x)) \in C_P\} \\ \qquad\qquad \dot{x} \in F(x) := F_P(x, \kappa_c(x)), \\ x \in D := \{x \in \mathbb{R}^n : (x, \kappa_d(x)) \in D_P\} \\ \qquad\qquad x^+ \in G(x) := G_P(x, \kappa_d(x)) \end{cases} \tag{11.31}
$$

is such that the set K enjoys the forward invariant property of interest. The case when $\mathcal{H}_K$ is a generic hybrid controller, having both continuous and discrete dynamics, can be treated similarly.

A design approach for the synthesis of a static state-feedback law $\kappa = (\kappa_c, \kappa_d)$ inducing a nominal invariance property of a given set $K \subset \mathbb{R}^n$ consists of applying Theorem 11.4 or Theorem 11.14 to the closed loop $\mathcal{H}$ in (11.31), and employing the sufficient conditions therein using barrier functions to choose the feedback law. Such a design produced is as follows.

Nominal design for forward pre-invariance of K**:** Given $\mathcal{H}_P = (C_P, F_P, D_P, G_P, \mathrm{Id})$ as in (2.7) and a closed set K, find functions κ_c and κ_d, and a barrier function candidate B with respect to K for the resulting hybrid closed-loop system $\mathcal{H} = (C, F, D, G)$ in (11.31) such that the following hold:

1. The feedback κ_c is such that

   a) The set $C := \{x \in \mathbb{R}^n : (x, \kappa_c(x)) \in C_P\}$ is closed.
   b) The flow map F defined as $F(x) := F_P(x, \kappa_c(x))$ for each $x \in C$ is outer semicontinuous and locally bounded relative to C, and $F_P(x, \kappa_c(x))$ is nonempty and convex for each $x \in C$.
   c) Condition (11.23) holds with C and F given in item 1a and item 1b above, respectively, and is $\mathcal{U}$ an open neighborhood of ∂K as in Definition 11.1.

2. The feedback κ_d is such that

   a) The set $D := \{x \in \mathbb{R}^n : (x, \kappa_d(x)) \in D_P\}$ is closed.
   b) The jump map G defined as $G(x) := G_P(x, \kappa_d(x))$ for each $x \in D$ is outer semicontinuous and locally bounded relative to D, and $G_P(x, \kappa_d(x))$ is nonempty for each $x \in D$.
   c) Condition (11.24) holds with D and G given in item 2a and item 2b above, respectively.

3. The feedbacks κ_c and κ_d are such that (11.25) holds.

When all of these conditions hold, item 1 in Theorem 11.14 implies that the set K is forward pre-invariant for $\mathcal{H}_P$ when controlled by the state-feedback pair (κ_c, κ_d) – that is, the set K is controlled forward pre-invariant.

The design procedure above can immediately be extended to forward invariance of K by enforcing the conditions in item 2 of Theorem 11.14. A design procedure using multiple barrier functions can also be formulated using the conditions in Theorem 11.9. Similarly, a design procedure to render K weakly invariant follows using the conditions in § 11.2.2. A design procedure for a feedback pair inducing forward invariance in the presence of disturbances can also be formulated by exploiting Theorem 11.18. Very importantly, following Chapter 10, it is also possible to extend the idea of choosing the state-feedback law pair from regulation maps and of constructing a pointwise minimum norm state-feedback pair for the purposes of forward invariance; see § 11.5.

11.4   EXERCISES

Exercise 76 (Properties of K). Given a hybrid closed-loop system $\mathcal{H} = (C, F, D, G)$ in $\mathbb{R}^n$ satisfying the hybrid basic conditions, a set $K \subset \mathbb{R}^n$, and a barrier function candidate B with respect to K for $\mathcal{H}$, show the following properties:

1. K is equal to (11.1).

2. K is closed.

3. $\partial K = (B^{-1}(0) \cup \partial C \cup \partial D) \cap K$.

Exercise 77 (Timer with resets). Consider a scalar hybrid closed-loop system $\mathcal{H} = (C, F, D, G)$ and a set $K = [0, 1]$. Using the notions in Definition 3.13, determine the invariance properties of K when

1. (C, F, D, G) is given by

$$F(x) := 1 \qquad \forall x \in C := [0, 1], \qquad G(x) := 0 \qquad \forall x \in D := \{1\}$$

2. (C, F, D, G) is given by

$$F(x) := 1 \qquad \forall x \in C := [0, 2], \qquad G(x) := 0 \qquad \forall x \in D := \{1\}$$

3. (C, F, D, G) is given by

$$F(x) := 1 \qquad \forall x \in C := [0, 1], \qquad G(x) := [0, 2] \qquad \forall x \in D := \{1\}$$

4. (C, F, D, G) is given by

$$F(x) := 1 \qquad \forall x \in C := [0, 1], \qquad G(x) := [0, 1] \qquad \forall x \in D := [1, 2]$$

Exercise 78 (Invariance in a planar system with finite escape times). Consider the hybrid closed-loop system $\mathcal{H} = (C, F, D, G)$ in $\mathbb{R}^2$ with data given by

$$F(x) := \begin{bmatrix} 1 + x_1^2 \\ 0 \end{bmatrix} \qquad \forall x \in C := \{x \in \mathbb{R}^2 : x_1 \in [0, \infty), x_2 \in [-1, 1]\}$$

$$G(x) := \begin{bmatrix} x_1 + \mathbb{B} \\ x_2 \end{bmatrix} \qquad \forall x \in D := \{x \in \mathbb{R}^2 : x_1 \in [0, \infty), x_2 = 0\}$$

1. Show that $K = C$ is weakly forward pre-invariant for $\mathcal{H}$.

2. Show that K is not forward pre-invariant for $\mathcal{H}$.

3. Replace G by $G(x) =: [2x_1 \ x_2]^\top$ and show that K is forward pre-invariant for $\mathcal{H}$.

4. Are each of these properties robust to some nonzero disturbance?

Exercise 79 (Weak invariance in a planar system). Consider the hybrid closed-loop system $\mathcal{H} = (C, F, D, G)$ in $\mathbb{R}^2$ with data given by

$$F(x) := \begin{cases} [1 \ 1]^\top & \text{if } x_2 > 1 - x_1 \\ \overline{\text{con}} \left\{ [1 \ 1]^\top, [-1 \ -1]^\top \right\} & \text{if } x_2 = 1 - x_1 \\ [-1 \ -1]^\top & \text{if } x_2 < 1 - x_1 \end{cases}$$
$$\forall x \in C := \{x \in \mathbb{R}^2 : x_1 \in [0, 1], x_2 \in [0, 1]\};$$

$$G(x) := \begin{cases} [\frac{1}{2} + \frac{1}{4}\mathbb{B} \ \frac{1}{2}]^\top & \text{if } x_2 \in \{0, 1\}, x_1 \subset (0, 1) \\ \{[\frac{1}{2} + \frac{1}{4}\mathbb{B} \ \frac{1}{2}]^\top, [\frac{1}{2} \ \frac{1}{2} + \frac{1}{4}\mathbb{B}]^\top\} & \text{if } x \in \{(0, 0), (0, 1), (1, 1), (1, 0)\} \\ [\frac{1}{2} \ \frac{1}{2} + \frac{1}{4}\mathbb{B}]^\top & \text{if } x_1 \in \{0, 1\}, x_2 \in (0, 1) \end{cases}$$
$$\forall x \in D := \partial C = (\{0, 1\} \times [0, 1]) \bigcup ([0, 1] \times \{0, 1\}).$$

Show that the set $K = [\frac{1}{2}, 1] \times [\frac{1}{2}, 1]$ is weakly forward invariant for $\mathcal{H}$.

Exercise 80 (Stabilization of a point on the unit circle). Consider the system given in §1.2.3 and the hybrid controller given in Exercise 15 for global stabilization of a point on the unit circle. Show the following properties for the hybrid closed-loop system therein:

1. The set $K = C$ is forward invariant.

2. The set $K := C_{K,2} \times \{1\}$ is weakly forward invariant.

3. For each c_1' and c_2' such that $c_2 < c_2' < c_1' < 0$ and $c_1 < c_1'$, the set $K := \{(z, q) \in \mathbb{S}^1 \times \{1\} : z_1 \geq c_2'\} \cup \{(z, q) \in \mathbb{S}^1 \times \{0\} : z_1 \leq c_1'\}$ is forward invariant.

Exercise 81 (Robust invariance for a simplified juggling system). Consider the simplification in Example 11.2 of the one-degree-of-freedom juggling system defined in Example 2.32 but with the addition of a disturbance w_d at jumps: the jump map G is replaced by $G(x, w_d) := [x_1 \quad -ex_2 + w_d]^\top$. Define the jump set D, which includes the allowed disturbances w_d, such that jumps are still triggered when $x_1 = 0$ and $x_2 \leq 0$ and the set K in Example 11.2 is robustly forward invariant for the resulting hybrid closed-loop system $\mathcal{H}_w$. Make sure that disturbances w_d other than the zero disturbance are allowed.

Exercise 82 (Invariance-based control for a thermostat). The evolution of the temperature of a room with a heater can be modeled by a linear-time invariant system with state z denoting the temperature of the room and with input $u = (u_1, u_2)$, where u_1 denotes whether the heater is turned on ($u_1 = 1$) or turned off ($u_1 = 0$) while u_2 denotes the temperature outside the room. With these definitions and the models in Example 11.11, the evolution of the temperature is given by

$$\dot{z} = -\alpha z + \begin{bmatrix} z_{\text{heat}} & 1 \end{bmatrix} \begin{bmatrix} u_1 \\ u_2 \end{bmatrix}$$

subject to $(z, u) \in \{(z, u) \in \mathbb{R} \times \mathbb{R}^2 : u_1 \in \{0, 1\}\}$, where the constant $\alpha > 0$ represents the decay rate of the temperature and z_{heat} is a constant representing the heater capacity. Note that the model in Example 11.11 has u_2 equal to the constant z_{out} denoting the outside temperature.

1. Given constants $z_{\min} < z_{\max}$, design a static state-feedback law assigning u_1 and determine conditions on the constants α, z_{heat} and on the allowed values for u_2 rendering forward invariant a closed set K defined such that, when projected onto the state space of the temperature, it is equal to $[z_{\min}, z_{\max}]$.

2. Validate your design in item 1 numerically.

3. Suppose that the measurements of z are affected additively by measurement noise $m \in \delta \mathbb{B}$, $\delta > 0$. Redesign the controller in item 1 and determine a nonzero value $\delta > 0$ such that the set K therein is robustly forward invariant.

Exercise 83 (Invariance in obstacle avoidance with target). Consider the problem of globally stabilizing an autonomous vehicle to a target while avoiding an obstacle in Example 4.13. Show that $K = C$ is forward invariant for the hybrid closed-loop system proposed therein.

Exercise 84 (Forward invariance for the DC/AC inverter). Perform the following analysis for the DC/AC converter in Example 1.1:

1. Show Lemma 2.38. Hint: consider whether V therein is a barrier function candidate.

2. Show that when the parameters are such that $\mathcal{T} \subset \Gamma$, the set $\mathcal{T}$ is forward invariant.

11.5   NOTES

Forward invariance has a broad range of applications that go beyond those illustrated in this chapter. Forward invariance plays a key role in the solution to problems emerging in air traffic management [200], obstacle avoidance in vehicular networks [201], threat assessment in semi-autonomous cars [202], network control systems [203], and building control [204].

Forward invariance appears in the literature under different names. In [205], a forward invariance-type notion is referred to as flow-invariance, in [206] as positive invariance, and in [207] as viability. The latter property corresponds to the weak invariance notion in Definition 3.13. A robust forward invariance notion that is comparable to the one in Definition 11.17 was used in [204].

Tools to verify invariance of a set for continuous-time and discrete-time systems are available in the literature. The survey article [206] and the book [207] summarize these and other analysis results for forward invariance of sets in continuous-time and discrete-time systems. In the seminal article [208], the so-called *Nagumo Theorem* is established to determine forward invariance (and weak forward invariance) of a locally compact set K for continuous-time systems with unique solutions. Given a locally compact set K that is to be rendered forward invariant and a continuous-time system with a continuous vector field, the Nagumo Theorem requires that, at each point in the boundary of K, the vector field belongs to the tangent cone to K; see also [207, Theorem 1.2.1]. This result has been revisited and extended in several directions. In [209], conditions for weak invariance as well as invariance for closed sets are provided – a result guaranteeing finite-time weak invariance is also presented. In particular, a result therein shows that a closed set K is forward invariant for a continuous-time system with unique solutions if and only if the vector field (and minus the vector field) are subtangential to K at each point in it. A similar result is known as the Bony-Brezis theorem, which, instead of involving a condition on the tangent vectors, requires the vector field to have a nonpositive inner product with any (exterior) normal vector to the set K [210, 211].

Taking advantage of convexity and linearity of the objects considered, [212] provides necessary and sufficient conditions for forward invariance of convex polyhedral sets for linear time invariant discrete-time systems. Essentially, the condition for forward invariance in [212] consists of requiring that the new value of the state after every iteration of the discrete-time system belongs to the set that is to be rendered forward invariant. This condition is very natural, it is necessary, and can be interpreted as the discrete-time counterpart of the condition in the Nagumo Theorem. For the case of time-varying continuous-time systems, [205] provides conditions guaranteeing forward invariance properties of K given by a sublevel set of a Lyapunov-like function; see also [213, 214, 215].

The analysis of forward invariance of a set for systems under the effect of perturbations has also been studied in the literature; see [216] for the case when K is a cone, [217, 218] when K is a polyhedral, to just list a few [202, 201]. The notion of robust controlled forward invariance has also been studied in the literature; see, e.g., [204, 219, 220].

The study of forward invariance in hybrid systems is not as mature as for continuous-time and discrete-time systems. When the continuous dynamics of a hybrid system are discretized, the methods for purely discrete-time systems mentioned above are applicable or can be extended without significant effort for certain classes of hybrid models in discrete time; see, in particular, the results for a class of piecewise affine discrete-time systems in [221]. "Pure" results for systems of hybrid nature include the results for invariance in impulsive differential inclusions in [42]. In particular, conditions to guarantee (weak – or viability – and strong) forward invariance of closed sets and a numerical algorithm to generate invariant kernels are proposed in [42]. These results build from a vast literature on the study of viability and invariance for differential inclusions; see, e.g., [207, 42, 222], to just list a few.

Some of the results presented in this chapter are extensions of results in [223, 224, 225] to the case of hybrid plants $\mathcal{H}_P$ and hybrid closed-loop systems $\mathcal{H}$. However, the main source with the basic ideas used to develop the results presented in this section is the book [207]. The work therein, which is for general differential inclusions without constraints, makes an important observation that is exploited in this chapter:

Forward invariance of a set K for a continuous-time system is a property that depends on the dynamics of the system only outside the set.

This is why, unlike results in the literature, the results in this chapter provide sufficient conditions that, for the continuous dynamics of the system, enforce properties only in a neighborhood of the boundary ∂K (relative to the complement of K). A version of the notion of barrier function candidate in Definition 11.1 was presented in [226]. The construction of the set K in (11.1) was also introduced therein; see [226, Definition 7]. Other definitions of barrier functions are available in the literature. In particular, [227] considers a scalar barrier function candidate that is positive and locally bounded on the interior of the set K, and that, in addition, approaches infinity as its argument of the barrier function converges the boundary of the set. When using such barrier function notion, solutions starting from the interior of the set to render invariant are not allowed to reach its boundary.

Similar sufficient conditions as those in Theorem 11.4 for forward pre-invariance and forward invariance using a continuously differentiable scalar barrier function appeared in [226, Theorem 1]. Along with the notion of multiple barrier function candidate in Definition 11.8, the case of multiple barrier functions captured in Theorem 11.9 appeared in [228], also for continuously differentiable barrier functions – see [229] for conditions on multiple barrier functions assuring weak invariance. More details on the case when the barrier function candidate that is only locally Lipschitz, which is given in Theorem 11.14, are in [230]. The locally Lipschitz case was also studied in [231], for a different notion of barrier function and in the context of obstacle avoidance problems. Very importantly, it should be noted that the majority of the conditions required by the results in this chapter to guarantee forward invariance are also necessary; see [232].

The notions in Definition 3.13 and in Definition 11.17 appeared in [233] and [234], respectively, in the context of sufficient conditions for invariance using tangent cone conditions. Due to space constraints, the sufficient conditions for forward invariance proposed in this chapter are solely using barrier functions, and can be seen as alternative versions to those in [233] and [234]. As outlined in § 11.1, the conditions proposed in this chapter exploit the fact that, in the general case, the set K is the intersection of the zero sublevel sets of scalar functions, for which the "variation" of the barrier function candidate can be assessed without computing solutions, which is prohibitive, or tangent cones, which at times is expensive.

Chapter Twelve

Temporal Logic

In most control problems, design specifications are first given in high-level terms before they are translated into fundamental properties. For instance, high-level design specifications associated to the obstacle avoidance with target problem shown in the cover photo of this book are as follows:

$$\textit{Eventually reach the target and always avoid the obstacle.} \qquad (12.1)$$

As described in Example 4.13, this specification requires the system to satisfy two different fundamental properties, simultaneously: avoidance of the obstacle at all times and convergence to the target after some finite time. *Temporal logic* allows the formulation of such specifications requiring the system – or better said, its solutions – to satisfy diverse properties over time. Temporal logic can be efficiently employed to determine *safety* – namely, "something bad never happens" – and *liveness* – "something good eventually happens." This language is powered by the combination of operators allowing for the formulation of statements using boolean logic and time, with functions of the state of the system, called *atomic propositions*.

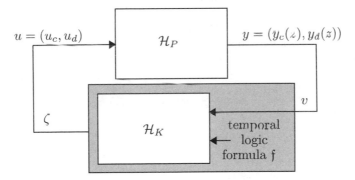

Figure 12.1: Hybrid closed-loop system in the context of satisfaction of temporal logic specifications.

This chapter introduces the operators and semantics required to formulate temporal logic specifications for systems with hybrid dynamics. As Figure 12.1 depicts, the presentation focuses on hybrid closed-loop systems $\mathcal{H}$, which, as introduced in Chapter 2, might emerge from the plant or the controller being hybrid; see § 2.3.1. In this way, specifications given in terms of temporal logic formulas can be satisfied by a closed-loop solution either when the solution evolves continuously or when it jumps. The relationship between specifications stated in the temporal logic language and dynamical properties of the system are established and conditions guaranteeing the satisfaction of the formulas are presented.

12.1 OVERVIEW

A key step in control design is the formulation of specifications in a manner that they are amenable to the tools to be employed. In a wide range of control problems, such specifications can be converted into stability, attractivity, and robustness properties, for which Lyapunov theorems, invariance principles, and perturbation analysis tools can be applied. However, some design problems include specifications that cannot be translated into one of those properties due to involving multiple objectives and time, like the one in (12.1). The approach presented in this chapter consists of using Linear Temporal Logic (LTL) to handle such specifications. LTL is an expressive, high-level language that permits the definition of specifications of *linear temporal logic formulas*. LTL formulas are given by linear combinations of operators and atomic propositions. The operators that can be used in the formulas include boolean operators, such as the well-known *and* and *not*, and also temporal operators, such as *always* and *eventually*, introduced later in this chapter. With such operators at hand, design specifications requiring the satisfaction of multiple properties following an order that involves time can be captured via temporal logic formulas.

For instance, using the notation in Example 4.13, a temporal logic formula capturing (12.1) can be defined as follows. Define the atomic proposition $\mathfrak{a}_1$ to be true when the state z of the autonomous vehicle therein is at the target location z^*, and to be false otherwise. Also, define the atomic proposition $\mathfrak{a}_2$ to be true when z belongs to the region $\mathcal{O}$ where the obstacle is located, and to be false when it is not in that region. Then, the specification in (12.1) is captured by the formula

$$\mathfrak{f} = (\Diamond \mathfrak{a}_1) \wedge (\Box \neg \mathfrak{a}_2)$$

where, as formally defined in § 12.2, $\Diamond$ represents the *eventually* operator, $\wedge$ the logic *and* operator, $\Box$ the *always* operator, and $\neg$ the logic negation. This formula is true when the following two expressions are true:

- $\Diamond \mathfrak{a}_1$: This expression requires $\mathfrak{a}_1$ to "eventually" become true; i.e., it requires the existence of a time at which $\mathfrak{a}_1$ is true. It captures the requirement in (12.1) of reaching the target since when $\mathfrak{a}_1$ is true, the vehicle state z is at the target location.

- $\Box \neg \mathfrak{a}_2$: This expression requires $\neg \mathfrak{a}_2$ to "always" be true; i.e., it requires that $\mathfrak{a}_2$ is false at all times. Since $\mathfrak{a}_2$ is defined as false when z is not in $\mathcal{O}$, then it assures that the vehicle is never in the obstacle region.

Similar formulas can be constructed for more complex specifications, using a variety of boolean and temporal operators.

LTL formulas require properties to occur at certain times. In the formula above, $\Box \neg \mathfrak{a}_2$ asks for $\neg \mathfrak{a}_2$ to be true all the time, while the expression $\Diamond \mathfrak{a}_1$ requires proposition $\mathfrak{a}_1$ to be satisfied at some finite-time instant. This is in contrast to the attractivity property in Definition 3.1, in which convergence may not occur at a specific finite time, but rather, only as (hybrid) time tends to infinity. As a consequence, analysis and design tools that rely solely on the construction of Lyapunov functions for pre-asymptotic stability may not be applicable. On the other hand, the fact that such tools neither require solving for the solutions to the system nor discretization makes them very appealing.

To satisfy specifications given in terms of LTL formulas, the approach proposed in this chapter exploits the power of Lyapunov-based tools by recasting such specifications as fundamental dynamical properties of suitably defined sets. To illustrate the approach, consider the expression $\Diamond \mathfrak{a}_1$ in the formula above and the simple vehicle model (4.36) with state $z \in \mathbb{R}^2$. Treating $\mathfrak{a}_1$ as a function of the state, a solution $t \mapsto z(t)$ to (4.36) satisfies $\Diamond \mathfrak{a}_1$ if there exists $t \in \operatorname{dom} z$ such that $\mathfrak{a}_1(z(t))$ is true. This requirement can be recast as the property that the solution $t \mapsto z(t)$ reaches after a finite amount of time the set of points on which $\mathfrak{a}_1$ is true. Such a set is given by

$$\left\{ z \in \mathbb{R}^2 \: : \: \mathfrak{a}_1(z) \text{ is true} \right\}$$

Hence, to certify $\Diamond \mathfrak{a}_1$, it suffices to show that solutions reach this set in finite time. This dynamical property is called *finite-time attractivity* which, as shown in § 12.3, is a necessary property for $\Diamond \mathfrak{a}_1$. Conveniently, it is possible to formulate Lyapunov conditions that guarantee such finite-time convergence property, without having to solve for the solutions to the system explicitly, or relying on discretization.

Readers of Chapter 11 may have already noticed that the expression $\Box \neg \mathfrak{a}_2$ associated to the specification in (12.1) can already be certified using the tools presented in that chapter. The corresponding set to be rendered forward (pre-)invariant is given by

$$\left\{ z \in \mathbb{R}^2 \: : \: \mathfrak{a}_2(z) \text{ is false} \right\}$$

In fact, every solution $t \mapsto z(t)$ to the vehicle model in (4.36) has to stay away from the obstacle, which is assured when $\mathfrak{a}_2(z(t))$ is false for all time. In this case, the dynamical property of interest is *forward pre-invariance* of the set above. Forward pre-invariance is not only necessary for $\Box \neg \mathfrak{a}_2$ to be satisfied but, very importantly, justifies the need for the initial value of the state z to be so that $\mathfrak{a}_2$ is false, at least in the case where the specification must hold at the initial time.

As alluded in the previous paragraph, in addition to capturing properties that have to be satisfied at specific times, one can additionally require that an LTL formula as a whole is satisfied at particular time instances. For example, the requirement that the LTL expression $\Box \neg \mathfrak{a}_2$ holds at *the initial time* is equivalent to forward pre-invariance, but the requirement that it holds *after some finite time* requires an "eventual" version of that notion. In the latter case, to satisfy $\Box \neg \mathfrak{a}_2$, one could first show that the set $\left\{ z \in \mathbb{R}^2 \: : \: \mathfrak{a}_2(z) \text{ is false} \right\}$ is reached in finite time – say, exactly at flow time t' – and, after that time, solutions have to remain in the said set. Then, using LTL lingo, one can say that $\Box \neg \mathfrak{a}_2$ is satisfied at $t = t'$ since from t' onwards, $\neg \mathfrak{a}_2$ is satisfied.

Similarly, one can say that the expression $\Diamond \mathfrak{a}_1$ is satisfied at the initial time (assumed to be $t = 0$), if there exists $t' \geq 0$ such that $\mathfrak{a}_1(z(t'))$ is true. Note that it might be the case that $\Diamond \mathfrak{a}_1$ is not satisfied for any future $t > t'$. However, one may want to enforce that $\Diamond \mathfrak{a}_1$ is satisfied after $t > t'$ by requiring that $\Diamond \mathfrak{a}_1$ is satisfied at each $t \geq 0$. A way to reason about this requirement is as follows. First, note that if $\Diamond \mathfrak{a}_1$ is true at t', then it is true at each $t \in [0, t']$: in fact, for each $t \in [0, t']$, there exists a time $\tilde{t} \in [t, t']$ such that $\mathfrak{a}_1(z(\tilde{t}))$ is true – in particular, $\mathfrak{a}_1$ is true at $\tilde{t} = t'$. Then, $\Diamond \mathfrak{a}_1$ being satisfied at each $t \geq 0$ requires the following: for each $t \geq 0$, there exists $\tilde{t} \geq t$ such that $\mathfrak{a}_1(z(\tilde{t}))$ is true. In other words, $\mathfrak{a}_1$ has to be recurrently satisfied when the formula $\Diamond \mathfrak{a}_1$ is to be satisfied for each $t \geq 0$. Certainly, such a property holds trivially when the solution satisfies $\mathfrak{a}_1$ for all times after some finite time, but that is not a necessary condition for $\Diamond \mathfrak{a}_1$ to be satisfied for each $t \geq 0$.

The approach outlined above extends to the case of solutions to hybrid closed-loop systems $\mathcal{H}$ as in (2.19). Given an LTL formula $\mathfrak{f}$ capturing a particular specification and a solution $(t, j) \mapsto x(t, j)$ to $\mathcal{H}$, the goal is to determine whether $\mathfrak{f}$ is satisfied, at the initial time $(t, j) = (0, 0)$ only or at each $(t, j) \in \operatorname{dom} x$. For example, if the particular formula $\mathfrak{f} = \Diamond \mathfrak{a}_1$ is to be satisfied at $(t, j) = (0, 0)$, then the goal is to show that there exists $(t', j') \in \operatorname{dom} x$ such that $\mathfrak{a}_1(x(t', j'))$ is true. Using the natural ordering of elements in a hybrid time domain mentioned in Definition 2.26, if $\mathfrak{f}$ is to be satisfied at $(t, j) \in \operatorname{dom} x$, the question is whether there exists $(t', j') \in \operatorname{dom} x$, $t' + j' \geq t + j$, such that $\mathfrak{a}_1(x(t', j'))$ is true.

12.2   LTL SEMANTICS

The syntax for the language associated with LTL can be defined recursively using *atomic propositions*. For a hybrid system $\mathcal{H}$ with state x, an atomic proposition is defined as a *statement* given in terms of the state x that, for each possible value of x, is either True or False. Atomic propositions are treated as functions of the state.

Definition 12.1 (Atomic proposition). *An atomic proposition $\mathfrak{a}$ is a function depending on the system state x. For each x, the value of the atomic proposition $\mathfrak{a}$ at x is either equal to True or equal to False, that is, $\mathfrak{a} : \mathbb{R}^n \to \{\text{True}, \text{False}\}$.*

The atomic propositions $\mathfrak{a}_1$ and $\mathfrak{a}_2$ introduced in § 12.1 in terms of the state are defined as follows: with $x = z$ and for each $x \in \mathbb{R}^2$,

$$\mathfrak{a}_1(x) := \begin{cases} \text{True} & \text{if } x = z^* \\ \text{False} & \text{otherwise} \end{cases}$$

$$\mathfrak{a}_2(x) := \begin{cases} \text{True} & \text{if } x \in \mathcal{O} \\ \text{False} & \text{otherwise} \end{cases}$$

Definition 12.2 (Turnstile notation). *Let $\mathfrak{a}$ be an atomic proposition and $x \in \mathbb{R}^n$.*

- *The proposition $\mathfrak{a}$ being True at x implies that $\mathfrak{a}(x) = \text{True}$, which is denoted*

$$x \vDash \mathfrak{a} \tag{12.2}$$

- *The proposition $\mathfrak{a}$ being False at x implies that $\mathfrak{a}(x) = \text{False}$, which is denoted*

$$x \nvDash \mathfrak{a} \tag{12.3}$$

The double turnstile symbol $\vDash$ is a binary relation that, here, indicates that the expression on the left satisfies the sentence(s) on the right.

Boolean logic operators can be employed in the formulation of LTL formulas. These operators enable the combination of atomic propositions in the construction of formulas. When applied to atomic propositions, these operators are required to hold at each value of x, or, when a solution $(t, j) \mapsto x(t, j)$ to $\mathcal{H}$ is given, at each (t, j) in the domain of x.

Definition 12.3 (Logic operators). *The logic operators are as follows.*

- *The logic operator $\neg$ denotes negation or, simply, not.*

- *The logic operator $\vee$ denotes disjunction or, simply, or.*

- *The logic operator $\wedge$ denotes conjunction or, simply, and.*

- *The logic operator $\Rightarrow$ denotes implication or, simply, then.*

- *The logic operator $\Leftrightarrow$ denotes equivalence or, simply, if and only if.*

The logic operators can be combined with propositions using standard boolean logic. For instance, given atomic propositions $\mathfrak{a}_1$ and $\mathfrak{a}_2$, the expression $\mathfrak{a}_1 \vee \neg \mathfrak{a}_2$ is True when either $\mathfrak{a}_1$ is True or $\mathfrak{a}_2$ is False. The meaning of these operators when applied to a solution to $\mathcal{H}$ is formalized in the definition of the semantics of LTL, which is given in Definition 12.5.

As outlined in §12.1, temporal operators expand the LTL language by allowing propositions to hold at certain times on the domain of a solution. The *always* and *eventually* operators mentioned in §12.1 are examples of temporal operators in LTL. In addition, the *next* operator enforces that an atomic proposition is true at the next time instance. More interestingly, the (weak and strong) *until* operator can be employed to capture specifications that require that a proposition is true up to the time that another proposition is true.

Definition 12.4 (Temporal operators). *The temporal operators are as follows.*

- *The next operator is denoted $\bigcirc$.*

- *The always operator is denoted $\square$.*

- *The eventually operator is denoted $\Diamond$.*

- *The strong until operator is denoted $\mathcal{U}_s$.*

- *The weak until operator is denoted $\mathcal{U}_w$.*

With the logic and temporal operators as defined above, the semantics for the LTL language are introduced as follows. These semantics formalize the meaning of a solution to $\mathcal{H}$ satisfying an atomic proposition.

Definition 12.5 (LTL semantics). *Let $\mathfrak{a}$, $\mathfrak{a}_1$, and $\mathfrak{a}_2$ be atomic propositions. Given a solution x to $\mathcal{H}$ and $(t, j) \in \mathrm{dom}\, x$*

$$(x, (t, j)) \vDash \mathfrak{a} \quad \Leftrightarrow \quad x(t, j) \vDash \mathfrak{a} \tag{12.4a}$$

$$(x, (t, j)) \vDash \neg \mathfrak{a} \quad \Leftrightarrow \quad (x, (t, j)) \nvDash \mathfrak{a} \tag{12.4b}$$

$$(x, (t, j)) \vDash \mathfrak{a}_1 \vee \mathfrak{a}_2 \quad \Leftrightarrow \quad (x, (t, j)) \vDash \mathfrak{a}_1 \; or \; (x, (t, j)) \vDash \mathfrak{a}_2 \tag{12.4c}$$

$$(x, (t,j)) \vDash \mathfrak{a}_1 \wedge \mathfrak{a}_2 \quad \Leftrightarrow \quad (x, (t,j)) \vDash \mathfrak{a}_1 \ \textit{and} \ (x, (t,j)) \vDash \mathfrak{a}_2 \tag{12.4d}$$

$$(x, (t,j)) \vDash \bigcirc \mathfrak{a} \quad \Leftrightarrow \quad (t, j+1) \in \operatorname{dom} x \ \textit{and} \ (x, (t, j+1)) \vDash \mathfrak{a} \tag{12.4e}$$

$$(x, (t,j)) \vDash \square \mathfrak{a} \quad \Leftrightarrow \quad \begin{aligned} &(x, (t',j')) \vDash \mathfrak{a} \quad \forall (t',j') \in \operatorname{dom} x \\ &\textit{s.t.} \ t' + j' \geq t + j \end{aligned} \tag{12.4f}$$

$$(x, (t,j)) \vDash \diamond \mathfrak{a} \quad \Leftrightarrow \quad \begin{aligned} &\exists (t',j') \in \operatorname{dom} x, \\ &\quad t' + j' \geq t + j \ \textit{s.t.} \ (x, (t',j')) \vDash \mathfrak{a} \end{aligned} \tag{12.4g}$$

$$(x, (t,j)) \vDash \mathfrak{a}_1 \, \mathcal{U}_s \mathfrak{a}_2 \quad \Leftrightarrow \quad \begin{aligned} &\exists (t',j') \in \operatorname{dom} x, \\ &t' + j' \geq t + j \ \textit{s.t.} \ (x, (t',j')) \vDash \mathfrak{a}_2, \\ &\textit{and} \ \forall (t'',j'') \in \operatorname{dom} x \ \textit{s.t.} \\ &t + j \leq t'' + j'' < t' + j', (x, (t'',j'')) \vDash \mathfrak{a}_1 \end{aligned} \tag{12.4h}$$

$$(x, (t,j)) \vDash \mathfrak{a}_1 \, \mathcal{U}_w \mathfrak{a}_2 \quad \Leftrightarrow \quad \begin{aligned} &(x, (t',j')) \vDash \mathfrak{a}_1 \ \forall (t',j') \in \operatorname{dom} x \ \textit{s.t.} \\ &t' + j' \geq t + j, \textit{or} \ (x, (t,j)) \vDash \mathfrak{a}_1 \, \mathcal{U}_s \mathfrak{a}_2 \end{aligned} \tag{12.4i}$$

These semantics uniquely determine the meaning of a solution x satisfying an atomic proposition $\mathfrak{a}$ alone, or a *sentence* defined using the logic and temporal operators. Such sentences are usually referred to as *formulas*. Similar to atomic propositions, formulas are treated as functions of the state.

Definition 12.6 (LTL formula). *An LTL formula* $\mathfrak{f}$ *is a function defined as a linear combination of atomic propositions and LTL operators.*

For example, the expression in (12.4d) defines the formula $\mathfrak{f} = \mathfrak{a}_1 \wedge \mathfrak{a}_2$ and the expression in (12.4g) defines the formula $\mathfrak{f} = \diamond \mathfrak{a}_1$.

The equivalent notation introduced in Definition 12.2 and the semantics in Definition 12.5 for atomic propositions are also used for formulas. More precisely, an LTL formula $\mathfrak{f}$ being satisfied by a solution $(t,j) \mapsto x(t,j)$ at some time (t,j) is given by

$$(x, (t,j)) \vDash \mathfrak{f}$$

while $\mathfrak{f}$ not being satisfied at (t,j) is given by

$$(x, (t,j)) \nvDash \mathfrak{f}$$

Note that according to the semantics in (12.4e), the formula $\mathfrak{f} = \bigcirc \mathfrak{a}$ is satisfied by $x(t,j)$ – namely, $(x, (t,j)) \vDash \mathfrak{f}$ – when $x(t, j+1)$ satisfies $\mathfrak{a}$ with both (t,j) and $(t, j+1)$ belonging to $\operatorname{dom} x$. The next operator is the only operator that imposes properties on the hybrid time domain of the solution.

12.3   CHARACTERIZATION OF BASIC FORMULAS

Basic LTL formulas are those that involve a single temporal operator. Such basic formulas include

$$\bigcirc \mathfrak{a}, \qquad \Box \mathfrak{a}, \qquad \Diamond \mathfrak{a}, \qquad \mathfrak{a}_1 \mathcal{U}_s \mathfrak{a}_2, \qquad \mathfrak{a}_1 \mathcal{U}_w \mathfrak{a}_2$$

where $\mathfrak{a}$, $\mathfrak{a}_1$, and $\mathfrak{a}_2$ are atomic propositions. As outlined in § 12.1, the approach proposed in this chapter to guarantee the satisfaction of basic LTL formulas consists of constructing the set of states for which the atomic propositions are true. More precisely:

Given the atomic proposition $\mathfrak{a}$ and a hybrid closed-loop system $\mathcal{H} = (C, F, D, G)$ with $x \in \mathbb{R}^n$, the set associated to $\mathfrak{a}$ is defined as

$$K_{\mathfrak{a}} := \{x \in X \ : \ \mathfrak{a}(x) = \text{True}\} \tag{12.5}$$

where, for convenience, in this chapter, the set X is defined as all possible values for the solutions to $\mathcal{H}$, that is, $X := \overline{C} \cup D \cup G(D)$.

With a set associated to each atomic proposition defined as in (12.5), to show that a solution x to $\mathcal{H}$ satisfies a given basic formula $\mathfrak{f}$ at $(t, j) = (0, 0)$ – that is, $(x, (0,0)) \models \mathfrak{f}$ – the following must hold according to the semantics introduced in Definition 12.5:

- For $\mathfrak{f} = \bigcirc \mathfrak{a}$, $x(0, 1)$ belongs to $K_{\mathfrak{a}}$.

- For $\mathfrak{f} = \Box \mathfrak{a}$, $x(t, j)$ belongs to $K_{\mathfrak{a}}$ for all $(t, j) \in \operatorname{dom} x$.

- For $\mathfrak{f} = \Diamond \mathfrak{a}$, $x(t, j)$ belongs to $K_{\mathfrak{a}}$ for some $(t, j) \in \operatorname{dom} x$.

- For $\mathfrak{f} = \mathfrak{a}_1 \mathcal{U}_s \mathfrak{a}_2$, x belongs to $K_{\mathfrak{a}_1}$ until some $(t, j) \in \operatorname{dom} x$, at which $x(t, j)$ belongs to $K_{\mathfrak{a}_2}$.

- For $\mathfrak{f} = \mathfrak{a}_1 \mathcal{U}_w \mathfrak{a}_2$, $x(t, j)$ belongs to $K_{\mathfrak{a}_1}$ for all $(t, j) \in \operatorname{dom} x$, or $(x, (t, j)) \models \mathfrak{a}_1 \mathcal{U}_s \mathfrak{a}_2$.

The proposed approach consists of expressing these properties in terms of dynamical properties of $\mathcal{H}$. For instance, the property required for $\mathfrak{f} = \Box \mathfrak{a}$ can only hold if $x(0, 0)$ belongs to the set $K_{\mathfrak{a}}$ and stays in it for all future time, which resembles the forward pre-invariance notion in Definition 3.13. In this section, necessary and sufficient conditions for the satisfaction of basic LTL formulas that are expressed in terms of such dynamical properties are presented.

12.3.1   Properties of $\mathcal{H}$ for the Next Operator

According to (12.4e) in the definition of semantics, a solution x to $\mathcal{H}$ satisfies the formula

$$\mathfrak{f} = \bigcirc \mathfrak{a} \tag{12.6}$$

at $(t,j) \in \operatorname{dom} x$ if $x(t,j+1)$ satisfies the atomic proposition $\mathfrak{a}$. This property implies that for the solution x to satisfy $\mathfrak{f}$ at each $(t,j) \in \operatorname{dom} x$, $x(t,j+1)$ has to satisfy $\mathfrak{a}$ for each $(t,j) \in \operatorname{dom} x \setminus \{(0,0)\}$. Note that this implies that $\operatorname{dom} x = \{0\} \times \mathbb{N}$, namely, that the maximal solutions to $\mathcal{H}$ are discrete and complete. This observation leads to the following equivalence.

Proposition 12.7 (Equivalence conditions for $\mathfrak{f} = \bigcirc \mathfrak{a}$). *Given a hybrid closed-loop system $\mathcal{H} = (C, F, D, G)$ as in (2.19) and an atomic proposition $\mathfrak{a}$, let $K_{\mathfrak{a}}$ be given as in (12.5). The formula $\mathfrak{f} = \bigcirc \mathfrak{a}$ is satisfied for each maximal solution x to $\mathcal{H}$ at each $(t,j) \in \operatorname{dom} x$ if and only if the following properties hold:*

1. $\overline{C} \subset D$;

2. Every maximal solution to $\mathcal{H}$ does not flow; and

3. $G(D) \subset K_{\mathfrak{a}} \cap D$.

Proof. To show necessity, assume that $\mathfrak{f}$ is satisfied for each maximal solution x to $\mathcal{H}$ at each $(t,j) \in \operatorname{dom} x$. Then, by the semantics of the temporal operator $\bigcirc$ in (12.4e), $(x, (t, j+1)) \vDash \mathfrak{a}$ with $(t,j) = (0,0)$. Hence, $(0,1) \in \operatorname{dom} x$. Then, $x(0,0) \in D$, and, by definition of solution to $\mathcal{H}$ in Definition 2.33, item 1 is established. This fact automatically implies that flows are not possible. In fact, if flow is possible from a point $x_\circ \in C$, then there exists a solution x to $\mathcal{H}$ with $x(0,0) = x_\circ$ and $[0,\varepsilon) \times \{0\} \subset \operatorname{dom} x$ for some $\varepsilon > 0$. This implies that x does not satisfy $\mathfrak{f}$ at $(t,j) = (0,0)$. This is a contradiction, and item 2 follows. To show that item 3 holds, note that the semantics of $\bigcirc$, when each maximal solution x satisfies $\mathfrak{f}$ for each $(t,j) \in \operatorname{dom} x$, implies that $(t, j+1) \in \operatorname{dom} x$ and $x(t, j+1) \in K_{\mathfrak{a}}$ for each $(t,j) \in \operatorname{dom} x$. By definition of solution to $\mathcal{H}$, it implies that for each $(t,j) \in \operatorname{dom} x$, $x(t,j) \in D$ and $G(x(t,j)) \subset K_{\mathfrak{a}}$. Hence, $G(D) \subset K_{\mathfrak{a}}$ and item 3 holds. The property $G(D) \subset D$ follows easily by contradiction. Suppose that there exists $(t', j') \in \operatorname{dom} x$ such that $x(t', j')$ satisfies $\mathfrak{f}$ but that $x(t', j' + 1) \notin D$. Since $\mathfrak{f}$ is satisfied for each solution x to $\mathcal{H}$ at each $(t,j) \in \operatorname{dom} x$, in particular, at $(t,j) = (t', j' + 1) \in \operatorname{dom} x$, then it has to be the case that $x(t', j' + 1) \in D$. This is a contradiction implying that item 3 holds. The proof of sufficiency is left as an exercise – see Exercise 86. $\qquad\blacksquare$

Remark 12.8 (*On next operator*). Items 1 and 2 in Proposition 12.7 imply that every maximal solution to $\mathcal{H}$ can only jump. Hence, solutions can only start from D. In fact, if that is not the case, there would exist a trivial solution to $\mathcal{H}$, which, in turn, would imply that $\mathfrak{f} = \bigcirc \mathfrak{a}$ does not hold from a solution to $\mathcal{H}$. Though it is not necessary as the example below illustrates, one could require that the flow set C is just the empty set, since flows are not allowed. In such a case, $\mathcal{H}$ reduces to a discrete-time system. $\qquad\triangle$

The following academic example illustrates Proposition 12.7.

Example 12.9. *Consider the hybrid system $\mathcal{H}$ with scalar state x and data*[1]

$$F(x) := 1 \quad \forall x \in C := \mathbb{N}, \qquad , \qquad G(x) := \operatorname{sign}(x) \quad \forall x \in D := \mathbb{R} \qquad (12.7)$$

---

[1] The map sign is the set-valued sign map; see List of Symbols.

Let the atomic proposition $\mathfrak{a}$ *be defined as*

$$\mathfrak{a}(x) := \begin{cases} \text{True} & \text{if } |x| = 1 \\ \text{False} & \text{otherwise} \end{cases}$$

for each $x \in D$. *Then, the set* $K_\mathfrak{a}$ *in* (12.5) *is given by* $K_\mathfrak{a} = \{-1, 1\}$. *The sufficient conditions in Proposition 12.7 are used to show that the formula* $\mathfrak{f} = \bigcirc\mathfrak{a}$ *is satisfied for all maximal solutions to* $\mathcal{H}$. *Items 1 and 2 hold by inspection. For each* $x \in D$,

$$G(x) = -1, \qquad G(x) = 1, \qquad \text{or} \qquad G(x) = \{-1, 1\}$$

depending on whether $x < 0$, $x > 0$, *or* $x = 0$, *respectively. This implies that* $G(x) \subset K_\mathfrak{a}$. *Then, since* $D = \mathbb{R}$, *item 3 in Proposition 12.7 holds.*

12.3.2 Forward Invariance for the Always Operator

According to the semantics of the always operator $\square$ in (12.4f), the formula

$$\mathfrak{f} = \square\mathfrak{a} \tag{12.8}$$

is satisfied by a solution x to $\mathcal{H}$ at (t, j) when $x(t', j')$ satisfies $\mathfrak{a}$ for each $t' + j' \geq t + j$ such that $(t', j') \in \operatorname{dom} x$. For a solution x to $\mathcal{H}$ to satisfy this formula at each $(t, j) \in \operatorname{dom} x$, the solution needs to start and stay in the set $K_\mathfrak{a}$ in (12.5). In light of these observations, the following equivalence between forward pre-invariance – as in Definition 3.13 – and the always operator holds.

Proposition 12.10 (Equivalence conditions for $\mathfrak{f} = \square\mathfrak{a}$ at $(t, j) = (0, 0)$). *Given a hybrid closed-loop system* $\mathcal{H} = (C, F, D, G)$ *as in* (2.19) *and an atomic proposition* $\mathfrak{a}$, *let* $K_\mathfrak{a}$ *be given as in* (12.5). *The formula* $\mathfrak{f} = \square\mathfrak{a}$ *is satisfied for each solution* x *to* $\mathcal{H}$ *with* $x(0, 0) \vDash \mathfrak{a}$ *at* $(t, j) = (0, 0)$ *if and only if* $K_\mathfrak{a}$ *is forward pre-invariant for* $\mathcal{H}$.

Proof. First, necessity is shown. With $\mathfrak{f}$ satisfied at $(t, j) = (0, 0)$ for each solution x to $\mathcal{H}$, since $x(0, 0)$ satisfies $\mathfrak{a}$, each solution x to $\mathcal{H}$ satisfies $\operatorname{rge} x \subset K_\mathfrak{a}$. Then, according to Definition 3.13, $K_\mathfrak{a}$ is pre-forward invariant for $\mathcal{H}$. Sufficiency is immediate: $K_\mathfrak{a}$ being forward pre-invariant for $\mathcal{H}$ implies that each solution to $\mathcal{H}$ that starts from $K_\mathfrak{a}$ stays in $K_\mathfrak{a}$ for all future time. Then, by definition of $K_\mathfrak{a}$ in (12.5), $x(0, 0) \vDash \mathfrak{a}$ and $x(t, j) \vDash \mathfrak{a}$ for all $(t, j) \in \operatorname{dom} x$. $\square$

Remark 12.11. Proposition 12.10 establishes a straightforward relationship between the formula $\mathfrak{f} = \square\mathfrak{a}$ and forward pre-invariance of $K_\mathfrak{a}$. Proposition 12.10 is stated for $(t, j) = (0, 0)$, but immediately extends to each (t, j) in the domain of the solution: $\mathfrak{f} = \square\mathfrak{a}$ being satisfied at each $(t, j) \in \operatorname{dom} x$ for each solution x is equivalent to $K_\mathfrak{a}$ being forward pre-invariant. This result is left as an exercise – see Exercise 88. $\triangle$

When $K_\mathfrak{a}$ is not forward pre-invariant, then the formula might be satisfied but only after some time. The following example illustrates this situation.

Example 12.12 (Solutions satisfying $\mathfrak{f} = \square\mathfrak{a}$ after some time). *Let the atomic proposition* $\mathfrak{a}$ *be given by*

$$\mathfrak{a}(x) := \begin{cases} 1 & \text{if } x \in [0, 1] \\ 0 & \text{otherwise} \end{cases}$$

*for each $x \in \mathbb{R}$, and consider the hybrid closed-loop system $\mathcal{H}$ with data (C, F, D, G)
given by*

$$F(x) := 0 \qquad\qquad \forall x \in C := \left[0, \tfrac{1}{2}\right]$$

$$G(x) := \begin{cases} 2 & \text{if } x = 1 \\ 0 & \text{if } x = 2 \end{cases} \qquad \forall x \in D := \{1\} \cup \{2\}$$

The set $K_\mathfrak{a}$ resulting from (12.5) is given by $[0, 1]$. Now, pick $x(0, 0) = 1$. Though $x(0, 0)$ satisfies $\mathfrak{a}$, the unique maximal solution x from that initial condition does not satisfy $\mathfrak{f} = \square \mathfrak{a}$ since after the first jump, $x(0, 1) = 2 \notin K_\mathfrak{a}$. However, $x(t, 2) = 0$ for all $t \geq 0$. Then, though $K_\mathfrak{a}$ is not forward pre-invariant for $\mathcal{H}$, x satisfies $\mathfrak{f}$ for all $(t, j) \in \operatorname{dom} x$ such that $t + j \geq 2$.

As illustrated by Example 12.12, when the initial condition does not satisfy $\mathfrak{a}$ or the set $K_\mathfrak{a}$ is not forward pre-invariant, the formula $\mathfrak{f} = \square \mathfrak{a}$ might be satisfied after some hybrid time. A result that pertains to this situation is given below. It uses the following invariance notion.

Definition 12.13 (Eventually forward pre-invariance). *Given a hybrid closed-loop system $\mathcal{H}$ as in (2.19), a nonempty set $K \subset \mathbb{R}^n$ is said to be <u>eventually forward pre-invariant</u> for $\mathcal{H}$ if every maximal solution x to $\mathcal{H}$ is such that there exists $(t, j) \in \operatorname{dom} x$ such that $x(t', j') \in K_\mathfrak{a}$ for all $(t', j') \in \operatorname{dom} x$, $t' + j' \geq t + j$.*

Proposition 12.14 (Equivalence conditions for $\mathfrak{f} = \square \mathfrak{a}$ at some (t, j)). *Given a hybrid closed-loop system $\mathcal{H} = (C, F, D, G)$ as in (2.19) and an atomic proposition $\mathfrak{a}$, let $K_\mathfrak{a}$ be given as in (12.5). The formula $\mathfrak{f} = \square \mathfrak{a}$ is satisfied for each maximal solution x to $\mathcal{H}$ at some $(t, j) \in \operatorname{dom} x$ if and only if $K_\mathfrak{a}$ is eventually forward pre-invariant for $\mathcal{H}$.*

Proof. To show necessity, since each maximal solution x to $\mathcal{H}$ satisfies $\mathfrak{f} = \square \mathfrak{a}$ at some $(t, j) \in \operatorname{dom} x$, $x(t', j') \vDash \mathfrak{a}$ for all $(t', j') \in \operatorname{dom} x$ such that $t' + j' \geq t + j$. By the definition of $K_\mathfrak{a}$, this property implies that $x(t', j') \in K_\mathfrak{a}$ for all such (t', j')'s. According to Definition 12.13, this implies that $K_\mathfrak{a}$ is eventually forward pre-invariant for $\mathcal{H}$. The proof of sufficiency follows from similar steps to the proof of Proposition 12.10. $\qquad\square$

12.3.3 Finite-Time Attractivity for the Eventually Operator

The semantics for the eventually operator $\Diamond$ in (12.4g) state that for a solution $(t, j) \mapsto x(t, j)$ to $\mathcal{H}$ to satisfy

$$\mathfrak{f} = \Diamond \mathfrak{a} \tag{12.9}$$

at $(t, j) \in \operatorname{dom} x$, there has to exist a hybrid time $(t', j') \in \operatorname{dom} x$, $t' + j' \geq t + j$, at which $x(t', j')$ satisfies $\mathfrak{a}$. Using the construction of the set $K_\mathfrak{a}$ in (12.5) associated to $\mathfrak{a}$, the satisfaction of this formula requires the distance between the solution x and the set $K_\mathfrak{a}$ to converge to zero after a finite amount of flow or after finitely many jumps. Very importantly, unlike $\mathfrak{f} = \square \mathfrak{a}$, after the distance becomes zero, this distance may become nonzero again, indicating that the solution may leave $K_\mathfrak{a}$.

To guarantee that every solution satisfies $\mathfrak{f}$ in (12.9) at each $(t, j) \in \operatorname{dom} x$, the following finite-time attractivity notion for hybrid systems is introduced. In this notion, the amount of time required for a solution x to reach a (closed) set $K_\mathfrak{a}$ is captured by the settling-time functional $t_{K_\mathfrak{a}}$ whose argument is the solution x and its value is a positive number determining the amount of hybrid time needed to converge to $K_{\mathfrak{a}}$. In other words, given a solution x, $t_{K_{\mathfrak{a}}}(x)$ is the time to reach $K_{\mathfrak{a}}$.

Definition 12.15 (Finite-time attractivity). *Given a hybrid closed-loop system* $\mathcal{H} = (C, F, D, G)$ *as in* (2.19), *a nonempty set* $K \subset \mathbb{R}^n$ *is said to be finite-time attractive for* $\mathcal{H}$ *if each maximal solution* x *to* $\mathcal{H}$ *satisfies* $\sup_{(t,j)\in\operatorname{dom} x} t+j \geq t_K(x)$ *and*

$$\lim_{(t,j)\in\operatorname{dom} x \,:\, t+j\nearrow t_K(x)} |x(t,j)|_K = 0 \tag{12.10}$$

The following equivalence between this notion and the eventually operator holds.

Proposition 12.16 (Equivalence conditions for $\mathfrak{f} = \Diamond\mathfrak{a}$). *Given a hybrid closed-loop system* $\mathcal{H} = (C, F, D, G)$ *as in* (2.19) *and an atomic proposition* $\mathfrak{a}$, *let* $K_{\mathfrak{a}}$ *be given as in* (12.5). *The formula* $\mathfrak{f} = \Diamond\mathfrak{a}$ *is satisfied for each maximal solution* x *to* $\mathcal{H}$ *at* $(t, j) = (0, 0)$ *if and only if* $K_{\mathfrak{a}}$ *is finite-time attractive for* $\mathcal{H}$.

Proof. Necessity follows directly from the semantics of the eventually operator $\Diamond$ in (12.4g). Since $\mathfrak{f}$ is satisfied for every maximal solution x to $\mathcal{H}$ at $(t, j) = (0, 0)$, there exists $(t', j') \in \operatorname{dom} x$ such that $x(t', j') \vDash \mathfrak{a}$. As a consequence, $K_{\mathfrak{a}}$ is finite-time attractive for $\mathcal{H}$ with $t_{K_{\mathfrak{a}}}(x) = t' + j'$. To show sufficiency, note that finite-time attractivity of $K_{\mathfrak{a}}$ implies that each maximal solution x to $\mathcal{H}$ is such that there exists $(t, j) \in \operatorname{dom} x$ such that $x(t, j) \in K_{\mathfrak{a}}$. This implies that $\mathfrak{f} = \Diamond\mathfrak{a}$ is satisfied for every maximal solution x to $\mathcal{H}$ at $(t, j) = (0, 0)$. $\qquad\square$

12.3.4 Properties of $\mathcal{H}$ for the Until Operator

Using the approach proposed in this chapter, the satisfaction of the formula

$$\mathfrak{f} = \mathfrak{a}_1 \mathcal{U}_s \mathfrak{a}_2 \tag{12.11}$$

requires solutions to $\mathcal{H}$ to start from or reach the set $K_{\mathfrak{a}_2}$ in finite time, but before then, stay in the set $K_{\mathfrak{a}_1}$. In fact, according to (12.4h), to guarantee that a solution x satisfies $\mathfrak{f}$ in (12.11) at $(t, j) = (0, 0)$, the solution needs to start and stay in the set $K_{\mathfrak{a}_1}$ at least until convergence to the set $K_{\mathfrak{a}_2}$ occurs; or the solution needs to start from the set $K_{\mathfrak{a}_2}$. This property of solutions is characterized in the following result.

Proposition 12.17 (Equivalence conditions for $\mathfrak{f} = \mathfrak{a}_1 \mathcal{U}_s \mathfrak{a}_2$). *Given a hybrid closed-loop system* $\mathcal{H} = (C, F, D, G)$ *as in* (2.19) *and atomic propositions* $\mathfrak{a}_1$ *and* $\mathfrak{a}_2$, *let* $K_{\mathfrak{a}_1}$ *and* $K_{\mathfrak{a}_2}$ *be constructed as in* (12.5) *with* $\mathfrak{a}$ *replaced by* $\mathfrak{a}_1$ *and by* $\mathfrak{a}_2$, *respectively. The formula* $\mathfrak{f} = \mathfrak{a}_1 \mathcal{U}_s \mathfrak{a}_2$ *is satisfied for each maximal solution* x *to* $\mathcal{H}$ *at* $(t, j) = (0, 0)$ *with* $x(0, 0) \vDash \mathfrak{a}_1$ *or* $x(0, 0) \vDash \mathfrak{a}_2$ *if and only if for each* $x_\circ \in K_{\mathfrak{a}_1} \cup K_{\mathfrak{a}_2}$, *each maximal solution* x *to* $\mathcal{H}$ *from* $x_\circ$ *satisfies the following property: there exists* $(t, j) \in \operatorname{dom} x$ *such that*

1. $|x(t, j)|_{K_{\mathfrak{a}_2}} = 0$; *and*

2. $x(t', j') \in K_{\mathfrak{a}_1}$ *for all* $(t', j') \in \operatorname{dom} x$ *such that* $t' + j' < t + j$.

Proof. *Sufficiency follows directly from the semantics of the strong until operator in* (12.4h). *To show necessity, with* $\mathfrak{f}$ *satisfied for each maximal solution* x *at* $(t, j) = (0, 0)$ *with* $x(0, 0) \vDash \mathfrak{a}_1$ *or* $x(0, 0) \vDash \mathfrak{a}_2$, *by the semantics of* $\mathcal{U}_s$, *there exists* $(t, j) \in$

$\mathrm{dom}\,x$ *such that*

   i) $x(t, j)$ *satisfies* $\mathfrak{a}_2$*; and*

   ii) $x(t', j')$ *satisfies* $\mathfrak{a}_1$ *for all* $(t', j') \in \mathrm{dom}\,x$ *such that* $t' + j' < t + j$.

Item i implies that $x(t, j) \in K_{\mathfrak{a}_2}$*. Hence,* $|x(t, j)|_{K_{\mathfrak{a}_2}} = 0$ *and item 1 holds. Item ii implies that* $x(t', j') \in K_{\mathfrak{a}_1}$ *for all* $(t', j') \in \mathrm{dom}\,x$ *such that* $t' + j' < t + j$*. Hence, item 2 holds.* □

12.4 SUFFICIENT CONDITIONS

The characterizations in §12.3 reveal key dynamical properties needed from $\mathcal{H}$ for its solutions to satisfy basic LTL formulas. In this section, these characterizations are exploited to formulate sufficient conditions for formulas involving the always, the eventually, or the until operator. The proposed sufficient conditions do not require computing the solutions to the system to assure that the formula is satisfied. In particular, these conditions certify invariance and finite-time attractivity of the set associated with the given formula. Conveniently, the results in Chapter 11 are employed to guarantee forward pre-invariance for formulas involving the always operator. Lyapunov conditions that guarantee finite-time attractivity of a set are presented for formulas that use the eventually operator. These conditions resemble those in Theorem 3.19 for attractivity in the limit and, as required, assure that the solutions belong to the set of interest after a finite amount of flow time or finitely many jumps.

12.4.1 Sufficient Conditions for the Always Operator

In § 12.3.2, Proposition 12.10 establishes that $\mathfrak{f} = \square \mathfrak{a}$ is satisfied at $(t, j) = (0, 0)$ for solutions to $\mathcal{H}$ from initial conditions satisfying $\mathfrak{f}$ when the associated set $K_{\mathfrak{a}}$ is forward pre-invariant for $\mathcal{H}$ – see Remark 12.11 and Exercise 87 for the case of the formula satisfied for each (t, j). The sufficient conditions for forward pre-invariance of general sets given in Chapter 11 are applied to assure that such formula holds. More precisely, the conditions in terms of a barrier function in Theorem 11.4 are exploited to formulate conditions guaranteeing the satisfaction of $\mathfrak{f} = \square \mathfrak{a}$.

Theorem 12.18 (Sufficient conditions for $\mathfrak{f} = \square \mathfrak{a}$ using barrier functions). *Given a hybrid closed-loop system* $\mathcal{H} = (C, F, D, G)$ *as in (2.19) and an atomic proposition* $\mathfrak{a}$*, suppose the following properties hold:*

1. $\mathcal{H}$ satisfies the hybrid basic conditions;

2. The set $K_{\mathfrak{a}}$ in (12.5) is closed and a subset of $C \cup D$;

3. There exists a barrier function candidate[2] with respect to $K_{\mathfrak{a}}$ for $\mathcal{H}$ such that

$$\mathfrak{a}(x) = \begin{cases} \texttt{True} & \text{if } B(x) \leq 0, x \in C \cup D \\ \texttt{False} & \text{otherwise} \end{cases} \qquad \forall x \in \mathbb{R}^n$$

---

[2]See Definition 11.1.

4. There exists an open neighborhood $\mathcal{U}$ of $\partial K_{\mathfrak{a}}$ such that

$$\dot{B}(x) \leq 0 \qquad \forall x \in (\mathcal{U} \backslash K_{\mathfrak{a}}) \cap C \qquad (12.12)$$

5. For each $x \in D \cap K_{\mathfrak{a}}$,

$$B^{+}(x) \leq 0 \qquad (12.13)$$

Then, the formula $\mathfrak{f} = \Box\mathfrak{a}$ is satisfied for each solution x to $\mathcal{H}$ and for each $(t, j) \in \operatorname{dom} x$ if $x(0,0) \vDash \mathfrak{a}$.

Theorem 12.18 follows from an application of item 1 of Theorem 11.4 with $K = K_{\mathfrak{a}}$. The proof is left as an exercise. Next, Theorem 12.18 is illustrated in examples. Though rather simple, the first example pinpoints a few subtleties involved in applying Theorem 12.18 to certify a temporal logic formula.

Example 12.19 ($\Box\mathfrak{a}$ for a hybrid system with timer and logic variables). *Consider a hybrid closed-loop system $\mathcal{H} = (C, F, D, G)$ with state $x = (\tau, q) \in [0, T^*] \times \{0, 1\}$ given by*

$$F(x) := \begin{bmatrix} 1 \\ 0 \end{bmatrix} \qquad \forall x \in C := [0, T^*] \times \{0, 1\}$$

$$G(x) := \begin{bmatrix} 0 \\ 1 - q \end{bmatrix} \qquad \forall x \in D := \{T^*\} \times \{0, 1\}$$

where τ denotes a timer and q is a logic variable. The parameter T^* is the period of the timer. The evolution of τ and q is as follows. During flows, the timer counts ordinary time and the logic variable remains constant. When the timer reaches T^*, then the value of the logic variable is flipped: it is changed from zero to one, or from one to zero. Now, consider the atomic proposition $\mathfrak{a}$ defined as

$$\mathfrak{a}(x) := \begin{cases} \text{True} & \text{if } \tau \in [0, T^*] \\ \text{False} & \text{otherwise} \end{cases}$$

for each $x \in \mathbb{R}^2$. Trivially, $\mathfrak{a}$ is true for each solution to $\mathcal{H}$ since those are such that the τ component remains in $[0, T^*]$ for all time. Then, every solution to $\mathcal{H}$ satisfies $\mathfrak{f} = \Box\mathfrak{a}$. Note that the system does not have solutions from points that are not in C.

Next, the same conclusion is obtained by an application of Theorem 12.18 with a barrier function candidate B that is identically zero.

- *Items 1-3 hold since C and D are closed sets, and F and G are continuous functions. Hence, $\mathcal{H}$ satisfies the hybrid basic conditions. Furthermore, the set $K_{\mathfrak{a}}$ is equal to C, which is closed and a subset of $C \cup D$.*

- *Items 4 and 5 hold since B is identically zero.*

Therefore, Theorem 12.18 certifies that the formula $\mathfrak{f} = \Box\mathfrak{a}$ is satisfied for each solution x to $\mathcal{H}$ at each $(t, j) \in \operatorname{dom} x$.

Now, consider the more interesting case when the flow map of $\mathcal{H}$ is given by $F(x) = (q, 0)$ and the atomic proposition $\mathfrak{a}$ is defined as

$$\mathfrak{a}(x) := \begin{cases} \texttt{True} & \text{if } \tau = 0, q = 0 \\ \texttt{False} & \text{otherwise} \end{cases}$$

for each $x \in \mathbb{R}^2$. In this case, the set $K_\mathfrak{a}$ in (12.5) is given by $K_\mathfrak{a} = \{(0, 0)\}$. To show that $\mathfrak{f} = \Box\mathfrak{a}$ is satisfied for each solution to $\mathcal{H}$ from $K_\mathfrak{a}$, consider the barrier function candidate $B(x) = \tau + q$. Note that if $x \in C \cup D = [0, T^*] \times \{0, 1\}$ and $B(x) \leq 0$, then $\tau = q = 0$. It is straightforward to check that items 1-3 hold.

- *Pick $\mathcal{U}$ small enough so that $\mathcal{U} \cap (\mathbb{R} \times \{1\})$ is empty. Since each $x = (\tau, q) \in (\mathcal{U} \backslash K_\mathfrak{a}) \cap C$ is such that $q = 0$, it follows that*

$$\dot{B}(x) = q + 0 \leq 0$$

Then, item 4 holds.

- *Item 5 holds since $K_\mathfrak{a} \cap D$ is empty.*

Therefore, Theorem 12.18 certifies that the formula $\mathfrak{f} = \Box\mathfrak{a}$ is satisfied for each solution x to $\mathcal{H}$ with $x(0, 0) \vDash \mathfrak{a}$ at each $(t, j) \in \text{dom } x$. Note that solutions from $x(0, 0)$ arbitrarily close to $K_\mathfrak{a}$ are constant and never satisfy the formula.

The following example is a variation of the system in Example 11.2.

Example 12.20 ($\Box\mathfrak{a}$ for a planar hybrid system). *Consider a hybrid closed-loop system $\mathcal{H} = (C, F, D, G)$ with state $x = (x_1, x_2) \in \mathbb{R}^2$ given by*

$$\begin{aligned} F(x) &:= \begin{bmatrix} 0 & 1 \\ -1 & 0 \end{bmatrix} x & \forall x \in C &:= \{x \in \mathbb{R}^2 : x_2 \geq 0\} \\ G(x) &:= \lambda x & \forall x \in D &:= \{x \in \mathbb{R}^2 : x_2 = 0\} \end{aligned} \tag{12.14}$$

where $|\lambda| \leq 1$. Let the atomic proposition $\mathfrak{a}$ be defined as

$$\mathfrak{a}(x) := \begin{cases} \texttt{True} & \text{if } |x|^2 \leq r, x_2 \geq 0 \\ \texttt{False} & \text{otherwise} \end{cases} \tag{12.15}$$

for each $x \in \mathbb{R}^2$, where $r \geq 0$. Then, the set $K_\mathfrak{a}$ in (12.5) is given as

$$K_\mathfrak{a} = \{x \in C \cup D : B(x) \leq 0\} = \{x \in \mathbb{R}^2 : |x|^2 \leq r, x_2 \geq 0\}$$

with

$$B(x) := x^\top x - r \qquad \forall x \in \mathbb{R}^2$$

As shown next, the conditions in Theorem 12.18 hold for each $r \geq 0$:

- *Items 1-3 hold by construction. In fact, $\mathcal{H}$ satisfies the hybrid basic conditions. Moreover, the set $K_\mathfrak{a}$ is closed and a subset of $C \cup D = \mathbb{R} \times \mathbb{R}_{\geq 0}$.*

- *Item 4 holds since, with B being continuously differentiable,*

$$\langle \nabla B(x), F(x) \rangle = 0 \qquad \forall x \in C$$

- *Item 5 is satisfied since*

$$B(G(x)) = \lambda^2 x^\top x - r \le |x|^2 - r \le 0 \qquad \forall x \in D \cap K_{\mathfrak{a}}$$

and $|\lambda| \le 1$.

Then, for each $r \ge 0$ *and* $|\lambda| \le 1$, $\mathfrak{f} = \Box\mathfrak{a}$ *is satisfied for each solution* x *to* $\mathcal{H}$ *with* $x(0,0) \vDash \mathfrak{a}$.

12.4.2 Sufficient Conditions for the Eventually Operator

As shown in § 12.3.3, finite-time attractivity of the set $K_{\mathfrak{a}}$ is mandatory for the satisfaction of the formula $\mathfrak{f} = \Diamond\mathfrak{a}$. Attractivity in the limit, as implied by conditions in Theorem 3.19, is not enough for solutions x to satisfy such a formula. Indeed, having $(t,j) \mapsto |x(t,j)|_{K_{\mathfrak{a}}}$ converge to zero as $t+j$ tends to infinity does not imply the existence of $(t',j') \in \operatorname{dom} x$ at which $x(t',j') \in K_{\mathfrak{a}}$. While at times it might be enough to converge to a neighborhood of $K_{\mathfrak{a}}$ in finite (hybrid) time, which is guaranteed for complete solutions when $K_{\mathfrak{a}}$ is attractive, the satisfaction of $\mathfrak{f} = \Diamond\mathfrak{a}$ requires convergence to $K_{\mathfrak{a}}$ after finite time. Lyapunov-based conditions that assure such finite-time attractivity of $K_{\mathfrak{a}}$ are presented in the following result.

Theorem 12.21 (Sufficient conditions for $\mathfrak{f} = \Diamond\mathfrak{a}$ at $(t,j) = (0,0)$). *Given a hybrid closed-loop system* $\mathcal{H} = (C, F, D, G)$ *as in* (2.19) *and an atomic proposition* $\mathfrak{a}$, *suppose the set* $K_{\mathfrak{a}}$ *in* (12.5) *is closed and that there exists a set* $\mathcal{U} \subset X$ *containing an open neighborhood of* $K_{\mathfrak{a}}$ *such that* $G(\mathcal{U}) \subset \mathcal{U}$. *Furthermore, suppose that one of the following conditions hold:*

1. *There exist a continuous function* $V : \mathbb{R}^n \to \mathbb{R}_{\ge 0}$, *locally Lipschitz on an open neighborhood of* $\overline{C} \cap \mathcal{U}$, *and constants* $c_1 > 0, c_2 \in [0,1)$ *such that*

   a) *For each* $x_\circ \in X \cap \mathcal{U}$ *such that* $\mathfrak{a}(x_\circ) = \mathtt{False}$, *each maximal solution* x *to* $\mathcal{H}$ *from* $x_\circ$ *satisfies*

   $$V(x_\circ)^{1-c_2} \le c_1(1-c_2) \sup_{(t,j) \in \operatorname{dom} x} t \qquad (12.16)$$

   b) *There exist functions* $\alpha_1, \alpha_2 \in \mathcal{K}_\infty$ *such that*

   $$\alpha_1(|x|_{K_{\mathfrak{a}}}) \le V(x) \le \alpha_2(|x|_{K_{\mathfrak{a}}}) \qquad \forall x \in X \cap \mathcal{U} \qquad (12.17)$$

   and

       i. *For each* $x \in C \cap \mathcal{U}$ *such that* $\mathfrak{a}(x) = \mathtt{False}$,

   $$\dot{V}(x) \le -c_1 V(x)^{c_2}$$

       ii. *For each* $x \in D \cap \mathcal{U}$ *such that* $\mathfrak{a}(x) = \mathtt{False}$,

   $$\Delta V(x) \le 0$$

or

2. *There exist a continuous function* $V : \mathbb{R}^n \to \mathbb{R}_{\geq 0}$, *locally Lipschitz on an open neighborhood of* $\overline{C} \cap \mathcal{U}$, *and a constant* $c > 0$ *such that*

   a) *For each* $x_\circ \in X \cap \mathcal{U}$ *such that* $\mathfrak{a}(x_\circ) = \text{False}$, *each maximal solution* x *to* $\mathcal{H}$ *from* $x_\circ$ *satisfies*

   $$ceil\left(\frac{V(x_\circ)}{c}\right) \leq \sup_{(t,j) \in \text{dom}\, x} j \qquad (12.18)$$

   b) *There exist functions* $\alpha_1, \alpha_2 \in \mathcal{K}_\infty$ *satisfying* (12.17) *and*
      i. *For each* $x \in C \cap \mathcal{U}$ *such that* $\mathfrak{a}(x) = \text{False}$,

      $$\dot{V}(x) \leq 0$$

   ii. *For each* $x \in D \cap \mathcal{U}$ *such that* $\mathfrak{a}(x) = \text{False}$,

   $$\Delta V(x) \leq -\min\{c, V(x)\}$$

where the functions $\dot{V}$ *and* ΔV *are defined in* (3.13) *and* (3.16), *respectively.*[3] *Then, the formula* $\mathfrak{f} = \Diamond \mathfrak{a}$ *is satisfied for each maximal solution* x *to* $\mathcal{H}$ *from* $L_V(r) \cap X$ *at* $(t,j) = (0,0)$, *where* $L_V(r)$ *with* $r \in [0, \infty]$ *is a sublevel set of* V *contained in* $\mathcal{U}$. *Moreover, each maximal solution* x *from* $x_\circ \in L_V(r) \cap X$ *is such that the first time* $(t', j') \in \text{dom}\, x$ *for which* $x(t', j') \vDash \mathfrak{a}$ *satisfies*

$$t' + j' = t_{K_\mathfrak{a}}(x) \qquad (12.19)$$

where $t_{K_\mathfrak{a}}(x)$ *is upper bounded as follows:*

3. *If item 1 holds, then* $t_{K_\mathfrak{a}}(x)$ *is upper bounded by* $t^\star(x_\circ) + j^\star(x_\circ)$, *where* $t^\star(x_\circ) = \frac{V(x_\circ)^{1-c_2}}{c_1(1-c_2)}$ *and* $j^\star(x_\circ)$ *is such that* $(t^\star(x_\circ), j^\star(x_\circ)) \in \text{dom}\, x$.

4. *If item 2 holds, then* $t_{K_\mathfrak{a}}(x)$ *is upper bounded by* $t^\star(x_\circ) + j^\star(x_\circ)$, *where* $j^\star(x_\circ) = ceil(\frac{V(x_\circ)}{c})$ *and* $t^\star(x_\circ)$ *is such that* $(t^\star(x_\circ), j^\star(x_\circ)) \in \text{dom}\, x$ *and* $(t^\star(x_\circ), j^\star(x_\circ) - 1) \in \text{dom}\, x$.

Proof. Only item 1 is shown – the proof of item 2 is left as an exercise. Let x be maximal solution to $\mathcal{H}$ with $x(0,0) = x_\circ \in L_V(r) \cap X \cap \mathcal{U}$ such that $\mathfrak{a}(x_\circ) = \text{False}$. Since $L_V(r) \subset \mathcal{U}$, items 1b(i) and 1b(ii) imply that the solution x remains in $\mathcal{U}$ at least until it satisfies $\mathfrak{a}$. Pick any $(t,j) \in \text{dom}\, x$ and let $0 = t_0 \leq t_1 \leq \cdots \leq t_{j+1} = t$ satisfy

$$\text{dom}\, x \cap ([0,t] \times \{0, 1, \ldots, j\}) = \bigcup_{i=0}^{j} ([t_i, t_{i+1}] \times \{i\}) \qquad (12.20)$$

For each $i \in \{0, 1, \ldots, j\}$ and all $s \in (t_i, t_{i+1})$, $x(s,i) \in C \cap \mathcal{U}$. Condition 1b(i) implies that, for each $i \in \{0, 1, \ldots, j\}$ and for almost all $s \in [t_i, t_{i+1}]$ such that $x(s,i) \in (C \cap \mathcal{U}) \setminus K_\mathfrak{a}$,

$$\tfrac{d}{ds} V(x(s,i)) \leq -c_1 V(x(s,i))^{c_2} \qquad (12.21)$$

---

[3]Since $\mathcal{H}$ does not necessarily satisfy the hybrid basic conditions, max in the expressions in (3.13) and (3.16) are to be replaced by sup.

which implies that

$$V(x(s,i))^{-c_2} \, dV(x(s,i)) \le -c_1 ds \tag{12.22}$$

Integrating over $[t_i, t_{i+1}]$ both sides of this inequality yields

$$\frac{1}{1-c_2} \left(V(x(t_{i+1},i))^{1-c_2} - V(x(t_i,i))^{1-c_2} \right) \le -c_1(t_{i+1} - t_i) \tag{12.23}$$

Similarly, for each $i \in \{1,2,\ldots,j\}$, condition 1b(ii) implies that if $x(t_i, i-1) \in (D \cap \mathcal{U}) \setminus K_\mathfrak{a}$ then

$$V(x(t_i,i)) \le V(x(t_i, i-1)) \tag{12.24}$$

The two inequalities in (12.23) and (12.24) imply that, for each $(t,j) \in \operatorname{dom} x$,

$$\frac{1}{1-c_2} \left(V(x(t,j))^{1-c_2} - V(x_\circ)^{1-c_2} \right) \le -c_1 t \tag{12.25}$$

Using $G(\mathcal{U}) \subset \mathcal{U}$, the lower bound on the function V, and the fact that $c_2 \in (0,1)$, gives

$$\alpha_1^{1-c_2}(|x(l,j)|_{K_\mathfrak{a}}) \le V(x(t,j))^{1-c_2} \le V(x_\circ)^{1-c_2} - c_1(1-c_2)t \tag{12.26}$$

Then, it follows that

$$|x(t,j)|_{K_\mathfrak{a}} \le \alpha_1^{-1} \left((V(x_\circ)^{1-c_2} - c_1(1-c_2)t)^{\frac{1}{1-c_2}} \right) \tag{12.27}$$

By setting this upper bound to zero and solving for t, the settling-time functional evaluated at the solution x satisfies

$$t_{K_\mathfrak{a}}(x) \le t^\star(x_\circ) + j^\star(x_\circ) \tag{12.28}$$

where $t^\star(x_\circ) = \frac{V(x_\circ)^{1-c_2}}{c_1(1-c_2)}$, and $j^\star(x_\circ)$ is such that $(t^\star(x_\circ), j^\star(x_\circ)) \in \operatorname{dom} x$. Note that $t^\star(x_\circ) < \sup_{(t,j) \in \operatorname{dom} x} t$ given by item 1a), the existence of $(t^\star(\xi), j^\star(\xi)) \in \operatorname{dom} x$ is guaranteed. $\qquad\qquad\square$

The settling-time functional $t_{K_\mathfrak{a}}$ in Theorem 12.21 determines an upper bound on the amount of (hybrid) time it would take for a solution to satisfy the formula $\mathfrak{f} = \Diamond \mathfrak{a}$. Conveniently, the upper bounds provided in Theorem 12.21 do not require the computation of solutions to $\mathcal{H}$. In fact, they depend only on the function V and the constants involved in the conditions therein. The price to pay is finding V.

Remark 12.22. The conditions in item 1 in Theorem 12.21 require V to strictly decrease during flows but only requires V to be nonincreasing at jumps; i.e., V is weak. The conditions in item 2 require the opposite. It is immediate to show that the same result therein holds when V decreases both during flows and at jumps, via the satisfaction of the conditions in item 1 and in item 2. Though finding a function V with such a property might not always be easy, the upper bound on the settling-time functional $t_{K_\mathfrak{a}}$ can be significantly improved when such a function is available. Note that the only conditions requiring information about solutions are those in (12.16) and (12.18). These conditions hold for free when maximal solutions are complete with its domain of definition unbounded in the t direction or in the j direction, respectively. $\qquad\qquad\triangle$

The following examples illustrate Theorem 12.21.

Example 12.23 ($\Diamond\mathfrak{a}$ for a hybrid system with timer). *Consider a hybrid closed-loop system* $\mathcal{H} = (C, F, D, G)$ *with state* $x = (z, \tau) \in \mathbb{R} \times [0, T^*] =: X$ *and data*

$$
\begin{aligned}
F(x) &:= \begin{bmatrix} -k|z|^\alpha sgn(z) \\ 1 \end{bmatrix} & \forall x \in C := \mathbb{R} \times [0, T^*] \\
G(x) &:= \begin{bmatrix} -z \\ 0 \end{bmatrix} & \forall x \in D := \mathbb{R} \times \{T^*\}
\end{aligned}
\tag{12.29}
$$

where $\alpha \in (0, 1)$ *and* $k > 0$*. Consider the atomic proposition* $\mathfrak{a}$ *defined as*

$$
\mathfrak{a}(x) := \begin{cases} \text{True} & \text{if } z = 0 \\ \text{False} & \text{otherwise} \end{cases} \qquad \forall x \in X
$$

To show that maximal solutions to $\mathcal{H}$ *satisfy* $\mathfrak{f} = \Diamond\mathfrak{a}$*, consider the function* $V :$ $\mathbb{R}^2 \to \mathbb{R}_{\geq 0}$ *given by* $V(x) = \frac{1}{2}z^2$ *for each* $x \in \mathbb{R}^2$*. With* $K_\mathfrak{a}$ *given in (12.5), for each* $x \in C \setminus K_\mathfrak{a}$*,*

$$
\langle \nabla V(x), F(x) \rangle = -k|z|^{1+\alpha} = -2^{\frac{1+\alpha}{2}} k V(x)^{\frac{1+\alpha}{2}}
$$

and, for each $x \in D \setminus K_\mathfrak{a}$*,*
$$
V(G(x)) - V(x) = 0
$$

Therefore, the conditions in item 1b) in Theorem 12.21 hold with $\mathcal{U} = \mathbb{R} \times \mathbb{R}$*,* $c_1 = 2^{\frac{1+\alpha}{2}} k > 0$*, and* $c_2 = \frac{1+\alpha}{2} \in (0, 1)$*. In addition, item 1a) therein also holds since every maximal solution to* $\mathcal{H}$ *is complete with its domain of definition unbounded in the* t *direction. Thus, according to Theorem 12.21, the formula* $\mathfrak{f} = \Diamond\mathfrak{a}$ *is satisfied for each maximal solution* x *to* $\mathcal{H}$ *at* $(t, j) = (0, 0)$*. In addition, since* $K_\mathfrak{a}$ *is forward pre-invariant for* $\mathcal{H}$*, this formula is satisfied at each* $(t, j) \in \operatorname{dom} x$*.*

Example 12.24 ($\Diamond\mathfrak{a}$ for the bouncing ball). *Consider the hybrid system model* $\mathcal{H} = (C, F, D, G)$ *in Example 11.2, which is the well-known bouncing ball system – a simplified version of the one-degree-of-freedom juggling system introduced in Example 2.32. Every maximal solution to this system is Zeno. Define an atomic proposition* $\mathfrak{a}$ *as follows: for each* $x \in \mathbb{R}^2$*,*

$$
\mathfrak{a}(x) := \begin{cases} \text{True} & \text{if } x_2 \leq 0 \\ \text{False} & \text{otherwise} \end{cases}
$$

With $K_\mathfrak{a}$ *in (12.5) and* $\mathcal{U} = \mathbb{R}^2$*, let* $V(x) = |x_2|$ *for all* $x \in \mathbb{R}^2$*. This function is continuously differentiable on the open set* $X \setminus (\mathbb{R} \times \{0\})$*. It follows that for all* $x \in C \setminus K_\mathfrak{a}$*,*
$$
\dot{V}(x) = \langle \nabla V(x), F(x) \rangle = -\gamma
$$

Then, item 1b(i) in Theorem 12.21 holds with $c_1 = \gamma$ *and* $c_2 = 0$*. Since* $D \setminus K_\mathfrak{a} = \emptyset$*, item 1b(ii) holds vacuously. While it can be shown that every maximal solution to* $\mathcal{H}$ *is complete but with its domain of definition unbounded in the* j *direction only – this is due to solutions to* $\mathcal{H}$ *being Zeno – direct integration of the continuous dynamics show that each maximal solution* x *evolves continuously for* $|x_2(0,0)|/\gamma$ *seconds. Then, item 1a) holds. Therefore, the formula* $\mathfrak{f} = \Diamond\mathfrak{a}$ *is satisfied for each maximal solution to* $\mathcal{H}$ *at* $(t, j) = (0, 0)$*. Since every maximal solution from* $K_\mathfrak{a}$*, after some time, jumps from* $K_\mathfrak{a}$ *and then converges to* $K_\mathfrak{a}$ *again in finite time, the formula* $\mathfrak{f} = \Diamond\mathfrak{a}$ *holds for every* (t, j) *in the domain of each solution to* $\mathcal{H}$*.*

The following example illustrates Theorem 12.21 in a hybrid system with maximal solutions that jump out of $K_\mathfrak{a}$ but, after finite hybrid time, jump back to $K_\mathfrak{a}$, recurrently.

Example 12.25 ($\Diamond\mathfrak{a}$ for impulse-coupled oscillators). *Consider a hybrid closed-loop system $\mathcal{H} = (C, F, D, G)$ modeling two impulsive oscillators, each with a resettable timer state, τ_1 and τ_2, respectively. In between impulses, the oscillators count time, linearly, according to $\dot{\tau}_1 = \dot{\tau}_2 = \gamma$, where $\gamma > 0$ is a constant. The impulses are triggered when any of the timers reaches the threshold $T^* > 0$. At such events, any timer that has reached the threshold is reset to zero and, when only $\tau_1 = T^*$, the timer τ_2 is incremented in value according to $\tau_2 + \widetilde{\varepsilon}\tau_2$ where $\widetilde{\varepsilon} > 0$ is a constant. Similarly, when only $\tau_2 = T^*$, then τ_1 is reset according to $\tau_1 + \widetilde{\varepsilon}\tau_1$. This mechanism can be modeled by a hybrid system with state $x = (\tau_1, \tau_2) \in [0, T^*] \times [0, T^*] =: X$ and data given by*

$$
F(x) := \begin{bmatrix} \gamma \\ \gamma \end{bmatrix} \qquad \forall x \in C := X
$$

$$
G(x) := \begin{bmatrix} \widetilde{G}((1 + \widetilde{\varepsilon})\tau_1) \\ \widetilde{G}((1 + \widetilde{\varepsilon})\tau_2) \end{bmatrix} \quad \forall x \in D := \{x \in C : \max\{\tau_1, \tau_2\} = T^*\}
$$

$$(12.30)$$

where the set-valued map $\widetilde{G}$ is given by

$$
\widetilde{G}(\tau) = \begin{cases} \tau & \text{if } \tau \in [0, T^*) \\ \{0, T^*\} & \text{if } \tau = T^* \\ 0 & \text{if } \tau \in (T^*, \infty) \end{cases} \qquad \forall \tau \in \mathbb{R}_{\geq 0}
$$

Now, consider the atomic proposition $\mathfrak{a}$ defined as

$$
\mathfrak{a}(x) := \begin{cases} \text{True} & \text{if } \tau_1 = \tau_2 \\ \text{False} & \text{otherwise} \end{cases}
$$

for each $x \in X$ and the formula $\mathfrak{f} = \Diamond\mathfrak{a}$. This specification corresponds to eventual synchronization of the two timers; that is, synchronization after a finite amount of hybrid time. To show that this formula is satisfied for all maximal solutions to $\mathcal{H}$ from a properly defined set $\mathcal{U}$, let $k = \frac{\widetilde{\varepsilon}T^*}{2+\widetilde{\varepsilon}}$ and $m^* = \frac{(1+\widetilde{\varepsilon})T^*}{2+\widetilde{\varepsilon}}$. Then, define

$$
V(x) := \min\{|\tau_1 - \tau_2|, T^* + k - |\tau_1 - \tau_2|\} \qquad \forall x \in X
$$

This function is locally Lipschitz and, at points x in $\mathcal{X} \setminus K_\mathfrak{a}$, is continuously differentiable, where $K_\mathfrak{a}$ is given as in (12.5) and $\mathcal{X}$ is given by the open set

$$
\mathcal{X} := \{x \in X : V(x) < m^*\} = \{x \in X : |\tau_1 - \tau_2| \neq m^*\}
$$

Pick $m \in (0, m^*)$ and define

$$
\mathcal{U} := \{x \in X : V(x) \leq m\}
$$

By the definition of V, it follows that

$$
\dot{V}(x) = 0 \qquad \forall x \in (C \cap \mathcal{U}) \setminus K_\mathfrak{a}
$$

For each $x \in (D \cap \mathcal{U}) \setminus K_{\mathfrak{a}}$, exploiting the symmetry in the definition of V, without loss of generality, only the case $x = (T^*, \tau_2) \in (D \cap \mathcal{U}) \setminus K_{\mathfrak{a}}$ with $\tau_2 \in [0, T^*] \setminus \{\frac{T^*}{2+\tilde{\varepsilon}}\}$ is considered. At such points, G is single valued and the following hold:

$$V(x) = \min\{T^* - \tau_2, k + \tau_2\}$$
$$V(G(x)) = \min\{\widetilde{G}((1 + \tilde{\varepsilon})\tau_2), T^* + k - \widetilde{G}((1 + \tilde{\varepsilon})\tau_2)\}$$

When $\widetilde{G}((1 + \tilde{\varepsilon})\tau_2) = 0$, it follows that $V(G(x)) = 0$. When $\widetilde{G}((1 + \tilde{\varepsilon})\tau_2) = (1 + \tilde{\varepsilon})\tau_2$, there are two cases based on the value of τ_2 relative to $\frac{T^*}{2+\tilde{\varepsilon}}$:

a) If $\tau_2 < \frac{T^*}{2+\tilde{\varepsilon}}$, then $V(x) = k + \tau_2 > (1 + \tilde{\varepsilon})\tau_2 \geq V(G(x))$;

b) If $\tau_2 > \frac{T^*}{2+\tilde{\varepsilon}}$, then $V(x) = T^* - \tau_2 \geq V(G(x))$.

Thus, $\Delta V(x) \leq 0$ for all $x \in (D \cap \mathcal{U}) \setminus K_{\mathfrak{a}}$. An application of Proposition 2.34 implies that every maximal solution to $\mathcal{H}$ is complete. Moreover, given $\tilde{\varepsilon} > 0$, for $\varepsilon = \frac{\tilde{\varepsilon}}{1+\tilde{\varepsilon}}$ and m such that $(K_{\mathfrak{a}} + \varepsilon \mathbb{B}) \cap C \subset C \cap \mathcal{U}$, $G(x) = 0 \in K_{\mathfrak{a}}$ for all $x \in D \cap \mathcal{U} \cap (K_{\mathfrak{a}} + \varepsilon \mathbb{B})$. Therefore, it follows from Theorem 12.21 that maximal solutions to $\mathcal{H}$ from $\mathcal{U}$ converge to $K_{\mathfrak{a}}$ in finite hybrid time. Note that maximal solutions starting from $K_{\mathfrak{a}}$ jump out of $K_{\mathfrak{a}}$ but they converge back to $K_{\mathfrak{a}}$ after a finite amount of hybrid time. In fact, from $D \cap K_{\mathfrak{a}}$, which is the singleton $\{(T^*, T^*)\}$, the jump map G is given by the set $\{0, T^*\} \times \{T^*, 0\}$, allowing for a solution that jumps to $\{(0, T^*)\}$, which is in D but not in $K_{\mathfrak{a}}$. However, for this solution, flow after such a jump is not possible, and for it to be maximal, it has to jump. The value of the solution after that second jump is in $K_{\mathfrak{a}}$. Then, the formula $\mathfrak{f} = \Diamond \mathfrak{a}$ is satisfied for each maximal solution x to $\mathcal{H}$ from $\mathcal{U}$ and for each $(t, j) \in \mathrm{dom}\, x$.

12.4.3 Sufficient Conditions for the Until Operator

Given two atomic propositions, $\mathfrak{a}_1$ and $\mathfrak{a}_2$, for the formula $\mathfrak{f} = \mathfrak{a}_1 \mathcal{U}_s \mathfrak{a}_2$ to be satisfied, the semantics of the strong until operator ($\mathcal{U}_s$) requires the satisfaction of $\mathfrak{a}_2$ at some hybrid time, and, until then, the satisfaction of $\mathfrak{a}_1$. This fact is captured by Proposition 12.17, precisely, by item 1 and item 2, respectively. Building from the conditions in the previous sections, the following result provides sufficient conditions for the formula $\mathfrak{f} = \mathfrak{a}_1 \mathcal{U}_s \mathfrak{a}_2$ to be satisfied for all maximal solutions x to $\mathcal{H}$, both at $(t, j) = (0, 0)$ and at any $(t, j) \in \mathrm{dom}\, x$.

Theorem 12.26 (Sufficient conditions for $\mathfrak{f} = \mathfrak{a}_1 \mathcal{U}_s \mathfrak{a}_2$ at $(t, j) = (0, 0)$). *Given a hybrid closed-loop system $\mathcal{H} = (C, F, D, G)$ as in (2.19) and atomic propositions $\mathfrak{a}_1$ and $\mathfrak{a}_2$, let $K_{\mathfrak{a}_1}$ and $K_{\mathfrak{a}_2}$ be constructed as in (12.5) with $\mathfrak{a}$ replaced by $\mathfrak{a}_1$ and by $\mathfrak{a}_2$, respectively. Suppose there exists a set $\mathcal{U} \subset X$ containing an open neighborhood of $K_{\mathfrak{a}_2}$ such that $G(\mathcal{U}) \subset \mathcal{U}$. Then, the formula $\mathfrak{f} = \mathfrak{a}_1 \mathcal{U}_s \mathfrak{a}_2$ is satisfied for every maximal solution x to $\mathcal{H}$ at $(t, j) = (0, 0)$ if the following conditions hold:*

1. $K_{\mathfrak{a}_2}$ is closed;

2. Either item 1 or item 2 in Theorem 12.21 is satisfied with $\mathfrak{a}$ therein replaced by $\mathfrak{a}_2$, for some function V and constant $r \in [0, \infty]$ as required therein;

3. $x(0, 0) \in (K_{\mathfrak{a}_1} \cap L_V(r)) \cup K_{\mathfrak{a}_2}$; and

4. $(L_V(r) \cap X) \setminus K_{\mathfrak{a}_2} \subset K_{\mathfrak{a}_1}$;

where $L_V(r)$ is a sublevel set of V contained in $\mathcal{U}$. Moreover, an upper bound for the settling-time functional $t_{K_{\mathfrak{a}_2}}$ is given by item 3 or by item 4 in Theorem 12.21, depending on whether item 1 or item 2 in Theorem 12.21 holds, respectively. Furthermore, if

5. $G(K_{\mathfrak{a}_2} \cap D) \subset L_V(r) \cap X$

then the formula $\mathfrak{f} = \mathfrak{a}_1 \mathcal{U}_s \mathfrak{a}_2$ is satisfied for each maximal solution x to $\mathcal{H}$ at each $(t, j) \in \operatorname{dom} x$.

Proof. The proof follows from an application of Theorem 12.21 and the properties in the assumptions. With the assumptions in item 1 and item 2, Theorem 12.21 implies that $\Diamond \mathfrak{a}_2$ is satisfied for each maximal solution x to $\mathcal{H}$ from $L_V(r) \cap X$ since they reach $K_{\mathfrak{a}_2}$ in finite hybrid time. For each such solution to satisfy the given formula, the solution has to remain in $K_{\mathfrak{a}_1}$ at least until it reaches $K_{\mathfrak{a}_2}$ if it starts from a point $K_{\mathfrak{a}_1} \setminus K_{\mathfrak{a}_2}$. This property is assured by item 4: since V is nonincreasing along solutions, the solution remains in $L_V(r)$ if it starts from a point in $K_{\mathfrak{a}_1} \cap L_V(r)$, and is guaranteed to stay in $K_{\mathfrak{a}_1}$, at least until it reaches $K_{\mathfrak{a}_2}$. For solutions that start in $K_{\mathfrak{a}_2}$, the formula is automatically satisfied. Then, $\mathfrak{f} = \mathfrak{a}_1 \mathcal{U}_s \mathfrak{a}_2$ is satisfied for each maximal solution to $\mathcal{H}$ from $(K_{\mathfrak{a}_1} \cap L_V(r)) \cup K_{\mathfrak{a}_2}$ at $(t, j) = (0, 0)$. When item 5 holds, the formula is also satisfied at each (t, j) in the domain of the solution. Since item 5 implies that from points in $K_{\mathfrak{a}_2} \cap D$, the state is mapped to $(L_V(r) \cap (\overline{C} \cup D))$, solutions can only jump out of $K_{\mathfrak{a}_2}$ to a point in $K_{\mathfrak{a}_1}$, as item 4 guarantees. After such jumps, convergence to $K_{\mathfrak{a}_2}$ in finite hybrid time occurs again. Repeating this argument recurrently establishes the result. $\square$

Remark 12.27 (On exploiting forward invariance). The condition in item 4 in Theorem 12.26 is instrumental in guaranteeing that $\mathfrak{a}_1$ is satisfied at least until $\mathfrak{a}_2$ is satisfied. As argued in its proof, this condition combined with the properties of V certifying finite-time attractivity of $K_{\mathfrak{a}_2}$ at some (t', j') ensures the satisfaction of $\mathfrak{a}_1$ up to that hybrid time. Alternative and typically more restrictive sufficient conditions include those that require that the set $K_{\mathfrak{a}_1}$ is forward pre-invariant. Sufficient conditions for forward pre-invariance of $K_{\mathfrak{a}_1}$ can be obtained via Theorem 11.4, for example. Less restrictive conditions can be obtained by relaxing such forward invariance property to the following property:

$\star$) Solutions either start and stay in $K_{\mathfrak{a}_1}$ or if they leave it, they remain in the set $K_{\mathfrak{a}_2}$ after they leave it.

This property is a special case of *conditional invariance*. A set K_2 is said to be conditionally invariant for $\dot{x} = f(x)$ with respect to a set $K_1 \subset K_2$ if for each $x_\circ \in K_1$, each solution $t \mapsto x(t)$ to $\dot{x} = f(x)$ starting from $x_\circ$ satisfies $x(t) \in K_2$ for all $t \in \operatorname{dom} x$. Then, property $\star$ corresponds to conditional invariance with $K_1 = K_{\mathfrak{a}_1}$ and $K_2 = K_{\mathfrak{a}_1} \cup K_{\mathfrak{a}_2}$. $\triangle$

Example 12.24 featuring the bouncing ball system is used to illustrate Theorem 12.26.

Example 12.28 (Strong until specification for a juggling system). *Consider the bouncing ball system model in Example 11.2 and Example 12.24 with data as in*

(11.2)-(11.3). *The atomic propositions* $\mathfrak{a}_1$ *and* $\mathfrak{a}_2$ *are defined as*

$$\mathfrak{a}_1(x) := \begin{cases} \texttt{True} & \text{if } x_2 \geq 0 \\ \texttt{False} & \text{otherwise} \end{cases}, \qquad \mathfrak{a}_2(x) := \begin{cases} \texttt{True} & \text{if } x_2 \leq 0 \\ \texttt{False} & \text{otherwise} \end{cases}$$

for each $x \in \mathbb{R}^2$. *Let the sets* $K_{\mathfrak{a}_1}$ *and* $K_{\mathfrak{a}_2}$ *be given as in* (12.5) *with* $\mathfrak{a}_1$ *and* $\mathfrak{a}_2$ *in place of* $\mathfrak{a}$, *respectively. It follows that* $K_{\mathfrak{a}_1} = \mathbb{R} \times [0, \infty)$ *and* $K_{\mathfrak{a}_2} = \mathbb{R} \times (-\infty, 0]$, *and by construction, item 1 of Theorem 12.26 holds. As shown in Example 12.24, item 2 in Theorem 12.26 is satisfied with* $\mathcal{U} = \mathbb{R}^2$ *and* $r = \infty-$ *in fact, item 1 in Theorem 12.21 holds. Since* $r = \infty$, *the set in item 3 is equal to* $\mathbb{R}^2$. *It follows that*

$$(L_V(r) \cap X) \setminus K_{\mathfrak{a}_2} = ([0, \infty) \times \mathbb{R}) \setminus (\mathbb{R} \times [0, \infty))$$

is contained in $K_{\mathfrak{a}_1}$, *which implies that item 4 holds. Also, it is easy to see that item 5 holds as well. Then, Theorem 12.26 implies that every maximal solution* x *to* $\mathcal{H}$ *satisfies* $\mathfrak{f} = \mathfrak{a}_1 \mathcal{U}_s \mathfrak{a}_2$ *at each* $(t, j) \in \operatorname{dom} x$. *The intuition here is simple. Every solution to* $\mathcal{H}$ *converges to* $K_{\mathfrak{a}_1}$ *after finite hybrid time. Also, every solution from* $K_{\mathfrak{a}_1}$ *jumps to* $K_{\mathfrak{a}_2}$, *from where it converges back to* $K_{\mathfrak{a}_1}$ *in finite hybrid time. Moreover, from the definition of* $K_{\mathfrak{a}_1}$ *and* $K_{\mathfrak{a}_2}$, *if a solution does not belong to one of these sets, then it belongs to the other set. In fact, for this system,* $K_{\mathfrak{a}_1} \cup K_{\mathfrak{a}_2} = X$. *A temporal logic specification using the strong until operator for which* $K_{\mathfrak{a}_1} \cup K_{\mathfrak{a}_2}$ *is a strict subset of* X *is in Problem 93.*

The following result captures the rather simple situation seen in Example 12.28, in which the union of $K_{\mathfrak{a}_1}$ and $K_{\mathfrak{a}_2}$ cover the state space, and provides sufficient conditions for the satisfaction of a formula using the weak until operator. A similar result holds for the strong until operator when $K_{\mathfrak{a}_2}$ is finite-time attractive.

Theorem 12.29 (Sufficient conditions for $\mathfrak{f} = \mathfrak{a}_1 \mathcal{U}_w \mathfrak{a}_2$). *Given a hybrid closed-loop system* $\mathcal{H} = (C, F, D, G)$ *as in* (2.19) *and atomic propositions* $\mathfrak{a}_1$ *and* $\mathfrak{a}_2$, *let* $K_{\mathfrak{a}_1}$ *and* $K_{\mathfrak{a}_2}$ *be constructed as in* (12.5) *with* $\mathfrak{a}$ *replaced by* $\mathfrak{a}_1$ *and by* $\mathfrak{a}_2$, *respectively. Suppose that for each* $x \in \mathbb{R}^n$, *either* $\mathfrak{a}_1(x) = \texttt{True}$ *or* $\mathfrak{a}_2(x) = \texttt{True}$, *and that every maximal solution to* $\mathcal{H}$ *is complete. Then, the formula* $\mathfrak{f} = \mathfrak{a}_1 \mathcal{U}_w \mathfrak{a}_2$ *is satisfied for every maximal solution* x *to* $\mathcal{H}$ *at each* $(t, j) \in \operatorname{dom} x$.

Proof. Let x be a maximal solution to $\mathcal{H}$. If $x(0, 0)$ is such that $\mathfrak{a}_2(x(0, 0)) = \texttt{True}$, then $\mathfrak{f}$ is satisfied at $(t, j) = (0, 0)$. To show that the formula is also satisfied at each $(t, j) \in \operatorname{dom} x$, proceeding by contradiction, suppose there exists $(t', j') \in \operatorname{dom} x$ such that $\mathfrak{a}_1(x(t', j')) = \texttt{False}$ and $\mathfrak{a}_2(x(t', j')) = \texttt{False}$. Then, since on X, either $\mathfrak{a}_1$ or $\mathfrak{a}_2$ is true, then $x(t', j') \notin X$. Hence, the solution is not complete, which is a contradiction. Next, let $x(0, 0)$ be such that $\mathfrak{a}_1(x(0, 0)) = \texttt{True}$. To show that $\mathfrak{f}$ is satisfied at $(t, j) = (0, 0)$ and at any $(t, j) \in \operatorname{dom} x$, also proceeding by contradiction, suppose there exists $(t', j') \in \operatorname{dom} x$ such that $\mathfrak{a}_1(x(t', j')) = \texttt{False}$ and $\mathfrak{a}_2(x(t', j')) = \texttt{False}$. A contradiction ensues due to completeness of the solution x using the argument above. $\qquad \square$

The sufficient conditions presented in this section can also be used to certify formulas that involve more than one operator. See Section 12.6 for a discussion on such extension. Some of the exercises given next illustrate how to use the proposed conditions for control design.

12.5   EXERCISES

Exercise 85 (Obstacle avoidance with target, pass one). Consider the point-mass model of a ground vehicle given by

$$\dot{z} = u \qquad z \in \mathbb{R}^2, u \in \mathbb{R}^2 \tag{12.31}$$

where z denotes its position and u its control input (force). Consider the following design specification:

> *Given regions $R_1, R_2,$ and R_3, guarantee that both R_1 and R_2 are eventually reached and that R_3 is always avoided.*

where R_1 is the square region in $\mathbb{R}^2$ with vertices $\{(1.5, 10.5), (2.5, 10.5), (1.5, 12.5), (2.5, 12.5)\}$, R_2 has vertices $\{(1.2, 1.5), (1.3, 1.5), (1.2, 2.5), (1.3, 2.5)\}$, and R_3 has vertices $\{(0.5, 5), (6, 5), (0.5, 7.5), (6, 7.5)\}$.

1. Define atomic propositions and a formula capturing the given specification.

2. For zero initial condition for z, determine a path on the plane that would satisfy the formula.

3. Outline the main tasks that a control algorithm should perform to satisfy the given specification.

4. Explain how you would validate that your controller satisfies the given specification for every possible solution.

Exercise 86 (Sufficient conditions for $\mathfrak{f} = \bigcirc \mathfrak{a}$). Given an atomic proposition $\mathfrak{a}$, let $K_\mathfrak{a}$ be given as in (12.5).

1. Show the sufficiency part of Proposition 12.7, that is, show that if items 1-3 therein hold then the formula $\mathfrak{f} = \bigcirc \mathfrak{a}$ is satisfied for each maximal solution x to $\mathcal{H}$ at each $(t, j) \in \operatorname{dom} x$.

2. Relax the semantics of the temporal operator $\bigcirc$ to its weaker version denoted $\bigcirc_w$ and defined as

$$(x, (t, j)) \vDash \bigcirc_w \mathfrak{a} \quad \Leftrightarrow \quad (x, (t, j+1)) \vDash \mathfrak{a} \qquad \text{if } (t, j+1) \in \operatorname{dom} x$$

Formulate necessary and sufficient conditions so the formula $\mathfrak{f} = \bigcirc_w \mathfrak{a}$ to be satisfied for each solution x to $\mathcal{H}$ at each $(t, j) \in \operatorname{dom} x$.

Exercise 87 (Sufficient conditions for $\mathfrak{f} = \square \mathfrak{a}$). Given a hybrid system $\mathcal{H}$ with $x \in X$ and given an atomic proposition $\mathfrak{a}$, let $K_\mathfrak{a}$ be given as in (12.5). Show that the formula $\mathfrak{f} = \square \mathfrak{a}$ is satisfied for each solution x to $\mathcal{H}$ with $x(0, 0) \vDash \mathfrak{a}$ at each $(t, j) \in \operatorname{dom} x$ if and only if $K_\mathfrak{a}$ is forward pre-invariant for $\mathcal{H}$.

Exercise 88 ($\Diamond \mathfrak{a}$ for a planar hybrid system). Show that each maximal solution x to the planar hybrid closed-loop system $\mathcal{H}$ given in Example 12.20 is such that $\mathfrak{f} = \Diamond \mathfrak{a}$, with $\mathfrak{a}$ given in (12.15), is satisfied at some $(t, j) \in \operatorname{dom} x$ when $|\lambda| < 1$.

Exercise 89 ($\Box\mathfrak{a}$ via barrier functions). Show Theorem 12.18.

Exercise 90 (Temporal logic specification for a thermostat). Consider the model of the temperature of a room given in Exercise 82.

1. Design a control algorithm that assures that

$$\mathfrak{f} = \Diamond\Box\mathfrak{a}$$

   holds for each solution x of the hybrid closed-loop system at each $(t, j) \in \operatorname{dom} x$, where

$$\mathfrak{a}(x) := \begin{cases} 1 & \text{if } z \in [z_{\min}, z_{\max}] \\ 0 & \text{otherwise} \end{cases}$$

   with $z_{\min}$ and $z_{\max}$ defined as and satisfy the conditions in Exercise 82.

2. Validate your design in item 1 numerically.

Exercise 91 (Sufficient conditions for $\mathfrak{f} = \Diamond\mathfrak{a}$). For the sufficient conditions for $\mathfrak{f} = \Diamond\mathfrak{a}$ in Theorem 12.21, perform the following tasks:

1. Follow the steps in the proof of item 1 in Theorem 12.21 to show that the claim therein holds when the conditions in item 2 hold.

2. Formulate conditions requiring a strict decrease of V during flows and at jumps guaranteeing that $\mathfrak{f} = \Diamond\mathfrak{a}$ is satisfied. Propose an upper bound for the settling-time functional $t_{K_{\mathfrak{a}}}$ and compare it with those in Theorem 12.21.

Exercise 92 ($\Diamond\mathfrak{a}$ for impulse-coupled oscillators). For the hybrid system and temporal specification in Example 12.25 show the following:

1. Every maximal solution x to $\mathcal{H}$ from $\mathcal{U}$ satisfies the formula $\mathfrak{f} = \Diamond\mathfrak{a}$ for each $(t, j) \in \operatorname{dom} x$.

2. There exist maximal solutions to $\mathcal{H}$ such that $\mathfrak{f}$ therein is not satisfied. Characterize the family of such solutions.

3. Propose a variation of the model $\mathcal{H}$ in Example 12.25 such that every maximal solution x from $[0, T^*] \times [0, T^*]$ is complete and satisfies $\mathfrak{f}$ therein for each $(t, j) \in \operatorname{dom} x$.

Exercise 93 (Until for a juggling system). Consider the simplified juggling system in Example 11.2 and Example 12.24 with data as in (11.2)-(11.3), where $\lambda \in [0, 1)$, and, given $c \geq 0$, the atomic propositions $\mathfrak{a}_1$ and $\mathfrak{a}_2$ defined as

$$\mathfrak{a}_1(x) := \begin{cases} \texttt{True} & \text{if } \gamma x_1 + \frac{1}{2}x_2^2 = c \\ \texttt{False} & \text{otherwise} \end{cases} , \qquad \mathfrak{a}_2(x) := \begin{cases} \texttt{True} & \text{if } x \in D \\ \texttt{False} & \text{otherwise} \end{cases}$$

for each $x \in \mathbb{R}^2$. Consider the specification $\mathfrak{f} = \mathfrak{a}_1 \mathcal{U}_s \mathfrak{a}_2$.

1. Characterize the initial conditions for which, regardless of the parameters c and λ, each maximal solution to $\mathcal{H}$ satisfies $\mathfrak{f}$ at $(t, j) = (0, 0)$.

2. Determine if there exist a choice of parameters c and λ such that each maximal solution x to $\mathcal{H}$ satisfies $\mathfrak{f}$ for each $(t, j) \in \text{dom } x$.

Exercise 94 (Obstacle avoidance with target). Consider the obstacle avoidance with target problem in Section 12.1 (and also in Example 4.13 and Exercise 85), using a planar model with state $z = (z_1, z_2, z_3, z_4)$ for the vehicle, where (z_1, z_2) corresponds to planar position and (z_3, z_4) to planar velocity.

1. Consider a generic state-feedback hybrid controller $\mathcal{H}_K$ and define the hybrid closed-loop system, and denote it $\mathcal{H}$.

2. Define the atomic propositions involved in the problem, the associated sets as in (12.5), and a temporal logic formula capturing the desired goal.

3. Employ the sufficient conditions in § 12.4 to determine conditions on the hybrid controller so that the specification is satisfied for each maximal solution to the closed-loop system at $(t, j) = (0, 0)$.

4. Synthesize a hybrid controller so that the proposed conditions are satisfied.

5. Validate your design numerically.

12.6   NOTES

One of the initial articles – if not the first – about linear temporal logic is [235]. This article proposes a new approach for sequential and parallel program verification. A program is modeled as a discrete-time system with a state taking values from a discrete set and whose evolution over discrete time is determined by a *transition relation*. Such systems can be modeled as a hybrid system with empty flow set, arbitrary flow map, discrete jump set, and jump map given by the transition relation; see the dynamical models of finite state machines in [236]. For this class of systems, the work in [235] proposes a formal approach in which the basic concept is the time dependence of the events on the executions – or, equivalently, solutions. For this purpose, two formal systems are presented for providing a basis for temporal reasoning: a formalization of the method of intermittent assertions and an adaptation of a tense logic system. The latter is found to be suitable for reasoning about concurrent programs. Building from followup work in [237], numerous contributions pertaining to modeling, analysis, design, and verification of LTL specifications for dynamical systems have appeared in the literature in recent years. The majority of the results in the literature are for systems with discrete dynamics, either obtained directly from modeling or after discretization of their continuous-time dynamics; see [238], [239], [240], and [241], to just list a few. In some settings, the discrete-time model has a state with some components that are continuous valued and others that are discrete valued; see, e.g., [240] and [241].

Some of the results for the satisfaction of temporal logic formulas in this chapter appeared in [236] and [242] – see also [243]. The idea of constructing the set in (12.5) collecting all state values for which an atomic proposition is true was introduced in [236, Section III.D] and further developed in [242]. The equivalence conditions in § 12.3 have been reported in [243]. The sufficient conditions in § 12.4, along with several other special cases and examples appeared in [242] and in [243]. As mentioned therein, the conditions in Chapter 11 are derived using results in § 11.2.1 – specifically, Theorem 11.4. The latter result using barrier functions, along with numerous other results that are useful to certify formulas using the always operator, appeared in [226]; see also [232, 228]. The sufficient conditions guaranteeing finite-time attractivity of a set therein are inspired by those in [244] – in fact, Theorem 12.21 borrows the ideas from [244, Theorem 3.5] and [244, Theorem 3.9]. The idea of employing dynamical properties and Lyapunov-like conditions to certify temporal logic formulas was also pursued in [245] for alternating-time temporal logic specifications in continuous-time systems.

Table 12.1 summarizes the sufficient conditions in this chapter for basic formulas with a single temporal operator. In addition to regularity properties on the data of the system, these conditions require finite-time attractivity or invariance of sets like the one defined in (12.5). Furthermore, these properties can be asserted by using certificates in terms of Lyapunov functions [244] and barrier functions [226, 232, 228] for hybrid closed-loop systems $\mathcal{H}$. The case of logic operators can be treated similarly by using intersections, unions, and complements of the sets where the propositions hold. For instance, sufficient conditions for $\square(\mathfrak{a}_1 \wedge \mathfrak{a}_2)$ can immediately be derived from the sufficient conditions given in this chapter by instead considering the set

$$K_{\mathfrak{a}_1 \wedge \mathfrak{a}_2} := \{x \in X : \mathfrak{a}_1(x) = \texttt{True}\} \cap \{x \in X : \mathfrak{a}_2(x) = \texttt{True}\} \qquad (12.32)$$

The results for forward invariance of intersections of sets in [228] provide sufficient conditions that only involve properties of the individual sets, which leads to a compositional method to certify the satisfaction of such formula.

$\mathfrak{f}$	Sufficient Conditions
$\bigcirc \mathfrak{a}$	Properties of the data of $\mathcal{H}$ – Proposition 12.7
$\square \mathfrak{a}$	Forward invariance – Theorem 12.18 (and Theorems 11.4, 11.9, 11.14)
$\diamondsuit \mathfrak{a}$	Finite-time attractivity – Theorem 12.21
$\mathfrak{a}_1 \mathcal{U}_s \mathfrak{a}_2$	Forward invariance and finite-time attractivity – Theorem 12.26
$\mathfrak{a}_1 \mathcal{U}_w \mathfrak{a}_2$	Forward invariance or finite-time attractivity – Theorem 12.29

Table 12.1: Sufficient conditions for $\bigcirc$, $\square$, $\diamondsuit$, $\mathcal{U}_s$, and $\mathcal{U}_w$.

The conditions in Table 12.1 can be combined to certify formulas that involve more than one boolean and temporal. Several such combinations requiring forward invariance and finite-time attractivity appeared in [243, Section 6]. A systematic methodology to satisfy general formulas within the considered language is part of current research and relies on the fact that formulas that combine more than one operator can be decomposed into simpler formulas for which our results for formulas with a single operator presented in this chapter can be applied. For instance, to decompose a general formula into several formulas with a single operator, one can employ the finite state automaton representation of an LTL formula [246, 247, 248]. See [243, Section 6] for more details.

Appendix A

Mathematical Review

This appendix introduces some of the basic concepts, objects, and notions used in this book. It also presents supporting results used in some of the proofs.

A.1 MODELS

A mathematical model of a control system describes a relationship between its inputs, state, and its output. The *state* of a control system is defined by the internal variables that, together with the inputs, are needed to determine the output of the system. A widely used mathematical model of a control system is given by a differential equation of the form

$$\dot{z} = F_P(z, u), \qquad y = h(z) \tag{A.1}$$

where z is the state, u is the control input, and y is the output. The function F_P defines the velocity of the state and is referred to as *the right-hand side* of the control system. The function h defines the output of the system and is called *the output map*.

The system in (A.1) is typically called *the plant*.[1] It is said to be a *continuous-time system* due to the velocity $\dot{z}$ being the variation of z with respect to ordinary time $t \in \mathbb{R}_{\geq 0}$. Given an initial condition $z_\circ$ for the state and a time function u defining the control input, the state satisfies

$$z(t) = z_\circ + \int_0^t F_P(z(s), u(s)) \, ds \tag{A.2}$$

This function of time is said to be a *state trajectory* of the plant. The state trajectory provides a solution to the differential equation (A.1) for the given initial condition and input. Indeed, it solves the following *initial value problem*:

(IVP) Given an initial condition $z_\circ$ and a time function u, find a state trajectory z satisfying
$$\dot{z}(t) = F_P(z(t), u(t)), \qquad z(0) = z_\circ \tag{A.3}$$

Due to this property, the state trajectory is said to be *a solution* to the plant (A.1).

---

[1] The use of the term "plant" dates back to the 1920's, when control rooms emerged in power plants and other major factories. Human operators in control rooms monitored charts and, to make corrections to the processes, manually opened or closed valves and turned switches on or off.

Given a solution to (A.1) and the associated input u, the evolution of the output of the plant is given by

$$y(t) = h(z(t)) \tag{A.4}$$

It is typically desired that the solution is defined for each $t \in \mathbb{R}_{\geq 0}$. However, that is not always possible when F_P is nonlinear. Examples are provided in § 2.1.

In most control problems, the control input to the plant is assigned via a function of its output. When this input is assigned via

$$u = \kappa(y) \tag{A.5}$$

it is said that κ is a *static output-feedback control law*. In the case that u is assigned via

$$u = \kappa(z) \tag{A.6}$$

then κ is said to be a *static state-feedback control law*. In either case, the resulting system is *the closed-loop system*. The closed-loop system obtained from using the state-feedback law in (A.6) is given by

$$\dot{z} = F_P(z, \kappa(z)) \tag{A.7}$$

Since its right-hand side depends on the state only, this system is said to be *autonomous*.

A discussion about stability, attractivity, and asymptotic stability for a closed-loop system like the one in (A.7) is in § 3.1.

Existence and uniqueness of solutions to (A.1) has been thoroughly studied in the literature of differential equations, see, e.g., [249]. These properties have also been studied in the context of continuous-time systems with a nonlinear right-hand side, namely, *nonlinear systems*. Chapter 3 in [95] provides an in-depth presentation of those properties; see also § 2.3.4. In simple words, existence and uniqueness of solutions depends on the regularity properties of the right-hand side of the system, in particular, continuity and Lipschitzness, as defined in § A.2. For the particular case of F_P and h given by

$$F_P(z, u) = Az + Bu, \qquad h(z) = Mz \tag{A.8}$$

where A, B, and M are matrices of appropriate dimensions, solutions to (A.1) exist and are unique. Indeed, for the initial condition $z_\circ$ and a time function u, the solution to (A.1) is given by

$$z(t) = \exp(At)z_\circ + \int_0^t \exp(A(t-s))Bu(s)\,ds \tag{A.9}$$

and the output is equal to $Mz(t)$. Both the state trajectory z and the output trajectory y are functions of time that are defined for each $t \in \mathbb{R}_{\geq 0}$. The term $\exp(At)$ is the exponential of the matrix At. A plant with a right-hand side as in (A.8) enjoys linearity and time-invariance properties, and is said to be a *linear time-invariant system*; see § 1.1. Several textbooks in the literature cover the theory of linear time-invariant systems in great detail. One such textbook is [250].

The mathematical models provided above are *nominal* in the sense that they do not include disturbances. Extending the differential equation model in (A.1) to the case when disturbances are present amounts to making F_P depend on the disturbance. The evolution of the state for a plant with a generic disturbance input w is given by

$$\dot{z} = F_P(z, u, w) \tag{A.10}$$

A model of a hybrid plant with disturbances is presented §2.3.5.

A powerful way to handle the effect of the disturbance is to define a new mathematical model that includes all solutions that could emerge due to any possible allowed disturbance. Denoting by $\mathcal{W}$ the set of possible values for the disturbance w, a mathematical model capturing all possible solutions to (A.10) for a given control input u is given by the *differential inclusion*

$$\dot{z} \in F_P(z, u, \mathcal{W}) \tag{A.11}$$

where $F_P(z, u, \mathcal{W}) = \{F_P(z, u, w) : w \in \mathcal{W}\}$; see List of Symbols. Note that when evaluated at (z, u), $F_P(z, u, \mathcal{W})$ may return more than one value. This nondeterminism allows several choices for the velocity of z, in this way, allowing to generate every possible solution that could emerge from the disturbance w taking values from $\mathcal{W}$. Indeed, the differential inclusion model in (A.11) rewritten as follows suggests that the effect of w has been captured by the multiple values of its right-hand side:

$$\dot{z} \in F(z, u) := F_P(z, u, \mathcal{W}) \tag{A.12}$$

Examples using this reformulation in terms of an inclusion are given throughout this book. The reader might find Example 2.14 and Example 9.1 insightful.

The map F in (A.12) is a *set-valued map*. Set-valued maps are introduced in §A.2. The interested reader is referred to Chapter 5 of [251] and Chapter 1 of both [85] and [222] for a concise exposition to set-valued maps, and to [85] for an in-depth presentation of the theory of differential inclusions.

Differential inclusions also emerge when closed-loop systems have a right-hand side that is discontinuous. Nonsmoothness of the right-hand side of the plant or of the feedback law can lead to closed-loop systems that are not robust, or to closed-loop systems for which no solutions exist from certain initial conditions; see §1.2.4 and §2.1. When the right-hand side is discontinuous, a differential inclusion can be defined via a regularization process. In particular, Example 2.3 and Example 9.5 employ the process in [58] to build the so-called *Filippov system*. The model associated with this system consists of a differential inclusion with a set-valued right-hand side that has enough regularity to guarantee robustness and existence of solutions. This issue is studied in-depth in [58] for nonlinear continuous-time systems and in Chapter 4 of [1] in the context of hybrid systems; see also [8].

In addition to continuous-time models, discrete-time models are widely used in the literature. A discrete-time model of a closed-loop system is introduced in §2.1. Such systems may represent the closed-loop system resulting from assigning the control input to a function of the state. Similar constructions to the ones given in this section can be formulated for a plant and a closed-loop system given in terms of a discrete-time model.

A.2   MAPS

A *function* F that maps each point in $\mathbb{R}^m$ to a point in $\mathbb{R}^n$ is denoted $F : \mathbb{R}^m \to \mathbb{R}^n$. The set $\mathbb{R}^m$ is the *domain* of the function and the set $\mathbb{R}^n$ is its *codomain*. Given $x \in \mathbb{R}^m$, $F(x)$ is the value of the function at x, or, equivalently, the evaluation of the function F at x. This should be emphasized: $F(x)$ is not the function itself, but rather its value at x.[2] The range (or image) of the function is the set of values that the function takes on its domain $\mathbb{R}^m$, namely, $\{F(x) : x \in \mathbb{R}^m\}$.

> In this book, a function is *always* single valued. Namely, given $F : \mathbb{R}^m \to \mathbb{R}^n$ and a point $x \in \mathbb{R}^m$, the function F evaluated at x is a point. Equivalent terms for "function" are *single-valued map* and *single-valued mapping*.

There are many ways to define a function $F : \mathbb{R}^m \to \mathbb{R}^n$. In this book, functions are defined in a variety of ways, as the following examples illustrate:

1. The function F corresponding to the identity map is defined as

$$F(x) := x \qquad \forall x \in \mathbb{R}^n$$

In this case, $m = n$. Equivalently, this function might be defined as

$$F = \mathrm{Id}$$

2. The function F corresponding to the maximum absolute value of the components of $x \in \mathbb{R}^n$ is defined as

$$x \mapsto F(x) := \max_{i \in \{1,2,\ldots,m\}} |x_i|$$

This notation does not explicitly define the domain and codomain of the function. That information is typically provided separately, is clear from context, or is irrelevant.

Note that the arrow used in item 2 is different from the arrow used in $F : \mathbb{R}^m \to \mathbb{R}^n$. In fact, the arrow in item 2 denotes how x is mapped by F, while the arrow in $F : \mathbb{R}^m \to \mathbb{R}^n$ indicates that F maps points from $\mathbb{R}^m$ to points in $\mathbb{R}^n$.

> A *set-valued map* associates every point in its domain to a set in its codomain. A set-valued map is distinguished from a single-valued map through the use of double arrow notation, namely, $F : \mathbb{R}^m \rightrightarrows \mathbb{R}^n$ indicates that F is a set-valued map. Equivalent terms for a set-valued map are *set-valued mapping* and *multi-valued map*.

The definition of domain and of range of a set-valued map given next come directly from [251]. See also Section 2.1 of Chapter 2 and Chapter 5 in [1].

---

[2]A reader that understands this difference would find the statement *"The function $F(x)$ is defined as x^2"* informal and perhaps confusing. A formal version of such statement is *"The function F is defined, for each $x \in \mathbb{R}$, as $F(x) := x^2$"*.

Definition A.1 (Domain). *The domain of a set-valued map $F : \mathbb{R}^m \rightrightarrows \mathbb{R}^n$ is denoted* $\operatorname{dom} F$ *and is given by the points in $\mathbb{R}^m$ at which F is nonempty, namely,*

$$\operatorname{dom} F = \{x \in \mathbb{R}^m : F(x) \neq \emptyset\}$$

Remark A.2. The domain of the function $F : \mathbb{R}^m \rightarrow \mathbb{R}^n$ is $\operatorname{dom} F = \mathbb{R}^m$ since a function maps single points onto single points, and a point is a nonempty set. One could interpret a single-valued map as a special case of a set-valued map, and in this way, allow for the domain of F to be a strict subset of $\mathbb{R}^m$. Such interpretation is not used in this book as it is ambiguous. Instead, when a particular domain – say, S – for a function is desired for a single-valued map F, $F : S \rightarrow \mathbb{R}^n$ is written. Recall that a set S_1 is a strict (or proper) subset of a set S_2 if $S_2 \setminus S_1$ is nonempty. $\triangle$

Definition A.3 (Range). *The range of a set-valued map $F : \mathbb{R}^m \rightrightarrows \mathbb{R}^n$ is denoted* $\operatorname{rge} F$ *and is given by the set of points in $\mathbb{R}^n$ that F can attain, namely,*

$$\operatorname{rge} F = \{y \in \mathbb{R}^n : \exists x \in \mathbb{R}^m \text{ such that } y \in F(x)\}$$

Equivalently, the range of F is given by $F(\operatorname{dom} F)$.

The codomain and the range of a (single-valued or set-valued) map are not necessarily the same. The codomain is the set of points that the map is allowed to take values from. On the other hand, the range (or, equivalently, the image) is the set of values that the map actually attains. For instance, the (scalar) saturation function is given by the single-valued map $F : \mathbb{R} \rightarrow \mathbb{R}$ defined as $F(x) := \operatorname{sign}(x) \max\{|x|, 1\}$ for all $x \in \mathbb{R}$. Its codomain is $\mathbb{R}$ but its range is $\operatorname{rge} F = [-1, 1]$.

The following definition introduces the graph of a set-valued map. In simple words, the graph of a map collects the input and output "pairs" that are related through the map.

Definition A.4 (Graph). *The graph of a set-valued map $F : \mathbb{R}^m \rightrightarrows \mathbb{R}^n$ is denoted* $\operatorname{gph} F$ *and is given by the set*

$$\operatorname{gph} F = \{(x, y) \in \mathbb{R}^m \times \mathbb{R}^n : y \in F(x)\}$$

A.3 SETS

The following properties of sets are used throughout this book.

Definition A.5 (Closed set in $\mathbb{R}^n$). *A set $S \subset \mathbb{R}^n$ is closed if for each convergent sequence of points $x_i \in S$, its limit x satisfies $x = \lim_{i \rightarrow \infty} x_i \in S$.*

For example, the set $S = \mathbb{B}$ denotes the unit closed ball centered at the origin of $\mathbb{R}^n$. Namely, $\mathbb{B} = \{x \in \mathbb{R}^n : |x| \leq 1\}$. Every convergent sequence of points in $\mathbb{B}$ either converges to a point in $\operatorname{int} \mathbb{B} = \{x \in \mathbb{R}^n : |x| < 1\}$ or $\partial \mathbb{B} = \{x \in \mathbb{R}^n : |x| = 1\}$.

The notion in Definition A.5 is relative to $\mathbb{R}^n$. A set S is closed relative to a set O if $S = \overline{S} \cap O$.

Definition A.6 (Bounded and compact sets in $\mathbb{R}^n$). *A set $S \subset \mathbb{R}^n$ is bounded if there exists $\rho \in \mathbb{R}_{\geq 0}$ such that $S \subset \rho\mathbb{B}$. A set $S \subset \mathbb{R}^n$ is compact if it is closed and bounded.*

Definition A.7 (Neighborhood in $\mathbb{R}^n$). *A neighborhood $\mathcal{U}$ of a point or a set $S \subset \mathbb{R}^n$ is a subset of $\mathbb{R}^n$ that contains an open set $\mathcal{V}$ strictly containing S, namely, $\mathcal{U}$ is such that there exist an open set $\mathcal{V}$ and $\epsilon > 0$ such that $S + \epsilon\mathbb{B}^\circ \subset \mathcal{V} \subset \mathcal{U}$.*

Note that this definition of neighborhood does not insist on $\mathcal{U}$ or S to be open or closed.

Tangent cones play an important role in the analysis of hybrid systems. The following tangent cone, which is also known as the *contingent cone* or the *Bouligand cone*, is used in results pertaining to existence of solutions and invariance of sets; see Proposition 2.34. More details about the tangent cone and its use in nonsmooth analysis is in [100].

Definition A.8 (Tangent cone). *The tangent cone to a set $S \subset \mathbb{R}^n$ at a point $x \in \mathbb{R}^n$, denoted $T_S(x)$, is the set of all vectors $w \in \mathbb{R}^n$ for which there exist sequences $x_i \in S$, $\tau_i > 0$ with $x_i \to x$, $\tau_i \searrow 0$, and $w = \lim_{i \to \infty} \frac{x_i - x}{\tau_i}$.*

The following result is a version of the well-known Nagumo Theorem. It provides a useful characterization of the tangent cone. It was reported in [223, Lemma 2.20]; see also [230, Appendix, Corollary 3].

Lemma A.9. *Given closed sets $M \subset \mathbb{R}^n$ and $K \subset M$, let $B : \mathbb{R}^n \to \mathbb{R}$ be such that $K = \{x \in M : B(x) \leq 0\}$ and let $x \in \partial K \cap \operatorname{int} M$ be such that B is continuously differentiable on an open neighborhood around x and $\nabla B(x) \neq 0$. Then,*

$$T_K(x) = \{\chi \in \mathbb{R}^n : \langle \nabla B(x), \chi \rangle \leq 0\}$$

The distance from a point to a generic set is defined as follows.

Definition A.10 (Distance to a set). *Given $x \in \mathbb{R}^n$ and a nonempty set $S \subset \mathbb{R}^n$, the distance from x to S is denoted $|x|_S$ and is defined by*

$$|x|_S := \inf_{\chi \in S} |x - \chi|$$

A.4   REGULARITY

The notions of boundedness, continuity, and Lipschitzness of functions are standard and available in real analysis textbooks. The book [252] provides an elementary introduction to real analysis. The book [83] is more advanced and provides an in-depth presentation of such concepts. For completeness, these notions are included here.

Definition A.11 (Bounded function). *A function $F : S \to \mathbb{R}^n$ is said to be locally bounded at $x \in S$ if there exists a neighborhood $\mathcal{U}$ of x such that the set $F(\mathcal{U} \cap S)$*

is bounded. The function F is said to be locally bounded if it is locally bounded at each $x \in S$. Given a set $S' \subset S$, the function F is said to be bounded on S' if the set $F(S')$ is bounded.

Definition A.12 (Continuous function – first pass). *A function $F : S \to \mathbb{R}^n$ is said to be continuous at $x \in S$ if for each $\varepsilon > 0$ there exists $\delta > 0$ such that $|F(\chi) - F(x)| \leq \varepsilon$ for all $\chi \in (x + \delta\mathbb{B}) \cap S$. A function $F : S \to \mathbb{R}^n$ is said to be continuous if it is continuous at each $x \in S$.*

Definition A.12 is the so-called ε-δ definition of continuity. An equivalent definition in terms of limits is as follows.[3]

Definition A.13 (Continuous function – second pass). *A function $F : S \to \mathbb{R}^n$ is said to be continuous at $x \in S$ if for each sequence $x_i \in S$ converging to x, $\lim_{x_i \to x} F(x_i) = F(x)$.*

Functions that are continuous on a closed subset of a Euclidean space can always be continuously extended to the entire Euclidean space; see, e.g., [253].

Remark A.14 (On the range of continuous functions). Any continuous function $F : \mathbb{R}^m \to \mathbb{R}^n$ evaluated on a bounded set $S \subset \mathbb{R}^m$ is such that its range $F(S)$ is bounded. When the function is not defined on the entirety of $\mathbb{R}^m$, then the situation is more delicate. For example, the function F defined as $F(x) = 1/x$ only for each $x > 0$ is such that its image $F(S)$ for $S = (0, 1)$ is $[1, \infty)$, which is unbounded. Another important property of a continuous function $F : \mathbb{R}^m \to \mathbb{R}^n$ is that its image on any compact set $S \subset \mathbb{R}^m$ is compact – the function actually attains its bounds at values of its argument that are within the compact set. This property does not extend to general closed (not bounded) sets though. For instance, the continuous function F defined as $F(x) = \exp(-x)$ for each $x \in \mathbb{R}$ is such that the range $F(S)$ with $S = [0, \infty)$ closed is equal to $(0, 1]$, which is not closed. $\triangle$

Remark A.15 (On continuity of the distance to a set). The distance function introduced in Definition A.10 is continuous. Showing this property is left as an exercise; see Exercise 96. $\triangle$

The following classes of functions are useful in establishing bounds in the context of pre-asymptotic stability; see Chapter 3.

Definition A.16 (Class-$\mathcal{K}$ functions). *A function $\alpha : \mathbb{R}_{\geq 0} \to \mathbb{R}_{\geq 0}$ is a class-$\mathcal{K}$ function, also written $\alpha \in \mathcal{K}$, if α is zero at zero, continuous, and strictly increasing.*

Definition A.17 (Class-$\mathcal{K}_\infty$ functions). *A function $\alpha : \mathbb{R}_{\geq 0} \to \mathbb{R}_{\geq 0}$ is a class-$\mathcal{K}_\infty$ function, also written $\alpha \in \mathcal{K}_\infty$, if α is zero at zero, continuous, strictly increasing, and unbounded; i.e., $\alpha \in \mathcal{K}$ and unbounded.*

Definition A.18 (Class-$\mathcal{KL}$ functions). *A function $\beta : \mathbb{R}_{\geq 0} \times \mathbb{R}_{\geq 0} \to \mathbb{R}_{\geq 0}$ is a class-$\mathcal{KL}$ function, also written $\beta \in \mathcal{KL}$, if it is nondecreasing in its first argument, nonincreasing in its second argument, $\lim_{r \searrow 0} \beta(r, s) = 0$ for each $s \in \mathbb{R}_{\geq 0}$, and $\lim_{s \to \infty} \beta(r, s) = 0$ for each $r \in \mathbb{R}_{\geq 0}$.*

---

[3]The topological characterization of continuity of a function F is not used in this book, but it certainly useful: F is said to be continuous if for each open set $\mathcal{U}$, $\{x : F(x) \in \mathcal{U}\}$ is open.

Definition A.19 (Proper indicator). *A proper indicator $\omega : \mathbb{R}^n \to \mathbb{R}_{\geq 0}$ of $\mathcal{A}$ on $\mathcal{U}$ is a continuous function such that the following hold: i) $\omega(x_i) \to \infty$ when $i \to \infty$ if either $|x_i| \to \infty$ or the sequence $\{x_i\}_{i=1}^{\infty}$ converges to the boundary of $\mathcal{A}$, and ii) $\omega(x) = 0$ if and only if $x \in \mathcal{A}$.*

The following notion is used in the definition of solutions to a hybrid system; see § 2.3.3.

Definition A.20 (Absolutely continuous function). *A function $s : [a, b] \to \mathbb{R}^n$ is said to be absolutely continuous if for each $\varepsilon > 0$ there exists $\delta > 0$ such that for each countable collection of disjoint subintervals $[a_k, b_k]$ of $[a, b]$ such that $\sum_k (b_k - a_k) \leq \delta$, it follows that $\sum_k |s(b_k) - s(a_k)| \leq \varepsilon$. A function $s : \mathbb{R}_{\geq 0} \to \mathbb{R}^n$ is said to be locally absolutely continuous if $r \mapsto s(r)$ is absolutely continuous on each compact subinterval of $\mathbb{R}_{\geq 0}$.*

Definition A.21 (Locally Lipschitz function). *A function $F : S \to \mathbb{R}^n$ is said to be locally Lipschitz (or, equivalently, locally Lipschitz continuous) if for each $x \in S$ there exist $\delta > 0$ and $L \geq 0$ such that $|F(x_1) - F(x_2)| \leq L|x_1 - x_2|$ for all $x_1, x_2 \in (x + \delta \mathbb{B}) \cap S$.*

Absolutely continuous functions are also continuous, uniformly continuous, and of bounded variation. Furthermore, any Lipschitz function is also absolutely continuous. Very importantly, the classical time derivative $\frac{d}{dt} s(t)$ of a locally absolutely continuous function $s : [a, b] \to \mathbb{R}^n$ exists and is finite for almost every point in $[a, b]$ – indeed, $r \mapsto s(r)$ is differentiable almost everywhere; see the discussion above Definition 2.27.

The following result about properties of solutions to a hybrid closed-loop system along flows follows from [1, Lemma 5.26]. It appeared in [118, Lemma 5].

Lemma A.22 (Flows within tangent cone). *Suppose the flow set C and the flow map F defining the data (C, F, D, G) of the hybrid closed-loop system $\mathcal{H}$ satisfy (A1) and (A2) in Definition 2.20. Let x be a solution to $\mathcal{H}$. Then, for each $j \in \mathbb{N}$ such that $I_x^j = \{t \in \operatorname{dom} x : (t, j) \in \operatorname{dom} x\}$ has a nonempty interior, the solution x satisfies*

$$\frac{d}{dt} x(t, j) \in F(x(t, j)) \cap T_C(x(t, j)) \qquad \text{for almost all } t \in I_x^j$$

Upper semicontinuity for a function is defined as follows.

Definition A.23 (Upper semicontinuous function). *A scalar function $F : S \to \mathbb{R}$ is said to be upper semicontinuous at $x \in S$ if for each sequence $x_i \in S$ converging to x, $\limsup_{i \to \infty} F(x_i) \leq F(x)$.*

The following result establishes that the functions $\dot{V}$ and ΔV defined in Chapter 3 are upper semicontinuous. It is essentially [9, Lemma 4.6].

Lemma A.24 (Upper semicontinuity of $\dot{V}$ and ΔV). *If V is continuous on $C \cup D \cup G(D)$ and locally Lipschitz on a neighborhood of $\overline{C}$, then the functions*

$$u_C(x) := \begin{cases} \dot{V}(x) & \text{if } x \in C \\ -\infty & \text{otherwise} \end{cases} \qquad u_D(x) := \begin{cases} \Delta V(x) & \text{if } x \in D \\ -\infty & \text{otherwise} \end{cases}$$

defined for each $x \in \mathbb{R}^n$ with $\dot{V}$ given[4] in (3.13) and ΔV in (3.16), respectively, are upper semicontinuous.

In addition to functions with certain regularity as defined above, functions with values having a particular sign are employed in this book. The following definitions introduce positive definite functions.

Definition A.25 (Positive definite functions – first pass). *A function $\rho : \mathbb{R}_{\geq 0} \to \mathbb{R}_{\geq 0}$ is said to be positive definite, also written $\rho \in \mathcal{PD}$, if $\rho(s) > 0$ for all $s > 0$ and $\rho(0) = 0$.*

Definition A.26 (Positive definite functions – second pass). *Given a nonempty set $\mathcal{A} \subset \mathbb{R}^n$, a function $V : \mathbb{R}^n \to \mathbb{R}_{\geq 0}$ is said to be positive definite with respect to $\mathcal{A}$, also written $V \in \mathcal{PD}(\mathcal{A})$, if $V(x) > 0$ for all $x \in \mathbb{R}^n \setminus \mathcal{A}$ and $V(\mathcal{A}) = 0$.*

The property of a scalar-valued function having compact sublevel sets plays a key role on the size of the basin of attraction of pre-asymptotic stability. Such a property is guaranteed when the function is *radially unbounded*, as defined next.

Definition A.27 (Radial unboundedness). *A function $V : \mathbb{R}^n \to \mathbb{R}_{\geq 0}$ is said to be radially unbounded if*

$$\lim_{|x| \to \infty} V(x) = \infty$$

Given a nonempty set $\mathcal{A} \subset \mathbb{R}^n$, a function $V : \mathbb{R}^n \to \mathbb{R}_{\geq 0}$ is said to be radially unbounded with respect to $\mathcal{A}$ if

$$\lim_{|x|_{\mathcal{A}} \to \infty} V(x) = \infty$$

Linear (affine) growth of the values of functions and set-valued maps enables ruling out finite escape times in dynamical systems.

Definition A.28 (Linear growth). *A set-valued map $F : \mathbb{R}^m \rightrightarrows \mathbb{R}^n$ is said to have linear growth on the set $S \subset \mathbb{R}^m$ if there exists $k \geq 0$ such that $\sup_{\chi \in F(x)} |\chi| \leq k(1 + |x|)$ for each $x \in S$.*

A notion closely related to upper semicontinuity as in Definition A.23 for set-valued maps is outer semicontinuity. This notion is defined as in [251].

Definition A.29 (Outer semicontinuity). *A set-valued map $F : \mathbb{R}^m \rightrightarrows \mathbb{R}^n$ is outer semicontinuous (osc) at $x \in \mathbb{R}^m$ if for each sequence of points x_i converging to x and each convergent sequence of points $y_i \in F(x_i)$ it follows that $y \in F(x)$, where $\lim_{i \to \infty} y_i = y$. The map F is outer semicontinuous if it is outer semicontinuous at each $x \in \mathbb{R}^m$. Given a set $S \subset \mathbb{R}^m$, $F : \mathbb{R}^m \rightrightarrows \mathbb{R}^n$ is outer semicontinuous relative to S if the set-valued map from $\mathbb{R}^m$ to $\mathbb{R}^n$ defined by $F(x)$ for $x \in S$ and $\emptyset$ for $x \notin S$ is outer semicontinuous at each $x \in S$.*

---

[4]The functions u_C and u_D are extensions to $\mathbb{R}^n$ of the definitions of $\dot{V}$ and ΔV in (3.13) and ΔV in (3.16), respectively.

In Definition A.13, continuity of a function is defined in terms of limits. Similarly, outer semicontinuity of a set-valued map can be defined in terms of the outer limit of a sequence of sets, as follows.

Definition A.30 (Outer semicontinuity). *A set-valued map $F : \mathbb{R}^m \rightrightarrows \mathbb{R}^n$ is outer semicontinuous (osc) at $x \in \mathbb{R}^m$ if for each sequence of points x_i converging to x,*

$$\limsup_{i \to \infty} F(x_i) \subset F(x)$$

where, defining the sequence of sets $S_i := F(x_i)$, $\limsup_{i \to \infty} S_i$ is the outer limit of the sequence S_i. The outer limit of S_i is given by the set of points $x \in \mathbb{R}^n$ for which there exist a subsequence $S_{i_k} \subset S$ and points $x_{i_k} \in S_{i_k}$, $k \in \mathbb{N}$, such that $\lim_{k \to \infty} x_{i_k} = x$.

Outer semicontinuity follows from the graph of a map being closed.

Lemma A.31 (OSC and closed graph). *A set-valued map $F : \mathbb{R}^m \rightrightarrows \mathbb{R}^n$ is outer semicontinuous if and only if $\operatorname{gph} F$ is closed. More generally, given a set $S \subset \mathbb{R}^m$, a set-valued map $F : \mathbb{R}^m \rightrightarrows \mathbb{R}^n$ is outer semicontinuous relative to S if and only if the set*

$$\{(x, y) \in \mathbb{R}^m \times \mathbb{R}^n : x \in S, \ y \in F(x)\}$$

is relatively closed in $S \times \mathbb{R}^n$.

Local boundeness of set valued maps is defined as follows.

Definition A.32 (Locally bounded set-valued map). *A set-valued map $F : \mathbb{R}^m \rightrightarrows \mathbb{R}^n$ is said to be locally bounded at $x \in \mathbb{R}^m$ if there exists a neighborhood $\mathcal{U}$ of x such that $F(\mathcal{U})$ is bounded. The map F is said to be locally bounded if it is locally bounded at each $x \in \mathbb{R}^m$. Given a set $S \subset \mathbb{R}^m$, the map F is said to be locally bounded relative to S if the set-valued map from $\mathbb{R}^m$ to $\mathbb{R}^n$ defined by $F(x)$ for $x \in S$ and $\emptyset$ for $x \notin S$ is locally bounded at each $x \in S$.*

The construction presented in the following result is used in this book several times. This result appeared in [114, Appendix, Lemma 2]. As the result shows, when the individual maps are outer semicontinuous and locally bounded, the map constructed therein has those properties as well.

Lemma A.33 (Outer semicontinuity and locally boundedness of the union of maps). *Given closed sets $D_1 \subset \mathbb{R}^m$ and $D_2 \subset \mathbb{R}^m$ and set-valued maps $G_1 : D_1 \rightrightarrows \mathbb{R}^n$ and $G_2 : D_2 \rightrightarrows \mathbb{R}^n$ that are outer semicontinuous and locally bounded relative to D_1 and D_2, respectively, the set-valued map $G : D \rightrightarrows \mathbb{R}^n$, given by*

$$G(x) := G_1(x) \cup G_2(x) = \begin{cases} G_1(x) & \text{if } x \in D_1 \backslash D_2 \\ G_2(x) & \text{if } x \in D_2 \backslash D_1 \\ G_1(x) \cup G_2(x) & \text{if } x \in D_1 \cap D_2 \end{cases} \tag{A.13}$$

for each $x \in D := D_1 \cup D_2$, is outer semicontinuous and locally bounded relative to the closed set D.

A notion widely used in the literature of set-valued maps is upper semicontinuity of a set-valued map.

Definition A.34 (Upper semicontinuous set-valued map). *The set-valued map $F : \mathbb{R}^m \rightrightarrows \mathbb{R}^n$ is upper semicontinuous at x if for every $\varepsilon > 0$ there exists $\delta > 0$ such that $\chi \in x + \delta\mathbb{B}$ implies $F(\chi) \subset F(x) + \varepsilon\mathbb{B}$. The map F is said to be upper semicontinuous if it is upper semicontinuous at each $x \in \mathbb{R}^m$.*

The following lemma relates outer semicontinuity and upper semicontinuity of a set-valued map.

Lemma A.35 (OSC vs. upper semicontinuity of set-valued maps). *Let $F : \mathbb{R}^m \rightrightarrows \mathbb{R}^n$ be a set-valued map. Let $x \in \mathbb{R}^m$ be such that $F(x)$ is closed. If F is upper semicontinuous at x then F is outer semicontinuous at x. If F is locally bounded at x, then the reverse implication is true.*

Another continuity property of set-valued maps that is used in this book is inner (or lower) semicontinuity; see [251, Chapter 5.B].

Definition A.36 (Inner semicontinuous set-valued map). *The set-valued map $F : \mathbb{R}^m \rightrightarrows \mathbb{R}^n$ is inner semicontinuous at $x \in \mathbb{R}^m$ if*

$$\liminf_{x_i \to x} F(x_i) \supset F(x)$$

where

$$\liminf_{x_i \to x} F(x_i) = \{z : \forall x_i \to x, \exists z_i \to z \text{ s.t. } z_i \in F(x_i)\}$$

is the so-called inner limit of F. The map F is said to be inner semicontinuous if it is inner semicontinuous at each $x \in \mathbb{R}^m$.

A set-valued map with domain that is not necessarily open is said to be inner semicontinuous if its trivial extension proposed in the following result is inner semicontinuous. This construction, which is denoted F_2 therein, appeared in [192, Lemma 4.2] and, as the result states, under suitable assumptions, leads to an inner semicontinuous map.

Lemma A.37. *Suppose the set-valued map $F_1 : \mathbb{R}^n \rightrightarrows \mathbb{R}^m$ is inner semicontinuous. Furthermore, suppose F_1 has nonempty and convex values on a closed set $K \subset \mathbb{R}^n$. Then, the set-valued map defined for each $x \in \mathbb{R}^n$ as $F_2(x) := F_1(x)$ if $x \in K$, $F_2(x) := \mathbb{R}^m$ otherwise, is inner semicontinuous with nonempty and convex values.*

The following result appeared in [191, Corollary 2.13]. It is used in the application of the selection theorems given below. See also [254, Proposition 4.4].

Corollary A.38. *Given an inner semicontinuous set-valued map W and an upper semicontinuous function w, the set-valued map defined as*

$$\mathcal{T}^\circ(x) := \{\chi \in W(x) : w(x, \chi) < 0\}$$

is inner semicontinuous at each x where W is nonempty and w is defined. Furthermore, at each x where $\mathcal{T}^\circ$ is inner semicontinuous, the closure of $\mathcal{T}^\circ(x)$ satisfies $\overline{\mathcal{T}^\circ(x)} = \{\chi \in W(x) : w(x, \chi) \leq 0\}$.

The selection theorem due to Michael provides conditions guaranteeing that a continuous selection from a set-valued map exists; see [255].

Theorem A.39 (Continuous selection). *Given an inner semicontinuous set-valued map $F : \mathbb{R}^m \rightrightarrows \mathbb{R}^n$ with nonempty, convex, and closed values, there exists a continuous selection $f : \mathbb{R}^m \rightarrow \mathbb{R}^n$, that is, the function f is continuous and satisfies*

$$f(x) \in F(x) \qquad \forall x \in \mathbb{R}^m$$

The following proposition presents a result providing a way to make a selection from a set-valued map that has minimum pointwise norm. This result is [191, Proposition 2.19] and is used in Chapter 10.

Proposition A.40. *(Minimum norm selection) Given an inner semicontinuous set-valued map $F : \mathbb{R}^m \rightrightarrows \mathbb{R}^n$ with nonempty, convex, and closed values, the function $f : \mathbb{R}^m \rightarrow \mathbb{R}^n$ defined as $f(x) := \arg\min \{|z| : z \in F(x)\}$ for each $x \in \mathbb{R}^m$ is continuous and* gph f *is closed.*

The following result appeared in [192, Theorem 4.5] and is used in the proof of Theorem 10.10.

Theorem A.41 (Existence of stabilizing feedback). *Under the conditions of Theorem 10.10, if*

1. *The set-valued map $\mathcal{T}_c'$ is inner semicontinuous at each $z \in \Pi_c(C_P) \cap \mathcal{I}(0)$;*

2. *The set-valued map $\mathcal{T}_d'$ is inner semicontinuous at each $z \in \Pi_d(D_P) \cap \mathcal{I}(0)$;*

then $\mathcal{A}$ is pre-asymptotically stabilizable for $\mathcal{H}_P$ by a continuous state-feedback pair (κ_c, κ_d), namely, there exists a pair of continuous functions (κ_c, κ_d) such that the resulting hybrid closed-loop system in (2.24) has the set $\mathcal{A}$ pre-asymptotically stable.

A.5   EXERCISES

Exercise 95 (Properties of continuous functions). Show that for continuous functions $F_1 : \mathbb{R}^m \rightarrow \mathbb{R}^n$ and $F_2 : \mathbb{R}^r \rightarrow \mathbb{R}^m$ the following hold:

1. $F_1(S)$ is bounded if $S \subset \mathbb{R}^m$ is bounded.

2. $F_1(S)$ is compact if $S \subset \mathbb{R}^m$ is compact.

3. $x \mapsto F(x) := F_1(F_2(x))$ defined on $\mathbb{R}^r$ is continuous.

4. S is closed if $F_1(S)$ is closed.

Is $F_1(S)$ closed if F_1 is continuous and S is closed?

Exercise 96 (Continuity of the distance to a set). Show that the distance function introduced in Definition A.10 is continuous. Hint: use the triangle inequality.

Exercise 97 (Regularity of set-valued maps). Given set-valued maps $F_1 : \mathbb{R}^m \rightrightarrows \mathbb{R}^n$ and $F_2 : \mathbb{R}^m \rightrightarrows \mathbb{R}^n$, and a function $F_3 : \mathbb{R}^p \rightarrow \mathbb{R}^m$, show the following:

1. If F_1 is outer semicontinuous and F_3 is continuous, then the set-valued map F defined as $F(x) := F_1(F_3(x))$ for all $x \in \mathbb{R}^p$ is outer semicontinuous.

2. If F_1 and F_2 are outer semicontinuous, then the set-valued map F defined as $F(x) := F_1(x) \cup F_2(x)$ for all $x \in \mathbb{R}^m$ is outer semicontinuous.

Exercise 98 (Bounds on positive definite functions). Given a continuous function $V : \mathbb{R}^n \to \mathbb{R}_{\geq 0}$ that is positive definite with respect to a nonempty compact set $\mathcal{A} \subset \mathbb{R}^n$ show the following:

1. There exist $r > 0$ and class-$\mathcal{K}_\infty$ functions α_1^r and α_2^r such that

$$\alpha_1^r(|x|_\mathcal{A}) \leq V(x) \leq \alpha_2^r(|x|_\mathcal{A}) \qquad \forall x \in \mathcal{A} + r\mathbb{B}$$

2. If, in addition, V has compact sublevel sets – or, equivalently, V is radially unbounded with respect to $\mathcal{A}$ – then there exist class-$\mathcal{K}_\infty$ functions α_1 and α_2 such that

$$\alpha_1(|x|_\mathcal{A}) \leq V(x) \leq \alpha_2(|x|_\mathcal{A}) \qquad \forall x \in \mathbb{R}^n \qquad \text{(A.14)}$$

Exercise 99 (Properties of class-$\mathcal{K}_\infty$ functions). Given class-$\mathcal{K}_\infty$ functions α_1 and α_2, show the following:

1. $s \mapsto \alpha_1(\alpha_2(s))$ defined on $\mathbb{R}_{\geq 0}$ is a class-$\mathcal{K}_\infty$ function.

2. α_1^{-1} defined as $\alpha_1^{-1}(\alpha_1(s)) = s$ for all $s \in \mathbb{R}_{\geq 0}$ is a class-$\mathcal{K}_\infty$ function.

Exercise 100 (Closedness of sets and maps). Perform the following tasks after solving Exercise 95:

1. Given a function $F : \mathbb{R}^m \to \mathbb{R}$ and a nonempty set $S \subset \mathbb{R}^m$, show that if F is continuous and the set S is closed then, for each $r \in F(S)$, the set $\{x \in S : F(x) = r\}$ is closed. Relate this property to the level set of a Lyapunov function candidate as in Definition 3.17.

2. Given a function $F : \mathbb{R}^m \to \mathbb{R}$ and a nonempty set $S \subset \mathbb{R}^m$, show that if F is continuous and the set S is closed then, for each $r \in F(S)$, the set $\{x \in S : F(x) \leq r\}$ is closed. Relate this property to the sublevel set of a Lyapunov function.

3. Given a function $F : \mathbb{R}^n \to \mathbb{R}^n$ and a nonempty set $S \subset \mathbb{R}^n$, show that if F is continuous and the set S is closed then the set $\{x \in \mathbb{R}^n : F(x) \in S\}$ is closed. Relate this property to the preimage of a function.

4. Given a nonempty set $S \subset \mathbb{R}^m$, $m = m_1 + m_2$, with m_1 and m_2 being positive integers, show that if S is closed then the map defined as $F(x) := \{x_2 \in \mathbb{R}^{m_2} : (x_1, x_2) \in S\}$ for each $x \in \mathbb{R}^n$ has closed values in its domain.

Proof of the Hybrid Lyapunov Theorem

This appendix provides a proof of item 1 and item 2 in Theorem 3.19. Proofs of item 3 can be found in [1, Chapter 3] (see also [9]), from where the proof of item 4 follows; see also [98] and [160].

B.1 PROOF OF STABILITY OF $\mathcal{A}$

The proof of stability of $\mathcal{A}$ in item 1 follows the ideas used for the proof of [91, Theorem 7.6]. The proof proceeds in two steps. In the first step, it is shown that there exists a sublevel set of V that is contained in $\mathcal{U}$, is compact, and, according to item 1 in Definition 3.13, is forward pre-invariant for $\mathcal{H}$. In the second step, relying on forward pre-invariance of the sublevel set of V, stability of $\mathcal{A}$ is shown by, for any given $\varepsilon > 0$, finding $\delta > 0$ such that item 1 in Definition 3.1 holds. The first step is established in the following lemma. As defined in Theorem 3.19, X denotes $C \cup D \cup G(D)$.

Lemma B.1 (Properties of a sublevel set of V). *Suppose the assumptions required in item 1 in Theorem 3.19 hold. Given $\varepsilon' > 0$ such that $\widetilde{\mathcal{A}} + 2\varepsilon'\mathbb{B} \subset \mathcal{U}$, where $\widetilde{\mathcal{A}} = \mathcal{A} \cap X$, there exists $r' > 0$ such that the set*

$$K := \left\{ x \in (\widetilde{\mathcal{A}} + \varepsilon'\mathbb{B}) \cap X \,:\, V(x) \le r' \right\} \tag{B.1}$$

is a compact forward pre-invariant set for $\mathcal{H}$.

Proof. Compactness of K follows from the fact that $(\widetilde{\mathcal{A}} + \varepsilon'\mathbb{B}) \cap X$ is compact and V is continuous; see Exercise 100. To show forward pre-invariance of K, since, according to Definition 3.17, V is positive definite on X with respect to $\mathcal{A}$, note that there exists $r'_\epsilon > 0$ such that

$$x \in (\widetilde{\mathcal{A}} + 2\varepsilon'\mathbb{B}) \cap X, \qquad V(x) \le r'_\epsilon$$

implies

$$x \in (\widetilde{\mathcal{A}} + \varepsilon'\mathbb{B}) \cap X$$

Using (3.19) and the fact that $\widetilde{\mathcal{A}} \subset \mathcal{U}$, since $\Delta V(x) \le 0$ for all $x \in \widetilde{\mathcal{A}}$ and V is positive definite on X with respect to $\mathcal{A}$, from the definition of ΔV in (3.16) it follows that each $\chi \in G(x)$ is such that $V(\chi) = 0$. Since[1] V is positive definite on

---

[1] When V is positive definite only on $X \cap \mathcal{U}$ with respect to $\mathcal{A}$, only when $\chi \in \mathcal{U}$ one has that $\chi \in \mathcal{A}$. In such a case, only $G(\widetilde{\mathcal{A}}) \cap \mathcal{U} \subset \widetilde{\mathcal{A}}$ can be asserted.

X with respect to $\mathcal{A}$, then $\chi \in \mathcal{A}$. It follows that

$$G(\widetilde{\mathcal{A}}) \subset \widetilde{\mathcal{A}}$$

Now, by outer semicontinuity and local boundedness of G from the hybrid basic conditions in Definition 2.20, using Lemma A.35, G is upper semicontinuous according to Definition A.34. It follows that, for each $x \in \widetilde{\mathcal{A}}$ and for the chosen ε', there exists $\varepsilon'' > 0$ such that[2] $G(x + \varepsilon''\mathbb{B}) \subset G(x) + \varepsilon'\mathbb{B}$. Then,

$$G(\widetilde{\mathcal{A}} + \varepsilon''\mathbb{B}) \subset \widetilde{\mathcal{A}} + \varepsilon'\mathbb{B}$$

Using positive definiteness of V with respect to $\mathcal{A}$, there exists $r''_\epsilon > 0$ such that

$$x \in (\widetilde{\mathcal{A}} + 2\varepsilon'\mathbb{B}) \cap X, \qquad V(x) \le r''_\epsilon$$

imply

$$x \in (\widetilde{\mathcal{A}} + \varepsilon''\mathbb{B}) \cap X$$

Combining these properties for r'_ϵ and for r''_ϵ, it follows that, with $r_\epsilon := \min\{r'_\epsilon, r''_\epsilon\}$,

$$x \in (\widetilde{\mathcal{A}} + 2\varepsilon'\mathbb{B}) \cap X, \qquad V(x) \le r_\epsilon \tag{B.2}$$

implies

$$x \in (\widetilde{\mathcal{A}} + \varepsilon'\mathbb{B}) \cap X, \quad G(x) \subset (\widetilde{\mathcal{A}} + \varepsilon'\mathbb{B}) \cap X \tag{B.3}$$

This property is next used to show that K is forward pre-invariant for $\mathcal{H}$ with $r' = r_\epsilon$. Let $(l, j) \mapsto x(t, j)$ be a solution to $\mathcal{H}$ from K. By definition of K in (B.1), $x(0, 0)$ belongs to $(\widetilde{\mathcal{A}} + \varepsilon'\mathbb{B}) \cap X$ and satisfies $V(x(0, 0)) \le r'$. In addition, $x(0, 0)$ satisfies (B.2).

- If $(0, 1) \in \text{dom}\, x$, then $x(0, 1) \in G(x(0, 0))$ and, from (B.3), $x(0, 1) \in (\widetilde{\mathcal{A}} + \varepsilon'\mathbb{B}) \cap X$. Moreover, since ΔV is nonincreasing due to (3.19), $V(x(0, 1)) \le r'$ and, hence, $x(0, 1) \in K$.

- Suppose there exists $T > 0$ such that $[0, T] \times \{0\} \subset \text{dom}\, x$, and for some $t' \in (0, T]$, $x(t', 0) \notin K$. Since, via (3.18), V is nonpositive on $C \cap \mathcal{U}$, $t \mapsto V(x(t, 0))$ is nonincreasing over $[0, T]$. Then, according to the definition of K, to have $x(t', 0) \notin K$, the solution has to leave the set $(\widetilde{\mathcal{A}} + \varepsilon'\mathbb{B}) \cap X$. Hence, to leave K, by continuity of $t \mapsto x(t, 0)$, there has to exist $t'' \in (0, t']$ such that $x(t'', 0) \in ((\widetilde{\mathcal{A}} + 2\varepsilon'\mathbb{B}) \setminus (\widetilde{\mathcal{A}} + \varepsilon'\mathbb{B})) \cap X$. Since $x(t'', 0)$ satisfies (B.2), which in turn implies (B.3), it has to be the case that $x(t'', 0) \in (\widetilde{\mathcal{A}} + \varepsilon'\mathbb{B}) \cap X$. This property further implies that $x(t'', 0) \in K$. This is a contradiction. Then, the solution x cannot leave the set K by flowing.

Since solutions to $\mathcal{H}$ from K cannot leave K by jumping or flowing, the set K is forward pre-invariant for $\mathcal{H}$. $\qquad\qquad\qquad\qquad\qquad\qquad\qquad\qquad\qquad\qquad\qquad$ $\square$

Note that, by construction, K in (B.1) is a subset of $\mathcal{U}$.

To show stability of $\mathcal{A}$, let $\varepsilon > 0$ be given. Pick $\varepsilon' \in (0, \varepsilon]$. Let $r' > 0$ come from Lemma B.1. From the definition of K, using continuity of V, there exists $\delta \in (0, \varepsilon')$

---

[2]At points where G is empty, this property still holds.

such that $x \in (\widetilde{\mathcal{A}} + \delta \mathbb{B}) \cap X$ implies $V(x) \leq r'$. Since $\delta < \varepsilon'$, using the fact that K is forward pre-invariant for $\mathcal{H}$, every solution x to $\mathcal{H}$ with $|x(0,0)|_{\mathcal{A}} \leq \delta$ remains in K; in particular, each such solution satisfies $|x(t,j)|_{\mathcal{A}} \leq \varepsilon' \leq \varepsilon$ for all $(t,j) \in \operatorname{dom} x$. Then, $\mathcal{A}$ is stable for $\mathcal{H}$.

B.2   PROOF OF PRE-ASYMPTOTIC STABILITY OF $\mathcal{A}$

Proofs of the claims in each of the items in item 2 for the case where $\mathcal{A}$ is compact are provided next. The proof of the claims in item 3 in Theorem 3.19 for the case where $\mathcal{A}$ is closed follow the same steps in the proofs of [1, Theorem 3.18, Proposition 3.24, Proposition 3.27, Proposition 3.29, and Proposition 3.30]. The proof of item 4 in Theorem 3.19 follows directly from such proofs, by using the expressions of α_1, α_2, ρ_c, and ρ_d in [1, (3.3)].

B.2.1   Proof of Item 2a of Theorem 3.19

Stability of $\mathcal{A}$ is shown in § B.1. From that proof, using Lemma B.1, there exists $\mu > 0$ such that every solution x to $\mathcal{H}$ with $|x(0,0)|_{\mathcal{A}} \leq \mu$ is bounded and such that $x(t,j) \in \mathcal{U}$ for all $(t,j) \in \operatorname{dom} x$. Since $\mathcal{H}$ satisfies the hybrid basic conditions, the Hybrid Invariance Principle in Theorem 3.23 characterizes convergence of such solutions, when complete. Since $\dot{V}$ satisfies (3.20) and ΔV satisfies (3.21), then

$$\dot{V}^{-1}(0) = \mathcal{A} \cap C, \qquad \Delta V^{-1}(0) = \mathcal{A} \cap D$$

Then, by item 1 of Theorem 3.23, every complete solution x to $\mathcal{H}$ with $|x(0,0)|_{\mathcal{A}} \leq \mu$ – which as stated above is bounded – converges to the largest weakly invariant set in (3.32), which reduces to

$$V^{-1}(r) \cap \mathcal{U} \cap \left[\dot{V}^{-1}(0) \cup \left(\Delta V^{-1}(0) \cap G(\Delta V^{-1}(0)) \right) \right]$$
$$= (\mathcal{A} \cap X) \cap \mathcal{U} \cap [(\mathcal{A} \cap C) \cup ((\mathcal{A} \cap D) \cap G(\mathcal{A} \cap D))] \subset \mathcal{A} \cap X$$

and r is zero. Then, the set $\mathcal{A}$ is pre-attractive for $\mathcal{H}$.

B.2.2   Proof of Item 2b of Theorem 3.19

Stability of $\mathcal{A}$ can be established as in § B.1. Then, following the steps in § B.2.1, there exists $\mu > 0$ such that every solution x to $\mathcal{H}$ with $|x(0,0)|_{\mathcal{A}} \leq \mu$ is bounded and such that $x(t,j) \in \mathcal{U}$ for all $(t,j) \in \operatorname{dom} x$. Then, following [9, Definition 3.2], the ω-limit set of the solution x satisfies $\Omega(x) \subset K \subset \mathcal{U}$. Given any $\xi' \in \Omega(x)$, since $\Omega(x)$ is weakly invariant according to [9, Lemma 4.1], let x' be any solution to $\mathcal{H}$ from ξ' verifying forward invariance of $\Omega(x)$; i.e., $\operatorname{rge} x' \subset \Omega(x)$. By [9, Lemma 4.1], V is constant along x'. Suppose that $V(x'(t,j)) =: r^* > 0$ for all $(t,j) \in \operatorname{dom} x'$. Then, in particular, $\Omega(x) \cap \mathcal{A} = \emptyset$. If item 2b(i) of the assumptions holds, then x' is discrete since $V(x'(t,j)) = r^* > 0$ for all $(t,j) \in \operatorname{dom} x'$. Hence, by item 2b(ii) it converges to $\mathcal{A}$. But this contradicts V being constant along x'. Hence, x converges to $\mathcal{A}$.

B.2.3   Proof of Item 2c of Theorem 3.19

The proof follows the one in § B.2.2. The only part that changes is when the assumptions are invoked. If item 2c(i) of the assumptions holds, then x' is continuous since $V(x'(t,j)) = r^* > 0$ for all $(t,j) \in \operatorname{dom} x'$. Hence, by item 2c(ii) it converges to $\mathcal{A}$. Since this contradicts V being constant along x', x converges to $\mathcal{A}$.

B.2.4   Proof of Item 2d of Theorem 3.19

Stability of $\mathcal{A}$ also follows from § B.1. As in § B.2.1, there exists $\mu > 0$ such that every solution x to $\mathcal{H}$ with $|x(0,0)|_{\mathcal{A}} \leq \mu$ is bounded and such that $x(t,j) \in \mathcal{U}$ for all $(t,j) \in \operatorname{dom} x$. Note that, without loss of generality, μ can be chosen small enough so that $V(x(0,0)) \in (0, r^*)$. Applying the Hybrid Invariance Principle in Theorem 3.23, every complete solution x to $\mathcal{H}$ with $|x(0,0)|_{\mathcal{A}} \leq \mu$ converges to the largest weakly invariant set in (3.32) for some r. Since V does not increase along solutions, r has to be in $[0, r^*)$. Since for each $x_\circ \in \mathcal{U}$ such that $r = V(x_\circ) > 0$, there is no complete solution to $\mathcal{H}$ from $x_\circ$ that remains in $V^{-1}(r) \cap \mathcal{U}$, then r in the application of the Hybrid Invariance Principle has to be zero. Since, using the definition of $V^{-1}(0)$ in (3.29), $V^{-1}(0) \subset \mathcal{A} \cap X$, then $\mathcal{A}$ is pre-attractive.

B.2.5   Proof of Item 2e of Theorem 3.19

Following the ideas in the proof of Lemma B.1, pick any $\varepsilon' > 0$ such that $\widetilde{\mathcal{A}} + 2\varepsilon'\mathbb{B} \subset \mathcal{U}$, where $\widetilde{\mathcal{A}} = \mathcal{A} \cap X$. Let

$$r' = \min\left\{V(x) : x \in \operatorname{dom} V, |x|_{\widetilde{\mathcal{A}}} - \varepsilon'\right\} \tag{B.4}$$

which is positive since $\varepsilon' > 0$ and V is positive definite with respect to $\mathcal{A}$. Pick $r_1 \in (0, r')$ small enough so that

$$Z := \{x \in \operatorname{dom} V \cap X : V(x) \leq r_1\}$$

is compact. Note that Z is in the interior of $\widetilde{\mathcal{A}} + 2\varepsilon'\mathbb{B}$. Hence, $Z \subset \mathcal{U}$. Given M as in (3.25), let $r_0 > 0$ be such that $\exp(M)r_0 \leq r_1$. Pick any solution x to $\mathcal{H}$ with $x(0,0)$ such that $V(x(0,0)) \leq r_0$. It follows that $x(t,j) \subset \mathcal{U}$ for all $(t,j) \in \operatorname{dom} x$. In fact, using (3.23), (3.24), and (3.25) in (3.5) leads to

$$V(x(t,j)) \leq \exp(\lambda_c t + \lambda_d j)V(x(0,0)) \leq \exp(M)\exp(-\gamma(t+j))V(x(0,0)) \tag{B.5}$$

for all $(t,j) \in \operatorname{dom} x$. Since γ is positive and $\exp(M)\exp(-\gamma(t+j))V(x(0,0)) \leq r_1$ for all (t,j) and $r_1 \leq r'$, then x remains in $\mathcal{U}$.

Now, given $\varepsilon > 0$, let r' be given as in (B.4) with $\varepsilon' \in (0, \varepsilon/2)$. Pick $r_1 \in (0, r')$ small enough so that Z is compact. Then, choose $\delta \in (0, \varepsilon')$ such that

$$\exp(M)\max\left\{V(x) : x \in \operatorname{dom} V, |x|_{\widetilde{\mathcal{A}}} \leq \delta\right\} \leq r_1$$

It follows that for each solution x to $\mathcal{H}$ with $|x(0,0)|_{\widetilde{\mathcal{A}}} \leq \delta$, since $\exp(M)V(x(0,0)) \leq r_1$ implies $V(x(t,j)) \leq r_1$ via (B.5), the solution x remains in Z. Since Z is a subset of $\widetilde{\mathcal{A}} + 2\varepsilon'\mathbb{B}$, which is a subset of $\widetilde{\mathcal{A}} + \varepsilon\mathbb{B}$, then $|x(t,j)|_{\mathcal{A}} \leq \varepsilon$ for each $(t,j) \in \operatorname{dom} x$. Hence, $\mathcal{A}$ is stable for $\mathcal{H}$. Pre-attractivity follows using $\mu = \delta$ with δ as just defined, by invoking (B.5) and compactness of Z.

Bibliography

[1] R. Goebel, R. G. Sanfelice, and A. R. Teel. *Hybrid Dynamical Systems: Modeling, Stability, and Robustness.* Princeton University Press, New Jersey, 2012.

[2] J. Chai and R. G. Sanfelice. A robust hybrid control algorithm for a single-phase dc/ac inverter with variable input voltage. In *Proceedings of the 2014 American Control Conference*, pages 1420–1425, 2014.

[3] D. Morin. *Introduction to Classical Mechanics: With Problems and Solutions.* Cambridge University Press, 2008.

[4] J. W. Grizzle, G. Abba, and F. Plestan. Asymptotically stable walking for biped robots: Analysis via systems with impulse effects. *IEEE Transactions on Automatic Control*, 46(1):51–64, 2001.

[5] R. Goebel, J. P. Hespanha, A. R. Teel, C. Cai, and R. G. Sanfelice. Hybrid systems: generalized solutions and robust stability. In *Proc. 6th IFAC Symposium in Nonlinear Control Systems*, page 1–12, 2004.

[6] R. Goebel and A. R. Teel. Solutions to hybrid inclusions via set and graphical convergence with stability theory applications. *Automatica*, 42(4):573–587, 2006.

[7] C. Cai, A. R. Teel, and R. Goebel. Smooth Lyapunov functions for hybrid systems - Part I: Existence is equivalent to robustness. *IEEE Transactions on Automatic Control*, 52(7):1264–1277, July 2007.

[8] R. G. Sanfelice, R. Goebel, and A.R. Teel. Generalized solutions to hybrid dynamical systems. *ESAIM: Control, Optimisation and Calculus of Variations*, 14(4):699–724, 2008.

[9] R. G. Sanfelice, R. Goebel, and A. R. Teel. Invariance principles for hybrid systems with connections to detectability and asymptotic stability. *IEEE Transactions on Automatic Control*, 52(12):2282–2297, 2007.

[10] C. Cai, A. R. Teel, and R. Goebel. Smooth Lyapunov functions for hybrid systems. Part II: (Pre)Asymptotically stable compact sets. *IEEE Transactions on Automatic Control*, 53(3):734–748, April 2008.

[11] R. Goebel, R. G. Sanfelice, and A. R. Teel. Hybrid dynamical systems. *IEEE Control Systems Magazine*, 29(2):28–93, April 2009.

[12] M. S. Branicky, V. S. Borkar, and S. K. Mitter. A unified framework for hybrid control: Model and optimal control theory. *IEEE Transactions on*

Automatic Control, 43(1):31–45, 1998.

[13] J. Lygeros, C. Tomlin, and S. S. Sastry. Controllers for reachability specifications for hybrid systems. *Automatica*, 35:349–370, 1999.

[14] C. Tomlin, J. Lygeros, and S. S. Sastry. A game theoretic approach to controller design for hybrid systems. *Proceedings of IEEE*, 88:949–970, 2000.

[15] C. Prieur and A. Astolfi. Robust stabilization of chained systems via hybrid control. *IEEE Transactions on Automatic Control*, 48(10):1768–1772, October 2003.

[16] M. S. Shaikh and P. E. Caines. On the hybrid optimal control problem: Theory and algorithms. *IEEE Transactions on Automatic Control*, 52:1587–1603, 2007.

[17] A. van der Schaft and H. Schumacher. *An Introduction to Hybrid Dynamical Systems*. Lecture Notes in Control and Information Sciences, Springer, 2000.

[18] W. S. Levine and D. Hristu-Varsakelis, editors. *Handbook of Networked and Embedded Control Systems*. Springer, 2005.

[19] W. M. Haddad, V. Chellaboina, and S. G. Nersesov. *Impulsive and Hybrid Dynamical Systems: Stability, Dissipativity, and Control*. Princeton University Press, 2006.

[20] J. Chai and R. G. Sanfelice. Hybrid feedback control methods for robust and global power conversion. In *Proceedings of the 5th Analysis and Design of Hybrid Systems*, pages 298–303, October 2015.

[21] B. Short and R. G. Sanfelice. A hybrid predictive control approach to trajectory tracking for a fully actuated biped. In *Proceedings of the American Control Conference*, pages 3526–3531, August 2018.

[22] R. G. Sanfelice and A. R. Teel. A "throw-and-catch" hybrid control strategy for robust global stabilization of nonlinear systems. In *Proc. 26th American Control Conference*, page 3470–3475, 2007.

[23] R. O'Flaherty, R. G. Sanfelice, and A. R. Teel. Robust global swing-up of the pendubot via hybrid control. In *Proc. 27th American Control Conference*, page 1424–1429, 2008.

[24] R. W. Brockett. *Differential Geometric Control Theory*, chapter Asymptotic stability and feedback stabilization, pages 181–191. Birkhauser, Boston, MA, 1983.

[25] E. P. Ryan. On brockett's condition for smooth stabilizability and its necessity in a context of nonsmooth feedback. *SIAM Journal on Control and Optimization*, 32(6):1597–1604, 1994.

[26] J. P. Hespanha and A. S. Morse. Stabilization of nonholonomic integrators via logic-based switching. *Automatica*, 35(3):385–393, 1999.

[27] R. Goebel, C. Prieur, and A. R. Teel. Smooth patchy control Lyapunov functions. *Automatica*, 45(3):675–683, 2009.

[28] S. P. Bhat and D. S. Bernstein. A topological obstruction to continuous global stabilization of rotational motion and the unwinding phenomenon. *Systems & Control Letters*, 39(1):63–70, 2000.

[29] E. D. Sontag. *Mathematical Control Theory: Deterministic Finite Dimensional Systems*. Springer-Verlag, New York, 1990.

[30] R. G. Sanfelice, A. R. Teel, and R. Goebel. Supervising a family of hybrid controllers for robust global asymptotic stabilization. In *Proc. 47th IEEE Conference on Decision and Control*, page 4700–4705, 2008.

[31] A. R. Teel, R. G. Sanfelice, and R. Goebel. *Hybrid Control Systems*. Springer, 2009.

[32] R. G. Sanfelice, M. J. Messina, S. E. Tuna, and A. R. Teel. Robust hybrid controllers for continuous-time systems with applications to obstacle avoidance and regulation to disconnected set of points. In *Proc. 25th American Control Conference*, page 3352–3357, 2006.

[33] S. Phillips and R. G. Sanfelice. Robust distributed synchronization of networked linear systems with intermittent information. *Automatica*, 105:323–333, July 2019.

[34] A. R. Teel and N. Kapoor. Uniting global and local controllers. In *Proc. European Control Conference*, 1997.

[35] P. Morin, R. M. Murray, and L. Praly. Nonlinear rescaling of control laws with application to stabilization in the presence of magnitude saturation. In *Proc. 4th IFAC Symposium on Nonlinear Control Systems*, Enschede, the Netherlands, 1998.

[36] Z. Pan, K. Ezal, A. J. Krener, and P.V. Kokotovic. Backstepping design with local optimality matching. *IEEE Transactions on Automatic Control*, 46(7):1014–1027, 2001.

[37] V. Andrieu and C. Prieur. Uniting two control Lyapunov functions for affine systems. *IEEE Transactions on Automatic Control*, 55(8):1923–1927, 2010.

[38] C. Prieur and A. R. Teel. Uniting local and global output feedback controllers. *IEEE Transactions on Automatic Control*, 56(7):1636–1649, 2011.

[39] D. D. Bainov and P. S. Simeonov. *Systems with Impulse Effect: Stability, Theory, and Applications*. Ellis Horwood Limited, Chichester England and New York, 1989.

[40] V. Lakshmikantham, D. D. Bainov, and P. S. Simeonov. *Theory of Impulsive Differential Equations*, volume 6 of *Series in Modern Applied Mathematics*. World Scientific, Singapore, 1989.

[41] T. Yang. *Impulsive control theory*, volume 272 of *Lecture Notes in Control and Information Sciences*. Springer-Verlag, 2001.

[42] J.-P. Aubin, J. Lygeros, M. Quincampoix, S. S. Sastry, and N. Seube. Impulse differential inclusions: a viability approach to hybrid systems. *IEEE*

Transactions on Automatic Control, 47(1):2–20, 2002.

[43] L. Tavernini. Differential automata and their discrete simulators. *Nonlinear Analysis, Theory, Methods & Applications*, 11(6):665–683, 1987.

[44] T. A. Henzinger. The theory of hybrid automata. In *Proc. 11th Annual Symp. on Logic in Comp. Science*, pages 278–292, 1996.

[45] J. Lygeros, K. H. Johansson, S. N. Simić, J. Zhang, and S. S. Sastry. Dynamical properties of hybrid automata. *IEEE Transactions on Automatic Control*, 48(1):2–17, 2003.

[46] P. J. Antsaklis, J. A. Stiver, and M. D. Lemmon. Hybrid systems modeling and autonomous control systems. In R. L. Grossman, A. Nerode, A. P. Ravn, and H. Rishel, editors, *Hybrid Systems*, volume 736. Lecture Notes in Computer Science, 1993.

[47] A. N. Michel, L. Wang, and B. Hu. *Qualitative Theory of Dynamical Systems*. Dekker, 2001.

[48] M. Broucke and A. Arapostathis. Continuous selections of trajectories of hybrid systems. *Systems & Control Lett.*, 47:149–157, 2002.

[49] J. J. Moreau. *Topics in Nonsmooth Mechanics*, chapter Bounded variation in time, pages 1–74. Birkhäuser Verlag, 1988.

[50] G. N. Silva and R. B. Vinter. Measure driven differential inclusions. *Journal of mathematical analysis and applications*, 202(3):727–746, 1996.

[51] B. Brogliato *Nonsmooth Mechanics Models, Dynamics and Control*. Springer, London, 1996.

[52] F. L. Pereira and G. N. Silva. Lyapunov stability of measure driven impulsive systems. *Differential Equations*, 40(8):1122–1130, 2004.

[53] R. I. Leine and N. van de Wouw. *Stability and Convergence of Mechanical Systems with Unilateral Constraints*, volume 36 of *Lecture Notes in Applied and Computational Mechanics*. Springer Verlag, Berlin, 2008.

[54] M. Bohner. *Dynamic equations on time scales: An introduction with applications*. Birkhauser, 2001.

[55] N. N. Krasovskii. *Problems of the Theory of Stability of Motion*. Stanford Univ. Press, 1963. Trans. of Russian edition, Moscow, 1959.

[56] H. Hermes. Discontinuous vector fields and feedback control. In J.K. Hale and J.P. LaSalle, editors, *Differential Equations and Dynamical Systems*, pages 155–165. Academic Press, New York, 1967.

[57] O. Hàjek. Discontinuous differential equations I. *Journal of Differential Equations*, 32:149–170, 1979.

[58] A. F. Filippov. *Differential Equations with Discontinuous Right-Hand Sides*. Kluwer, 1988.

[59] A. Bacciotti and F. Ceragioli. Stability and stabilization of discontinuous systems and nonsmooth Lyapunov functions. *ESAIM: Control, Optimisation and Calculus of Variations*, 4:361–376, 1999.

[60] W.P.M.H. Heemels, J. M. Schumacher, and S. Weiland. Linear complementarity systems. *SIAM J. Appl. Math.*, 60(4):1234–1269, 2000.

[61] A. Bemporad, F. Borrelli, and M. Morari. Piecewise linear optimal controllers for hybrid systems. In *Proceedings of the 2000 American Control Conference*, volume 2, pages 1190–1194. IEEE, 2000.

[62] W.P.M.H. Heemels, B. De Schutter, and A. Bemporad. Equivalence of hybrid dynamical models. *Automatica*, 37(7):1085–1091, July 2001.

[63] G. Ferrari-Trecate, F.A. Cuzzola, D. Mignone, and M. Morari. Analysis of discrete-time piecewise affine and hybrid systems. *Automatica*, 38(12):2139–2146, December 2002.

[64] F. Borrelli, A. Bemporad, and M. Morari. *Predictive control for linear and hybrid systems*. Cambridge University Press, 2017.

[65] P. Peleties and R. DeCarlo. Asymptotic stability of switched systems using Lyapunov functions. In *Proc. 31st IEEE Conference on Decision and Control*, volume 4, pages 3438–3439, 1992.

[66] A. Balluchi, L. Benvenuti, M. D. Di Benedetto, and A. L. Sangiovanni-Vincentelli. Design of observers for hybrid systems. In *International Workshop on Hybrid Systems: Computation and Control*, pages 76–89. Springer, 2002.

[67] A. V. Savkin and R. J. Evans. *Hybrid dynamical systems: controller and sensor switching problems*. Springer Science & Business Media, 2002.

[68] D. Liberzon. *Switching in Systems and Control*. Birkhauser, 2003.

[69] J. P. Hespanha. Uniform stability of switched linear systems: Extensions of LaSalle's invariance principle. *IEEE Transactions on Automatic Control*, 49(4):470–482, 2004.

[70] A. Bacciotti and L. Mazzi. An invariance principle for nonlinear switched systems. *Systems and Control Letters*, 54:1109–1119, 2005.

[71] J. L. Mancilla-Aguilar and R. A. Garcia. An extension of LaSalle's invariance principle for switched systems. *Systems Control Lett.*, 55:376–384, 2006.

[72] A. Tornambè. *Discrete-event system theory: an introduction*. World Scientific Pub. Co., 1995.

[73] P.J.G. Ramadge and W. M. Wonham. The control of discrete event systems. *Proceedings of the IEEE*, 77(1):81–98, 1989.

[74] L. E Holloway, B. H. Krogh, and A. Giua. A survey of petri net methods for controlled discrete event systems. *Discrete Event Dynamic Systems*, 7(2):151–190, 1997.

[75] C. G. Cassandras and S. Lafortune. *Introduction to discrete event systems*, volume 11. Kluwer Academic Publishers, 1999.

[76] R. G. Sanfelice. Interconnections of hybrid systems: Some challenges and recent results. *Journal of Nonlinear Systems and Applications*, 2(1-2):111–121, 2011.

[77] R. G. Sanfelice. Input-output-to-state stability tools for hybrid systems and their interconnections. *IEEE Transactions on Automatic Control*, 59(5):1360–1366, May 2014.

[78] R. Naldi and R. G. Sanfelice. Passivity-based control for hybrid systems with applications to mechanical systems exhibiting impacts. *Automatica*, 49(5):1104–1116, May 2013.

[79] R. G. Sanfelice. *Control of Hybrid Dynamical Systems: An Overview of Recent Advances*, pages 146–177. Wiley, April 2013.

[80] J. Chai, P. Casau, and R. G. Sanfelice. Analysis and design of event-triggered control algorithms using hybrid systems tools. In *Proceedings of the 2017 IEEE Conference on Decision and Control*, pages 6057–6062, 2017.

[81] P. Collins. A trajectory-space approach to hybrid systems. In *Proceedings of the 16th International Symposium on Mathematical Theory of Network and Systems*, 2004.

[82] R. Goebel and R. G. Sanfelice. How well-posedness of hybrid systems can extend beyond zeno times. In *Proceedings of the IEEE Conference on Decision and Control*, pages 598–603, 2016.

[83] W. Rudin. *Real and Complex Analysis*. McGraw-Hill, 1966.

[84] R. L. Allen and D. Mills. *Signal analysis: time, frequency, scale, and structure*. John Wiley & Sons, 2004.

[85] J.-P. Aubin and A. Cellina. *Differential Inclusions*. Springer-Verlag, 1984.

[86] I. Natanson. *Theory of Functions of a Real Variable*. Frederick Ungar Publishing Co., New York, 1961.

[87] C. Cai and A. R. Teel. Characterizations of input-to-state stability for hybrid systems. *Syst. & Cont. Letters*, 58:47–53, 2009.

[88] J. Chai and R. G. Sanfelice. Forward invariance of sets for hybrid dynamical systems (Part I). *IEEE Transactions on Automatic Control*, 64:2426–2441, June 2019.

[89] A. Zavala-Rio and B. Brogliato. On the control of a one degree-of-freedom juggling robot. *Dynamics and Control*, 9:67–90, 1999.

[90] R. Ronsse, P. Lefèvre, and R. Sepulchre. Rhythmic feedback control of a blind planar juggler. *IEEE Transactions on Robotics*, 23(4):790–802, 2007.

[91] R. G. Sanfelice, A. R. Teel, and R. Sepulchre. A hybrid systems approach to trajectory tracking control for juggling systems. In *Proc. 46th IEEE Confer-*

ence on Decision and Control, page 5282–5287, New Orleans, LA, 2007.

[92] X. Tian, J. H. Koessler, and R. G. Sanfelice. Juggling on a bouncing ball apparatus via hybrid control. In *Proceedings of the IEEE/RSJ International Conference on Intelligent Robots and Systems*, page 1848–1853, 2013.

[93] R. G. Sanfelice and A. R. Teel. Dynamical properties of hybrid systems simulators. *Automatica*, 46(2):239–248, 2010.

[94] R. G. Sanfelice, D. A. Copp, and P. Nanez. A toolbox for simulation of hybrid systems in Matlab/Simulink: Hybrid Equations (HyEQ) Toolbox. In *Proceedings of Hybrid Systems: Computation and Control Conference*, page 101–106, 2013.

[95] H. K. Khalil. *Nonlinear Systems*. Prentice Hall, 3rd edition, 2002.

[96] C. H. Edwards. *Advanced calculus of several variables*. Courier Corporation, 2012.

[97] A.R. Teel and L. Praly. A smooth Lyapunov function from a class-$\mathcal{KL}$ estimate involving two positive semidefinite functions. *ESAIM: Control, Optimisation and Calculus of Variations*, 5:313–367, 2000.

[98] A. R. Teel, F. Forni, and L. Zaccarian. Lyapunov-based sufficient conditions for exponential stability in hybrid systems. *IEEE Transactions on Automatic Control*, 58(6):1591–1596, 2012.

[99] J. Chai and R. G. Sanfelice. Forward invariance of sets for hybrid dynamical systems (Part II). *To appear in IEEE Transactions on Automatic Control*, 2020.

[100] F. H. Clarke. *Optimization and Nonsmooth Analysis*. SIAM's Classic in Applied Mathematics, Philadelphia, 1990.

[101] C. G. Mayhew, R. G. Sanfelice, and A. R. Teel. Quaternion-based hybrid controller for robust global attitude tracking. *IEEE Transactions on Automatic Control*, 56(11):2555–2566, November 2011.

[102] R. G. Sanfelice and A. R. Teel. On singular perturbations due to fast actuators in hybrid control systems. *Automatica*, 47(4):692–701, April 2011.

[103] R. G. Sanfelice, A. R. Teel, R. Goebel, and C. Prieur. On the robustness to measurement noise and unmodeled dynamics of stability in hybrid systems. In *Proc. 25th American Control Conference*, page 4061–4066, 2006.

[104] R. G. Sanfelice and A. R. Teel. Lyapunov analysis of sample-and-hold hybrid feedbacks. In *Proc. 45th IEEE Conference on Decision and Control*, page 4879–4884, 2006.

[105] B. Altin and R. G. Sanfelice. On robustness of pre-asymptotic stability to delayed jumps in hybrid systems. In *Proceedings of the American Control Conference*, pages 2204–2209, 08/2018 2018.

[106] P. Morin, R.M. Murray, and L. Praly. Nonlinear rescaling of control laws with application to stabilization in the presence of magnitude saturation. In

Proc. 4th IFAC Symposium on Nonlinear Control Systems, Enschede, the Netherlands, 1998.

[107] C. Prieur and L. Praly. Uniting local and global controllers. In *Proceedings of the 38th IEEE Conference on Decision and Control*, volume 2, pages 1214–1219. IEEE, 1999.

[108] C. Prieur. Uniting local and global controllers with robustness to vanishing noise. *Math. Control Signals Systems*, 14:143–172, 2001.

[109] D. V. Efimov. Uniting global and local controllers under acting disturbances. *Automatica*, 42:489–495, 2006.

[110] D. V. Efimov, A. Loria, and E. Panteley. Multigoal output regulation via supervisory control: Application to stabilization of a unicycle. In *Proc. 2009 American Control Conference*, 2009.

[111] D. V. Efimov, A. Loria, and E. Panteley. Robust output stabilization: Improving performance via supervisory control. *International Journal of Robust and Nonlinear Control*, 21(10):1219–1236, 2011.

[112] D. Hustig-Schultz and R. G. Sanfelice. A robust hybrid heavy ball algorithm for optimization with high performance. In *Proceedings of the American Control Conference*, pages 151–156, July 2019.

[113] R. G. Sanfelice and C. Prieur. Robust supervisory control for uniting two output-feedback hybrid controllers with different objectives. *Automatica*, 49(7):1958–1969, July 2013.

[114] J. Chai, P. Casau, and R. G. Sanfelice. Analysis and design of event-triggered control algorithms using hybrid systems tools. *International Journal of Robust and Nonlinear Control*, April 2020.

[115] P. Tabuada. Event-triggered real-time scheduling of stabilizing control tasks. *IEEE Transactions on Automatic Control*, 52(9):1680–1685, 2007.

[116] F. Ferrante, F. Gouaisbaut, R. G. Sanfelice, and S. Tarbouriech. State estimation of linear systems in the presence of sporadic measurements. *Automatica*, 73:101–109, November 2016.

[117] W. Wang, D. Nešić, and R. Postoyan. Emulation-based stabilization of networked control systems implemented on flexray. *Automatica*, 59:73–83, 2015.

[118] R. Postoyan, P. Tabuada, D. Nešić, and A. Anta. A framework for the event-triggered stabilization of nonlinear systems. *IEEE Transactions on Automatic Control*, 60(4):982–996, 2015.

[119] D. P. N. Borgers and W.P.M.H. Heemels. Event-separation properties of event-triggered control systems. *IEEE Transactions on Automatic Control*, 59(10):2644–2656, 2014.

[120] F. Ferrante, F. Gouaisbaut, R. G. Sanfelice, and S. Tarbouriech. Observer-based control design for linear systems in the presence of limited measurement streams and intermittent input access. In *Proceedings of the American Control*

Conference, pages 4689–4694, June 2015.

[121] M. C. F. Donkers and W.P.M.H. Heemels. Output-based event-triggered control with guaranteed Linfinity-gain and improved event-triggering. In *49th IEEE Conference on Decision and Control (CDC)*, pages 3246–3251, 2010.

[122] W.P.M.H. Heemels, K. H. Johansson, and P. Tabuada. An introduction to event-triggered and self-triggered control. In *51st IEEE Conference on Decision and Control (CDC)*, pages 3270–3285, 2012.

[123] N. Marchand, S. Durand, and J. F. Guerrero-Castellanos. A general formula for event-based stabilization of nonlinear systems. *IEEE Transactions on Automatic Control*, 58(5):1332–1337, 2013.

[124] D. Theodosis and J. Tsinias. Sufficient lie algebraic conditions for sampled-data feedback stabilization. In *54th IEEE Conference on Decision and Control (CDC)*, pages 6490–6495, 2015.

[125] B. Boisseau, S. Durand, J. J. Martinez-Molina, T. Raharijaona, and N. Marchand. Attitude control of a gyroscope actuator using event-based discrete-time approach. In *IEEE International Conference on Event-based Control, Communication, and Signal Processing (EBCCSP)*, pages 1–6, 2015.

[126] J. F. Guerrero-Castellanos, J. J. Téllez-Guzmán, S. Durand, N. Marchand, J. U. Alvarez-Muñoz, and V. R. Gonzalez-Diaz. Attitude stabilization of a quadrotor by means of event-triggered nonlinear control. *Journal of Intelligent & Robotic Systems*, 73(1-4):123–135, 2014.

[127] T. Liu and Z. Jiang. Event-based control of nonlinear systems with partial state and output feedback. *Automatica*, 53:10–22, 2015.

[128] C. De Persis, R. Sailer, and F. Wirth. Parsimonious event-triggered distributed control: A Zeno free approach. *Automatica*, 49(7):2116–2124, 2013.

[129] A. Seuret and C. Prieur. Event-triggered sampling algorithms based on a Lyapunov function. In *50th IEEE Conference on Decision and Control and European Control Conference*, pages 6128–6133, 2011.

[130] T. Glück, A. Eder, and A. Kugi. Swing-up control of a triple pendulum on a cart with experimental validation. *Automatica*, 49(3):801–808, 2013.

[131] R. R. Burridge, A. A. Rizzi, and D. E. Koditschek. Toward a systems theory for the composition of dynamically dexterous robot behaviors. In *International Symposium on Robotics Research*, volume 7, pages 149–161. MIT Press, 1996.

[132] R. G. Sanfelice. Robust hybrid control systems. Ph.D., University of California, Santa Barbara, 2007.

[133] R. G. Sanfelice and E. Frazzoli. A hybrid control framework for robust maneuver-based motion planning. In *Proc. 27th American Control Conference*, page 2254–2259, 2008.

[134] E. Frazzoli, M. A. Dahleh, and E. Feron. Maneuver-based motion planning

for nonlinear systems with symmetries. *Robotics, IEEE Transactions on [see also Robotics and Automation, IEEE Transactions on]*, 21:1077–1091, 2005.

[135] R. Shvartsman, A. R. Teel, D. Oetomo, and D. Nešić. System of funnels framework for robust global non-linear control. In *2016 IEEE 55th Conference on Decision and Control*, pages 3018–3023. IEEE, 2016.

[136] I. Fantoni, R. Lozano, and M. W. Spong. Energy based control of the pendubot. *IEEE Transactions on Automatic Control*, 45:725–729, 2000.

[137] J. Rubi, A. Rubio, and A. Avello. Swing-up control problem for a self-erecting double inverted pendulum. *IEE Proc.- Control Theory Applications*, 149:169–175, 2002.

[138] P.-A. Absil and R. Sepulchre. A hybrid control scheme for swing up acrobatics. In *European Conference on Control ECC*, 2001.

[139] A. S. Morse, D. Q. Mayne, and G. C. Goodwin. Applications of hysteresis switching in parameter adaptive control. *IEEE Transactions on Automatic Control*, 37(9):1343–1354, 1992.

[140] R. H. Middleton, G. C. Goodwin, D. J. Hill, and D. Q. Mayne. Design issues in adaptive control. *IEEE Transactions on Automatic Control*, 33(1):50–58, January 1988.

[141] P. Peleties and R. A. DeCarlo. Asymptotic stability of m-switched systems using Lyapunov-like functions. In *Proc. American Control Conference*, pages 1679–1684, 1991.

[142] M. S. Branicky. Multiple Lyapunov functions and other analysis tools for switched and hybrid systems. *IEEE Transactions on Automatic Control*, 43(4):475–482, April 1998.

[143] J. P. Hespanha and A. S. Morse. Scale-independent hysteresis switching. In *Lecture Notes in Computer Science*, volume 1569, pages 117–122. Springer Berlin, Heidelberg, 1999.

[144] J. Malmborg, B. Berhardsson, and K. J. Åström. A stabilizing switching scheme for multi-controller systems. In *Proceedings of the Triennial IFAC World Congress*, volume F, pages 229–234, 1996.

[145] J. Malmborg and J. Eker. Hybrid control of a double tank system. In *Proceedings of the IEEE International Conference on Control Applications*, pages 133–138, 1997.

[146] R. Fierro, F. L. Lewis, and A. Lowe. Hybrid control for a class of underactuated mechanical systems. *IEEE Transactions on Systems, Man and Cybernetics, Part A: Systems and Humans*, 29(6):649–654, November 1999.

[147] H. Nakamura, Y. Yamashita, and H. Nishitani. Minimum projection method for nonsmooth control Lyapunov function design on general manifolds. *Systems & Control Letters*, 58(10–11):716–723, October 2009.

[148] H. Nakamura, Y. Fukui, N. Nakamura, and H. Nishitani. Multilayer minimum

projection method for nonsmooth strict control Lyapunov function design. *Systems & Control Letters*, 59(9):563–570, September 2010.

[149] C. G. Mayhew, R. G. Sanfelice, and A. R. Teel. Synergistic Lyapunov functions and backstepping hybrid feedbacks. In *Proc. 30th American Control Conference*, pages 3203–3208, 2011.

[150] C. G. Mayhew, R. G. Sanfelice, and A. R. Teel. Further results on synergistic Lyapunov functions and hybrid feedback design through backstepping. In *Proc. Joint Conference on Decision and Control and European Control Conference*, pages 7428–7433, 2011.

[151] C. G. Mayhew, R. G. Sanfelice, and A. R. Teel. Hybrid feedback design by backstepping synergistic Lyapunov functions. http://arxiv.org/abs/2009.03815, 2020.

[152] C. G. Mayhew, R. G. Sanfelice, and A. R. Teel. On path-lifting mechanisms and unwinding in quaternion-based attitude control. *IEEE Transactions on Automatic Control*, 58(5):1179–1191, May 2013.

[153] C. G. Mayhew and A. R. Teel. Hybrid control of spherical orientation. In *Proceedings of the 49th IEEE Conference on Decision and Control*, pages 4198–4203, 2010.

[154] C. G. Mayhew and A. R. Teel. Global asymptotic stabilization of the inverted equilibrium manifold of the 3D pendulum by hybrid feedback. In *Proceedings of the 49th IEEE Conference on Decision and Control*, pages 679–684, 2010.

[155] C. G. Mayhew and A. R. Teel. Hybrid control of planar rotations. In *Proceedings of the American Control Conference*, pages 154–159, 2010.

[156] C. G. Mayhew and A. R. Teel. Synergistic potential functions for hybrid control of rigid-body attitude. In *Proceedings of the American Control Conference*, pages 875–880, 2011.

[157] C. G. Mayhew and A. R. Teel. Hybrid control of rigid-body attitude with synergistic potential functions. In *Proceedings of the American Control Conference*, pages 287–292, 2011.

[158] P. Casau, R. G. Sanfelice, R. Cunha, and C. Silvestre. A globally asymptotically stabilizing trajectory tracking controller for fully actuated rigid bodies using landmark-based information. *International Journal of Robust and Nonlinear Control*, 25:3617–3640, 2015.

[159] P. Casau, R. G. Sanfelice, R. Cunha, D. Cabecinhas, and C. Silvestre. Robust global trajectory tracking for a class of underactuated vehicles. *Automatica*, 58:90–98, August 2015.

[160] P. Casau, C. G. Mayhew, R. G. Sanfelice, and C. Silvestre. Exponential stabilization of a vectored-thrust vehicle using synergistic potential functions. In *Proceedings of American Control Conference*, pages 6042–6047, 2016.

[161] P. Casau, C. G. Mayhew, R. G. Sanfelice, and C. Silvestre. Global exponential stabilization on the n-dimensional sphere. In *Proceedings of the American*

Control Conference, pages 3218–3223, June 2015.

[162] P.J.G. Ramadge. Some tractable supervisory control problems for discrete-event systems modeled by Buchi automata. *IEEE Transactions on Automatic Control*, 34(1):10–19, Jan. 1989.

[163] F. Lin. Robust and adaptive supervisory control of discrete event systems. *IEEE Transactions on Automatic Control*, 38(3):1848–1852, May 1993.

[164] A. S. Morse. Supervisory control of families of linear set-point controllers - Part 1: Exact matching. *IEEE Transactions on Automatic Control*, 41:1413–1431, 1996.

[165] A. S. Morse. Supervisory control of families of linear set-point controllers - Part 2: Robustness. *IEEE Transactions on Automatic Control*, 42:1500–1515, 1997.

[166] J. P. Hespanha, D. Liberzon, and A. S. Morse. Supervision of integral-input-to-state stabilizing controllers. *Automatica*, 38:1327–335, 2002.

[167] X. D. Koutsoukos, P. J. Antsaklis, J. A. Stiver, and M. D. Lemmon. Supervisory control of hybrid systems. *Proc. IEEE*, 88(7):1026–1049, July 2000.

[168] C. Prieur, R. Goebel, and A. R. Teel. Hybrid feedback control and robust stabilization of nonlinear systems. *IEEE Transactions on Automatic Control*, 52(11):2103–2117, November 2007.

[169] F. Ancona and A. Bressan. Patchy vector fields and asymptotic stabilization. *ESAIM: Control, Optimisation and Calculus of Variations*, 4:445–471, 1999.

[170] R. Sepulchre, M. Jankovic, and P. Kokotovic. *Constructive Nonlinear Control*. Springer, 1997.

[171] A. Isidori. *Nonlinear Control Systems II*. Communication and Control Engineering Series. Springer–Verlag, London, 1998.

[172] A. van der Schaft. *L2-Gain and Passivity Techniques in Nonlinear Control*. Springer, 2000.

[173] J. C. Willems. Dissipative dynamical systems. I. General theory. *Archive for Rational Mechanics and Analysis*, 45(5):321–352, 1972.

[174] P. Kokotovic and H. Sussman. A positive real condition for global stabilization of nonlinear systems. *Systems & Control Letters*, 13(2):125–133, 1989.

[175] R. Ortega, A. van der Schaft, I. Mareels, and B. Maschke. Putting energy back in control. *IEEE Control Systems Magazine*, 21(2):18–33, 2001.

[176] W. Lin and C. I. Byrnes. Passivity and absolute stabilization of a class of discrete-time nonlinear systems. *Automatica*, 31(2):263–267, 1995.

[177] R. Ortega and E. Garcia-Canseco. Interconnection and damping assignment passivity-based control: A survey. *European Journal of Control*, 10:432–450, 2004.

[178] A. Y. Pogromsky, M. Jirstrand, and P. Spangeous. On stability and passivity of a class of hybrid systems. In *Proc. 37th IEEE Conference on Decision and Control*, pages 3705–3710, Tampa, Florida, 1998.

[179] M. Zefran, F. Bullo, and M. Stein. A notion of passivity for hybrid systems. In *Proc. 49th IEEE Conference on Decision and Control*, pages 768–773, Orlando, Florida, 2001.

[180] J. Zhao and D. J. Hill. Passivity and stability of switched systems: A multiple storage function method. *Systems & Control Letters*, 57(2):158–164, 2008.

[181] B. Brogliato, R. Lozano, and O. Egeland. *Dissipative Systems Analysis and Control*. Springer, 2007.

[182] M. W. Spong, J. K. Holm, and D. Lee. Passivity-based control of bipedal locomotion. *IEEE Robotics & Automation Magazine*, 14(2):30–40, 2007.

[183] E. R. Westervelt, C. Chevallereau, J. H. Choi, B. Morris, and J. W. Grizzle. *Feedback control of dynamic bipedal robot locomotion*. CRC press, 2007.

[184] A. R. Teel. Asymptotic stability for hybrid systems via decomposition, dissipativity, and detectability. In *Proc. 49th IEEE Conference on Decision and Control*, pages 7419–7424, Atlanta, Georgia, 2010.

[185] W. M. Haddad and V. Chellaboina. Dissipativity theory and stability of feedback interconnections for hybrid dynamical systems. In *Proc. American Control Conference*, pages 2688–2694, Chicago, Illinois, 2000.

[186] Y. Or and A. D. Ames. Stability and completion of zeno equilibria in lagrangian hybrid systems. *IEEE Transactions on Automatic Control*, 56(6):1322 –1336, june 2011.

[187] W. J. Stronge. *Impact Mechanics*. Cambridge, University Press, 2000.

[188] E. D. Sontag. A "universal" construction of Artstein's theorem on nonlinear stabilization. *Systems and Control Letters*, 13:117–123, 1989.

[189] F. H. Clarke, Yu. S. Ledyaev, L. Rifford, and R. J. Stern. Feedback stabilization and Lyapunov functions. *SIAM: Journal of Control and Optimization*, 39(1):25–48, 2000.

[190] E. D. Sontag and H. J. Sussmann. General classes of control-Lyapunov functions. In *Stability Theory: Hurwitz Centenary Conference, Centro Stefano Franscini, Ascona, 1995.* 1996.

[191] R. A. Freeman and P. V. Kokotovic. *Robust Nonlinear Control Design: State-Space and Lyapunov Techniques*. Birkhauser, 1996.

[192] R. G. Sanfelice. On the existence of control Lyapunov functions and state-feedback laws for hybrid systems. *IEEE Transactions on Automatic Control*, 58(12):3242–3248, December 2013.

[193] S. Di Cairano, W.P.M.H. Heemels, M. Lazar, and A. Bemporad. Stabilizing dynamic controllers for hybrid systems: a hybrid control Lyapunov function approach. *IEEE Transactions on Automatic Control*, 59(10):2629–2643, 2014.

[194] R. G. Sanfelice. Robust asymptotic stabilization of hybrid systems using control Lyapunov functions. In *Proceedings of the 19th International Conference on Hybrid Systems: Computation and Control*, pages 235–244, April 2016.

[195] E. Moulay and W. Perruquetti. Stabilization of nonaffine systems: A constructive method for polynomial systems. *IEEE Transactions on Automatic Control*, 50(4):520–526, 2005.

[196] R. G. Sanfelice. Pointwise minimum-norm control laws for hybrid systems. In *Proceedings of the IEEE Conference on Decision and Control*, page 2665–2670, 2013.

[197] M. Krstic and H. Deng. *Stabilization of nonlinear uncertain systems.* Springer-Verlag, New York, 1998.

[198] R. G. Sanfelice. Clf-based control for hybrid dynamical systems. http://arxiv.org/abs/2009.03819, 2020.

[199] R. G. Sanfelice. A computationally tractable implementation of pointwise minimum norm state-feedback laws for hybrid systems. In *Proceedings of American Control Conference*, pages 4257–4262, 2016.

[200] C. Tomlin, G. J. Pappas, and S. Sastry. Conflict resolution for air traffic management: A study in multiagent hybrid systems. *IEEE Transactions on automatic control*, 43(4):509–521, 1998.

[201] X. Qi, D. Theilliol, D. Song, and J. Han. Invariant-set-based planning approach for obstacle avoidance under vehicle dynamic constraints. In *Proceedings of the IEEE International Conference on Robotics and Biomimetics*, pages 1692–1697, 2015.

[202] P. Falcone, M. Ali, and J. Sjoberg. Predictive threat assessment via reachability analysis and set invariance theory. *IEEE Transactions on Intelligent Transportation Systems*, 12(4):1352–1361, 2011.

[203] G. Pin and T. Parisini. Stabilization of networked control systems by nonlinear model predictive control: a set invariance approach. In *Nonlinear Model Predictive Control*, pages 195–204. Springer, 2009.

[204] P. Meyer, A. Girard, and E. Witrant. Robust controlled invariance for monotone systems: application to ventilation regulation in buildings. *Automatica*, 70:14–20, 2016.

[205] M. Fernandes and F. Zanolin. Remarks on strongly flow-invariant sets. *Journal of Mathematical Analysis and Applications*, 128(1):176–188, 1987.

[206] F. Blanchini. Set invariance in control. *Automatica*, 35(11):1747–1767, 1999.

[207] J.-P. Aubin. *Viability Theory.* Birkhauser, 1991.

[208] M. Nagumo. Über die lage der integralkurven gewöhnlicher differentialgleichungen. 24:551–559, 1942.

[209] J. A. Yorke. Invariance for ordinary differential equations. *Theory of Computing Systems*, 1(4):353–372, 1967.

[210] J. Bony. Principe du maximum, inégalité de harnack et unicité du probleme de cauchy pour les opérateurs elliptiques dégénérés. In *Annales de l'institut Fourier*, volume 19(1), pages 277–304, 1969.

[211] H. Brezis. On a characterization of flow-invariant sets. *Communications on Pure and Applied Mathematics*, 23(2):261–263, 1970.

[212] G. Bitsoris. On the positive invariance of polyhedral sets for discrete-time systems. *Systems & Control Letters*, 11(3):243–248, 1988.

[213] F. Zanolin. Bound sets, periodic solutions and flow-invariance for ordinary differential equations in $\mathbb{R}^n$: some remarks. *Rendiconti dell'Istituto di Matematica dell'Università di Trieste. An International Journal of Mathematics*, 19:76–92, 1988.

[214] V. Chellaboina, A. Leonessa, and W. M. Haddad. Generalized Lyapunov and invariant set theorems for nonlinear dynamical systems. *Systems & Control Letters*, 38(4):289–295, 1999.

[215] A. N. Gorban, I. Tyukin, E. Steur, and H. Nijmeijer. Lyapunov-like conditions of forward invariance and boundedness for a class of unstable systems. *SIAM Journal on Control and Optimization*, 51(3):2306–2334, 2013.

[216] S. Tarbouriech and C. Burgat. Positively invariant sets for constrained continuous-time systems with cone properties. *IEEE Transactions on Automatic Control*, 39(2):401–405, 1994.

[217] E. B. Castelan and J.C. Hennet. On invariant polyhedra of continuous-time linear systems. In *Proceedings of the 30th IEEE Conference on Decision and Control*, pages 1736–1741, 1991.

[218] B. Milani and C. Dórea. On invariant polyhedra of continuous-time systems subject to additive disturbances. *Automatica*, 32(5):785–789, 1996.

[219] H. Li, L. Xie, and Y. Wang. On robust control invariance of boolean control networks. *Automatica*, 68:392–396, 2016.

[220] S. Sadraddini and C. Belta. A provably correct mpc approach to safety control of urban traffic networks. *arXiv preprint arXiv:1602.01028*, 2016.

[221] S. V. Rakovic, P. Grieder, M. Kvasnica, D. Q. Mayne, and M. Morari. Computation of invariant sets for piecewise affine discrete time systems subject to bounded disturbances. In *Proceedings of the 43rd IEEE Conference on Decision and Control*, volume 2, pages 1418–1423. IEEE, 2004.

[222] J.-P. Aubin and H. Frankowska. *Set-valued analysis*. Springer Science & Business Media, 2009.

[223] S. Prajna, A. Jadbabaie, and G. J. Pappas. A framework for worst-case and stochastic safety verification using barrier certificates. *IEEE Transactions on Automatic Control*, 52(8):1415–1428, 2007.

[224] L. Dai, T. Gan, B. Xia, and N. Zhan. Barrier certificates revisited. *Journal of Symbolic Computation*, 80:62–86, 2017. SI: Program Verification.

[225] H. Kong, F. He, X. Song, W.N.N. Hung, and M. Gu. Exponential-condition-based barrier certificate generation for safety verification of hybrid systems. In *Comput. Aided Ver.*, pages 242–257, Springer Berlin, Heidelberg, 2013.

[226] M. Maghenem and R. G. Sanfelice. Barrier function certificates for invariance in hybrid inclusions. In *Proceedings of the 2018 IEEE Conference on Decision and Control*, pages 759–764, December 2018.

[227] A. D. Ames, X. Xu, J. W. Grizzle, and P. Tabuada. Control barrier function based quadratic programs for safety critical systems. *IEEE Transactions on Automatic Control*, 62(8):3861–3876, 2017.

[228] M. Maghenem and R. G. Sanfelice. Multiple barrier function certificates for forward invariance in hybrid inclusions. In *Proceedings of the American Control Conference*, pages 2346–2351, July 2019.

[229] M. Maghenem and R. G. Sanfelice. Multiple barrier function certificates for weak forward invariance in hybrid inclusions. In *Proceedings of the 2019 IEEE Conference on Decision and Control*, December 2019.

[230] M. Maghenem and R. G. Sanfelice. Sufficient conditions for forward invariance and contractivity in hybrid inclusions using barrier functions. *To appear in Automatica*, 2020. https://arxiv.org/pdf/1908.03980.pdf.

[231] P. Glotfelter, J. Cortés, and M. Egerstedt. Nonsmooth barrier functions with applications to multi-robot systems. *IEEE Control Systems Letters*, 1(2):310–315, 2017.

[232] M. Maghenem and R. G. Sanfelice. Characterizations of safety in hybrid inclusions via barrier functions. In *Proceedings of the Hybrid Systems: Computation and Control*, July 2019.

[233] J. Chai and R. G. Sanfelice. On notions and sufficient conditions for forward invariance of sets for hybrid dynamical systems. In *Proceedings of the 54th IEEE Conference on Decision and Control*, pages 2869–2874, December 2015.

[234] J. Chai and R. G. Sanfelice. On robust forward invariance of sets for hybrid dynamical systems. In *Proceedings of the American Control Conference*, pages 1199–1204, 2017.

[235] A. Pnueli. The temporal logic of programs. In *18th Annual Symposium on Foundations of Computer Science, 1977*, pages 46–57. IEEE, 1977.

[236] R. G. Sanfelice. *Analysis and Design of Cyber-Physical Systems: A Hybrid Control Systems Approach*, pages 3–31. CRC Press, 2015.

[237] Z. Manna and A. Pnueli. *The Temporal Logic of Reactive and Concurrent Systems*, volume 16. Springer, 1992.

[238] S. Karaman, R. G. Sanfelice, and E. Frazzoli. Optimal control of mixed logical dynamical systems with linear temporal logic specifications. In *Proc. 47th IEEE Conference on Decision and Control*, page 2117–2122, 2008.

[239] G. E. Fainekos, A. Girard, H. Kress-Gazit, and G. J. Pappas. Temporal logic

motion planning for dynamic robots. *Automatica*, 45(2):343–352, 2009.

[240] E. M. Wolff, U. Topcu, and R. M. Murray. Optimization-based trajectory generation with linear temporal logic specifications. In *Robotics and Automation (ICRA), 2014*, pages 5319–5325. IEEE, 2014.

[241] S. Saha and A. A. Julius. An MILP approach for real-time optimal controller synthesis with metric temporal logic specifications. In *American Control Conference (ACC), 2016*, pages 1105–1110. IEEE, 2016.

[242] H. Han and R. G. Sanfelice. Sufficient conditions for temporal logic specifications in hybrid dynamical systems. In *Proceedings of the 6th Analysis and Design of Hybrid Systems*, volume 51, pages 97–102, July 2018.

[243] H. Han and R. G. Sanfelice. Linear temporal logic for hybrid dynamical systems: Characterizations and sufficient conditions. *Nonlinear Analysis: Hybrid Systems*, 36, May 2020.

[244] Y. Li and R. G. Sanfelice. Finite time stability of sets for hybrid dynamical systems. *Automatica*, 100:200–211, February 2019.

[245] R. Dimitrova and R. Majumdar. Deductive control synthesis for alternating-time logics. In *Proceedings of the 14th International Conference on Embedded Software*, page 14. ACM, 2014.

[246] P. Wolper. Constructing automata from temporal logic formulas: A tutorial. In *School organized by the European Educational Forum*, pages 261–277. Springer, 2000.

[247] T. Babiak, M. Křetínský, V. Řehák, and J. Strejček. Ltl to büchi automata translation: Fast and more deterministic. In *International Conference on Tools and Algorithms for the Construction and Analysis of Systems*, pages 95–109. Springer, 2012.

[248] C. Belta, B. Yordanov, and E. A. Gol. *Formal methods for discrete-time dynamical systems*, volume 89. Springer, 2017.

[249] P. Hartman. *Ordinary Differential Equations*. Birkhauser, 1982.

[250] C-T. Chen. *Linear System Theory and Design*. Oxford University Press, 1999.

[251] R.T. Rockafellar and R. J-B Wets. *Variational Analysis*. Springer, Berlin, Heidelberg, 1998.

[252] K. A. Ross. *Elementary Analysis: The Theory of Calculus*. Springer Science & Business Media, 2013.

[253] H. Whitney. Analytic extensions of differentiable functions defined in closed sets. *Transactions of the American Mathematical Society*, 36(1):63–89, 1934.

[254] R. Freeman and P. V. Kokotovic. Inverse optimality in robust stabilization. *SIAM Journal of Control and Optimization*, 34:1365–1391, 1996.

[255] E. Michael. Continuous selections, I. *The Annals of Mathematics*, 63(2):361–382, 1956.

Index